Fuels and Petrochemicals Division

AA003143

Presentations at the 2010 AIChE Annual Meeting

Salt Lake City, Utah, USA
7-12 November 2010

ISBN: 978-1-61782-161-5

Printed from e-media with permission by:

Curran Associates, Inc.
57 Morehouse Lane
Red Hook, NY 12571

Some format issues inherent in the e-media version may also appear in this print version.

Printed by Curran Associates, Inc. (2011)

For permission requests, please contact AIChE
at the address below.

AIChE
3 Park Avenue
New York, NY 10016-5991

Phone: (203) 702-7660
Fax: (203) 775-5177

www.aiche.org

Additional copies of this publication are available from:

Curran Associates, Inc.
57 Morehouse Lane
Red Hook, NY 12571 USA
Phone: 845-758-0400
Fax: 845-758-2634
Email: curran@proceedings.com
Web: www.proceedings.com

TABLE OF CONTENTS

(15a) Condensation Reactions of Propanal Over Ceria-Zirconia Mixed Oxide Catalysts

Anirudhan Gangadharan, Chemical, Biological and Materials Engineering, University of Oklahoma, Norman, OK, Tawan Sooknoi, Chemistry, King Mongkut Institute of Technology Ladkrabang, Bangkok, Thailand, Lance L. Lobban, School of Chemical, Biological and Materials Engineering, Center for Biomass Refining, University of Oklahoma, Norman, OK, Daniel E. Resasco, Chemical, Bological and Materials Engineering, University of Oklahoma, Norman, OK and Richard G. Mallinson, School of Chemical, Biological and Materials Engineering, University of Oklahoma, Norman, OK

Vapor phase aldol condensation and ketonization reactions of propanal were investigated over $Ce_xZr_{1-x}O_2$ mixed oxides as a model reaction to produce gasoline range molecules from short aldehydes found in bio-oil mixtures. Several operating parameters were investigated. They include the type of carrier gas used (H_2 or He) and the incorporation of acids and water in the feed. Propanal is converted to higher carbon chain oxygenates on $Ce_xZr_{1-x}O_2$ by two pathways, aldol condensation and ketonization. The major products of these condensation reactions include 3-pentanone, 2-methyl-2-pentenal, 2-methylpentanal, 3-heptanone and 4-methyl-3-heptanone. It is proposed that the primary intermediate for the ketonization path is a surface carboxylate. The presence of acids in the feed inhibits the aldol condensation pathway by competitive adsorption that reduces the aldehyde conversion. Water also promotes ketonization and inhibits aldol condensation by increasing the concentration of surface hydroxyl groups that enhance the formation of surface carboxylates with the aldehyde. Hydrogen enhances cracking and production of light oxygenates and hydrocarbons. The light oxygenates may in turn be reincorporated into the reaction path, giving secondary products. However, the hydrocarbons do not react further. The analysis of the fresh and spent catalysts by XPS showed varying degrees of reduction of the oxide under different operating conditions and was in agreement with the reaction results. Changing the proportion of the parent oxides showed that increased Zr favored formation of aldol products while increased Ce favored ketonization. This occurred by shifting the balance of the acid-base properties of the active sites.

See more of this Session: Catalytic Biofuels Refining
See more of this Group/Topical: Fuels and Petrochemicals Division

(15b) Effect of Pt Loading On Enhancing Aldehyde Hydrogenation for One-Step Hydrogenation/Esterification: Model Reaction for Bio-Oil Upgrading

Yang Tang[1], Shaojun Miao[2], Brent H. Shanks[3] and Xiaoming Zheng[1], (1)Institute of catalysis, Department of chemistry, Zhejiang University, Hangzhou, China, (2)Department of Chemical and Biological Engineering, Iowa State University, Ames, IA, (3)Chemical & Biological Engineering, Iowa State University, Ames, IA

ABSTRACT: Bio-oil is acknowledged as a renewable, sustainable and promising energy alternative for fossil fuels, but there are some issues to deal with before its subsequent use. One is its corrosive nature due to its high content of organic acid and another is its reactive nature due to many unsaturated components such as aldehydes. Combined one-step hydrogenation/esterification (OHE) has been demonstrated a possible approach for bio-oil upgrading, by directly converting the two main reactive components, aldehydes and organic acids, to stable and combustible esters. Bifunctional mesoporous organic-inorganic hybrid silica involving platinum and propylsulfonic acid group (Pt/SBA15-PrSO3H) has been revealed feasible and superior in catalyzing OHE of acetaldehyde and acetic acid, which was employed as a model reaction for catalytic upgrading of bio-oil. But there are still potential and necessity to enhance the catalytic performance of this bifunctional catalyst. In this work, the enhancements of hydrogenation activity over metal sites of the mesoporous hybrid silicas were focused on. It is found that aqueous reduction-deposition by the relatively less efficient reducing agent (HCHO) is promising to prepare Pt nanoparticles into the pores of SBA15 with high dispersion. And further modifications, herein, high temperature treatments with air or H2 were tested. Results showed that both air and H2 treatments enhanced acetaldehyde hydrogenation activity over Pt/SBA15, especially for H2 treatment. Characterization results demonstrated that these treatments lead to more metallic Pt sites and activate hydrogen more efficiently, which were deemed as the main reasons for acetaldehyde hydrogenation activity enhancement. Consequently, it provided guideline for the further enhancement of the bifunctional catalysts (SBA15 functionalized with platinum and sulfonic acid groups). Finally, a new bifunctionalization procedure for this catalyst was proposed for future work aiming at improving the catalytic activity of one-step hydrogenation/esterification.

See more of this Session: Catalytic Biofuels Refining
See more of this Group/Topical: Fuels and Petrochemicals Division

(15c) Selective Deoxygenation of Furfuraldehyde On Bimetallic Catalysts

Daniel E. Resasco[1], Surapas Sitthisa[2], Lance Lobban[2], Tawan Sooknoi[2] and Richard Mallinson[2], (1) Chemical, Bological and Materials Engineering, University of Oklahoma, Norman, OK, (2)University of Oklahoma, Norman, OK

Furfuraldehyde is commonly found in bio-oil, and this remarkably reduces fuel properties and stability during storage. Deoxygenation is typically needed to produce the bio-oil more fungible with standard gasoline and diesel. This can be achieved by hydrotreating using various noble metal catalysts. The reaction mainly involves decarbonylation and hydrogenation of the carbonyl group. In the present work, the silica supported Pd-Cu and Fe-Ni bimetallics were used for hydrodeoxygenation of fufuraldehyde. The reactions were performed at the temperature between 210-230oC under atmospheric pressure of hydrogen. In the case of Pd/SiO2 catalysts, TPR suggests that the addition of Cu to supported Pd catalyst results in formation of Pd-Cu bimetallic alloy. The activity and selectivity of the catalysts were largely influenced by the Cu content. The yield of furan from decarboylation was significantly reduced when Cu was incorporated. However, increasing the hydrogenation activity to produce furfuryl alcohol can be observed. For Ni/SiO2 catalyst, the main products were furan and butane which is derived from ring opening reaction. 2-Methyl furan was the minor product on Ni catalyst. However, the addition of Fe to Ni catalyst dramatically increased the yield of 2-methyl furan while reducing the yield of furan. It is clearly seen that the addition of Cu into Pd catalyst and Fe into Ni catalyst can suppress the decarbonylation activity while enhancing the hydrogenation activity.

See more of this Session: Catalytic Biofuels Refining
See more of this Group/Topical: Fuels and Petrochemicals Division

(15d) Bio-Oil Refining Using Nanohybrid Catalysts That Stabilize Emulsions in Bi-Phasic Liquid Systems

Daniel E. Resasco[1], Paula Zapata[2], Min Shen[2], Pilar Ruiz[2] and Jimmy Faria[2], (1)Chemical, Bological and Materials Engineering, University of Oklahoma, Norman, OK, (2)University of Oklahoma, Norman, OK

Pyrolysis oil is a complex mixture of oxygenated compounds with up to ~30-40% of water. Depending on the cooling process used in the condensation of the pyrolysis vapors, the crude bio-oil can generate a biphasic system, in which molecules are distributed, between the two phases, depending on their solubility. It is desirable to conduct reactions at the water/oil interface that can both enhance the fuel value of the molecules and effect phase migration, based on solubility, avoiding fractionation by heating, which is known to negatively impact bio-oil. We have developed a new reaction/separation concept based on a family of recoverable nano-hybrid catalysts that simultaneously stabilize emulsions in biphasic bio-oil. These nanostructured solid particles exhibit a unique advantage in streamlining biomass refining, where the immiscibility and thermal instability of crude bio-oil greatly complicates purification procedures. These novel catalyst/emulsifier hybrids can catalyze reactions with high ?phase-selectivity? either in the aqueous or organic phases. These catalysts are obtained by fusing carbon nanotubes with metal-oxide particles, which results in two-faced nanohybrid solids that are able to stabilize water/oil emulsions by forming a rigid film at liquid-liquid interface of the droplets and increasing the apparent viscosity of the system . The metal oxide in the nanohybrid acts both, as the hydrophylic side of the emulsifier, and as a catalyst for condensation reactions in the aqueous phase. Accordingly, small oxygenates soluble in water, with low fuel value, condense via aldol-condensation, ketonization, or etherification resulting in products, which are no longer watersoluble molecules and therefore migrate to the organic phase. The oxide used can vary in acid/base characteristics. Some of the metal-oxides tested are; MgO, SiO2, TiO2 and ZnO. In the organic phase, transition metals such as Pd, Ni and Cu have been deposited onto the hydrophobic carbon nanotube of the nanohybrids to catalyze deoxygenation reactions including TCS2010 Symposium on Thermal and Catalytic Sciences for Biofuels and Biobased Products -- Submit Abstract 3/20/10 10:51 PM http://www.ucs.iastate.edu/mnet/tcs2010/form1/respond.html Page 2 of 2 nanohybrids to catalyze deoxygenation reactions including hydrogenation, hydrogenolysis, or decarbonylation occurring on the oil side of the emulsion. We have accomplished biphasic hydrodeoxygenation and condensation catalysis in high yields, using these nono-hybrid catalysts for several systems of interest in biomass refining. Reactions were conducted in a semi-batch reactor with the liquid composed of three layers oil/emulsion/water and with continuous flow of hydrogen, at temperatures in the range 80-250°C, and pressures (300-900 psi) high enough to maintain the water in the liquid state.

[1] M. Shen, D. E. Resasco, Langmuir, 25, 10843 (2009).

[2] S. Crossley, J. Faria, M. Shen and D. E. Resasco, Science, 327, 68 (2009).

See more of this Session: Catalytic Biofuels Refining
See more of this Group/Topical: Fuels and Petrochemicals Division

(15e) "Drop-in" Green Jet and Diesel Fuels From Renewable Oils

Edward N. Coppola, Process Development Group, Applied Research Associates, Inc., Panama City, FL

Current approaches for making jet fuel from alternate sources include the Fischer-Tropsch (FT) process and conventional hydrotreating and hydrocracking used to produce "Hydrotreated Renewable Jet" or HRJ. Both FT and HRJ-derived fuels have low density relative to the specification requirement, contain no aromatics, and must be blended with the petroleum-derived counterpart to meet current jet fuel specifications. Low density results in low volumetric energy content and other operational problems. Aromatic compounds are necessary to promote elastomer (O-ring) swelling and prevent fuel system leaks. The requirement to blend FT and HRJ fuels with the petroleum counterpart creates logistics problems. Therefore, a true "drop-in" alternative fuel is desired to meet current goals established by the Department of Defense and the U.S. Air Force. Applied Research Associates, Inc. (ARA) developed a novel hydrothermal process (patent pending) called "Catalytic Hydrothermolysis" (CH) and demonstrated the production of military jet fuel (JP-8) and diesel for Navy applications (F-76: Diesel, Naval Distillate) in a pilot-scale system. The hydrothermal process uses water to convert renewable plant and algal oils into pure hydrocarbon distillate fuels. The unique application of water minimizes hydrogen consumption and results in conversion of non-edible oils into ideal hydrocarbons for "drop-in" green jet and diesel fuels. In addition to the normal and branched paraffins that are formed by FT and conventional HRJ processes, the ARA CH process creates high density cycloparaffin and aromatic molecules in a single step, without blending, pyrolysis, catalytic reforming, catalytic hydroisomerization, or catalytic hydrocracking. ARA is under contract with the Air Force Research Laboratory (AFRL) to produce samples of specification-quality JP-8 for specification and "fit-for-purpose" testing. In addition to producing true "drop-in" fuels that meet density and aromatic requirements, the CH process eliminates catalytic hydrocracking and hydroisomerization operations, and reduces catalysts requirements, hydrogen consumption and carbon footprint. This paper provides the results of recent jet and diesel fuel production and characterization tests.

See more of this Session: Catalytic Biofuels Refining
See more of this Group/Topical: Fuels and Petrochemicals Division

(15f) Effect of Water On Esterification of Palmitic Acid Using ZSM-5 and Faujasite Zeolite Catalysts

Adebola T. Coker[1], Alexei V. Iretskii[2], Rafael Hernandez[1], Mark G. White[1] and Todd French[1], (1) Dave C. Swalm School of Chemical Engineering, Mississippi State University, Mississippi State, MS, (2)Lake Superior State University, Sault Ste. Marie, MI

Our research group has demonstrated the production of fatty acid methyl esters (biodiesel) using activated sludge generated from a wastewater treatment facility. One of the main operating costs of this process is drying of the sludge prior to oil extraction and reaction. A method of generating biodiesel from oil without transesterification is to convert the lipids to free fatty acids via hydrolysis, followed by esterification to obtain the fatty acid methyl esters (FAME). Thus, we have conducted a study to determine the maximum concentration of water that the esterification reaction would tolerate using zeolite catalysts.

This alternative method employs a solid acid to catalyze the esterification of the free fatty acid. Since the silica:alumina (SiO_2:Al_2O_3)of the zeolites affects its acidity and hydrophobicity, the influence of the SiO_2:Al_2O_3 ratio in these catalysts on the reactions will be analyzed. These catalysts were screened for highest activity in conversion of fatty acid to FAME and water tolerance. The effect of water on the reaction was examined at volume compositions of 5 %, 10%, 20%, and 50%.

Esterification was completed using palmitic acid to represent the free fatty acids present in the oils. Two types of zeolite catalysts – ZSM-5 and Faujasite (Y) zeolites with SiO_2:Al_2O_3 ratios of 80:1 and 30:1 were tested and compared. The reactants included 0.5 g fatty acid, 10 ml methanol, and 10 mass % catalyst, with reaction temperature of 65°C, and a duration of 3 hours. The effect of various concentrations of water on the esterification of palmitic acid using these catalysts, and the reaction kinetics were evaluated to determine their activity and usability in the presence of water. Reaction parameters such as rate constants, rates, and activation energy were obtained.

Preliminary results on the ZSM-5 catalyst with SiO_2:Al_2O_3 ratio of 30:1 shows the activation energy to be 37.5 KJ/mol which is comparable to literature results for strong acids on a similar system. The yield of FAME shows nonlinear dependency on water concentration.

See more of this Session: Catalytic Biofuels Refining
See more of this Group/Topical: Fuels and Petrochemicals Division

(15g) CFD ANALYSIS of the Heat Transfer Coefficient During Hydrotreatment of Palm Oil

Cesar C. Mendoza[1], Alejandro Molina Ochoa[1] and **Cesar A. Mahecha**[2], (1)Process and Energy School, National University of Colombia, Medellin Campus, Medellin, Colombia, (2)Colombian Petroleum Institute, ECOPETROL S.A., Piedecuesta, Colombia

ABSTRACT Hydrotreatment transforms vegetables oils into a zero-sulfur liquid with properties that resemble those of diesel fuel. In a hydrotreatment reactor, high pressure addition of hydrogen reduces the size of the fatty acids molecules through isomerization and thermal cracking at temperatures close to 350°C (see Figure 1). Figure 1. Figure 1. Proposed process for crude palm oil hydrotreatment Hydrotreatment, an exothermic reaction, is strongly affected by heat transfer as high temperatures in hot spots could degrade the fuel to undesired small-chain hydrocarbons. A preliminary analysis of the hydrotreatment process suggests that a packed bed reactor, similar to those used during hydrotreatment of regular mineral oil, should be preferred to a packed-tube reactor as heat transfer in the former, if correctly designed, is enough to prevent the formation of hot spots. However, this analysis strongly depends on the heat transfer coefficient in the packed bed reactor and at the walls. As hydrotreatment of palm oil for biodiesel production is a relative new technology, there are no correlations in the refereed literature for the heat transfer during hydrotreatment of palm oil in packed-bed reactors. This study considers the use of computational fluid dynamics (CFD) as an alternative for the evaluation of heat transfer in this process. A review of the scientific literature showed that, even though there are no references that deal with CFD simulation of hydrotreatment reactors, there are several studies of CFD simulation of packed-tube reactors. 2D and 3D CFD analyses yielded encouraging results when modeling external air flow through a sphere (see Figure 2), a cylinder and a bank of tubes. The Nusselt numbers (Nu) calculated from the CFD simulations are close to those obtained from commonly used correlations. This CFD analysis was extended to the calculation of the variation Nu with Re for packed spheres and cylinders at conditions that resemble those of the hydrotreatment of palm oil. These correlations, when applied to the original packed-bed reactor design, suggests that an external jacket in a packed-bed reactor could effectively remove the heat produced during hydrotreatment of palm oil. Figure 2. Figure 2. Comparison of the predicted variation of Nu with Re for external flow around a sphere obtained with CFD with four different turbulent models (( ): Spalart-Allmaras; ( ): Reynolds-Stress; ( ): Standard K-Omega; ( ): Standard K-epsilon) with that obtained with the Whitaker Correlation (---). Bold lines represent 15% below and above the value predicted with the Whitaker Correlation.

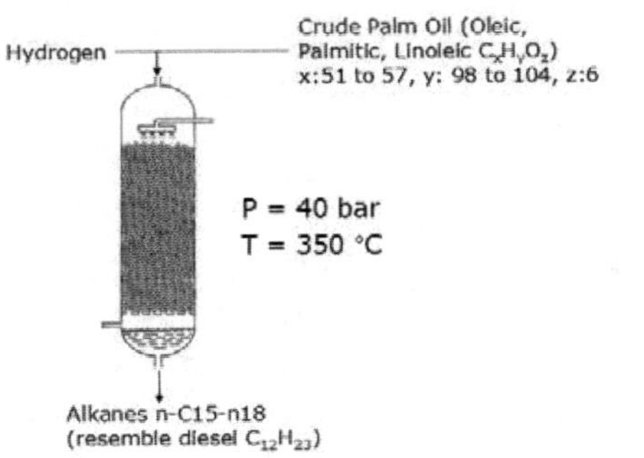

Figure 1. Proposed process for crude palm oil hydrotreatment

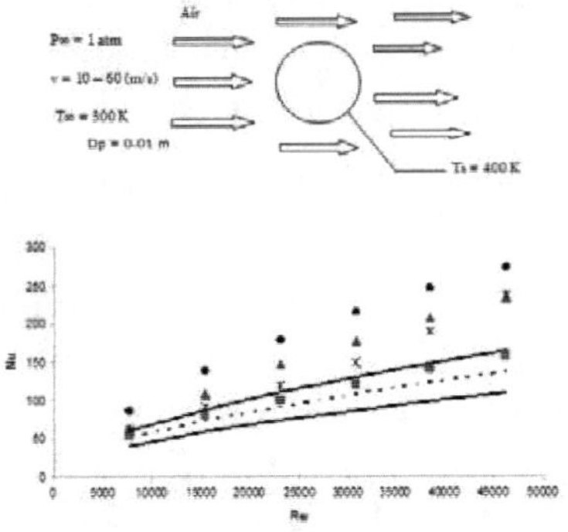

Figure 2. Comparison of the predicted variation of Nu with Re for external flow around a sphere obtained with CFD with four different turbulent models ((•): Spalart-Allmaras; (▲): Reynolds-Stress; (×): Standard K-Omega; (■): Standard K-epsilon) with that obtained with the Whitaker Correlation (---). Bold lines represent 15% below and above the value predicted with the Whitaker Correlation.

See more of this Session: Catalytic Biofuels Refining
See more of this Group/Topical: Fuels and Petrochemicals Division

Catalytic Cracking of a Rapeseed Oil for Production of Transportation Fuels and Chemicals: Yield Structure

Siauw H. Ng[1], Yu Shi[1], Lianhui Ding[2] and Song Chen[1], (1)CanmetENERGY Upgrading Program, Natural Resources Canada, Devon, AB, Canada, (2)Hydroprocessing Research, ConocoPhillips Bartlesville Technology Center, Bartlesville, OK

1. Introduction

Biofuels refer to a wide range of fuels derived from biological sources, especially biomass. Owing to considerations such as price escalation, energy security, and environmental concerns, biofuels, being a source of renewable energy, are gaining increased industrial and scientific interest. In 2008, biofuels provided 1.8% of the world's transportation fuels.[1] The European Union set targets for biofuel usage of 2% in 2005 and 5.75% in 2010.[2]

Canola is a cultivated variety of rapeseed, and canola oilseeds are rich in oil content (40–43%). The name "canola" was derived from "**Can**adian **O**ilseed, **L**ow-**A**cid" (referring to the erucic acid content). The canola plant produces seed-filled pods. The seeds are crushed to produce canola oil in addition to canola meal for animal feed.

Through transesterification, canola oil can be converted to biodiesel, which is known to produce fewer emissions and provide better lubrication than low-sulfur diesel from petroleum sources. The option to produce biogasoline from canola oil through thermal or catalytic cracking has been investigated by researchers in North America and Europe.[3–7]. Most of these studies dealt with cracking of canola oil using inert materials or pure zeolites in fixed-bed reactors at relatively low temperatures (<500°C) and low weight hourly space velocities (WHSV, <3.6 h^{-1}). In this study, a canola oil, a refinery FCC feed, and a 50/50 blend of the two were catalytically cracked over an equilibrium catalyst in a fluid-bed batch reactor at commercially relevant conditions.

2. Experimental

Edible-grade canola oil was obtained from a supermarket. This oil is similar in analysis to the bulk canola oil available from commercial suppliers except for the color (pale yellow versus green). The conventional FCC feed and equilibrium catalyst were supplied by a US refinery. Properties of the feeds and catalyst are given in Tables 1 and 2, respectively. The feeds were cracked in a fluid-bed microactivity test (MAT) unit at 540°C and 10–25 h^{-1} WHSV, with a constant oil injection time of 30 s. Details of the experiment have been reported elsewhere.[8]

Table 1. Feed Properties

Analysis	Unit	FCC Feed	Canola Oil	50/50 Blend
Density	g/mL	0.8928	0.9169	0.9072
Carbon	wt%	86.46	77.63	81.52
Hydrogen	wt%	13.20	11.92	12.51
Sulfur	wppm	687	<0.55	335
Nitrogen	wppm	370	0.71	190
Oxygen	wt%	0.31	11.58	6.52
Saturates	wt%	65.8	0.0	29.6
Aromatics	wt%	32.1	0.6	16.3
Polars	wt%	2.1	99.4	54.0
Asphaltenes	wt%	0.0	0.0	0.0

Table 2. Properties of Equilibrium Catalyst

Catalyst	Equilibrium Catalyst
Al_2O_3, wt %	35.9
SiO_2, wt %	51.0
Re_2O_3, wt %	0.28
TiO_2, wt %	1.27
Na_2O, wt %	0.52
Fe_2O_3, wt %	0.40

3. Results and Discussion

3.1. Conversion

Conversion is defined as the portion of the feed converted to 221°C– products, including gas and coke. Figure 1 shows that conversion increased with catalyst/oil (C/O) ratio. At a given ratio, canola oil gave the highest conversion, followed by the 50/50 blend and the conventional FCC feed. All curves tended to converge to the same conversion level at very high C/O ratio. At C/O ratio of 10, canola oil exhibited conversion ~5.5wt% higher than that of the FCC feed. It is important to determine whether the high conversion was due to high-value products such as gasoline and liquefied petroleum gas or low-value constituents such as dry gas and coke.

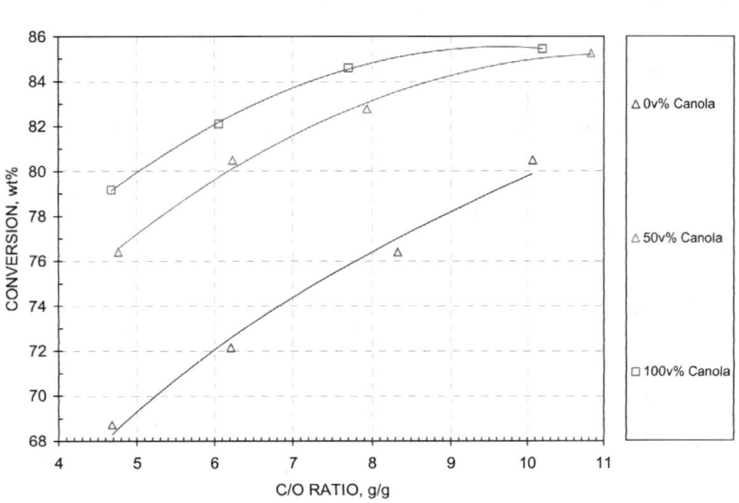

Figure 1. Relationship between Conversion and C/O Ratio

3.2. Dry Gas

Dry gas is composed of H_2, H_2S, CO, CO_2, and C_1–C_2 hydrocarbons. Figure 2 shows that dry gas yield increased exponentially with conversion. At a given conversion, canola oil and the blend gave higher dry gas yields than the FCC feed. The increase of dry gas yield with conversion was particularly strong for canola oil. Both CO and CO_2 were detected during cracking of the feeds containing canola oil, with CO yields at ~0.95 and ~2.55wt% and CO_2 yields at 1.3–1.8wt% and 1.3–2.2wt%, for the blend and pure canola oil, respectively.

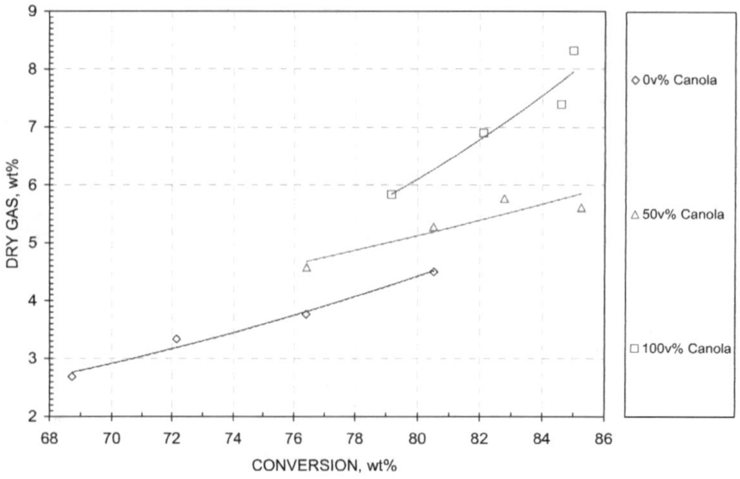

Figure 2. Relationship between Dry Gas Yield and Conversion

3.3. Liquefied Petroleum Gas

Liquefied petroleum gas (LPG) consists of C_3 and C_4 gaseous hydrocarbons. LPG is considered a valuable product since its components can be used as alkylation and petrochemical feedstocks. Figure 3 shows that, in contrast to the dry gas yields, the blend and canola oil exhibit lower LPG yields at a given conversion than the FCC feed.

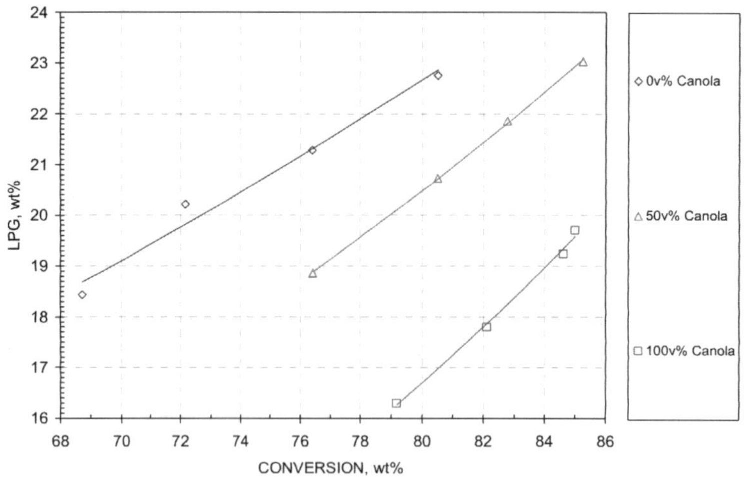

Figure 3. Variation of LPG Yield with Conversion

3.4 Gasoline

Gasoline (C_5–221°C boiling point) is the major and the most desirable product in FCC operation. Perhaps the most striking feature in cracking the canola-containing feeds was the significant drop in gasoline yield relative to that of the FCC feed. Figure 4 shows that, at 80wt% conversion, the blend lowered the gasoline yield by ~4.3wt% compared to the FCC feed whereas canola oil further reduced the yield by ~2.0wt% compared to the blend. It was believed that the loss of gasoline was mostly due to the formation of water, CO, and CO_2 from the canola component in the feeds. The water yield varied from 5.4 to 6.0wt% and from 10.2 to 11.0wt% for the blend and canola oil, respectively. Overcracking was observed for canola oil at 78–82wt% conversion.

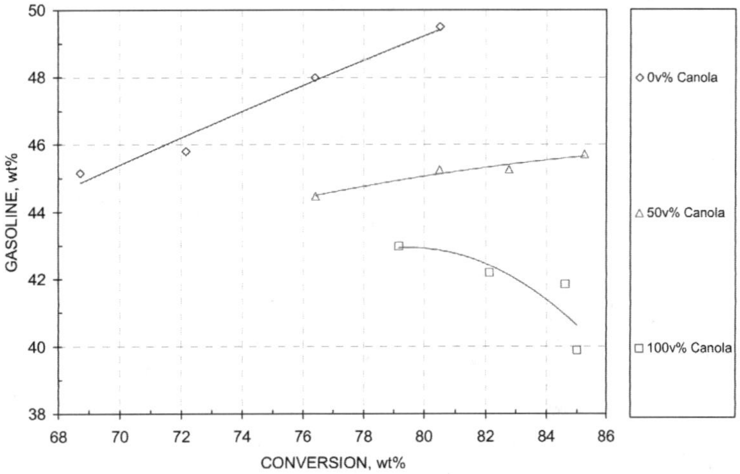

Figure 4. Relationship between Gasoline Yield and Conversion

3.5 Coke

In FCC operation, coke is necessary to supply heat for feed preheating and cracking. However, too much coke can seriously poison the catalyst and overload the air blower during catalyst regeneration, causing excessively high temperatures in the regenerator. Figure 5 shows that coke yield increased with conversion. The blend had a similar coke profile to that of the FCC feed but canola oil showed a sharp increase in coke yield at conversions higher than 80wt%.

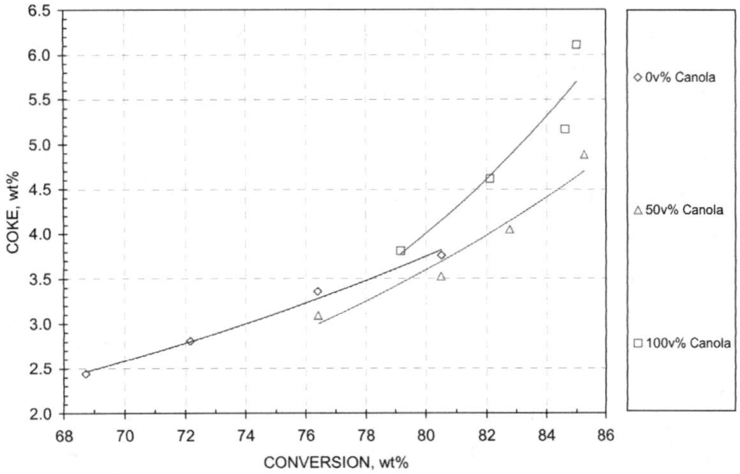

Figure 5. Relationship between Coke Yield and Conversion

3.5 Oxygen Balance

Table 3 gives the oxygen distribution in the gaseous and liquid products. It can be seen that, after cracking, most of the oxygen in the canola oil appears as H_2O (79.7–88.7 and 78.6–84.1wt% for the blend and canola oil, respectively), with the rest forming CO (8.5–9.3 and 12.2–12.8wt% for the blend and canola oil, respectively), or be retained in the TLP as phenols (about 2wt% for both feeds).

Table 3. Oxygen Balance in Cracking 50/50 Blend and Canola Oil

Run Number		3114D	3115D	3116D	3117D	3120D	3121D	3122D	3124D
Feed		50/50 Blend				Canola Oil			
Catalyst/Oil Ratio	g/g	4.761	6.234	7.931	10.828	4.675	6.058	7.699	10.271
WHSV	g/h/g	25.20	19.25	15.13	11.08	25.67	19.81	15.59	11.68
CO Yield	wt%	0.92	0.97	0.98	0.89	2.47	2.54	2.57	2.60
H2O Yield	wt%	5.41	5.71	5.85	6.02	10.24	10.61	10.96	10.97
Conversion	wt%	76.41	80.50	82.77	85.25	79.17	82.11	84.60	84.99
O2 conc in TLP	wt%	0.26	0.28	0.25	0.26	0.35	0.38	0.423	0.43
O2 balance (relative to feed oxygen)	wt%								
O2 as CO		8.7	9.2	9.3	8.5	12.2	12.5	12.7	12.8
O2 as H2O		79.7	84.1	86.1	88.7	78.6	81.4	84.0	84.1
O2 in TLP		2.4	2.4	2.0	1.9	1.8	1.8	1.9	1.9
Tot O2		90.9	95.8	97.4	99.1	92.6	95.7	98.6	98.8

4. Conclusions

- Catalytic cracking of feeds containing canola oil results in the formation of water, carbon monoxide, and carbon dioxide at the expense of gasoline production.
- Conversions of all feeds increase with C/O ratio. When the proportion of canola oil in the feed is higher, the conversion at a constant C/O ratio increases.
- For a given feed, as conversion increases, yields of unconverted products (i.e., diesel and heavy fuel oil) decrease while yields of other products increase, except for gasoline of which the increase may show a parabolic variation.
- At a given conversion, as the content of canola oil in the feed increases, yields of dry gas, diesel, coke, and water increase, while yields of LPG, gasoline, and heavy fuel oil decrease.
- After cracking, most of the oxygen in canola oil appears as H_2O, with the rest forming CO, CO_2, and oxygenates, such as phenols, in the TLP.

Acknowledgments

The authors wish to thank the analytical laboratory of CanmetENERGY.

References.

[1] "Towards Sustainable Production and Use of Resources: Assessing Biofuels". United Nations Environment Programme 2009-10-16.

[2] Selçuk, S., Kaygusuz, K., Sari, A., Energ. Sour. **2004,** *26* 1119–1129.

[3] Idem, R.O., Katikaneni, S.P.R., Bakhshi, N.N. "Thermal Cracking of Canola Oil: Reaction Products in the Presence and Absence of Steam", Energy Fuels **1996,** *10* 1150–1162.

[4] Katikaneni, S.P.R., Adjaye, J.D., Idem, R.O., Bakhshi, N.N. "Performance Studies of Various Cracking Catalysts in the Conversion of Canola Oil to Fuels and Chemicals in a Fluidized-Bed Reactor", JAOCS **1998,** *75*(3) 381–391.

[5] Sadrameli, S.M., Green, A.E.S. "Systematics of Renewable Olefins from Thermal Cracking of Canola Oil", Journal of Analytical and Applied Pyrolysis **2007**, *78*(2) 445-451.

[6] Katikaneni, S. P.R., Adjaye, J.D., Bakhshi, N.N. "Catalytic Conversion of Canola Oil to Fuels and Chemicals over Various Cracking Catalysts", CJChE **2009**, *73*(4) 484 – 497.

[7] Dupain, X., Costa, D., Schaverien, C.J., Makkee M., Moulijn, J. "Cracking of a Rapeseed Vegetable Oil under Realistic FCC Conditions", Applied Catalysis B: Environment 2007 *72* 44–61.

[8] Ng, S.H., Zhu, Y., Humphries, A., Zheng, L., Ding, F., Gentzis, T., Charland, J. P., Yui, S., *Energy Fuels* **2002,** *16*, 1196–1208.

.

(24a) Data Requirements for Heavy Oil Production - A Reservoir Fluid Perspective

Shawn D. Taylor and John Ratulowski, DBR Technology Center, Schlumberger, Edmonton, AB, Canada

There is a spectrum of methods available for the production and transportation of heavy oils from a variety of different environments (e.g. arctic, onshore, offshore, urban). The production methods may be grouped into one of three main categories; primary methods such as CHOPS, thermal methods including combustion and steam injection, and non-thermal methods such as VAPEX. These methods may be used on their own in or in combination, such as solvent assisted steam injection processes. In each of these cases, some level of understanding of the reservoir fluids is required to select and optimize the production process. As a general rule of thumb, techniques requiring alteration of the fluid properties during production require a greater understanding of the fluid properties.

From an industry perspective, there are a few key challenges that are common to understanding the heavy oil phase behavior and fluid properties for all production processes. The first challenge is deciding what samples to collect. The choice of sampling is determined to some extent on the type of processes (and the related data requirements) under consideration for a particular reservoir. In addition, there is a growing awareness that heavy oil fields often involve fluids with a range of fluid properties, where the variation in properties may be both vertical and aerial. Therefore, there is a need to develop new sample selection approaches that provide the necessary samples that meet the data generation requirements of the production processes of interest as well as to allow identification and quantification of potential fluid properties variations within a heavy oil field.

After selection of sampling locations, there is the challenge of actually acquiring the fluid samples. Ideally, samples would be acquired directly from the reservoir using fluid sampling tools. Unfortunately, due to the low mobility of heavy oil fluids in a reservoir, such tools are often ineffective and most fluid samples must be acquired at the surface when the wells are already flowing or through extraction of a small volume of fluid from a limited number of cores. As a result, thermodynamic models for reservoir simulation or transport will be developed using estimated fluid properties since key data may be not be measurable with the available samples, the data may not provided in time, or there is simply a lack of representative samples.

Once in the laboratory, there are challenges related to the handling of the samples and subsequent measurements. It most case the fluid samples contain some level of contamination, which may include drilling mud, fine solids, and emulsified water, that can alter the measured fluid properties. While there are options available to remove such contaminates, there is no reliable standard approach for 'cleaning' that ensures the samples remain representative. Thus, there is a need to develop a standard method or suite of standard methods for preparing 'representative' fluid samples for measurement. The final challenge is the ability to routinely and consistently characterize heavy oil fluids. While some data can be readily generated using existing standard methods, either directly or through minor modifications to conventional oil procedures, there are a number of key data needs where there is no 'standard' methodology. A simple example is the whole oil composition for heavy oils, where the options are often limited to methods that do not offer enough detail about the composition (e.g. SARA) or cannot resolve a significant portion of the composition (e.g. GC-based methods). A more detailed knowledge of the heavier components is needed for fluid modeling. A solution to this example may be a suite of standardized compositional approaches that address the requirements of each type of production process.

This presentation will provide a brief overview of the types of data required by different production processes and attempt to identify areas of potential technology development in the selection, acquisition, handling, and analysis of heavy oils.

See more of this Session: Heavy Oil and Other Challenging Resources I
See more of this Group/Topical: Topical D: Chemical Engineering in Oil and Gas Production and Other Complex Subsurface Processes

(24b) New Directions for the Oilsands and Heavy Oil Industry Enabled by Fundamental Research

John M. Shaw, Chemical and Materials Engineering, University of Alberta, Edmonton, AB, Canada

Hydrocarbon production and refining have entered their second century as global, industrial scale processes. There is a tendency to treat them as sunset industries with only limited research and development needs and potential, as large thermophysical property databases have been created, and a panoply of experimental techniques, and commercial computational tools have been available for some time. While it can be argued that for natural gas and light oil processing, this is so, even for these cases critical exceptions arise particularly with respect to phase behavior and physical property prediction quality that affect process safety and optimization [1]. For unconventional hydrocarbon resources (e.g.: heavy oil and bitumen), a growing segment of the hydrocarbon energy and petrochemical industries, this is certainly not the case. Customary experimental and computational tools are of limited value because bitumen and heavy oil include nano-dispersed and self-assembled phases, are opaque to visible light, include solid phases, and comprise large molecules that are ill-defined. Process identification, design and optimization present formidable challenges for engineers and researchers. Phase equilibria, and transport properties [2, 3] are not readily measured or predicted. Chemical reaction schemes are not readily optimized. Apparently simple measurements such as viscosity, a key transport property in production and pipelining, have become an uncertain and debated topic [3]. The number and nature of phases present and the enthalpy of mixing associated with bitumen + n-alkanes, a mixture class which arises in paraffinic deasphalting and blending, remain unresolved. Much work is conducted blind or by analogy and it is not clear whether existing processes, from production to separation and refining are appropriate or have been optimized with respect to capital or operating cost, greenhouse gas emission, efficiency or energy use. The molecular composition, phase behaviour and physical properties of these hydrocarbon resources are complex. Significant experimental and modeling challenges are present at all length scales. The nature and behaviours of these materials are only being discovered through creative and coordinated use of diverse and novel experimental and theoretical approaches targeting knowledge gaps at the molecular, supramolecular and bulk material length scales, and keeping track of what is really known at each step. One example, drawn from our recent work shows how to construct phase diagrams for bitumen and asphaltenes using a combination of calorimetry, rheometry, quantum mechanics, polarized light microscopy and photoacoustic infrared spectroscopy [4-6]! New data and knew knowledge are creating new opportunities in a challenged and challenging sector. 1. Fluid Phase Equilibria 264 (2008) 137-146; 2. Petroleum Science and Technology, 25 (2007) 773–790; 3. J. Chem. Eng. Data, 55 (2010) 1389-1397; 4. Fluid Phase Equilibria 268 (2008) 51–60; 5. Fluid Phase Equilibria 268 (2008), 134-141; 6. Fluid Phase Equilibria 272 (2008) 32-41.

See more of this Session: Heavy Oil and Other Challenging Resources I
See more of this Group/Topical: Topical D: Chemical Engineering in Oil and Gas Production and Other Complex Subsurface Processes

(24c) Towards a Modeling Framework for Heavy Oil Phase Behavior and Properties

Harvey Yarranton, Marco Satyro and Orlando Castellanos Diaz, Chemical and Petroleum Engineering, University of Calgary, Calgary, AB, Canada

Heavy oil and bitumen are a challenging resource to recover and process mainly due to their high viscosity. Typically, heat is used to reduce the viscosity; for example, thermal processes are used to recover in situ heavy oil and a hot water process is used to extract bitumen from mined oil sands. These methods are energy and water intensive. Solvent-based or solvent-assisted processes are a potential alternative but require the understanding of more complex phase behavior. Data, characterization methods, and modeling methodologies are all lacking for mixtures of heavy oil and solvents; hence, the development of these processes carries significant risk.

A comprehensive characterization and modeling methodology for these fluids is under development at the University of Calgary. The approach involves new characterization methods and an equation-of-state based phase behavior model supported by new property correlations. A synopsis of this approach is presented with a focus on characterization.

A key challenge in characterizing heavy oils is the large non-distillable fraction of the oil. Conventional methods are based on the distillable fraction and properties must be extrapolated over the non-distillable fraction. For heavy oils, liquid-liquid equilibrium calculations are sensitive to the characterization of the medium fraction of the oil and asphaltene precipitation is sensitive to the characterization of the heavy fraction. Therefore, phase behavior predictions require a more accurate fluid characterization than can be obtained from simple property extrapolations originally developed for distillation based methods.

An alternative to extrapolation is to more rigorously characterize the non-distillable fraction of heavy oils. The idea is to fractionate the non-distillable portion of the oil using solvent-based methods and to measure properties of these fractions including vapour pressure, molar mass, density, and heat capacity. This data will be combined with GC, distillation, SimDist, and possibly PNA analysis to create a more complete representation of the oil suitable for modeling with equations-of-state.

All of the measurements except for vapour pressure are straightforward. A deep vacuum apparatus was designed to measure the vapour pressures of low volatility heavy fractions and possible to aid in the fractionation of the lighter end of the non-distillable cut. Preliminary characterization results using this approach are presented and the sensitivity of VLLE predictions to the characterization is examined.

See more of this Session: Heavy Oil and Other Challenging Resources I
See more of this Group/Topical: Topical D: Chemical Engineering in Oil and Gas Production and Other

(24d) Measurement and Modeling of Phase Equilibrium in Heavy Oil / Propane Systems

Francesca Galimberti[1], Nicolas von Solms[2], Wei Yan[2], Birgit Elkjær Ascanius[2] and Erling H. Stenby[2], (1)Eni Spa, div.R&M, San Donato Milanese (MI), Italy, (2)Department of Chemical and Biochemical Engineering, Technical University of Denmark, Kgs. Lyngby, Denmark

In order to model phase equilibrium of heavy oil with propane, several experimental measurements were undertaken using four samples from two different heavy crude oils, two as natural feeds and two as products of previous treatments.

The phase equilibrium experiments were performed in a high pressure PVT cell consisting of heavy oil/ propane mixtures at a range of temperatures, pressures and solvent/oil ratios. The densities of the two resulting phases were measured at experimental conditions. The oil and the two phases resulting from the experiments (light and heavy) were further separated by SARA characterization (they were divided into Saturates, Aromatics, Resins and three asphaltenes fractions) and the mass, molecular weight, refractive index and densities of each of the six resulting SARA fractions were measured. These values were used in order to obtain critical properties of the fractions for phase equilibrium modeling with an equation of state (Peng-Robinson). About 70 equilibrium experiments were performed altogether.

An extensive modeling effort was also carried out using the Peng-Robinson equation of state, combined with an analysis of several correlations for critical properties, acentric factors and binary interaction parameters. The model gave results for oil yield, volume fraction of the phases as well as densities of the phases. Agreement was good in most cases, despite the complexity of the systems and the errors inherent in the various characterization methods.

See more of this Session: Heavy Oil and Other Challenging Resources I
See more of this Group/Topical: Topical D: Chemical Engineering in Oil and Gas Production and Other Complex Subsurface Processes

(24e) The Compositional Continuum of Petroleum: Detailed Molecular Characterization of Heavy Crude Oils and Asphaltenes by Ultrahigh Resolution FT-ICR Mass Spectrometry

Amy M. McKenna[1], Ryan P. Rodgers[1], Christopher L. Hendrickson[1], Gregory T. Blakney[1], Joshua T. Savory[1], Nathan T. Kaiser[1], Mmilili M. Mapolelo[1], Alan G. Marshall[2], Winston K. Robbins[3], Brandie M. Ehrmann[2] and Chang S. Hsu[4], (1)Ion Cyclotron Resonance Program, National High Magnetic Field Laboratory, Tallahassee, FL, (2)Ion Cyclotron Resonance Program, Florida State University, Tallahassee, FL, (3)Consultant, Brunswick, ME, (4)Ion Cyclotron Resonance Program, Consultant, Brunswick, ME

The world's current and future dependence on fossil fuels will shift the types of materials slated for production. Light, sweet crude will diminish as their reserves approach exhaustion and the production of heavier, more heteroatom-rich crudes will increase to meet world demand. As petroleum producers increase oil production to meet the world's growing energy needs, they will inevitably access an ever-widening quality range of petroleum materials (i.e., bitumen and heavy oils) and tap into deeper and deeper offshore reservoirs. Canada has continuously ramped up oil sand mining in Alberta over the past two decades to surpass the Middle East as the primary supplier to the United States. Richer in heteroatom content and present in severe environments (temperature and pressure), these materials pose major flow assurance challenges. Thus, detailed information on oil composition is paramount in understanding its reactivity and production behavior. Oil companies sell molecules and characterization at the molecular level is vital to determine how an oil will behave during production. However, the inherent complexity of crude oil makes it difficult to provide compositional information on even the most likely culprits in common production and refining problems. In fact, many of the most common characterization techniques are based on bulk properties or solubility. Fourier transform ion cyclotron resonance mass spectrometry (FT-ICR MS) allows detailed characterization of complex petroleum samples at the level of elemental composition assignment. Ultrahigh resolution (450,000-700,000 at m/z 500) enables identification of isobaric species that differ in mass by 3 milliDalton or less, and high mass accuracy (better than 300 ppb mass error) allows for unambiguous molecular formula assignment to each of more than ten to forty thousand peaks in each mass spectrum. Thus, it is now possible to identify, sort and monitor simultaneously thousands of elemental compositions as a function of boiling point as well as provide a comprehensive analysis of non-boiling species.

Proposed over 20 years ago, the Boduszynski model describes the evolution of petroleum composition as a function of atmospheric equivalent boiling point [1]. Based on his model, Boduszynski concluded that "most of petroleum components do not exceed a molecular weight of about 200." He acknowledged that the results are controversial, "These findings are significant because of the existing controversy over whether there is an appreciable concentration of molecules in petroleum having molecular weights greater than 2000 Da. Data show there is not." However, a definitive proof of Boduszynski's model requires direct, complete compositional characterization of complex distillate cuts unavailable at that time. If substantiated, the Boduszynski model would impose strict limits on molecular weight distribution for both distillable and nondistillables petroleum fractions that contradict many previously published assertions about petroleum molecular weight and composition.

Here, we present our cumulative efforts in heavy oil and asphaltene characterization by (+/-) electrospray and atmospheric pressure photoionization FT-ICR mass spectrometry; summarize class, type, and molecular weight trends for distillable/nondistillable petroleum species; and compare results to proposed asphaltene structural models. In addition, structural characterization of nickel and vanadyl porphyrins on from an unfractionated asphaltene will be presented in a single mass spectrum for the first time. Detailed characterization of an Arabian heavy crude oil distillation series defines maltene and asphaltene compositional space (aromaticity and carbon number), and calculation of the hydrogen to carbon ratio (H:C) bounds the distillable space for all identified classes, revealing that projection of the compositional continuum to high carbon number cannot account for bulk H:C ratios. Chromatographic fractionation of crude oils from various geographic regions with varying total acid number and heteroatom content followed by subsequent mass spectral characterization reveals compositional differences and provides further insight into the separation and characterization of nonpolar and polar species in crude oils. Characterization of entrained species in asphaltene aggregates will be discussed. Compositional information for distillables and nondistillables and its implications for asphaltene aggregation will be discussed in detail.

References [1] Boduszynski, M. M.; Altgelt, K. H., Composition and Analysis of Heavy Petroleum Fractions. CRC Press: New York, NY 1994.

Keywords: APPI, petroleum, Fourier Transform, Ion Cyclotron Resonance, ICR, FT-ICR, FTMS

See more of this Session: Heavy Oil and Other Challenging Resources I
See more of this Group/Topical: Topical D: Chemical Engineering in Oil and Gas Production and Other Complex Subsurface Processes

(24f) Understanding Asphaltene Aggregation and Deposition: A Necessity for the Future of Oil Research

Michael P. Hoepfner, Tabish Maqbool, Nasim Haji Akbari Balou and H. Scott Fogler, Chemical Engineering, The University of Michigan, Ann Arbor, MI

Heavy oil is an extensive but challenging source of energy. Many difficulties currently exist with traditional crude oil reserves and working with heavy oils only complicates the extraction, transportation and refining processes. This presentation will highlight why fundamental chemical engineering research on petroleum is tantamount to the efficient handling of heavy crude oils. Central to this area is the study of asphaltenes due to their large fraction in heavy oils and their strong tendency to aggregate and deposit with changes in temperature, pressure or composition. Our research group has recently shown that asphaltene aggregation occurs under conditions previously identified as stable, further complicating the understanding of an already bewildering fraction of oil [1]. We have also shown that in addition to aggregation occurring at low degrees of destabilization, asphaltene deposition also occurs under such conditions [2]. These recent findings question the validity of both previous experimental studies and many of the thermodynamic models currently used to predict asphaltene stability. Fundamental research on asphaltenes can reveal the true mechanism(s) for asphaltene stability that can be directly applied to the recovery, transportation and processing of heavy crude oil. The industrial implications of asphaltene stability will also be discussed.

[1] T. Maqbool, A.T. Balgoa, and H.S. Fogler, "Revisiting Asphaltene Precipitation from Crude Oils: A Case of Neglected Kinetic Effects," Energy & Fuels, vol. 23, Jul. 2009, pp. 3681-3686.

[2] M. Hoepfner, T. Maqbool, and H.S. Fogler, "Effects of Precipitant Concentration on Asphaltene Deposition," AIChE National Meeting, Nashville, TN, 2009.

See more of this Session: Heavy Oil and Other Challenging Resources I
See more of this Group/Topical: Topical D: Chemical Engineering in Oil and Gas Production and Other Complex Subsurface Processes

(82a) Influence of Thermal Treatment Temperature On Wood Structures

Ping Wang[1], Bret H. Howard[2], Bryan D. Morreale[2], Sheila Hedges[2], Dirk VanEssendelft[1] and David Berry[1], (1)U.S.DOE/NETL, Morgantown, WV, (2)U.S. DOE/NETL, Pittsburgh, PA

Utilization of biomass as a co-feed supplement with coal in various coal-fired/gasification energy conversion processes is typically thought to require size reduction of the biomass. It is difficult to reduce/pulverize biomass without pretreatment because biomasses are fibrous and compressible. Torrefaction is identified as a promising thermal pretreatment process. After torrefaction, biomass becomes brittle and friable and its grindability is improved. Biomass grindability relates to the mechanical properties of biomass, which is in turn related to cell wall structure and the thickness of the cell walls. The cell wall composition of biomass consists of cellulose, hemicelluloses and lignin. During thermal treatment such as torrefaction, hemicellulose is the most reactive of the three components and is preferentially decomposed. Thermal treatment causes chemical changes that can alter the structure of biomass and the treatment temperature used is probably the most important factor. The objectives of this study are to investigate the impacts of thermal treatment temperature on the hardwood microstructure using SEM and to understand the mechanism of how torrefaction can improve the grindability of wood biomass. Four pieces of hardwood cut from axial and radial systems will consecutively be treated at 200, 250, 300, 350, 450 and 600oC for 0.5 hr. After each thermal treatment the wood samples will be examined using SEM, and the sample cell wall compositions will be analyzed using TGA. The cell wall structure models reported in the literature will be used to elucidate the mechanism of the wood structure change and its component decomposition caused by thermal treatment at different temperatures.

See more of this Session: Biomass Gasification
See more of this Group/Topical: Fuels and Petrochemicals Division

(82b) Reactive Molecular Dynamics Simulations of Cellulose Gasification Chemistry

John W. Daily, Department of Mechanical Engineering, University of Colorado at Boulder, Boulder, CO and **Luke A. Erickson**, Mechanical Engineering, University of Colorado at Boulder, Boulder, CO

Two problems preventing the commercial viability of biomass gasification technologies are low product yields and high tar production. Both of these stem from a lack of understanding about the underlying chemistry. Detailed reaction mechanisms have not been developed because of the complexity of the thermal decomposition process and the difficulty observing all the short-lived intermediates in experiments. Computational methods, until recently, have been too either computationally expensive to model polymers or incapable of accurately describing bond breakage and formation. Reactive molecular dynamics may be a tool that will simulate biomass decomposition accurately enough to give insight into the reaction mechanisms. The ReaxFF force field [1] changes classical molecular dynamics by evaluating each pair of atoms at each timestep to determine the bond order from internuclear distance based on empirical relationships. Using empirical data decreases computation time by a factor of 10^3 over ab initio methods and also allows for molecular decomposition.

The current study has attempted to validate this approach by reproducing experimentally observed product distributions. These final product distributions have been well documented in the experimental literature. We expect to see five- to six-carbon rings early in the decomposition process followed by the production of gas phase molecules like hydroxyacetaldehyde, carbon oxides, water, and molecular hydrogen [2].

Simulations have been run on cellulose structures of 30 glucose base units for 1 ns in 0.2 fs timesteps. Temperature was varied between 900K and 2000K in constant temperature and constant energy simulations. Simulations in vacuum, nitrogen, and steam were considered. Results show production of physically realistic products including hydroxyacetaldehyde and water. However, depolymerization is not observed despite being widely evident in experiments. Further work will consider whether ReaxFF can account for unimolecular electron transfer reactions.

References: 1. Van Duin, A. C. T., et al., (2001). "ReaxFF: A Reactive Force Field for Hydrocarbons." J. Phys. Chem. A 105: 9396-9409. 2.Lin, Y.C., et al. (2009) "Kinetics and mechanism of cellulose pyrolysis" J Phys Chem C 113: 20097-20107.

See more of this Session: Biomass Gasification
See more of this Group/Topical: Fuels and Petrochemicals Division

(82c) Chemical Modeling of Biomass Gasification

Mark Nimlos[1], AnGayle K. Vasiliou[2], Adam. M Scheer[2], Mark Jarvis[1], Perrine Pepiot[1], David. J Robichaud[1], Katherine Gaston[1] and G. Barney Ellison[2], (1)National Bioenergy Center, National Renewable Energy Laboratory, Golden, CO, (2)Department of Chemistry and Biochemistry, University of Colorado at Boulder, Boulder, CO

The gasification of biomass produces a syngas that can be used for the production of renewable transportation fuels, but often requires cleaning and conditioning. Of particular importance is the formation of tars, or condensable organic compounds. These compounds can be removed by catalytic steam reforming, but this adds considerable cost and complexity. An understanding of the chemical mechanisms and kinetics of gasification allows design and optimization for minimized tar formation, thus reducing the demand on the steam reforming step. Of particular interest to gasification are Polynuclear Aromatic Hydrocarbons (PAHs), which are formed at typical gasification temperatures (700°C to 950°C) and are refractory. We have studied the pyrolysis of model compounds, biopolymers and biomass and have developed chemical models to describe the kinetics of PAH formation. Model compounds were pyrolyzed using a hyperthermal nozzle and products, including radicals, were measured with laser photoionization mass spectrometery and matrix isolation infrared spectroscopy. Biopolymers and biomass were studied in a Laminar Entrained Flow Reactor (LEFR) and tars were detected with a Molecular Beam Mass Spectrometer (MBMS). Kinetic models are being used in Computational Fluid Dynamics (CFD) modeling to develop comprehensive and predictive tools for gasifier performance. These tools will ultimately aid in improved gasifier design and operation for more cost effective fuel production.

See more of this Session: Biomass Gasification
See more of this Group/Topical: Fuels and Petrochemicals Division

(82d) Sawdust Pyrolysis and CO_2 Gasification of Petroleum Coke In a High Pressure Flat-Flame Burner

Aaron Lewis, Sam Goodrich, Kolbein Kolste and Thomas H. Fletcher, Chemical Engineering, Brigham Young University, Provo, UT

CO_2 gasification kinetics of petroleum coke were studied in a high pressure flat-flame burner from 1 to 15 atm and particle heating rate of 10^5 K/s. The rapid pyrolysis of sawdust was also studied using an atmospheric flat-flame burner. The high initial particle heating rates of the pressurized flat-flame burner more closely match industrial heating rates (10^6 K/s) than those provided by a drop tube reactor (10^4 K/s) or a TGA. Effects of temperature, pressure, and residence time on pyrolysis yields, char composition, mass release, and char morphology using SEM images were determined. All experiments were performed using a feed rate of ~1 g/hr of 45-75 micron particles to represent single particle behavior. Very low tar yields (< 6 wt%) were observed during sawdust pyrolysis from 1150-1450 K, primarily caused by secondary tar cracking. The CPD model, which was originally developed for coal, predicts reasonably well measured sawdust pyrolysis yields (tar, char, and light gas) when combined with a tar cracking model. SEM images of sawdust char reveal that char morphology continues to change following complete pyrolysis even though mass release asymptotes. SEM images of petroleum coke char show small cracks on the char surface where volatiles likely escaped. The fractures on the char surface were likely caused by high heating rate and not pressure since the cracks were observed at all pressures tested. The mass release during pyrolysis of sawdust at high heating rates exceeds the ASTM volatiles test by ~10 wt%. First-order kinetic rate constants were determined for CO_2 gasification of petroleum coke.

See more of this Session: Biomass Gasification
See more of this Group/Topical: Fuels and Petrochemicals Division

(82e) Effects of Feedstock and Particle Size On Biomass Gasification Tars

Katherine R. Gaston[1], Kim Magrini[2], Mark W. Jarvis[2], Kristin Smith[1] and Mark Nimlos[1], (1)National Bioenergy Center, National Renewable Energy Laboratory, Golden, CO, (2)National Renewable Energy Laboratory, Golden, CO

Particle size fractions (0.5 to 4.0 mm) of seven different feedstocks were studied in a biomass gasifier at 850°C. Light gases were measured and tars were analyzed by a molecular beam mass spectrometer; the results were correlated to the feedstock analyses.

See more of this Session: Biomass Gasification
See more of this Group/Topical: Fuels and Petrochemicals Division

ABSTRACT-EXTENDED

BIOMASS PYROLYSIS IN A THREE-STAGE GASIFIER

David L. Seidel and Thomas R. Marrero

SRADCO, LLC and University of Missouri

Columbia, MO

The purpose of this paper is to present results of experimental studies of the pyrolysis/gasification of various biomass materials. The experiments have been conducted over the last five years in one prototype system with three separate reaction chambers lined with refractory materials. In the feed chamber pyrolysis or primary gasification occurs; the second chamber serves as a partial combustor, and the outlet chamber is a high-temperature reactor. The prototype has sampling points in, after and in-between the three chambers from which a variety of biomass products--solid, liquid and gaseous--are extracted. Biomass feed materials have been: corn, corn stover, clover, grasses, milo stover, peanut shells, oak wood, wood sawdust, cow manure , horse manure , hog manure, compressed paper, grumb rubber, chipped rubber tires, whole tires, and other renewable wastes. These experiments have resulted in mass and energy balances for the pyrolysis/gasification system, operations with precise temperatures in each chamber, low air-emissions of criteria pollutants, and a relatively high thermal efficiency. The biomass derived products are high carbon content activated biochar, biocrude oils, and biofuel gases generated by a sustainable process. The properties of the biomass derived products were tested for purity, density, liquid and gas absorption, and adsorption, thermal conductivity, fuel characteristics, and other properties. Some of the results are an 85% carbon content biochar with high methane and nitrogen uptake properties, a 1250 BTU per cubic foot biogas, and a golden high viscosity lubricating oil, soil revitalizing amendment or fertilizer and various grades of biofuel oils and lubricants. The experiments yielded promising results for new techniques in the processing of activated carbon, nanocarbons, carbides, natural fertilizers, absorbents, and adsorbents, natural chemical solvents, oils, alternative fuels and gases. Thus, the significance of these prototype tests is the design and construction of a scaled-up system for commercial projects to manufacture carbon products for high-value applications utilizing low-value sustainable renewable biomass feed materials.

The gasification prototype was used in addition to pre-experiment dryers and electric ovens. The collection, analysis, and comparisons were conducted in accordance with ATSM and EPA testing methods. The basic analyses were conducted at the field sampling laboratory. No detailed chemical analyses were conducted during this studies period.

The following array of pictures is the photographic documentation of three phases of samples.

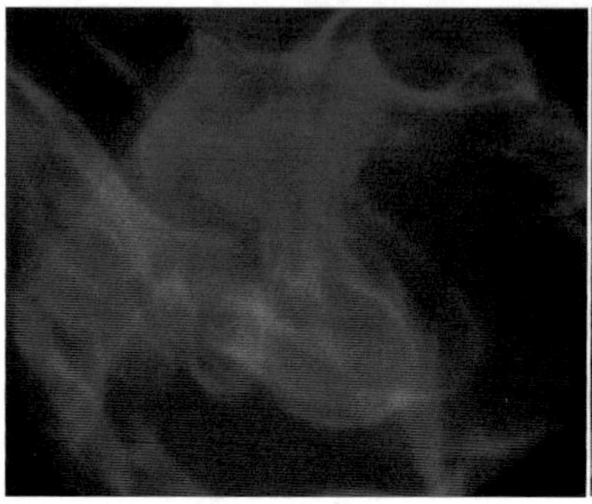

Flammable Gases

Bio-Liquids

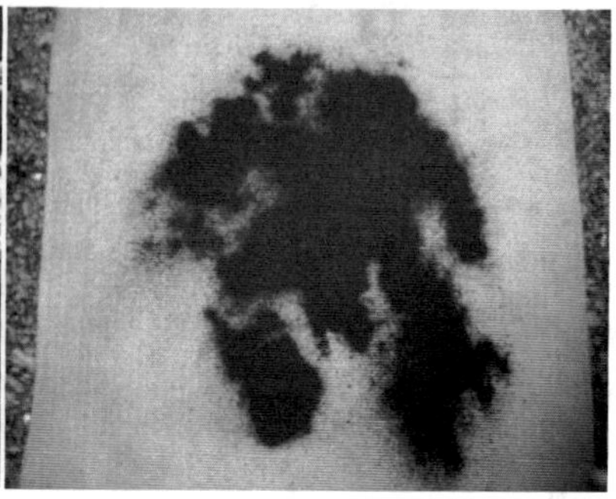

Pyrolysis Carbon Solids

The summary of the experimental studies are presented in the following matrix chart. The table rates the bio-product on the best results of a series of experimental tests with random samplings.

Sample	Moisture (%)	Carbonization	Final Carbon Content	Bio-Gas Weight %	Bio-Liquid Weight %	Adsorption Property	Thermal Conduct	Fuel Characteristics
Corn Seed	20	Poor	65	15	18	Low	Low	Fair
Corn Stover	35	Good	78	20	24	Medium	Medium	Good
Clover	27	Good	80	22	25	Medium	Medium	Good
Prairie Grass	32	Good	80	21	25	Medium	Medium	Good
Wheat Straw	24	Excellent	82	18	23	High	High	Good
Soybean Stover	30	Good	77	16	29	Medium	Medium	Good
Milo Stover	40	Good	68	22	27	Medium	Medium	Good
Peanut Shells	15	Excellent	89	22	25	High	High	Excellent
Oak Wood	37	Good	78	24	26	Medium	Medium	Excellent
Oak Sawdust	45	Poor	70	15	20	Medium	Medium	Good
Cow Manure	75	Good	65	18	32	Medium	Low	Good
Horse Manure	53	Good	66	17	28	Medium	Low	Good
Hog Manure	60	Good	64	16	30	Medium	Low	Good
Compressed Newspaper	9	Excellent	82	14	16	Medium	Medium	Excellent
Grumb Rubber	8	Excellent	82	15	18	Medium	High	Good
Tire Chips	6	Excellent	85	15	20	Medium	High	Excellent
Whole Tires	4	Excellent	85	15	20	Medium	High	Excellent

A Study on Modeling and Operational Optimization of Biomass Gasification Processes Using Neural Networks

Maurício B. de Souza Jr[1], *Amaro G. Barreto Jr*[2], *Leonardo Couceiro Nemer*[3]
Rio de Janeiro Federal University, Chemical Engineering Department, Centro de Tecnologia, bloco E, Ilha do Fundão, 21949-900, Rio de Janeiro – RJ, Brazil
E-mails: [1]*mbsj@eq.ufrj.br;* [2]*amaro@eq.ufrj.br;* [3]*leonardonemer@hotmail.com*

Patrícia Oliva Soares[4], *Cristina P. B. Quitete*[5]
PETROBRAS, R&D Center, R&D in Gas, Energy and Sustainable Development
Av. Horário de Macedo, 950, Cidade Universitária, 21941-915 Rio de Janeiro, Brazil
E-mails: [4]*oliva@petrobras.com.br;* [5]*crispontes@petrobras.com.br*

1. INTRODUCTION

Gasification processes are rather complex and difficult to model as they include gas-solid two-phase flow, mass and heat transfer, pyrolysis, homogeneous gas phase reactions and hetrogeneous gas-solid reactions. Modeling of these phenomena based on basic principles of conservation is still at an incipient stage of development (Guo et al., 2001). Additionally, most of the works have focused on coal gasification (e.g., Nougués et al., 2000; Guo et al., 1997). Consequently, the development of a mechanistic model demands that many idealizations and suppositions are made (Li et al., 2004), resulting in a very simplified model with little predictive capability.

Artificial neural networks (ANNs) are universal approximators (Haykin, 1999) and have received numerous applications (Himmelblau, 2000). The literature, as indicated by Xiao et al (2009), points out their ability to recognize highly nonlinear relations and organize disperse data in a nonlinear model in the context of empirical or hybrid modeling. These characteristics of the ANNs are very interesting and useful motivating their use in the modeling of biomass gasification processes.

Hence, the present work aims to investigate – through the use of artificial neural netorks (ANNs) and literature data – the correlation between the composition of the produced gas and the characteristics of different biomass for several operating conditions employed in fluidized bed gasifiers. Additionally, the neural network based developed model is employed to find conditions that maximize the yield of a given component of the produced gas.

This work is structured as follows. In section 2 fundamental aspects concerning biomass gasification and the modeling of the process are briefly reviewed. Section 3 focus on the modeling based on ANNs, while section 4 presents the optimization investigations using the developed models. Finally the main conclusions are presented in section 5.

2. BIOMASS GASIFICATION

2.1 Fundamental Aspects

Gasification is a process in which a solid or liquid fuel is converted into a gaseous fuel with or without contact with a gasfying agent. Coal, biomass, petroleum coke and other materials can be used in the process. The produced gas is mostly composed of hydrogen (H_2), carbon monoxide (CO), carbon dioxide (CO_2) methane (CH_4), traces of heavier hidrocarbons (as ethane and ethylene), water, nitrogen (when air is used as gasifying agent) and some contaminants. Besides the gaseous products, there are also subproducts as tar and solid non-converted residual carbon (char) (Basu, 2006).

The gas composition and the production of subproducts depend on several factors as: energy delivered to the process, type of gasifier, operating conditions and type of biomass employed. The gas produced can be used in different applications such as: gas turbines or internal combustion enginees, production of syngas and, after an adequate cleaning up and reforming, production of hydrogen or direct use on fuel cells (Chen et al, 2003, Baratieri et al, 2008). The reactions inside the gasifier can be divided in four stages according to the temperature (Basu, 2006): drying (> 150°C); pyrolysis (150 -700°C); combustion (700 - 1500°C) and reduction (800 -1100°C).

Some characterisitcs of the biomass have a significative effect on the performance of the gasifier. For this reason, proximate and ultimate analyses are used in order to characterize the biomass (Xiao et al., 2009).

Because of their flexibility for use with differenttypes of biomass (Basu, 2006), only fluidized bed (bubbling and circulating) gasifiers were studied in the present work. Different gasification agents can be considered, such as: air, oxygen, steam or a combination of them. In the case of use of air or oxygen, the heat released by the exothermic reactions between the oxygen and the fuel is used to keep the gasifier in the operating temperature and as heat source for the endothermic reactions. When steam is employed, it is necessary to use an external heat source (Basu, 2006).

2.2 Modeling of the Process

The availability of accuarate biomass gasification process models would help the operation and optimization of these processes. However, as noted by Nemtsov & Zabaniotou (2008), the majority of the works have been developed for coal. In comparison with coal, biomass is made up not only of lignin, but also of cellullose and hemicellulose, each one having its own thermal behavior, whst makes the biomass gasification even more difficult (Nemtsov & Zabaniotou, 2008).

Corella & Sanz (2005) commented the difficulties of developing a model based on the kinetic equations of the different reactions, together with the mass and energy balances and hydrodynamic considerations for a circulating fluidized bed biomass gasifier. In this work, the authors cite the objective of developing a model "as good as possible".

Previous literature papers employed neural networks to predict characteristics of combustion, pyrolisys or gasification processes. Guo et al. (1997) developed a hybrid gasification model using ANNs to estimate reactivity parameters of different types of coal with relative success. Dong et al. (2002) used ANNs to predict the emission of pollutant gases in the combustion process of a mixture of coal and urban solid residuals with a good agreement between experimental and predicted data.

In 2001, Guo et al. developed a hybrid model for the gasification of biomass in a fluidized reactor that employed steam as gasification agent. The authors used multilayer ANNs to estimate parameters of a phenomenological model. The hybrid model was used to determine the production rate of the gas and its composition in terms of H_2, CO, CO_2 e CH_4. However, the neural networks were trained for each biomass separately.

In 2009, Xiao et al. used ANNs to predict LHV (lower heating values) of the gas and of the gas with tar and char and gas yields using the following input variables: type of residual (paper, wood, kitchen garbage, plastic and textile materials), gasification temperature and equivalence ratio. The results indicate that ANNs are a viable alternative for the modeling of the studied process.

3. MODELING USING ANNs

3.1 Methodology

First, a literature search was carried out for experimental biomass gasification data. Data from 21 references for gasification of different biomass were collected (Doherty *et al.*, 2009, He *et al.*, 2009, Xiao *et al.* 2009, Baratieri *et al.* 2008, Nikoo *et al.*, 2008, Shen *et al.*, 2008, Souza-Santos 2008a,b, Siedlecki *et al.*, 2007, Cao *et al.*, 2006, Chen *et al.*, 2004, Li *et al.*, 2004, García-Ibañez *et al.*, 2004, Dellepiane *et al.*, 2003, Lv *et al.*, 2003, Schuster *et al.*, 2001, van der Drift *et al.*, 2001, Turn *et al.*, 1998, Faaij *et al.*, 1997, Boateng *et al.,* 1992, Ergundenler *et al.*, 1992).

The data search had three main information focuses: the gasification system (technical and operating aspects), type of biomass and the characteristics of the produced gas, as described in the following:

- Gasification system: type of gasifier as well as the dimensions of the reactor, the operating conditions and the gasification agent employed.
- Biomass: proximate and ultimate analysis data were collected. Additionally, when the heating value was not provided, its value was estimated based on the ultimate analysis data.

- Produced gas: the main information collected was the composition in terms of H_2, CO, CH_4 e CO_2.

Some characteristics of the gasification system (as the type of bed and the operation pressure) were restricted in the search for building the database that would be further used to train the neural networks. So, the ANNs were trained for fluidized bed gasifiers, using sand as bed and operated at atmospheric pressure. Only laboratory and pilot dimensions were used in this study. Initially, data for all the gasification agents were included in the training of the ANNs.

The database built had 181 input-ouput patterns (Nemer, 2010). In the following, some observations are presented regarding the collected data:

- The contents of ashes, volatile components and fixed carbon were determined on a dry matter basis.
- The variable S/B indicates the ratio between the values of steam and biomass feed mass flows.
- The variables C, H, N, O and S indicate the mass percentage of carbon, hydrogen, nitrogen, oxygen and sulfur, respectively in the biomass fuel.
- The composition of the produced gas is given in volumetric percentage.
- The variable C_2H_n indicates the sum of the hydrocarbons with two atom of carbon that are formed in the process (mostly, ethylene and ethane).

At first, the choice of the input variables for the ANNs model was made heuristically, considering the analysis of the studied problem and the influence of the input variables in the prediction of the composition of the produced gas. Later, a sensitivity analysis was also implemented in order to help in that task.

In this work, the Statistica Neural Networks – SNN (Statsoft®) sofware was used in order to train and validate the neural networks. The ANNs that presented the best performance were of the kind MLP (Multilayer Perceptron). MLP are feedforward, multilayered neural networks that typically present one input, at least one hidden layer and one output layer of computational nodes (the neurons). Details can be found in several references, including (Haykin, 1999). The MLPs employed in this work had onde hidden layer of hyperbolic tangent neurons.

3.2 Modeling Results using ANNs

ANNs were trained using partial information about the gasification system and the biomasses in order to make predictions about the produced gas (composition in terms of H_2, CO, CH_4, CO_2, C_2H_n). The following information concerning the operating conditions of the gasifier was used as input variables to the neural network: equivalence ratio, steam/biomass ratio (S/B), temperature of the gasifier (T) and gasification agent used (the categories: air, steam, air/steam, steam/oxygen). Additionally the following information about the biomass was also used as input variables to the neural network: proximate (specifically, moisture, ash and volatile contents) and ultimate (specifically, C, H and O percentages) analysis data.

It must be emphasized that only partial information was used in order to avoid a large number of input variables to the neural network and, consequently, a large number of neurons and of parameters, that could lead to an overdimensioned neural network, with little preditictive capability (Haykin, 1999), considering the limited amount of literature data used for training.

A very detailed study was carried out, concerning the design and comparison of multilayered (MLP) neural network models models. ANNs with multiple and individual outputs were trained. The inclusion of an input categorical variable that classified the gasifier as 'bubbling' or 'circulating fluidized bed' was also investigated. 'Specialized' neural networks developed for each gasification agent category and each gasifier type produced more parsimonious neural network models in terms of number of parameters.

High correlations values (ranging from 0.94 to 0.99) were obtained for these 'specialized' ANNs. This study is fully described in Nemer (2010). Here, due to space reasons, selected results are shown. The selection aimed to provide illustrative results of the application of ANNs in the modeling and optimization of gasification of idfferent type of biomass in fluidized gasifiers. In the following, results are shown for the neural network that predicts the hydrogen percentage in the produced gas of a bubbling fluidized gasifier, using steam as the gasification agent.

For the choice of this specific neural network, three hundred ANNs were compared employing 2/3 of the available patterns in the buit database for training and 1/3 for validation. The data for the input variables were between the following limits: *7.5 < Input1 (moisture) < 9.4; 71.02 < Input2 (volatile) < 82;36.57 < Input3 (C) < 48; 4.91 < Input4 (H) < 6.04; 39 < Input5 (O) < 45.43; 0.113 < Input6 (S/B) < 4.7 and* 650 < Input7 (T) < 900. The ANN selected was the one that presented the smaller errors both for training and validation.

The obtained MLP presents 7 linear neurons in the input layer (for the input variables: 1. moisture (%wt); 2. volatile content (%wt); 3. C (%wt); 4. H (%wt); 5. O (%wt); 6. S/B; 7. T (^{o}C)); 10 hyperbolic neurons in the hidden layer and 1 linear neuron in the outpout layer. It was trained using the backpropagation method during the 100 initial epochs and the conjugate gradient during the 127 last ones (Haykin, 1998, Nemer, 2010). This ANN will be further cited here as MLP 7:10:1.

Figure 1 illustrates the topology of MLP 7:10:1 (a) and its results (b). A very high correlation (0.99) between predicted ad observed value was obtained.

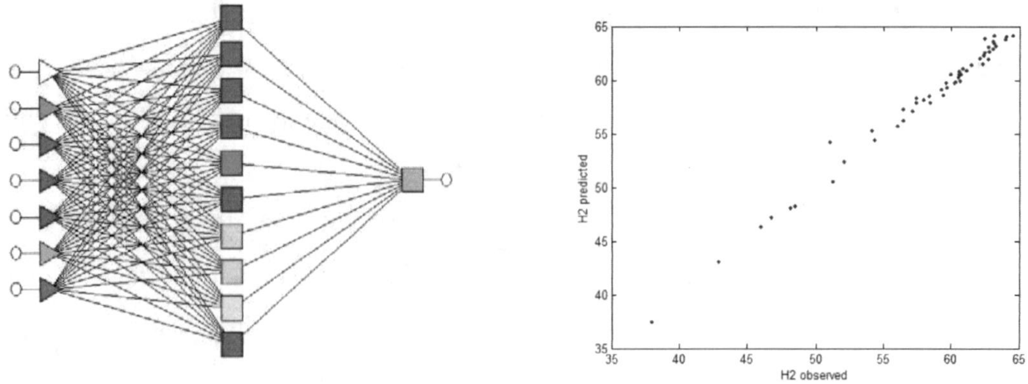

(a)

(b)

Figura 1 – MLP 7:10:1 for prediction of hydrogen in the produced gas (bubbling gasifier, gasificant agent: steam): (a) illustration; (b) predicted vs. experimental values

A sensitivity analysis was also carried out in order to evaluate the importance of each input variable to the predictive performance of the neural network. In this analysis, if one specific variable is considered 'unavailable' (that is, only its means value is used) the performance of the network should deteriorate and its Error should increase. Based on this fact, the sensitivity analysis, calculates the ratio between the Error and the Baseline Error (i.e. the error of the network if all variables are 'available'). This way, the higher the Ratio, the most important is that particular input variable to the performance of the ANN. The results of the sensitivity test are shown in Table 1, where the Rank lists the variables in order of importance. The results shown in Table 1 indicate that all input variables are important, being the hydrogen content in the biomass the most relevant, followed by the ratio steam/biomass.

Table 1 – MLP 7:10:1 for prediction of hydrogen in the produced gas (bubbling gasifier, gasificant agent: steam): sensitivity analysis

	Moisture	**Volatiles**	**C**	**H**	**O**	**S/B**	**T**
Ratio	8.00	4.45	4.85	8.877	8.160	8.27	6.49
Rank	4	7	6	1	3	2	5

4. PRELIMINARY RESULTS OF OPERATIONAL OPTIMIZATION

A preliminary investigation was also conducted of the optimization of the operation of a particular gasifier using the gasification model provided by the neural network. The neural model described in the previous section for a bubbling gasifier using steam as the gasificant agent was used.

For this study, the biomass was fixed and the operating conditions (in terms of T and S/B) were varied, according to the data present in the built database database (Nemer, 2010) in order to maximize the yield of a given component in the produced gas.

The results for wood and straw biomasses and maximization of the producction of hydrogen are described in the following to illustrate the procedure.

Initially, the response surfaces using the neural model and the data for each biomass were ploted as shown in Figure 2 (a) and (b) for wood and straw, respcctively. In Figure 2, the operating variables were varied considering the availability of data in that operating range. So, temperature was varied between 800 and 850 °C, for wood, and 650 and 900 °C, for straw. For the ratio S/B, the considered ranges were 1.1 < S/B < 4.7, for wood, and 0.4 < S/B < 0.9, for straw. It can be seen in Figure 2 that, if the operating values are restricted to those ranges, the maximization of H_2 in the produced gas demands higher S/B ratios and opposite directions for T (lower T for wood and higher T for straw). So, the model provided by the ANN provides information that could give the operator the right trends to maximize the production of a given product of interest.

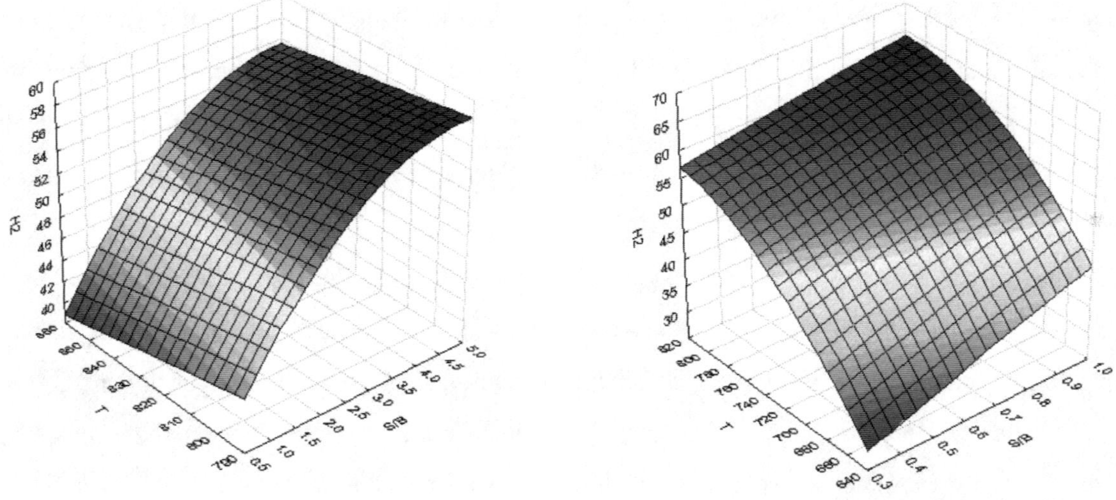

(a) (b)

Figure 2 - MLP 7:10:1 for prediction of hydrogen in the produced gas (bubbling gasifier, gasificant agent: steam): response surface for wood (a) and straw (b).

Advancing a further step in the optimization, just for the sake of a preliminary investigation, the ability of the neural network to generalize (interpolating the training data) was also evaluated.

For the bubbling gasifier, with steam as the gasificant agent, the database training data included the operating variables in the range 650 < T < 900 °C and 0.113 < S/B < 4.7 and three different biomasses (wood, straw and pine sawdust). A stochastic optimization method, the Particle

Swarm Optimization (PSO) (Kennedy & Eberhart, 1995), was applied, in order to find the optimum (maximum yield of H_2 in the produced gas), for a given biomass, considering the whole operating range in the training database. When that approach was applied, as an example, for straw biomass, the PSO algortihm found an optimum for hydrogen production of 81.07 for S/B = 4.7 and T = 700.95 ^oC. This result should be analysed very cautiously; the percentage of hydrogen seems very high – as literature report percentage of 72 % (Corella et al., 2008) – but it indicates for the operator a region that should be further examined experimentally in order to reach higher percentages of the component of interest in the produced gas. Additional results and details may be found in Nemer (2010).

5. CONCLUSIONS

ANNs are able to capture the latent chracteristics present in the experimental data used for training, including nonlinearities (Haykin, 1999). Hence, multilayer perceptron neuron networks were proposed here as an alternative tool for the empirical modeling of biomass gasification. Specifically, ANN models were developed to correlate operating conditions of the gasifier and biomass data with characteristics of the produced gas, using experimental data given in the literature. It was verified that the developed model showed a better performance when it was specifically built for a particular gasifier and a particular gasification agent (Nemer, 2010). Very high correlation rates between predictive and observed data were obtained.

The resulting trained ANN model is an algebric mapping between input-output data, demanding little computational time. That fact makes the use of neural netwok very attrative in real time control and/or optimization of the process. The preliminar optimization investigation carried out here showed that the ANNs may supply the operator with information of tendencies that should be further experimentally checked in order to reach the target of maximizing the amount of a given component in the produced gas.

Calibrition of the ANNs is easily performed, that is, whenever additional data are available, they may be added to the database and the ANN may be retrained, improving its predictive ability.

ACKNOWLEDGMENTS - This work was sponsored by Brazilian State Oil Company, PETROBRAS. Professor M. B. de Souza Jr acknowledges CNPq for a research fellowship.

REFERENCES

Basu, P. (2006). *Combustion and gasification in fluidized beds.* Taylor & Francis Group, USA.

Baratieri, M., Baggio, P., Fiori, L., & Grigiante, M. (2008). Biomass as an energy source: Thermodynamic constraints on the performance of the conversion process. *Bioresource technology* , 99, 7063-7073.

Boateng, A. A., Walawender, W. P., Fan, L. T., Chee, L. T. (1992). Fluidized-bed steam gasification of rice hull. *Bioresource Technology*, 40, 235-239.

Cao, Y., Wang, Y., Riley, J. T. *et al.* (2006). A novel biomass air gasification process for producing tar-free higher heating value fuel gas. *Fuel Processing Technology*, 87, 343-353.

Chen, G., Andries, J., Spliethoff, H. *et al.* (2004). Biomass gasification integrated with pyrolysis in a circulating fluidised bed. *Solar Energy*, 76, 1-3, 345-349.

Corella, J., & Sanz, A. (2005). Modeling circulating fluidized bed biomass gasifiers. A pseudo-rigorous model for stationary state. *Fuel Processing Technology 86* , 1021-1053.

Corella, J., Aznar, M. P., Caballero, M. A. et al. (2008). 140 g H_2/kg biomass d.a.f. by a CO-shift reactor downstream from a FB biomass gasifier and a catalytic steam reformer. *International Journal of Hydrogen Energy*, 3, 7, 1820-1826.

Dellepiane, D., Bosio, B., Arato, E. (2003). Clean energy from sugarcane waste: feasibility study of an innovative application of bagasse and barbojo. *Journal of Power Sources*, 122, 1, 47-56.

Doherty, W., Reynolds, A., Kennedy, D. (2009). The effect of air preheating in a biomass CFB gasifier using ASPEN Plus simulation, *Bioresource Technology*, 33, 9, 1158-1167.

Dong, C., Jin, B., Zhong, Z., & Lan, J. (2002). Tests on co-firing of municipal solid waste and coal in a circulating fluidized bed. *Energy Conversion and Management*, 43, 2189-2199.

Faaij, A., van Ree, R., Waldheim, L. *et al.* (1997), Gasification of Biomass Wastes and Residues for Electricity Production. *Biomass and Bioenergy*, 12, 6, 387-407.

García-Ibañez, P., Cabanillas, A., Sánchez, J. M. S. (2004). Gasification of leached orujillo (olive oil waste) in a pilot plant circulating fluidised bed reactor. Preliminary results. *Biomass and Bioenergy*, 27, 183-194.

Guo, B., Li, D., Cheng, C., Lü, Z.-a., & Shen, Y. (2001). Simulation of biomass gasification with a hybrid neural network model . *Bioresource Technology*, 76 , 77-83.

Guo, B., Shen, Y., Li, D., & Zhao, F. (1997). Modelling coal gasification with a hybrid neural network. *Fuel, Vol. 76, No. 12* , 1159-1164.

Haykin, S. (1999). *Neural Networks, A Compreensive Foundation*, 2nd ed., New Jersey, Prentice-Hall.

He, M., Xiao, B., Liu, S. *et al.* (2009). Hydrogen-rich gas from catalytic steam gasification of municipal solid waste (MSW): Influence of steam to MSW ratios and weight hourly space velocity on gas production and composition. *International Journal of Hydrogen Energy,* 2174-2183.

Himmelblau, D. H. (2000) Applications of Artificial Neural Networks in Chemical Engineering, *Korean Journal of Chemical Engineering*, 17, 4, 373-392.

Kennedy, J. and Eberhart R. (1995). Particle Swarm Optimization. *Proceedings of the IEEE International Conference on Neural Networks*, Perth, Australia, 1942-1948.

Li, X., Grace, J., Lim, C. *et al.* (2004). Biomass gasification in a circulating fluidized bed. *Biomass & Bioenergy*, 26, 171-193.

Lv, P., Chang, J., Xiong, Z et al. (2004), An experimental study on biomass air–steam gasification in a fluidized bed, Bioresource Technology, 95, 95–101.

Nemer, L. C. (2010), Modelagem do Processo de Gaseificação de Biomassa Empregando Redes Neuronais, Monograph, Escola de Química, UFRJ (*in portuguese*).

Nemtsov, D., Zabaniotou, A. (2008). Mathematical modelling and simulation approaches of agricultural residues air gasification in a bubbling fluidized bed reactor. *Chemical Engineering Journal* . 143, 1-3, 10-31.

Nikooa, M. B., Mahinpeya, N. (2008). Simulation of biomass gasification in fluidized bed reactor using ASPEN PLUS. *Biomass and Bioenergy*, 32, 12, 1245-1254.

Nougués, J.M., Pan, Y. G., Velo, E. *et al.* (2000). Identification of a pilot scale fluidised-bed coal gasification unit by using neural networks . *Applied Thermal Engineering, 20*, 1561-1575.

Schuster, G., Loffler, G., Weigl, K. *et al.* (2001), Biomass steam gasification – an extensive parametric modeling study. *Bioesource Technology*, 77, 71-79.

Shen, L., Gao, Y., Xiao, J. (2008), Simulation of hydrogen production from biomass gasification in interconnected fluidized beds. *Biomass and Bioenergy*, 32, 120-127.

Siedlecki, M., Simeone, E., De Jong, W. *et al.* (2007). Characterization of gaseous and condensable components in the product gas obtained during steam-oxygen gasification of biomass in a 100 KWth CFB gasifier. *15th European Biomass Conference and Exhibition - From Research to Market*, 1120-1127.

Souza-Santos, M. L. (2008a). Comprehensive Simulator (CSFMB) Applied to Circulating Fluidized Bed Boilers and Gasifiers. *The Open Chemical Engineering Journal*, 2, 106-118.

Souza-Santos, M. L. (2008b). CSFB applied to fluidized-bed gasification of special fuels. *Fuel*, 88, 5, 826-833.

Turn, S., Kinoshita, C., Zhang *et al.* (1998). An experimental investigation of hydrogen production from biomass gasification. *Int. J. Hydrogen Energy*, 23, 8 , 641-648.

van der Drift, A., van Doorn, J. and Vermeulen, J. W. (2001), Ten residual biomass fuels for circulating fluidized-bed gasification. *Biomass and Bioenergy*, 20, 45-56.

Xiao, G., Ni, M.-j., Chi, Y. *et al.* (2009). Gasification characteristics of MSW and an ANN prediction model. *Waste Management*, 29, 1, 240-244.

Combined Transport and Kinetic Modeling of Downdraft Biomass Gasifier

Pratik N Sheth[a], Amita U[b] and B V Babu[c1]

Birla Institute of Technology and Science (BITS), PILANI-333 031
Rajasthan, India

[a]Assistant Professor, Chemical Engineering Group, Email: pratik@bits-pilani.ac.in

[b]Student, Chemical Engineering Group,

[c]Dean-Student Welfare Division & Professor of Chemical Engineering
Phone: +91-1596-245073 Ext. 257; Fax: +91-1596-244183
Email: bvbabu@bits-pilani.ac.in; Homepage: http://discovery.bits-pilani.ac.in/~bvbabu

Abstract

Kinetics-free equilibrium models can predict the exit gas composition of the biomass gasifier, given the solid composition and the equilibrium temperature, but they cannot be used for reactor design. Hence, there is a need to develop a combined transport and kinetic model which takes into account of the kinetics of homogeneous and heterogeneous chemical reactions, transport of volatiles produced, heat and mass transfer between solid and gaseous phase and pyrolysis reactions. Taking into account of the importance of downdraft biomass gasifier and its commercial applications, it is essential to have a complete model for such a configuration. In the present study, a transient one-dimensional model is developed for the throated close-top downdraft biomass gasifier. The model takes into account of the drying, pyrolysis, secondary tar reactions, homogeneous gas reactions and heterogeneous combustion/gasification reactions. The experimental data obtained in our earlier study are used to validate the simulation results of the complete combined transport and kinetic model. The fraction of initial moisture content, air flow rate, temperature of the drying & pyrolysis zone, and chemical composition of the biomass are required as input data for the model to predict the composition of producer gas. The variation of molar fraction of producer gas components with time is predicted and compared with the experimental data. It is concluded from the present study that the developed model can predict the performance of the biomass gasifier, *a priori*. The results of this study are also useful in the design of a downdraft biomass gasifier.

1. Introduction

Bio-energy is now accepted as having the potential to provide the major part of the projected renewable energy provisions of the future. Gasification is one of the efficient

[1] Corresponding Author

technologies to convert the energy embedded in the biomass (Babu and Sheth, 2004). Gasification is a process of conversion of solid carbonaceous fuel into combustible gas by partial combustion. Downdraft gasification is comparatively cheap method of gasification that can produce a product gas with very low tar content (Giltrap et al., 2003). The understanding of the chemical and physical mechanisms during gasification is of fundamental importance for the optimal design of biomass gasifier (Babu and Chaurasia, 2004a-d). Modeling of biomass gasification implies the representation of chemical and physical phenomena constituting drying, pyrolysis, combustion, and reduction in the mathematical form (Babu and Sheth, 2005). The models of downdraft biomass gasification can be categorized into two groups: (1) Equilibrium models and (2) Combined transport and kinetic models. Equilibrium models are important in order to predict the thermodynamic limits of chemical reactions describing the gasification process. Kinetics-free equilibrium models can predict the exit gas composition, given the solid composition and the equilibrium temperature, but they cannot be used for reactor design (Sheth and Babu, 2009a). A transient one-dimensional model for the throated close-top downdraft biomass gasifier has been developed by Sheth and Babu (2009a). The model takes into account of the pyrolysis, secondary tar reactions, homogeneous gas reactions and heterogeneous combustion/gasification reactions. The drying and pyrolysis phenomena are modeled together as one zone. The drying zone is indirectly incorporated in the pyrolysis zone model. The pyrolysis model is divided into two subsystems, i.e., gas phase inside the bed and the individual particles. The model for single particle, developed by Babu and Chaurasia (2004c), is used in the complete gasifier model (Sheth and Babu, 2009a). The composition of volatiles is found using the experimental data of Boroson et al. (1989), which predicts the release of mainly water vapor from the pyrolyzing particle below 120 °C. Di Blasi (1998) also reported that at low temperatures the degradation rates are much slower than the drying rate. The presence of moisture is seen to delay wood pyrolysis and ignition. Therefore, drying and pyrolysis can be seen as two separate (sequential) processes in the biomass gasification (Di Blasi, 1998).

In the present study, the drying zone of the biomass gasifier is modeled independently and incorporated into the model proposed by Sheth and Babu (2009a). A model of moisture evaporation and transport phenomena is presented to account for the drying. The drying model is divided into two subsystems, i.e., the bulk phase in the bed and the individual particles. Bulk phase modeling is carried out by a transient mass balance for water vapor. Single particle model incorporates the conservation of moisture, water vapor and energy. The kinetic model of drying developed by Chan et al. (1985) for wood pyrolysis is used in the present study. The simulated model can predict the performance of the biomass gasifier, *a priori* which is useful in the design of a downdraft biomass gasifier.

2. Mathematical Model

A transient one dimensional model is developed for the throated close-top downdraft biomass gasifier. The model takes into account of the pyrolysis, secondary tar reactions, homogeneous gas reactions and heterogeneous combustion/gasification reactions. Eight gaseous species namely O_2, N_2, CO_2, CO, H_2O, H_2, CH_4 and tar are considered in the gas phase. In the pyrolysis and combustion zone, the solid phase is a biomass, whereas in the reduction zone it is a charcoal. The developed model is divided into four parts according to four different zones formed: (1) drying, (2) pyrolysis, (3) oxidation, and (4) reduction.

2.1 Drying

The first zone in which the feed comes in contact in biomass gasifier is the drying zone. Basically drying is a mass transfer operation resulting in the removal of water moisture by evaporation from a solid or semi-solid. In the drying zone, feed descend into the gasifier and moisture is removed using the heat generated in the zones below by evaporation. The water vapor flows downward in the gasifier. Part of it may be reduced to hydrogen in the reduction zone and the rest will end up as moisture in the gas. The rate of drying depends on the surface area of the fuel, the temperature difference between the feed and the hot gases, the re-circulation velocity, relative humidity of these gases, and the internal diffusivity of moisture within the fuel (Dogru et al., 2002). In this zone, no chemical reaction takes place and only the water removal is carried out.

Drying zone is modeled as a stack of particles in one dimension. The model presented here considers temperature gradient both in the bed and also inside the single particles. Thus the entire bed is divided into two subsystems, i.e., vapor phase inside the bed and the individual particles. The assumptions made in this model are: (1) the moisture released from biomass particles flow downward, (2) unsteady state operation, (3) velocity varies along the bed, (4) solid biomass does not move downward during the operation, and (5) porosity of the bed remains constant.

2.1.1 Transport Model for Vapor Phase in Drying Zone

The continuity equation for the water vapor in a packed bed is given by Eq. (1).

$$\varepsilon_{bed}\frac{\partial \rho_{w,bed}}{\partial t} = -\varepsilon_{bed}\left[v_{w,bed}\frac{\partial \rho_{w,bed}}{\partial z} + \rho_{w,bed}\frac{\partial v_{w,bed}}{\partial z}\right] + \frac{\partial}{\partial z}\left[\varepsilon_{bed} D_{w,eff,bed}\frac{\partial \rho_{w,bed}}{\partial z}\right] + a_p \dot{m}_{w,p} \tag{1}$$

where, ε_{bed} is the porosity of the bed, $v_{w,bed}$ is the velocity of the water vapor (m/s), $\rho_{w,bed}$ is the density of the water vapor in packed bed gas phase (kg/m^3), $\dot{m}_{w,p}$ is the mass transfer rate of water vapor from particle to the gas phase (kg/m^2 s), $D_{w,eff,bed}$ is the effective diffusivity of water vapor in packed bed gas phase, and a_p is the specific surface area of particle (m^2/m^3).

The initial and boundary conditions are given below by Eq. (2) and Eq. (3) respectively.

Initial Conditions:

at $t = 0$; $\rho_{w,\,bed} = 0$; $v_{w,\,bed} = 0$ (2)

Boundary Conditions:

at $z = 0$; $\rho_{w,\,bed} = 0$; $v_{w,\,bed} = 0$ for all $t > 0$ (3a)

at $z = L$; $\frac{\partial \rho_{w,bed}}{\partial z} = 0$ for all $t > 0$ (3b)

The water vapor generated from the biomass particle flows downward and its velocity is computed using Eq. (4).

$$v_{w,bed} = \left(\frac{\text{Flow rate}}{\text{Area}}\right) = \left(\frac{\rho_{w,bed} A\Delta z}{\text{Density of gaseous mixture}}\right)\left(\frac{1}{\Delta t}\right)\left(\frac{1}{\text{Area}}\right) \quad (4)$$

2.1.2Single Particle Model in Drying Zone

Drying of a particle is carried out using an unsteady state one dimensional variable property model of transport phenomena. It includes heat (conductive, convective and radiative modes) and mass (diffusive modes) transport of the water vapor released within the solid. An Arrhenius type kinetic reaction proposed by chan et al. (1985) and subsequently modified by Bryden et al. (2002) describing the mechanism of evaporation is considered in the model.

Equations (5) and (6) describe the reaction of moisture evaporation and kinetic rate expression respectively.

$$\text{Moisture } (M) \rightarrow \text{Watervapor } (w) \qquad \Delta H = -2.44 \text{ MJ/Kg} \quad (5)$$

$$k = A\exp\left(\frac{-E}{RT}\right) \qquad A = 5.13\times10^{10} \text{ s}^{-1} \qquad E = 88 \text{ KJ/Mol} \quad (6)$$

The equation for conservation of moisture is given by Eq. (7).

$$\frac{\partial C_M}{\partial t} = -kC_M \quad (7)$$

where, C_M = Density of moisture in solid phase

The conservation of water vapor is given by Eq. (8).

$$\frac{\partial(\varepsilon C_w)}{\partial t} = D_{ew}\left(\frac{b-1}{r}\frac{\partial C_w}{\partial r} + \frac{\partial^2 C_w}{\partial r^2}\right) + kC_M \quad (8)$$

The conservation of energy is given by Eq. (9).

$$\frac{\partial(\rho C_p T)}{\partial t} = k\left(\frac{b-1}{r}\frac{\partial T}{\partial r} + \frac{\partial^2 T}{\partial r^2}\right) - \left(D_{ew}\frac{\partial C_w}{\partial r}\right)C_{p,w}\frac{\partial T}{\partial r} + (-\Delta H)\left(-\frac{\partial \rho}{\partial t}\right) \quad (9)$$

The initial and boundary conditions are given by Eqs. (10-13).

At t =0;

$$C_M = C_{M0}; \quad C_w = 0; \quad T(r,0) = T_0 \quad (10)$$

Particle boundary condition:

At t >0,

At r = 0; $\frac{\partial C_w}{\partial t} = 0; \quad \left(\frac{\partial T}{\partial r}\right)_{r=0} = 0$ (11)

At r = R; $\left(D_{ew}\frac{\partial C_w}{\partial r}\right) = K_{m,w}(C_{w0} - C_{w1})$ (12)

At r = R; $K\left(\frac{\partial T}{\partial r}\right)_{r=R} = h(T_f - T) + 6\varepsilon(T_f^4 - T^4)$ (13)

2.2 Pyrolysis

Pyrolysis is a process by which a biomass feedstock is thermally degraded in the absence of oxygen/air to produce solid (charcoal), liquid (tar and other organics) and gaseous (H_2, CO_2, CO, etc.) products. Released volatiles from each biomass particle flow downward in packed pyrolysis bed. Rate of volatiles release depends on the particle size and temperature within a single particle. Pyrolysis bed is modeled as a stack of particles in one dimension. The model presented here considers temperature gradient both in the bed and also inside the single particles. Thus the entire bed is divided into two subsystems, i.e., gas phase inside the bed and the individual particles. The assumptions made in this model are: (1) the volatiles released from biomass particles flow downward, (2) unsteady state operation, (3) released volatiles is a mixture of CO, CH_4, CO_2, H_2, H_2O and tar, (4) velocity varies along the bed, (5) solid biomass does not move downward during the operation, and (6) porosity of the bed remains constant.

2.2.1 Transport Model for Gas Phase in Pyrolysis Zone

The continuity equation for the gas phase in a packed bed is given by Eq. (14).

$$\frac{\partial \varepsilon_{g,bed}\rho_{g,bed}}{\partial t} = -\frac{\partial}{\partial z}\left[\varepsilon_{g,bed}\rho_{g,bed}v_{g,bed}\right] + a_p\dot{m}_p \quad (14)$$

where, $\varepsilon_{g,bed}$ is the porosity of the bed, $v_{g,bed}$ is the gas velocity (m/s), $\rho_{g,bed}$ is the density of the gas phase (kg/m^3), $\dot{m}_p$ is the mass transfer rate from particle to the gas phase (kg/m^2 s), and a_p is the specific surface area of particle (m^2/m^3).

The species conservation equation is given by Eq. (15).

$$\varepsilon_{g,bed}\frac{\partial \rho_{ig,bed}}{\partial t} = -\varepsilon_{g,bed}\left[v_{g,bed}\frac{\partial \rho_{ig,bed}}{\partial z} + \rho_{ig,bed}\frac{\partial v_{g,bed}}{\partial z}\right] + \frac{\partial}{\partial z}\left[\varepsilon_{g,bed} D_{i,eff,bed}\frac{\partial \rho_{ig,bed}}{\partial z}\right] + a_p \dot{m}_{i,p} \tag{15}$$

where, $\rho_{ig,\,bed}$ is the density of the species i in packed bed gas phase, $D_{i,eff,bed}$ is the effective diffusivity of species i in packed bed gas phase, and $\dot{m}_{i,p}$ is the mass transfer rate of species i from particle to the gas phase (kg/m^2 s).

The initial and boundary conditions are given below by Eq. (16) and Eq. (17) respectively.

Initial Conditions:

at $t = 0$; $\rho_{\text{ig, bed}} = 0$; $v_{\text{g, bed}} = 0$ (16)

Boundary Conditions:

at $z = 0$; $\rho_{\text{ig, bed}} = 0$; $v_{\text{g, bed}} = 0$ for all $t > 0$ (17a)

at $z = L$; $\frac{\partial \rho_{ig,bed}}{\partial z} = 0$ for all $t > 0$ (17b)

Pyrolysed gas generated from the biomass particle flows downward and its velocity is computed using Eq. (4).

2.2.2 Single Particle

To describe the chemical process of pyrolysis in a single solid particle, an unsteady state one dimensional variable property model of transport phenomena is required. It should include heat (conductive, convective and radiative modes), mass (diffusive and convective modes) and momentum transport of the products formed within the solid (volatiles and gases). Babu and Chaurasia (2003 a-b; 2004 a-d) carried out exhaustive modeling and simulation studies on the pyrolysis of single solid particle. The model developed and modified by Babu and Chaurasia (2003 a-b; 2004 a-d) uses physically measurable parameters and practically explainable kinetic scheme, incorporating the convective and diffusion effects. Babu and Chaurasia (2004c) proposed three models for the pyrolysis of biomass under two categories namely, (1) Generalized reference model (Model-I), and (2) Simplified models (Model-II & Model-III). In most practical situations of industrial pyrolysis reactions, the contributions of the bulk motion of gases inside the pores of the particle are insignificant. Because the resistance offered by the pores in the solid particle is so high, the transport of these gases would take place essentially by a diffusion mechanism but not by bulk motion (i.e. convection). Taking this situation into consideration, Model –II is used in the present study, which is based on the assumption that there is no bulk motion contribution (i.e. convective transport is neglected) to the temperature profile and the product yield predictions.

2.3 Combustion

The biomass present in the oxidation zone reacts with oxygen to form carbon dioxide, which provides heat for the subsequent gasification reactions. In complete combustion, carbon present in biomass is completely converted to carbon dioxide while hydrogen is converted to water vapor. It is an exothermic reaction and yields temperatures in the range of 1000 °C to 1500 °C. In the present model, complete combustion of biomass is assumed which can be ensured by supplying excess air (usually around 20%) than stoichiometrical requirement. The combustion reaction therefore is represented as given by Eq. (18)

$$C_6H_{HB}O_{OB} + \left(6 + \frac{HB}{4} - \frac{OB}{2}\right) O_2 \longrightarrow 6CO_2 + \frac{HB}{2} H_2O \qquad (18)$$

The volatile products generated in the pyrolysis zone flow downwards and enter into the oxidation zone where a part of volatiles gets oxidized. It is assumed that the tar present in the pyrolysed gas mixture completely gets decomposed due to very high temperature present in the oxidation zone. The tar decomposition is represented by Eq. (19).

$$\mathrm{C_6H_{6.2}O_{0.2} + 7.45O_2 \longrightarrow 6CO_2 + 3.1H_2O} \qquad (19)$$

2.4 Reduction

The main components of the gaseous mixture leaving the combustion zone are carbon dioxide, water vapor, inert nitrogen, carbon monoxide, hydrogen and some amount of low molecular weight hydrocarbons such as methane, ethane, ethylene etc. In the reduction zone, the gaseous mixture passes through the hot porous charcoal bed resting above the grate. The reduction zone is often referred as gasification zone. Giltrap et al. (2003) developed a model of reduction zone of downdraft biomass gasifier to predict the composition of producer gas under steady state operation. The accuracy of the model is limited by the availability of data on the initial conditions at the top of the reduction zone. Moreover they assumed that the char reactivity factor (*CRF*) which represents the reactivity of char and the key variable in simulation is constant throughout the reduction zone. Giltrap's model (2003) is modified by incorporating the variation of *CRF* along the reduction zone of downdraft biomass gasifier in our earlier study (Babu and Sheth, 2006). It is assumed that *CRF* is exponentially increasing along the bed length of the reduction zone. Solid carbon in the form of char is assumed to be present throughout the reduction zone. Modeling equations of single particle model of pyrolysis (Babu and Chaurasia, 2004c) and reduction model proposed by Babu and Sheth (2006) are not presented in the present article to reduce the length of the article.

3. Experimental Studies

The main equipment of the experimental set up is an Imbert downdraft biomass gasifier. An Imbert downdraft biomass gasifier is the one, which has throated combustion zone and different diameter for pyrolysis and reduction zones unlike stratified downdraft biomass gasifier in which diameter of the gasifier is uniform through out (Reed and Das, 1988). Biomass is fed from top of the gasifier and air is introduced through nozzle in the combustion zone. The downdraft gasifier converts the solid biomass into a combustible gas, generally known as a producer gas. Biomass undergoes pyrolysis and gets oxidized in the combustion zone near air inlet. The pyrolysed gas mixture and the gases produced due to combustion passes over the charcoal bed resting above the grate and generate producer gas. The details of the experimental set up, procedure and operating variables are reported in our earlier study (Sheth and Babu, 2009b).

4. Results and Discussion

The experimental data reported in our earlier study (Sheth and Babu, 2009b) are used to validate the simulation results of the complete combined transport and kinetic model. The initial moisture content fraction, air flow rate, temperature of the pyrolysis zone, and chemical composition of the biomass are required as an input data for the model to predict the composition of producer gas. The combined transport and kinetic model is validated using the parametric values as given in Table-1.

Table 1 Details of the biomass gasification experiments

Experimental No	Air flow rate (m^3/h)	Initial moisture content (wt fraction, wet basis)	Biomass consumption Rate (kg/h)	Equivalence ratio(Φ)
1	2.7765	0.1145	2.10	0.2533
2	3.3935	0.0437	3.63	0.1791
3	1.8510	0.0437	2.12	0.1673
4	2.7765	0.0437	2.67	0.1992
5	2.7765	0.073	2.59	0.2054
6	1.8510	0.10	1.00	0.3546
7	2.7765	0.1518	2.20	0.2418
8	2.1595	0.07	1.488	0.278
9	2.1595	0.044	2.12	0.1951
10	2.1595	0.1167	1.1626	0.3558
11	2.1595	0.164	1.0424	0.3968

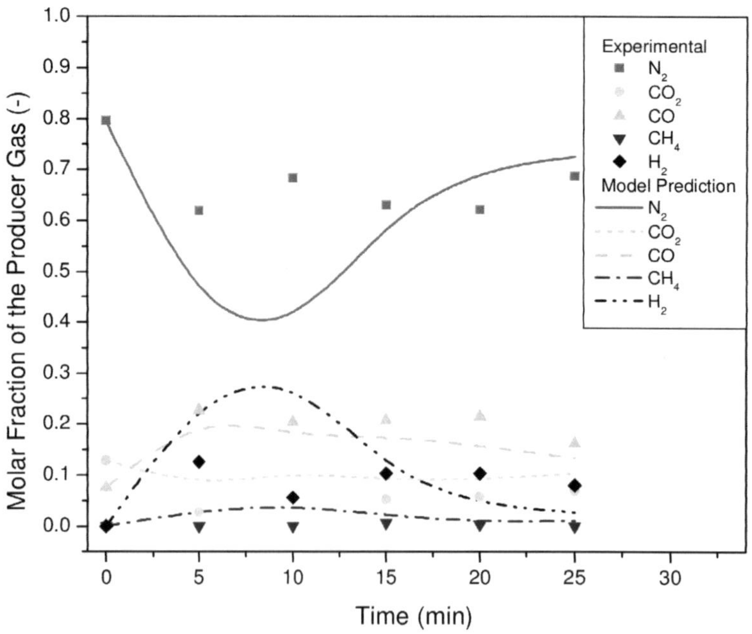

Fig. 1 Comparison of model predicted producer gas composition with experimental values of the present study (Φ = 0.2533)

The variation of molar fraction of producer gas components with time is predicted and compared with the experimental data. Fig. 1 show the comparison of model predicted molar fractions of producer gas components with those found experimentally for an equilibrium ratio of 0.2533. It is found that the model predicted molar fraction of nitrogen decreases first during the few initial minutes (5-10 min) of gasification. After that it increases and attains a steady value (after 10-15 min). The molar amount of nitrogen is constant for a particular flow rate of air as nitrogen acts as an inert but its composition varies due to the changes in molar amount of other components of gaseous mixture. It is found that the model predicted molar fractions of carbon monoxide and hydrogen increase first with time and after that it decreases and attains a steady value. It is observed that the model predicted molar fraction of methane is very less and almost remains constant. It is also found that the model predicted molar fraction of carbon dioxide decreases a little and attains a steady value.

The simulation results of the molar composition of various components of the producer gas match well with the experimental data of 10 minutes or higher from the start of run. For the experimental data of 5 min and 10 min, the simulation results differ more. This is because of the assumption taken in the model that all the gas generated in pyrolysis or reduction zone travel downwards in the gasifier. However, it is observed while carrying out the experiments that the gas produced in the pyrolysis zone first travels upwards and occupies the empty space above the biomass. After 5 - 10 minutes from the start of the run, the accumulated gas builds up a pressure and the producer gas gets started flowing downwards. Because of this, the model predicts higher concentration of hydrogen and carbon monoxide and lower concentration of nitrogen in comparison to the experimental data for initial 5 -10 minutes from the start of a particular experimental run. Fig. 2 shows the model predicted composition profile of gaseous phase across the gasifier at 5 min from the start of the experiment for the first run of the present study. Figs. 3 and 4 show the similar profiles at different times from the start of the experiment (run 1). The pyrolysis zone in the gasifier is represented by first 100 mm, oxidation zone by 100 to 140 mm and reduction zone by 140 to 270 mm in Figs. 2 to 4.

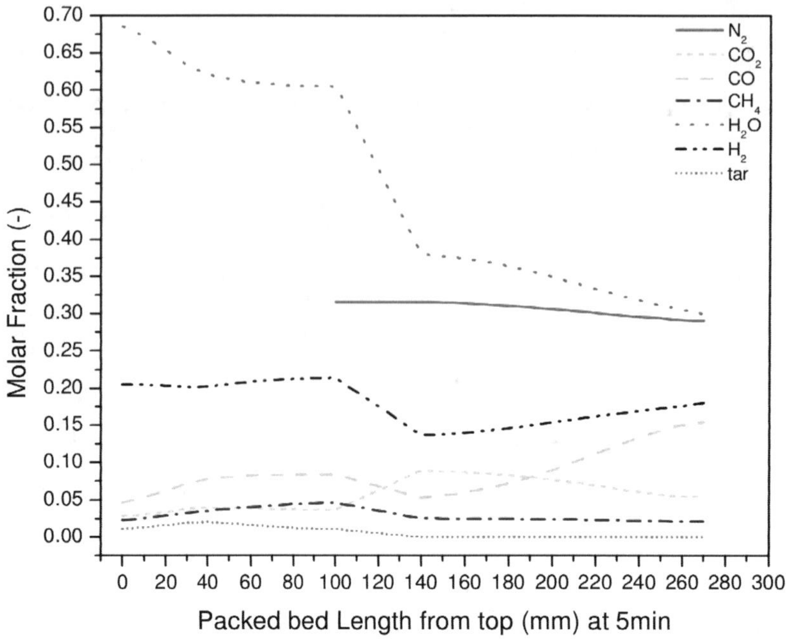

Fig. 2 Simulated composition profile across the gasifier at 5 min from the start of the experiment for the present study (*Φ* = 0.2533)

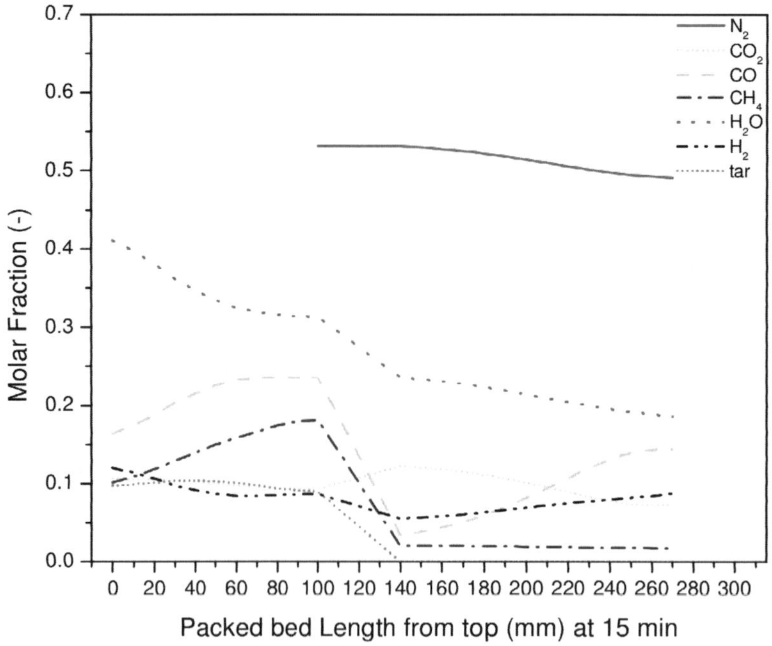

Fig. 3 Simulated composition profile across the gasifier at 15 min from the start of the experiment for the present study (*Φ* = 0.2533)

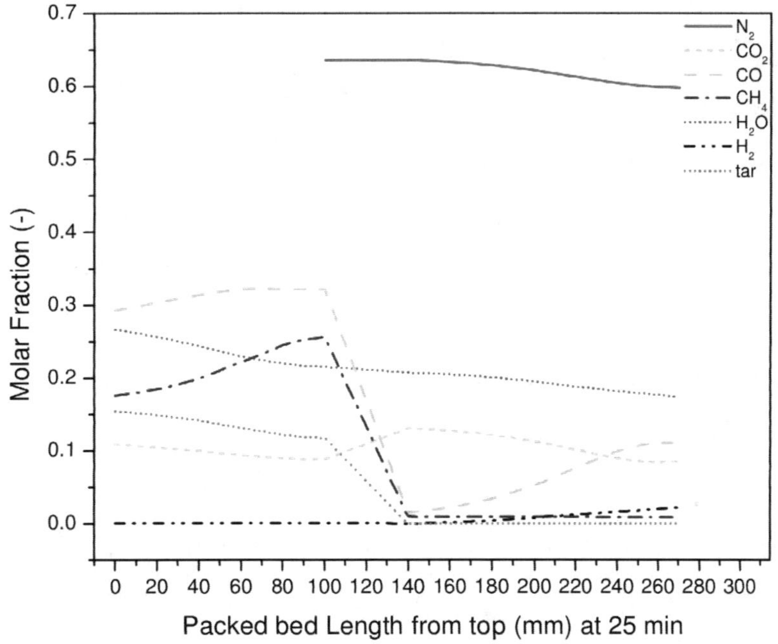

Fig. 4 Simulated composition profile across the gasifier at 25 min from the start of the experiment for the present study (Φ = 0.2533)

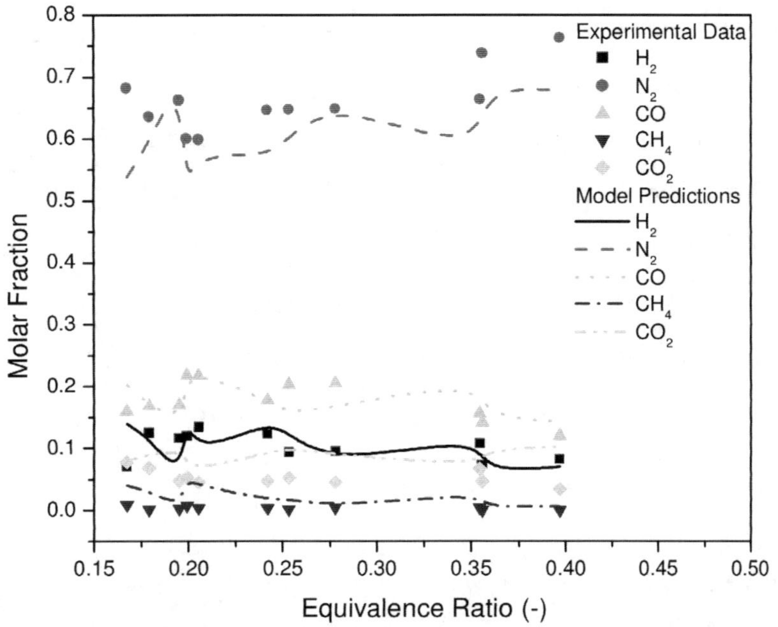

Fig. 5 Comparison of experimental data of present study with model predicted composition of producer gas

It shows the variation of molar fraction of the main components of the producer gas, i.e. nitrogen, carbon monoxide, carbon dioxide, methane, water vapor, hydrogen, and tar, with time in the gasifier. It is found that the composition of water vapor in the gaseous phase is the highest in pyrolysis zone (0-100 mm) and decreases in oxidation zone (100-140 mm) and reduction zone (140-270 mm) of the gasifier, when it travels through these zones (Figs. 2 to 4). It is also observed that the molar fraction of water vapor decreases with time in all zones of the gasifier. The molar fractions of CO and CH_4 in the gaseous phase increase in the pyrolysis zone as gas travels through it.

It is found that the molar fractions of CO, H_2, and CH_4 decrease in oxidation zone due to combustion. The molar fraction of CO_2 increases and that of tar decreases due to the high temperature oxidation. The molar fraction of nitrogen almost remains constant in the oxidation zone due to the constant supply of air. In the reduction zone of gasifier, the molar fractions of CO and H_2 increase and those of CO_2 and N_2 decrease. The molar flow rate of nitrogen does not change but due to the water gas reaction, methane formation reaction, and Boudouard reaction, formation of hydrogen and carbon monoxide occurs and the composition of nitrogen decreases in the reduction zone.

It is also found that the composition of carbon monoxide, methane, hydrogen and tar increases and that of water vapor decreases with time in the pyrolysis zone of the gasifier due to successive increase in the temperature of the pyrolysis zone. The simulated composition profiles shown in Figs. 2 to 4 at different times from the start of the experiment give proper insight of the gasifier. These simulated variations of the molar fraction of gaseous components with time match with the theory of downdraft biomass gasification. To study the variation of the simulated composition of the combined transport and kinetic model with equivalence ratio, the composition of producer gas is averaged over 25 minutes. Fig. 5.shows the comparison of the average composition of producer gas with those found experimentally as a function of equivalence ratio.

5. Conclusions

Based on the results obtained in the present study, the following conclusions are drawn:

1. The proposed combined transport and kinetic model is successfully validated with the experimental data reported in the literature.

2. The model predicted composition of producer gas matches very well with the experimental data reported in the literature.
3. The molar fractions of CO, CH_4 and tar increase and that of water vapor decreases with time in the pyrolysis zone of the gasifier due to a subsequent increase in the temperature of pyrolysis zone.
4. The molar fraction of water vapor in the gaseous phase is the highest in the pyrolysis zone and decreases as it travels through the oxidation and reduction zones of the gasifier.
5. The molar fractions of CO and CH_4 in the gaseous phase increase in the pyrolysis zone as gas travels through it. However, the molar fractions of CO, H_2, tar, and CH_4 decrease in the oxidation zone due to high temperature oxidation and hence the molar fraction of CO_2 increases.

References

Babu, B.V., Chaurasia A.S., 2004a. Parametric Study of Thermal and Thermodynamic Properties on Pyrolysis of Biomass in Thermally Thick Regime. Energy Conversion and Management, 45, 53-72.

Babu, B.V., Chaurasia A.S., 2004b. Dominant Design Variables in Pyrolysis of Biomass Particles of Different Geometries in Thermally Thick Regime. Chemical Engineering Science, 59, 611-622.

Babu, B.V., Chaurasia A.S., 2004c. Pyrolysis of Biomass: Improved Models for Simultaneous Kinetics & Transport of Heat, Mass, and Momentum. Energy Conversion and Management, 45, 1297-1327.

Babu, B.V., Chaurasia A.S., 2004d. Heat Transfer and Kinetics in the Pyrolysis of Shrinking Biomass Particle, Chemical Engineering Science, 59, 1999-2012.

Babu, B.V., Sheth, P.N., 2006. Modeling and Simulation of Reduction Zone of Downdraft Biomass Gasifier: Effect of Char Reactivity Factor, Energy Conversion and Management, 47, 2602-2611.

Babu, B.V., Sheth, P.N. 2004. Modeling and Simulation of Downdraft Biomass Gasifier, Proceedings of International Symposium & 57th Annual Session of IIChE in association with AIChE (CHEMCON-2004), Mumbai, December 27-30, (2004)

Babu, B.V., Sheth, P.N. 2005. Parametric Study of Factors affecting the Composition of Producer Gas, Proceedings of CHEMCON-2005, New Delhi.

Boroson, M.L., Howard, J.B., Longwell, J.P., Peters, W.A., 1989. Product Yields and Kinetics from the Vapor Phase Cracking of Wood Pyrolysis Tars. American Institute of Chemical Engineers Journal, 35, 120-128.

Bridgwater, A.V., 2002. Bio-Energy Research Group, Aston University, Birmingham B47ET, UK,(2002). http://www.icheme.org/.../conferences/gasi/Gasification% 20Conf%20Papers/ Session%202%20presentation-Bridgewater.

Bryden, K. M., K. W. Ragland, et al. (2002). "Modeling thermally thick pyrolysis of wood." Biomass and Bioenergy 22(1): 41-53.

Chan W.R., Kelbon M., Krieger B.B. 1985. Modeling and experimental verification of physical and chemical processes during pyrolysis of a large biomass particle, Fuel 64, 1505–13.

Di Blasi C., 1998. Multi-phase moisture transfer in the high temperature drying of wood particles, Chemical Engineering Science, 53, 353-366.

Di Blasi, C., 2000. Dynamic Behaviour of the Stratified Downdraft Gasifiers. Chemical Engineering Science, 55, 2931-2944.

Giltrap, D.L., McKibbin, R.; Barnes, G.R.G., 2003. A Steady State Model of Gas-Char Reactions in a Down Draft Biomass Gasifier. Solar Energy, 74, 85-91.

Sharma, A.K., 2007. Modeling Fluid and Heat Transport in the Reactive, Porous Bed of Downdraft (Biomass) Gasifier. International Journal of Heat and Fluid Flow, 28, 1518–1530.
Sheth, P.N., Babu, B.V., 2009a. Modeling and Simulation of Downdraft Biomass Gasifier. Proceedings of 2009 Annual Meeting of AIChE, Gaylord Opryland Hotel, Nashville, TN, USA, November 8-13, (2009)
Sheth, P.N., Babu, B.V., 2009b. Experimental studies on producer gas generation from wood waste in a downdraft biomass gasifier. Bioresource Technology 100(12): 3127-3133.
Wurzenberger, J.C., Wallner, S., Raupenstrauch, H., Khinst, J.G., 2002. Thermal Conversion of Biomass: Comprehensive Reactor and Particle Modeling, American Institute of Chemical Engineers Journal, 48, 2398-2411.

(135a) Biomass Pretreatment to Improve Bio-Oil Stability1

Najeeb M. Kuzhiyil[1], Dustin Dalluge[2] and Robert C Brown[2], (1)Mechanical Engineering, Iowa State University, Ames, IA, (2)Mechanical Engineeering, Iowa State University, Ames, IA

Storage stability is one of the most important properties of a fuel. Bio-oil produced from biomass by the fast pyrolysis process is known for its poor storage stability which makes it a less convenient fuel. The aldehydes and carboxylic acids contained in the bio-oil react with each other and with other species that causes the instability of the oil. Pretreatments are the first opportunity to improve the quality of bio-oil by selectively controlling the impurities in the feedstock that influence the pyrolysis reactions. Potential biomass pretreatment techniques were studied in order to understand their effect on the stability of the bio-oil produced from the treated feedstocks. Switchgrass was pretreated by washing with hot water and dilute acids separately, and also by infusing various acids at different concentrations. Pyrolysis-GC-MS set up was used to analyze the composition of the bio-oil produced from these treated feedstocks. For the purpose of predicting the stability of the oil produced, the pyrolysis products were categorized broadly into anhydrosugars, light oxygenates, furans and phenolics. Hot water and dilute acid wash improved the yield of levoglucosan and reduced the yield of carboxylic acids and aldehydes while the acid wash increased the yield of furans also. However, the overall effects were not good enough to justify the extra energy spent on drying the washed feedstock. In order to avoid the need for drying, acids were infused into switchgrass and tested. It was found that phosphoric acid at 2 wt% of the biomass produced the maximum yield of anhydrosugars and lowest yield of light oxygenates. Sulfuric acid was slightly better than phosphoric acid while nitric and hydrochloric acids were less effective. Acetic and formic acids were not effective at all. The oils produced from phosphoric and sulfuric acid treated switchgrass were found superior in terms of its storage stability when compared to the control oil.

1,This work was funded by the U.S. Department of Energy and the ConocoPhillips Company

See more of this Session: Biomass Pyrolysis
See more of this Group/Topical: Fuels and Petrochemicals Division

Process Design Data for Liquid Phase Pyrolysis of Biogenic Feedstock

Verena Mertlitz*, Nikolaus Schwaiger, Edgar Ahn, Matthäus Siebenhofer
BDI-BioDiesel International AG, Parkring 18, 8074 Grambach/Austria,
verena.mertlitz@bdi-biodiesel.com

This project focuses on pyrolytic conversion of CHO-feedstock such as fir chips under liquid phase operation conditions. The investigation of the reaction kinetics showed a change in the reaction mechanism at a reaction temperature of 350 [°C]. The activation energy for a reaction temperature below T = 350 [°C] was found to be 18 [kJ/mol]. At higher temperatures an activation energy of 105 [kJ/mol] was observed.
The temperature has a significant influence on the product distribution and the energy balance.

1. Introduction

Pyrolytic conversion of biogenic feedstock has become a high priority task due to declining crude oil reserves.
Subject of this project is the determination of design parameters for industrial scale plant design, including reaction mechanisms, thermodynamics and kinetics of liquid phase pyrolysis of CHO-feedstock. The results are needed for knowledge transfer from lab scale to technical scale.
From liquid phase pyrolysis improved convective heat transfer is expected, enabling fast pyrolysis conditions at ambient pressure. Operation temperature was limited to a range of less than 650 [K]. Investigations were carried out under isothermal conditions.

1.1 Experimental

The process was operated in a lab scale reactor in semi-batch mode. The rate of degradation for nearly monodisperse feed was recorded. The feed quantity was limited to an amount which causes a deviation of the reactor temperature of less than 2 [K] each feed cycle. The effect of temperature on product distribution was investigated. The influence of biomass composition on the mechanism and rate of pyrolytic degradation were analyzed. As feedstock wood, straw, pulp and bran were used.
The liquid and solid products were characterized by GC-MS, NMR-spectroscopy and elemental analysis. Reproducibility was checked with replicates.

1.2 Results and Discussion

The ratio of water formed during pyrolytic degradation increases with increasing temperature, passing a maximum, while formation of CHO-products linearly correlates

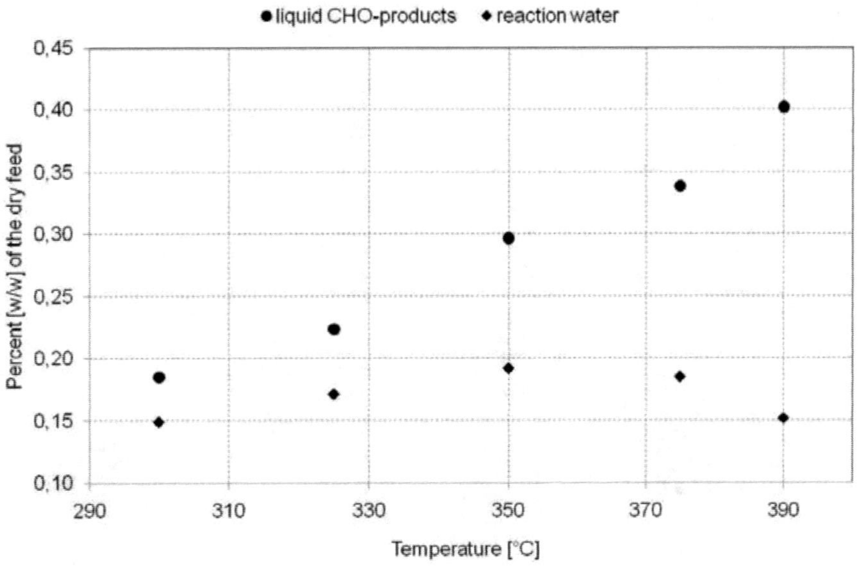

Figure 1: Temperature dependency of liquid products

with the operation temperature (Figure 1). Carbon transfer from feed to gaseous and liquid product phase also correlates linearly with the temperature.

Figure 2 shows the influence of the feedstock on the product distribution.

Figure 3 shows the formation of liquid products (liquid CHO-products plus water formed by chemical reaction) based on dry feed at different reaction temperatures. The feedstock was wood. The rate of formation of liquid products was best modelled by 2nd order kinetics. A shift in the reaction mechanism of the formation of liquid products was observed.

The activation energy for a reaction temperature below 350 [°C] was found to be 18 [kJ/mol]. Above 350 [°C] an activation energy of 105 [kJ/mol] was observed.

The reaction temperature also has an effect on the distribution of product fixed energy (Table 1). At lower temperatures (300 [°C], column 1) the energy of the biomass is mainly stored in the produced char.

Table 1: Energy content of products (based on the original biomass energy content) for different reaction temperatures

product	300 [°C]	350 [°C]	390 [°C]
Char	79%	53%	35%
Gas	1%	3%	12%
liquid CHO-products	19%	39%	47%
Loss	1%	5%	6%

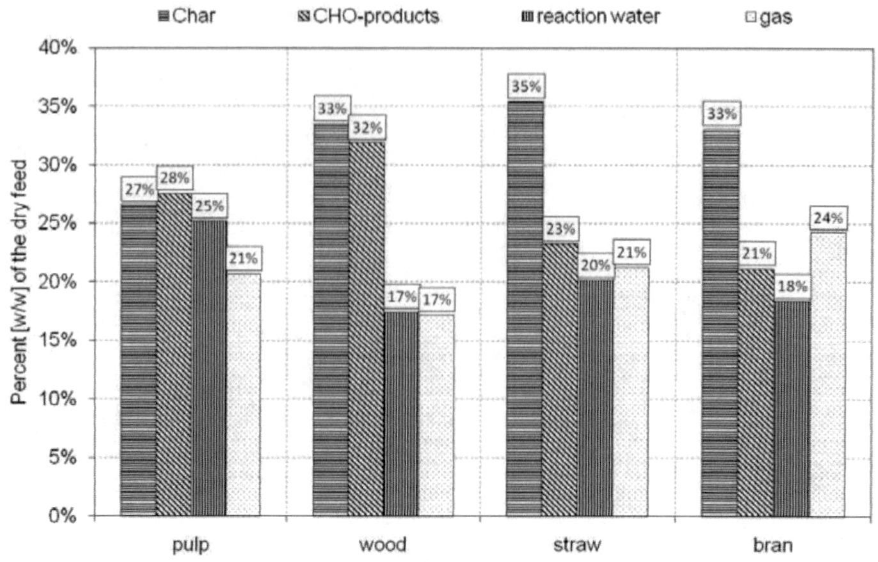

Figure 2: Influence of feedstock on the product distribution (Char, liquid CHO-products and reaction water).

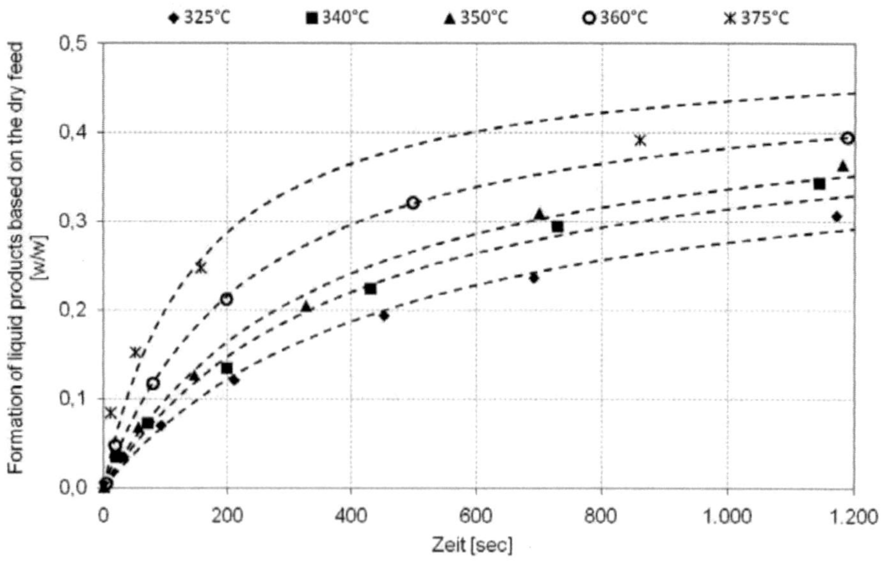

Figure 3: Formation of liquid CHO-products and reaction water based on dry feed.

At higher temperatures (390 [°C], column 3) the liquid-CHO-products are the fraction with the highest energy content. Different biogenic feedstock does not have a significant effect on energy distribution.
According to process thermodynamics liquid phase pyrolysis of wood based feedstock is a moderately exothermic process. The heat of reaction is less than the needed energy for vaporization. According to Figure 4 the overall process needs energy supply.

1.3 Conclusions
Based on the experimental results and thermodynamics it is possible to convert biogenic feedstock such as fir chips and straw into products with higher energy density by liquid phase pyrolysis.
The char left after separation of volatiles is highly reactive. Due to the elevated carbon content compared with the feed the char provides interesting application opportunities for further processing.
The macro-kinetics of liquid phase pyrolysis showed a shift in the reaction mechanism. Further investigation is needed to clarify the influence and interaction of the biomass composites cellulose, hemicelluloses and lignin on the reaction mechanism in detail.
Liquid phase pyrolysis of wood is moderately exothermic reaction energy cannot cover the demand for evaporation of condensable products.

1.4 Acknowledgments
This project is funded by the Climate and Energyfund Austria and is carried out under the program "Energy of the Future".
K. Zahel and A. Prettner are gratefully acknowledged for help and support. P. Pucher and G. Witek are acknowledged for helpful discussions.

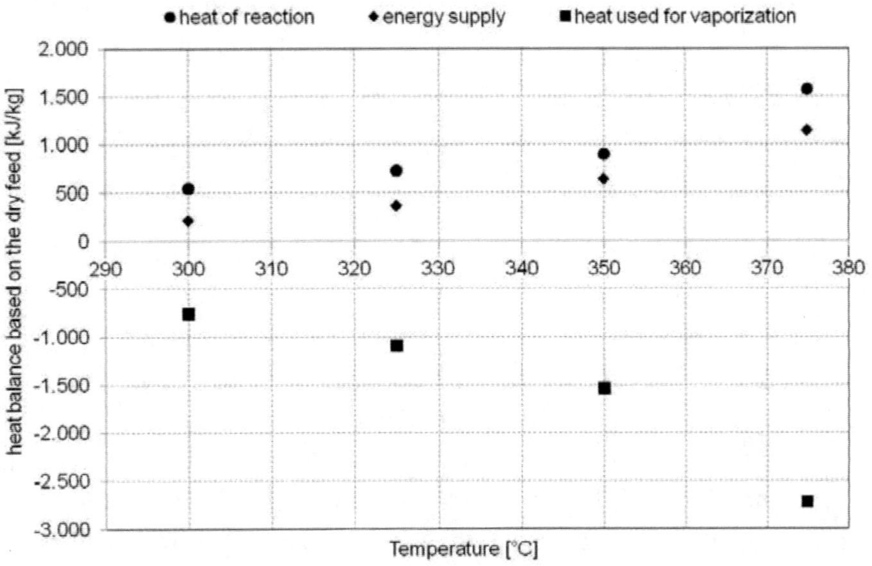

Figure 4: Results of the thermodynamic analysis

New method of total acid number (TAN) determination in biooils

FA Agblevor and J. Shao

Biological Systems Engineering, Virginia Polytechnic Institute and State University, Blacksburg, VA 24061

Abstract

Fast pyrolysis oil is one source of renewable liquid fuels, which is promising in satisfying environmental concerns over fossil fuels. The ASTM D664 method is used to determine the acidic constituents in petroleum products, lubricants, and biodiesel requiring mixtures of 2-propanol and toluene as solvent. The ASTM method also uses large quantities of solvents for the total acid number (TAN) determination. The ASTM D664 method has been adopted for the determination of TAN for bio-oils data published in literature. However, the properties of fast pyrolysis bio-oils are different from those of petroleum products and biodiesel, because bio-oils are soluble in polar solvents and partially soluble in water whereas petroleum products and biodiesel are insoluble in water and are more non-polar. An alternative method which is more suitable for TAN determination of pyrolysis bio-oil has been developed and validated. Aqueous potassium hydroxide (0.1 M KOH) was used as standard titrant instead of 0.1M KOH in alcoholic solution, and acetone was used as titration solvent instead of mixtures of toluene and 2-propanol as specified in the ASTM method. Four bio-oil samples were analyzed for determination of TAN with the new method and the ASTM method, respectively. The results showed that TAN of bio-oil samples obtained by the new method are in good agreement with the one obtained by the ASTM method. It was confirmed that the new method was simpler and relatively cheaper and more suitable for TAN determination of the pyrolysis oils than the ASTM D664 method.

Introduction

Pyrolysis of biomass is one of the most promising methods of producing liquid fuels from lignocellulosic biomass feedstocks. There are several variations of the pyrolitic conversion method including rapid pyrolysis, ultrafast pyrolysis, vacuum pyrolysis and fractional catalytic pyrolysis. All these methods produce liquid products that are acidic. The level of acidity varies with the pyrolysis method. The pH of liquids produced using these methods vary from 2 to 4. Because of the acidity of these oils, they could potentially cause corrosion in processing equipment or equipment using the oils either as fuel or other applications. The total acid number (TAN), which measures both the weak and strong acids in the oil is the traditional method of assessing the corrosivity of petroleum oils. The TAN is defined as the quantity of potassium hydroxide required to neutralize the acid in one gram of sample. The TAN is different from the pH in that whereas the pH measures the strength of the acidic or alkaline constituents, the TAN measures the concentration of the acidic constituents. TAN is also useful in detecting the presence of minor weak acids. For petroleum products, the ASTM standard method D664 is usually used. This method has been adapted to determine the TAN for biooils. However, both the physical and chemical properties of biooils are quite different from those of petroleum products. Whereas most petroleum products are hydrophobic, the biooils are hydrophilic and in some cases can dissolve substantial amounts of the biooil. Because of the hydrophobicity of the petroleum products, a mixture of solvents such as iso-propanol and toluene are used to ensure that the all the acidic components of the substrate are dissolved in the solvent prior to the titration. Large quantities of solvents are also normally required per determination of the TAN. A typical titration will require as much as 150 mL of solvent mixture; thus for a triplicate run, about 450 mL of solvent mixture will be required. For a large number of samples, the amount of solvent required becomes excessive, time consuming and not sustainable.

Biooils on the contrary are hydrophilic, easily dissolve in common solvents such methanol and acetone and are partially soluble in water. However in addition to weak organic acids such as acetic and propionic acids, biooils also contain phenols which are also weakly acidic. Because of the hydrophilicity of biooils, we proposed a new and simpler method for determining the acid number of the biooil. In this paper we report results of our new method of TAN determination for biooils.

Materials and methods

All chemicals used for these experiments (toluene, acetone, benzoic acid, potassium hydroxide, iso-propanol) were reagent grade obtained from Sigma- Aldrich (sigma-aldrich.com). the biooils used for the analysis were produced in our laboratory using both fractional catalytic pyrolysis method (Agblevor et al, 2010) and rapid pyrolysis using sand as the pyrolysis medium.

The biooils and benzoic acid standard were titrated using 798 MPT Titrino (Metrohm USA, Riverview, FL). One set of titration was carried out using the ASTM standard D 664 and another set was carried out using the new method. In the new method, 50 mL and 125 mL acetone were used to dissolve the biooil samples instead of the toluene iso-propanol mixture. The o.1M KOH solution was standardized with 0.1M hydrochloric acid solution. For each sample 0.1-0.5 g biooil was dissolved in 50 mL acetone in the first set of experiments and in the second set 125 mL of acetone was used to dissolve the biooil. The samples were titrated with 0.1 M KOH to the end point. Each sample was titrated in triplicate and the average and standard deviations calculated. Additionally each method was used to titrate benzoic acid standard to the end point. After the titration the total acid number was calculated using the following formula:

$$TAN = mgKOH/g = (A-B)*M*56.1/W$$

Where A = volume (mL) of KOH solution used to titrate the sample to end point; B = volume (mL) of KOH used for blank titration; M = molar concentration of KOH solution (mol/L); W = mass of biooil sample used.

Results and discussions

The TAN of benzoic acid standard using the two methods are similar to the theoretical calculations (Table 1). Clearly the accuracy of the new method is similar to that of the ASTM D 664 method. For the ASTM method, the volume of solvent mixture recommended was usually 125 mL. For the new method

50 mL and 125 mL of acetone were investigated. The results showed that there was no difference in the TAN using either 50 mL solvent or 125 mL. Thus, for subsequent analysis, only 50 mL solvent was used to dissolve the samples. In the ASTM, alcoholic KOH is used for the titration, but in the new method, 0.1 M KOH in water was adequate for the titration.

Table 1. Comparison of ASTM D 664 standard method with new acetone method

Method	ASTM D664	New acetone method
TAN (mg KOH/g benzoic acid)	456.81±2.61	461.39±3.52
Calculated TAN (mgKOH/g benzoic acid)	459.38	459.38
Repeatability	1.58%	2.11%
Accuracy	99.4%	100.4%

Repeatability= (2.77*SD)/(average)*100; Accuracy = Average value/calculated value

The comparison of the benzoic acid results showed similar TAN number, so the method was applied to both rapid pyrolysis biooil samples and catalytic pyrolysis biooil samples. The rapid pyrolysis biooil sample was supplied by NREL. The NREL biooil was prepared from hybrid poplar wood. The TAN number for bioooil samples determine by the two methods are shown in Table 2.

Table 2. Comparison of TAN for biooil samples using both ASTM D664 and new acetone method.

Biooil	ASTM D664, TAN (mgKOH/g)	New acetone method, TAN (mgKOH/g)
NREL (non-catalytic)	95.00±1.40	98.85±2.69
Poplar catalytic	86.92±0.55	89.93±1.81
Poplar catalytic	76.60±1.05	82.58±2.49
Pine catalytic	61.51±0.25	63.35±0.69

In all cases, the acetone method was slightly higher than the ASTM method. We attributed the differences to the improved solubility of the biooil in the acetone compared to the toluene/iso-propanol mixture. However, the TAN determined by the two methods are comparable. Since the new method is much simpler and does not require the use of large solvent volumes, and alcoholic potassium hydroxide, we recommend this approach as being more suitable for TAN determination in biooils.

(135d) Fractional Catalytic Pyrolysis of Biomass for Stable Pyrolysis Oils and Hydrocarbon Fuels Production

Foster A. Agblevor, Department of Biological Systems Engineering, Virginia Polytechnic Institute and State University, Blacksburg, VA, Ofei Mante, Biological Systsems Engineering, Virginia Polytechnic Institute and State University, Blacksburg, VA, S. Ted Oyama, Chemical Engineering, Virginia Polytechnic Institute and State University, Blacksburg, VA, Francine Battaglia, Mechanical Engineering, Virginia Polytechnic Institute and State University, Blacksburg, VA and Ron McClung, BASF Catalysts LLC, Iselin, NJ

Pyrolysis of biomass is developing rapidly and could play a very important role in the future renewable energy equation. However, rapid pyrolysis oils are usually acidic, unstable and have high viscosities that increase with storage time. We have developed fractional catalytic pyrolysis process that produces stable, low viscosity biomass pyrolysis oils that can be stored at ambient conditions without any significant increase in viscosity. The oils can be distilled at both atmospheric pressure and under vacuum without char or solid formation. Fractional catalytic pyrolysis oils produced from hybrid poplar wood was stored at ambient laboratory condition for more than 10 months and the dynamic viscosities changed from 11.2 ±0.45 cP to 12.7 cP ± 0.56 The oils were also characterized using the high temperature simulated distillation method and this showed 100% distillation at up to 640 oC (1200 F). About 15% of the stable pyrolysis oils were blended with Standard Gulf Coast gasoil and cracked in fluid catalytic cracking (FCC) unit into various fuel fractions. There was no difference in both conversion and yields of various products from the blend and the standard Gulf Coast gasoil. The pyrolysis oil/gasoil blend produced slightly less coke than the standard gasoil feedstock. 13C-NMR and FTIR analysis of the cracked fractions did not show any oxygenates in the products.

See more of this Session: Biomass Pyrolysis
See more of this Group/Topical: Fuels and Petrochemicals Division

IN SITU CATALYTIC UPGRADING OF BIOOIL WITH NOVEL CATALYSTS, FROM BENCH SCALE TO PILOT PLANT

*K.G. Kalogiannis[1], S. Stephanidis[1,2], S.S. Voutetakis[1], and A.A. Lappas[1]**

1. Chemical Process Engineering Research Institute (CPERI)/
Center for Research and Technology Hellas (CERTH)
2. Department of Mechanical Engineering, University of Western Macedonia

*Corresponding author: Tel: +0030-2310-498305, Fax: +0030-2310-498380,

email: angel@cperi.certh.gr, kkalogia@cperi.certh.gr

Introduction

Biomass flash pyrolysis is a very promising thermochemical process for the production of bio-liquids and/or chemicals [1, 2]. The catalytic pyrolysis of biomass has recently gained a lot of research interest for the production of bio-liquids from biomass flash pyrolysis in a single catalytic stage. However, early works have shown that the use of catalysts usually leads to additional water and coke production and to a decrease in the yield of organic products [3, 4]. Additionally, due to the fast catalyst deactivation during the biomass catalytic pyrolysis a catalyst regeneration step is necessary. Consequently, the fully circulating fluid bed (CFB) technology is the best technology for this process.

CPERI has extended experience in CFB technology for the FCC process. For biomass pyrolysis, CPERI has constructed a once through transported fluid bed unit. A detailed description of the unit and its operation has been given elsewhere [5]. The main disadvantage of the once through configuration is that it cannot run for a long time since spent catalyst should be reloaded manually into the regenerator. For the above reasons CPERI recently revamped this once through unit to a fully circulating system. CPERI has also experience in the catalytic pyrolysis of biomass for the production of an upgraded biooil. Catalyst evaluation takes place in a bench scale unit, while commercial catalysts have also been evaluated in the Pilot Plant unit to correlate the two units. This allows the fast screening of different catalytic materials.

The characterization of the produced biooil, the assessment of the achieved deoxygenation and the distribution of liquid products into different groups such as aromatic and aliphatic hydrocarbons, phenols, carbonylic compounds, furans, acids etc. is very important for both the understanding of the catalytic pyrolysis mechanisms and the evaluation of the effect of different catalytic materials on the end products. For these reasons several techniques have been developed that allow the determination of the elemental analysis of the produced biooil, its water yield and the qualitative and quantitative evaluation of the distribution of the different chemical groups. In this paper, the effect of several catalysts on the quality and yield of biooil will be investigated in the bench scale unit. Deoxygenation and GCMS analysis of the produced biooil are two important factors that will be thoroughly reviewed. In addition, data from the bench scale unit is compared to those received from the pilot plant unit so that a correlation between the two units is achieved.

Experimental

Prescreening experiments were performed on a bench scale fixed bed reactor made of stainless steel 316. The reactor was heated by a three zone furnace and was purged with nitrogen to ensure an oxygen-free atmosphere and to carry the pyrolysis vapours through the catalyst. The experimental setup is described in detail elsewhere [6]. Figure 1 presents a schematic diagram of the bench scale unit. The feed was Beechwood sawdust and the temperature of pyrolysis was 500 ^{0}C for all runs.

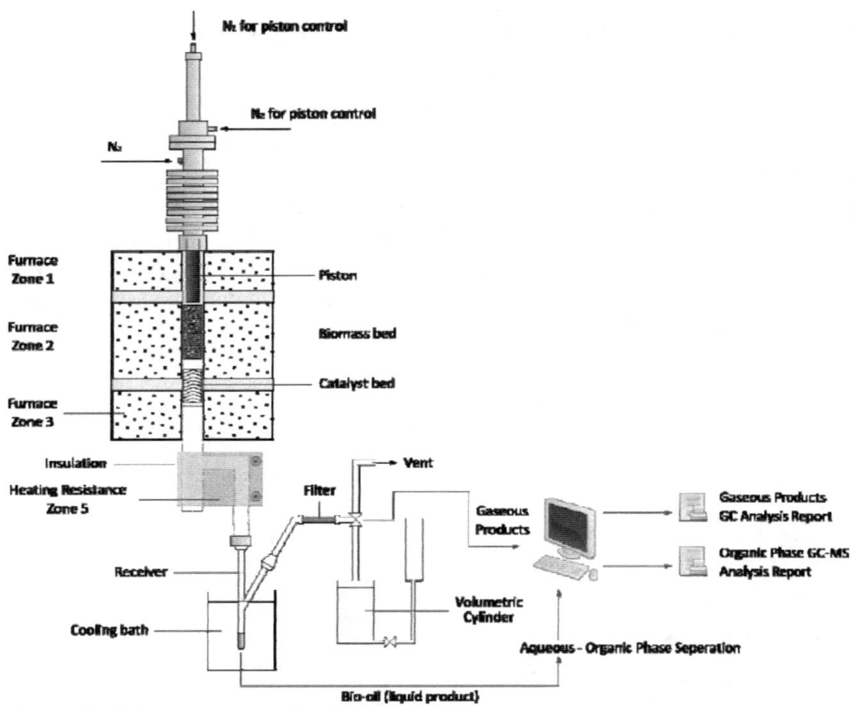

Figure 1. Bench scale biomass pyrolysis unit

In addition, some of the commercial catalysts were tested in the pilot plant unit. Mass balances were received in both pilot plant and bench scale units and the received biooils were analysed for oxygen and water content and with the help of GCMS a semiquantitative analysis for fuels and chemicals in the biooil took place.

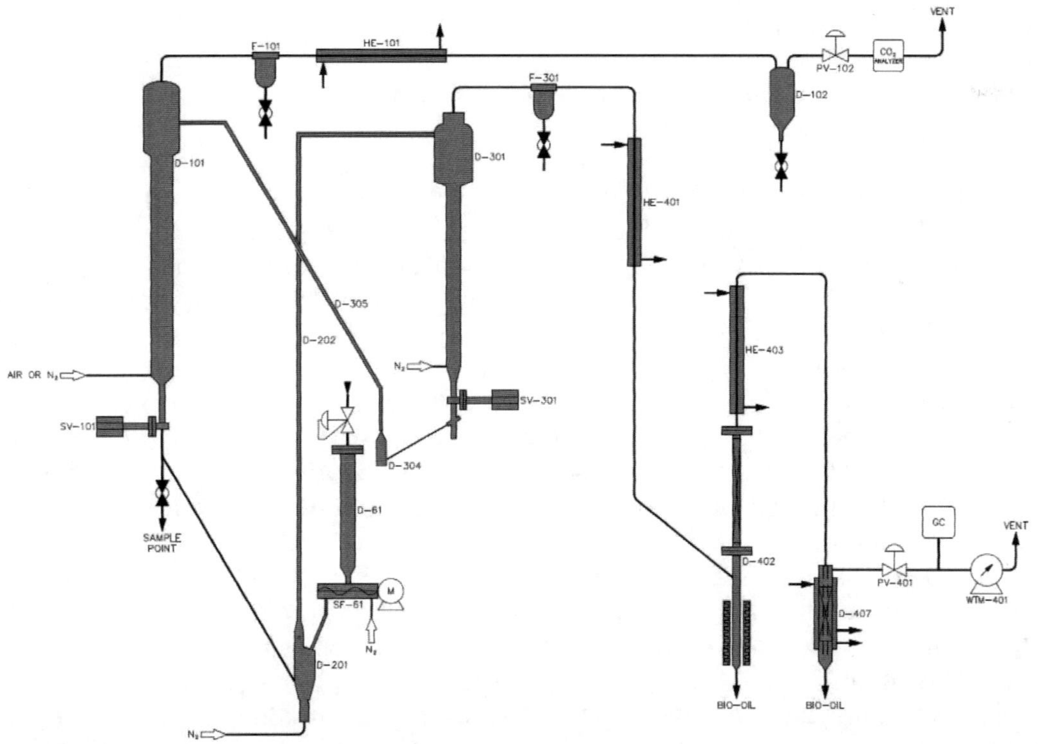

Figure 2. Schematic diagram of the biomass pyrolysis pilot plant in the fully circulating configuration

Figure 2 presents the biomass pyrolysis pilot plant unit in the fully circulating configuration. The unit can be continuously operated for long runs achieving mass balances always between 92-98 wt%. In the paper detailed results of tests will be presented.

Results and Discussion

All tests described in the current study for the bench scale unit were performed at similar experimental conditions Thus; the only independent variable of the work was the type of the catalyst. The mass balances of all tests which were satisfactory and within the range of 90-96 wt% have been normalized to 100%. Figure 3 presents the liquid, gas and coke yields of all the catalysts tested.

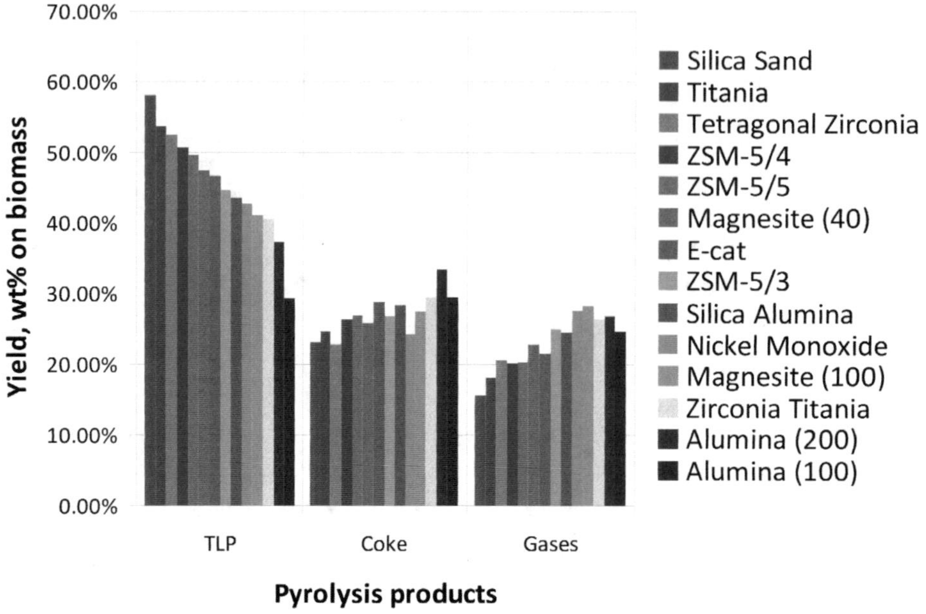

Figure 3. Liquid, Coke and Gases mass balance for catalytic pyrolysis runs of Beech sawdust at 500 ^{0}C in the bench scale unit.

Depending on the catalyst, different cracking takes place yielding different product yields and different biooil qualities. Figure 4 presents the organic biooil yield versus the oxygen content of the organic biooil. It is clear that different deoxygenation levels can be achieved with different catalysts. The lesser the oxygen content of the biooil the greater the decrease of the biooil yield. This is due to the fact that oxygen is removed from biooil in the form of CO, CO_2 and H_2O.

However, Figure 4 also reveals that different biooil yields may be achieved for a given deoxygenation level depending on the catalyst used. This way Silica-Alumina with a surface area of 400 m^2/g has an organic biooil yield of 16.5 wt% on biomass with an oxygen content of 30.5 wt% on biooil while the ZSM-5/3 has a yield of biooil 20.6 wt% on biomass with an oxygen content of 31.14 wt% on biooil respectively. Therefore, the ZSM-5/3 seems to be better suited for the catalytic upgrading of biooil compared to Silica-Alumina. The same is noted if a comparison between a USY Zeolite is made and ZSM-5/3 catalysts. Depending on the surface area of each catalyst different biooil yields and oxygen contents are achieved, however the ZSM-5 catalysts deoxygenate the biooil more effectively than the USY Zeolite. All catalysts produce less biooil of higher quality, less oxygen content, compared to the base cases of Silica sand that acts as inert material with no surface area and yields around 38 wt% organic biooil with an oxygen content of 38 wt%.

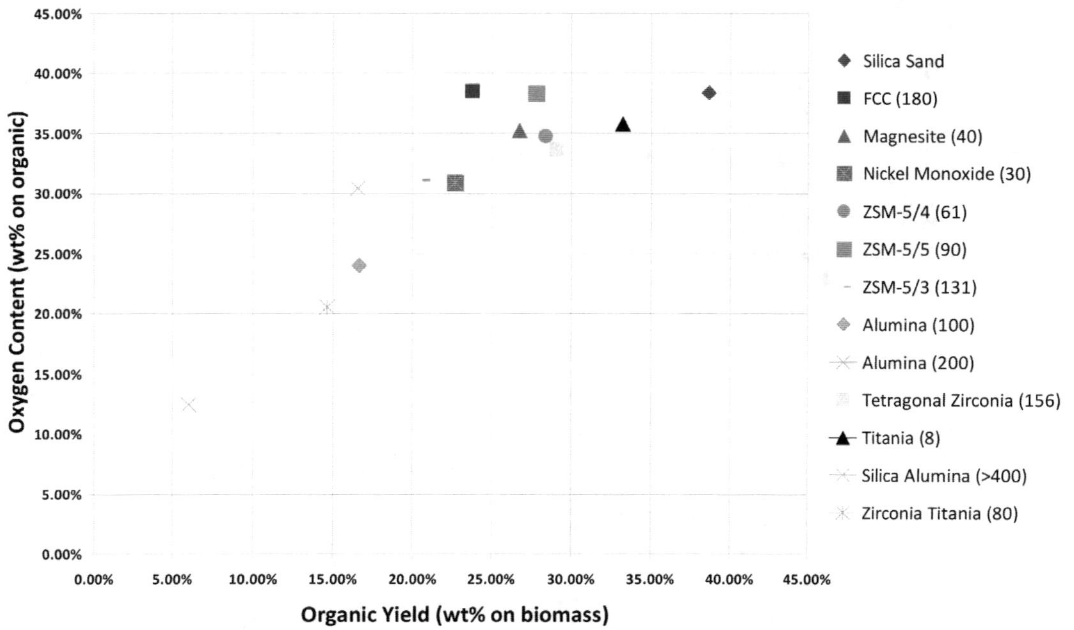

Figure 4. Oxygen content of organic biooil vs organic biooil yield.

GCMS analysis of the produced biooils is presented in table 1. The groups presented are Aromatic Hydrocarbons (AR), Aliphatic Hydrocarbons (ALI), Phenols (PH), Furans (FUR), Acids (AC), Esters (EST), Alcohols (AL), Ethers (ETH), Aldehydes (ALD), Ketones (KET), Polyaromatic Hydrocarbons (PAH), Sugars (SUG), Nitrogen compounds (NIT), Heavy compounds (HV) and Unidentified (UN).

Table 1. GCMS analysis of organic fraction (wt % on organic liquid) of biooils produced from different catalysts at 500 ^{0}C.

	Silica Sand	Tetr. Zirc.	FCC	ZSM-5/3	Magnesite (40)	Zirconia - Titania	Alumina (200)
AR	1.11	3.02	2.82	7.77	0.84	8.45	10.98
ALI	0.18	0.00	0.43	1.02	1.05	0.20	1.16
PH	29.62	32.03	36.15	37.02	39.12	32.76	25.76
FUR	1.04	0.37	1.47	1.67	0.83	1.43	1.91
AC	19.64	14.38	12.45	5.24	9.30	8.65	3.88
EST	4.18	3.29	0.64	1.12	6.12	3.91	0.89
AL	0.30	0.52	1.74	0.54	1.59	1.02	4.75
ETH	1.76	0.43	4.32	2.74	0.91	0.55	0.72
ALD	1.00	1.23	2.34	0.16	0.68	1.26	2.58
KET	13.62	11.42	8.41	6.03	14.36	10.65	4.18
PAH	0.00	0.00	1.79	2.74	0.23	4.87	10.48
SUG	1.77	0.50	0.00	0.00	0.00	0.45	0.00
NIT	4.58	3.50	3.60	6.10	1.84	2.39	4.05
HV	9.67	13.02	8.45	10.03	13.44	12.84	17.58
UN	11.53	16.29	15.39	17.82	9.69	10.57	11.08

From Table 1 it can be deduced that different catalysts enhance the production of different chemical groups. FCC, ZSM-5, Alumina and Zirconia-Titania produce more aromatics and PAHs while FCC, ZSM-5 and Magnesite also produce high Phenol yields. All catalysts reduce the Acids and Ketones yields that are largely responsible for the corrosivity and instability of biooils [7].

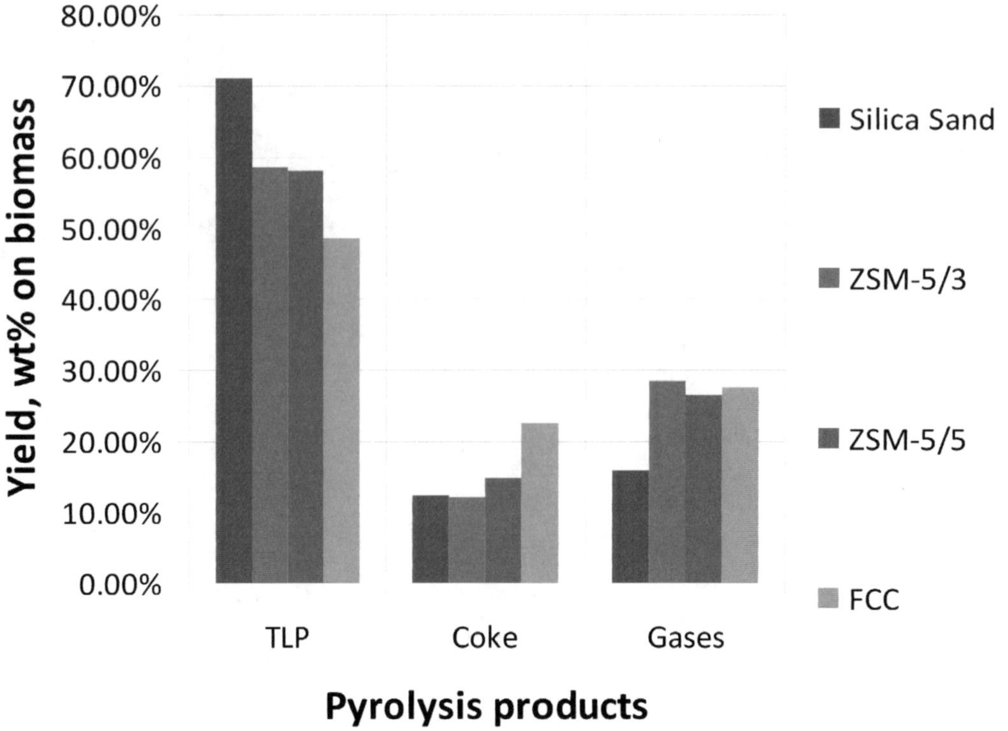

Figure 5. Pyrolysis product yields for in Pilot plant unit.

In the pilot plant two ZSM-5 additives with different surface areas (named ZSM-5/3 and ZSM-5/5), an FCC catalyst were also tested along with silica sand for the base case tests. Mass balances are presented in figure 5.

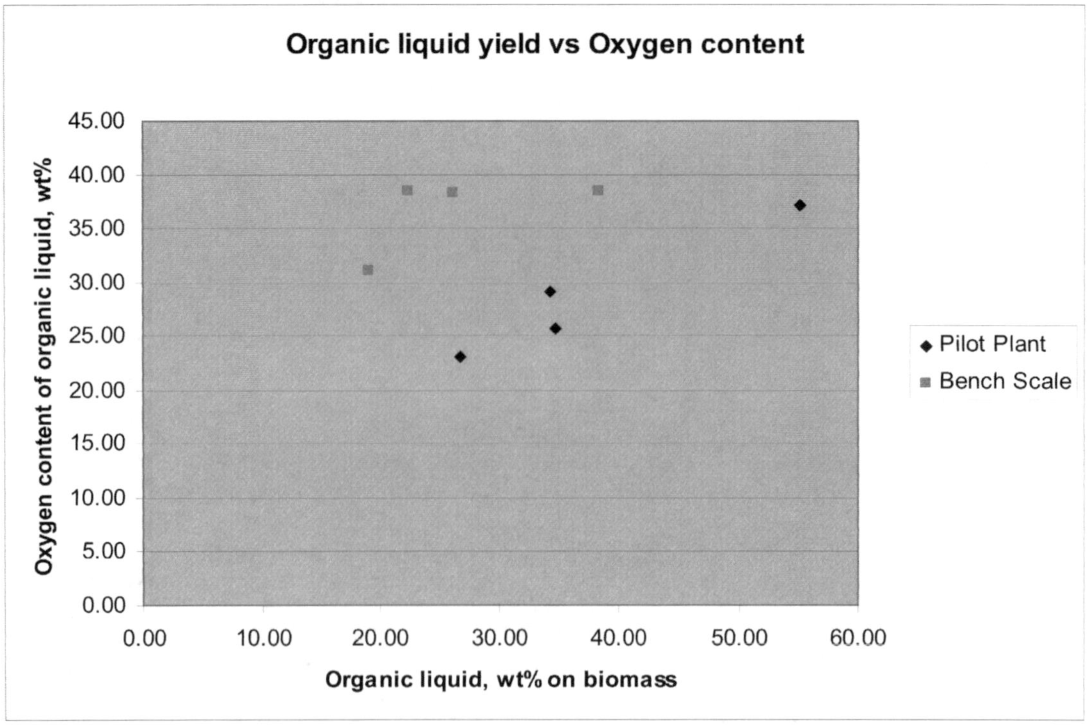

Figure 6. Comparison of oxygen content in biooils vs biooil yield produced in bench and pilot plant scale.

Figure 6 presents a correlation between the bench scale and the pilot plant data. In all cases, the pilot plant unit yields more organic biooil with less oxygen content. This is attributed to three main factors. Firstly, the heat transfer in the pilot plant unit where the biomass particles are directly mixed with hot catalyst results in the better mixing and cracking of the biomass. This way more liquid product and less coke and char are produced. In addition, the catalyst to biomass ratio in the pilot plant is 15:1, 30 times higher compared to the 1:2 ratio of the bench scale unit. The higher catalyst to biomass ratio along with the direct contact and better mixing of biomass particles with hot catalyst results in the further deoxygenation and cracking of the biomass. Finally, in the case of the pilot plant the catalyst is constantly regenerated by burning off coke in the regenerator vessel which allows it to retain its activity in contrast to the fixed bed catalyst that is deactivated due to coke formation on the catalyst. The general trend is that the pilot plant unit produces biooil in greater yields and with lower oxygen contents stressing the efficiency of the circulating fluid bed technology.

Conclusions

Catalytic biomass pyrolysis was investigated both at bench and pilot plant scale. Several new experimental and commercial catalysts were tested and the deoxygenation along with the selectivity for chemicals and fuels was evaluated. In addition a correlation between bench and pilot plant scale was achieved that allows for the fast screening of catalysts and the determination of the most suitable ones for the biomass catalytic cracking process with circulating fluid bed technology.

Acknowledgements

The work presented was funded by the ACENET HECABIO project contract no. ACE.07026

References

[1] A. Demirbas, Biomass resource facilities and biomass conversion processing for fuels and chemicals, Energy Convers. Managem. 42 (2001) 1357-1378..

[2] D. Meier, O. Faix, State of the art of applied fast pyrolysis of lignocellulosic materials- a review, Bioresource Technol. 68 (1999) 71-77.

[3] A. A. Lappas, M. C. Samolada, D. K. Iatridis, S. S. Voutetakis and I. A. Vasalos, Biomass pyrolysis in a circulating fluid bed reactor for the production of fuels and chemicals, Fuel, 81 (2002) 2087-2095.

[4] Adam, E. Antonakou, A. Lappas, M. Stöcker, M.H. Nilsen, A. Bouzga, J.E. Hustad, G. Øye, In situ catalytic upgrading of biomass derived fast pyrolysis vapours in a fixed bed reactor using mesoporous materials, Micropor. Mesopor. Mater. 96 (2006) 93-101.

[5] Lappas A., Dimitropoulos V., Antonakou E., Voutetakis S., Vasalos I., 7th World Congress of Chemical Engineering, (2005) 26.

[6] E. Antonakou, A. Lappas, M.H. Nilsen, A. Bouzga, M. Stöcker, Evaluation of various types of Al-MCM-41 materials as catalysts in biomass pyrolysis for the production of bio-fuels and chemicals, Fuel, 85 (2006) 2202.

[7] M.N. Islam, M.R.A. Beg, The fuel properties of pyrolysis liquid derived from urban solid wastes in Bangladesh, Bioresource Technol. 92 (2004) 181-186.

Title: Pyrolysis of algal biomass

Pyrolysis experiments with algal biomass were performed in a laboratory fixed-bed reactor. These experiments were carried out at 450, 500, 550 and 600°C to study the effects of temperature on biomass conversion to gas, oil and char products. These conditions were chosen based on thermo-gravimetric analyses of the feed stocks which showed that most of the thermal decomposition of algal biomass occurred within this temperature range. The reactor was purged with helium to maintain inert atmosphere for pyrolysis reactions. During our experiments, liquid products were collected by condensing with cold water; gas formation was monitored using mass flow meters and gas composition was measured using an online gas chromatograph equipped with a pulsed discharge helium ionization detector. Solid residues (or char) were recovered from the reactor at the end of the experiment. Proximate- (volatile matter, moisture content, ash and fixed carbon) and ultimate- analyses (C, H and N) of the feedstock and products were performed using various standard methods to close overall and element-specific mass balances. Additionally, bio-oil was analyzed using mass spectroscopy to identify specific compounds and the overall molecular weight distribution of this product. The heating values of biomass, bio-oil and bio-char were calculated. The results of these investigations will be presented and discussed.

Catalytic Upgrading of Switchgrass Pyrolysis Bio-oil Vapors

Jonathan E. Peters, Xiaohan Zhang, Trung T. Pham, Anirudhan Gangadharan, Xinli Zhu, Roberto Galiasso, Daniel E. Resasco and Richard G. Mallinson,
School of Chemical, Biological and Materials Engineering,
University of Oklahoma, Norman, OK, 73019, USA
Center for Biomass Refining,
University of Oklahoma, Norman, OK, 73019, USA

ABSTRACT

Catalytic upgrading of bio-oil from switchgrass pyrolysis was investigated over $Ce_{0.5}Zr_{0.5}O_2$ mixed oxides to produce higher carbon chain molecules for use as fuel additives from short aldehydes and carboxylic acids found in bio-oil mixtures. The experiments were carried out in a single two-stage process where the gaseous products from the pyrolysis reaction were fed directly to a fixed catalyst bed in a separate reactor in series with the pyrolysis reactor. The final products were condensed to the liquid phase and collected for analysis. GC by GC analysis of products shows evidence of catalytic activity. Acetic acid and propanoic acid found in the bio-oil were converted to acetone and 3-pentanone, respectively, among other products, when reacted over the catalyst. This is consistent with our work on model compounds that short aldehydes and carboxylic acids are converted to higher chain aldehydes and ketones by way of aldol condensation and ketonization over mixed oxides catalyst. Reacting over mixed oxides catalyst resulted in lower liquid yields from pyrolysis when compared to experiments without catalyst. The char yields remained the same while the gas yields, by difference, appear to have increased under these conditions. The coated monolith catalysts seem to have higher activity than the pelletized catalysts for the same amount of ceria and zirconia loading.

1. INTRODUCTION

There is strong motivation to produce renewable fungible fuels for transportation and power generation.[1,2] One way is by a high temperature thermo chemical breakdown of biomass called pyrolysis.[3] The liquid from pyrolysis is a complex mixture of oxygenated compounds called "bio-oil."[4] The light oxygenates in bio-oil can be upgraded to higher chain oxygenates for use as a fuel or fuel additive.[5] In this study, catalytic upgrading of bio-oil produced from switchgrass pyrolysis was investigated over $Ce_{0.5}Zr_{0.5}O_2$ mixed oxides to produce higher carbon chain molecules from short aldehydes and carboxylic acids found in bio-oil mixtures. This is consistent with our work on model compounds that short aldehydes and carboxylic acids are converted to higher chain aldehydes and ketones by way of aldol condensation and ketonization over mixed oxides catalyst.[5]

2. EXPERIMENTAL

2.1 Catalyst Preparation

A $Ce_{0.5}Zr_{0.5}O_2$ catalyst was prepared by a co-precipitation method with aqueous solutions of cerium (IV) ammonium nitrate and zirconium nitrate as described in the studies by Hori et al [6] and Noronha et al [7]. The aqueous solutions were mixed at a 1:1 ratio, co-precipitated with ammonia hydroxide, rinsed, dried, and calcined at 500°C for one hour.

2.2 Pyrolysis/Catalysis

The pyrolysis and the catalysis reactions were carried out in a single two-stage process in a nitrogen environment at atmospheric pressure. The pyrolysis stage consisted of a 0.870 in. I.D., stainless steel, counter-flow reactor with nitrogen fed in the bottom at 0.550 SLPM and pre-heated to 500 °C. The catalysis stage consisted of a 0.495 in. I.D., stainless steel, fixed-bed reactor maintained at 400 °C. The switchgrass, 180-250 micron diameter particles, was fed with a manual, auger-type feeder from the top of the first reactor by gravity. Later, some experiments were also carried out with 150-600 micron diameter switchgrass particles. The stages were operated in series where the nitrogen carried the pyrolysis vapor products to the catalytic reactor, and the catalysis products were condensed to the liquid phase for collection and analysis. Figure 1 shows the schematic of the two-stage process. Two sets of experiments were conducted; one set with catalyst and one without. In the first set, the catalytic reactor was charged with the $Ce_{0.5}Zr_{0.5}O_2$ mixed oxides fixed bed using pelletized catalyst with particle size 400-700 microns. The linear velocity of the gaseous products in the carrier gas is about 40 cm/s, and the catalyst bed is 1 cm long. The second set of experiments was conducted with ground glass of particle size 400-700 microns in place of the catalyst to show the changes from the catalyzed reactions while maintaining the fluid dynamics. The liquid products collected existed in two distinguishable phases, the aqueous fraction and the heavy organic fraction. In this study, the aqueous fraction was the focus and was analyzed separately by GCxGC, GC-MS, carbon NMR, free fatty acid number, and water content.

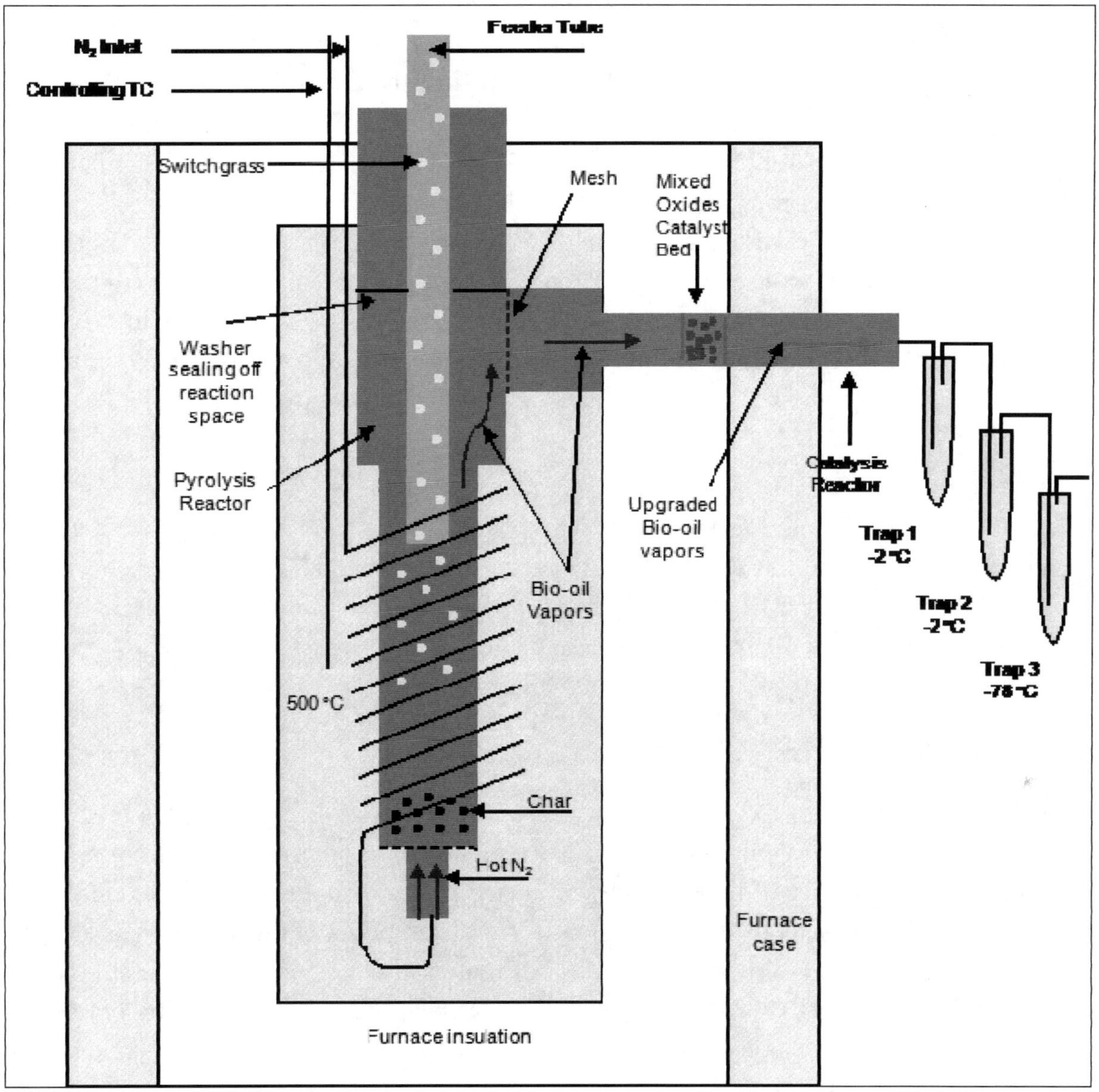

Figure 1: Pyrolysis-Catalysis Two-Stage Process Design Schematic

2.3 Two dimensional GCxGC

A GCxGC system was set up by modifying the GCMS-QP 5000 (Shimadzu), equipped with EI detector, with a two-stage thermal modulator (Zoex Corporation) and a secondary oven installed inside the main oven of the GCMS. A low temperature valve mounted outside the GCMS oven controls the cold jet from -80 °C to -120 °C, that produces narrow chemical pulses. A hot jet at 300°C releases the trapped solutes into the second column mounted inside the second oven. The modulator accumulates samples eluting from the first column for a period of 12 seconds, typically equal to one third of the duration of a chemical peak. However, some peaks may not completely elute within this modulation period, i.e. acetic acid, and showing smearing/tailing behavior in the 2D-chromatogram. An auto-sampler injects undiluted samples with a 10 micro-syringe (0.4-0.6 μl) and a split ratio of 25. The first column is for volatility separation with DB-1 30 m length, 0.25 mm I.D., 1 μm film thickness (Agilent). The second column is for polarity separation Carbowax 1 m length, 0.1 mm I.D., 0.36 μm film thickness (Agilent). The primary oven is set at 35°C ramping to 275°C with a ramp rate of 2.2°C/min. The secondary oven is set at 75°C ramping to 325°C with a ramp rate of 2.27 °C/min. The GC images are processed and analyzed using GC Image (TM) Version 2.0b4, provided by Zoex. In this contribution, the total response of the MS detector was obtained from the evolution of specific compounds, and thus providing a semi-quantitative analysis. On some samples, an HP 6890A equipped with an FID detector was used to analyze the sample compositions to compare the relative area ratios of reacted compounds with and without catalytic treatment.

2.4 NMR-C^{13} Spectroscopy

NMR-C^{13} analysis was performed using Varian Mercury VX-300 MHz spectrometer with 4-nuclei auto-switchable probe PFG at 75MHz magnet. Approximately 7 drops of each sample was added to 0.75 ml d- THF NMR solvent in 5mm NMR tube. For each sample, 16384 scans were collected in consecutive 14 hours. We could identify acetic acid peaks at 20ppm and 179ppm, and ketone range from 180ppm to 210ppm.

2.5 Total Acid Number

The total acid number of the sample was measured using TitroLine alpha plus Free Fatty Acid Analyzer (TitroLine alpha plus, SCHOTT Instruments) by potentiometric titration. The total acid number is the quantity of base, expressed in milligrams of potassium hydroxide that is required to neutralize all acidic constituents present in 1 g of sample. 0.5 mg of each sample was collected and weighed in a 150 ml beaker and 80 ml of toluene and ethanol, 1:1, solvent mixture was added to the sample. 0.1M KOH titration solvent was added gradually to reach the equilibrium point of titration reaction. A calibration titration with 80ml of titration solvent was performed for each set of samples.

2.6 Water content

Water content analysis (% wt) of each sample was carried out by TitroLine Karl-Fischer analyzer (TitroLine KF, SCHOOT Instruments), in which 0.018g of sample was injected using syringe and titrated with KF reagent for volumetric titration. The water content percentage was determined upon the KF reagent consumed. For each set of samples perform a calibration titration with 0.18g DI-water.

3. RESULTS AND DISCUSSION

Liquid and Char yields were measured by % wt of the switchgrass reacted in the pyrolysis reactor, and gas yields were calculated by difference. Figure 2 compares the product yields from identical experiments with and without catalyst, showing a decrease in liquid yields by 28% and an increase in gas yields by 35% when reacted with catalyst. This result is consistent with aldol condensation reactions and ketonization reactions that reduce the number of moles of liquid compounds and produce more carbon dioxide.

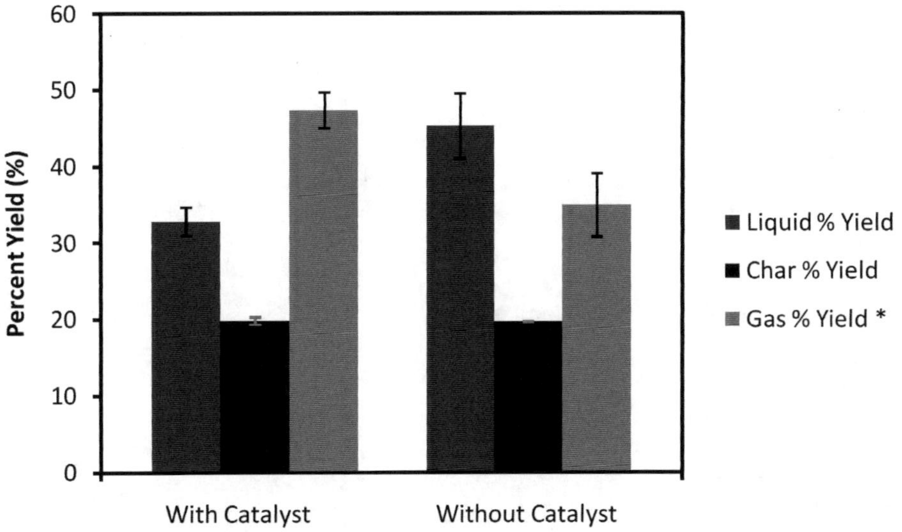

Figure 2: Product Yield Comparison
*Gas Yields were calculated by difference

The bubble graph in Figure 3 represents the peak volume response difference of the aqueous fraction of runs with and without catalyst. The positive responses are shown in the blue while the negatives are shown in red. A positive response means the volume/amount of a compound increases after catalyst treatment while the negative means the compound is consumed or reacted after treatment. The size of the bubble is equivalent to the magnitude of the difference. Note that a distinctively large red bubble at (12, 6) is acetic acid that is consumed after the catalyst treatment. Correspondingly, a large size blue bubble at (9.6, 7.4) is acetone being increased after the treatment. The acetone might be formed from ketonization reaction of acetic acid, shown as follows:

$$CH_3COOH + CH_3COOH \rightarrow CH_3COCH_3 + CO_2$$

However, the bubble size graph above is rather semi-quantitative due to the limitation of the mass spec detector. Thus, quantitative analysis is needed and can be carried out using GC with FID detector. Thus far, the main interest is focused on the conversion of the acid, relatively high concentration in pyrolysis oil, into a more stable product, in this case, acetone.

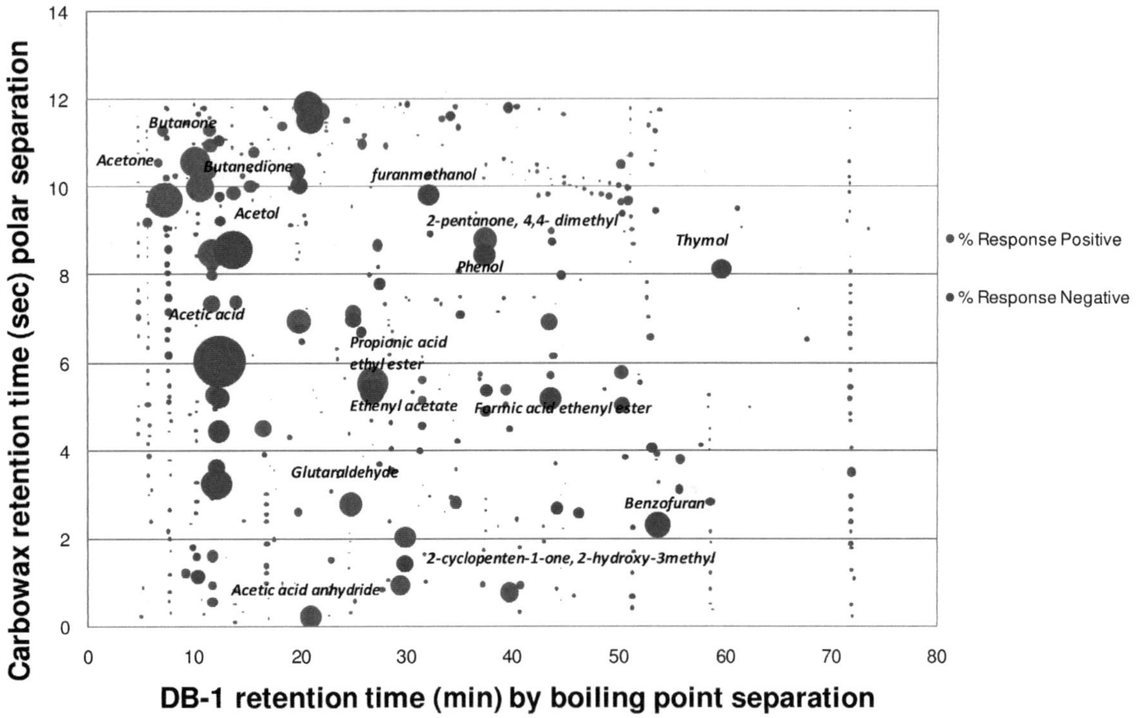

Figure 3: Positive and negative % Responses of peak volume with retention time from without catalyst 2D-GC and with catalyst 2D-GC samples.

Acetic acid and acetone standards were injected to confirm the retention times of these compounds in the aqueous fraction of the bio-oil samples. The injection of the samples with and without catalyst into GC FID showed that the area ratios of acetone/acetic acid are different, shown in Figure 4. Different amount of ceria and zirconia were coated onto inert silica alumina monoliths. The activity of these monoliths is much more stable compared to that of the pelletized catalysts, even for the same amount of loading.

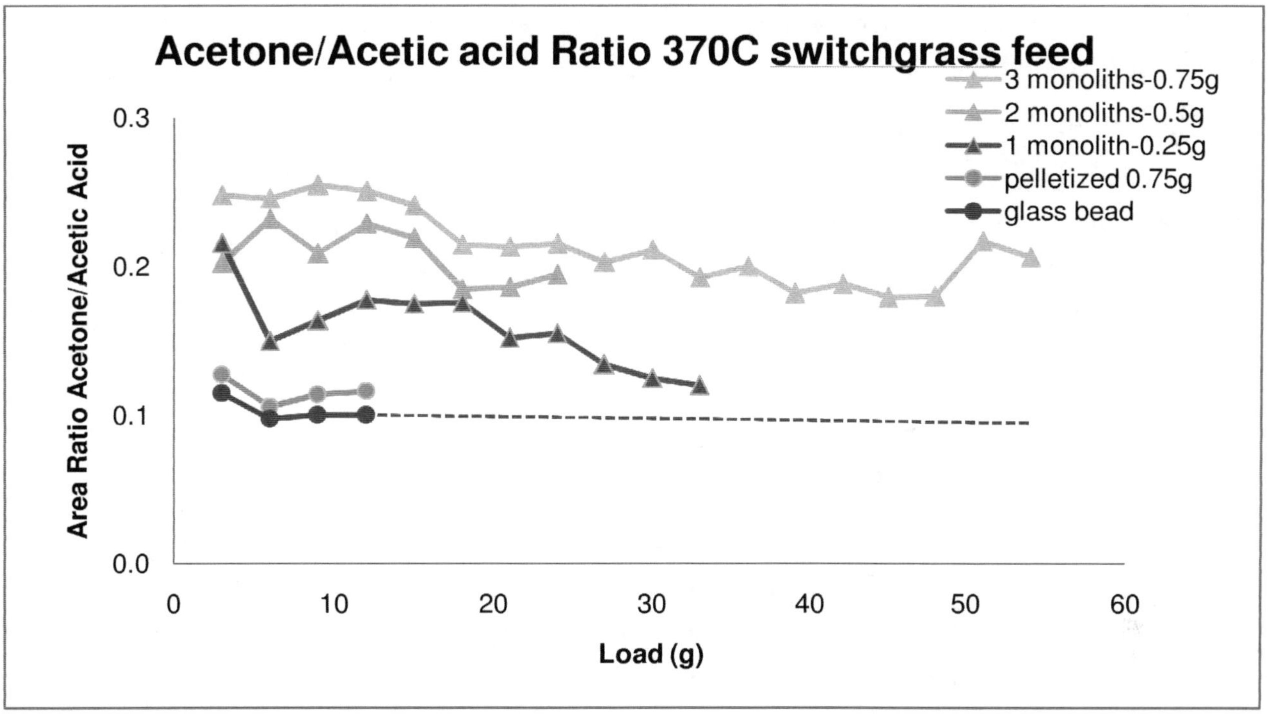

Figure 4: Acetone/Acetic Acid Area Ratio of feeding grass over coated monoliths vs. pelletized catalyst.

Another indicative parameter to investigate the activity of the catalyst is the CO_2/CO ratio by analyzing and quantifying outlet gas samples with various loads of fed grass. As ketonization reaction takes place, CO_2 is expected to be released more than when fed without catalysts. In Figure 5, for the 2 monoliths in series, the CO_2/CO molar ratio is about 1.4 to 1.8 and decreasing over time with gradual deactivation of the catalyst. However, without catalyst, this ratio stays low at about 1.1.

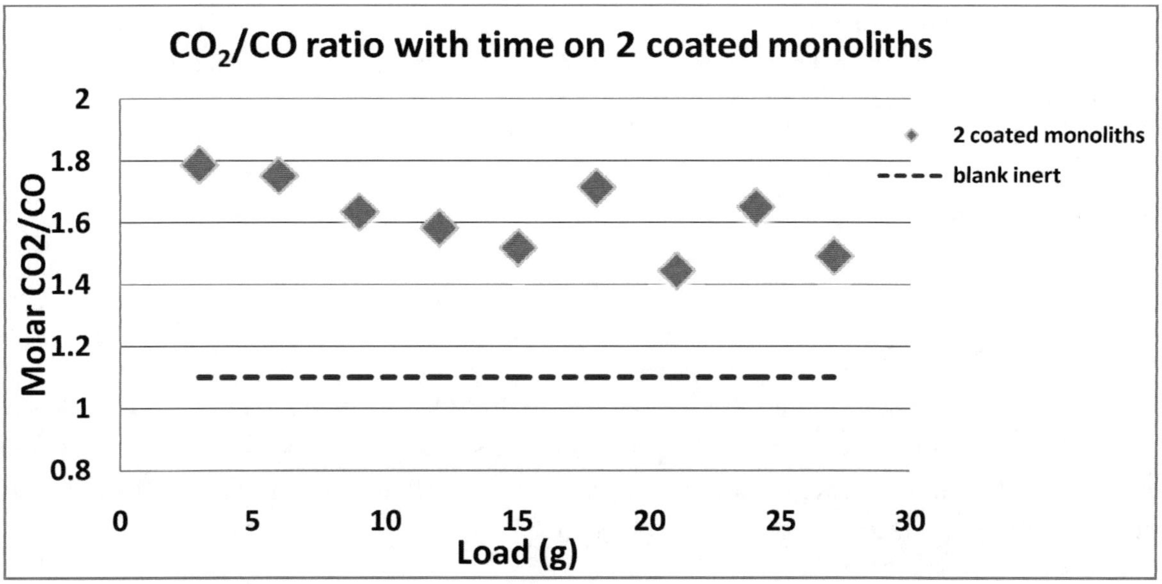

Figure 5: CO_2/CO molar ratio as indicator for catalyst activity. Herein shown ratio for 2 coated monoliths installed in series.

Besides acetic acid conversion, there are also other compounds being reacted with the presence of catalyst. In fact, propionic acid also converts in ketonization forming 3-pentanone, confirmed by GCMS. Cross ketonization of acetic acid and propionic acid results in the formation of butanone, that is also observed in the product spectrum change for with and without catalyst. Besides ketonization reactions, aldehydes in the bio-oil could also react in aldol-condensation to form large aldehydes. However, reactivity of the acids is much higher than that of the aldehydes [5]. There also exists competitive adsorption with other compounds, particularly the bulky molecules such as phenol, furfuraldehydes, etc., although these molecules might not react but rather take up sites on the surface of the catalyst.

After the catalyst upgrading, the total acid number drops approximately 10% due to the $Ce_{0.5}Zr_{0.5}O_2$ ketonization reaction which consumes acids to form ketones. The comparison of water content analysis between experiments with and without catalyst demonstrates that the percentage of water rises from 50% to 64% in average, which is also consistent with the ketonization reaction which produces water. Table 1 below summarizes the acid analysis results.

Table 1: Acid and Water Analysis of Aqueous Fraction of Bio-oil

Run#	Catalyst or Packing	Catalysis Temp. (°C)	TAN**	Water Content (%)
44c	Ce0.5Zr0.5O2	400	82.7	65.7%
48c	Ce0.5Zr0.5O2	400	94.2	63.6%
50c	Ce0.5Zr0.5O2	350	75.4	66.7%
53c1†	Ce0.5Zr0.5O2	400	76.3	53.4%
47	ground glass	400	97.5	53.4%
49	ground glass	400	89.2	50.3%
51	borosilicate	350	93.0	46.0%
52	borosilicate	350	104.6	50.1%

** Total Acid Number: the amount of KOH in mg needed to neutralize 1 g of the aqueous fraction.

†The catalyst in run 53c1 was used once before in run 44c

4. CONCLUSION

Upgrading switchgrass pyrolysis vapors over $Ce_{0.5}Zr_{0.5}O_2$ catalyst results in a decrease in liquid yields by 28% and an increase in gas yields by 35% under these conductions. The short aldehydes and carboxylic acids are converted to higher carbon chain aldehydes and ketones by way of aldol condensation and ketonization over the mixed oxides catalyst. This is marked by increasing the ratio of acetone to acetic acid in the aqueous products after reacting over the mixed oxides catalyst. While this only shows a part of the bigger picture, future studies will show more complete analysis of the total bio-oil.

REFERENCES

[1] Ragauskas, A. J. et al. *Science* **2006**, *311*, 484
[2] Huber, G. W.; Iborra, S.; Corma, A. *Chem. Rev.* **2006**, *106*, 4044
[3] Shafizadeh, F.; *J. Analytical Appl. Pyrolysis*, **1982**, *3*, 283
[4] Mohan, D.; Pittman, C. U. Jr.; Steele, P. H. *Energy & Fuels* **2006**, *20*, 848
[5] Gangadharan, A.; Resasco, D. E.; Mallinson, R. G. *Accepted* **2010**
[6]Hori, C. E.; Permana, H.; Ng, K. Y. S.; Brenner, A.; More, K.; Rahmoeller, K. M.; Belton, D. *Appl. Catal. B-Environ.* **1998**, *16*, 105
[7]Noronha, F. B.; Fendley, E. C.; Soares, R. R.; Alvarez, W. E.; Resasco, D. E. *Chem. Eng. J* **2001**, *82*, 21

(135h) Integrated Pyrolysis Quench Catalytic Cascade for Insitu Product Stabilization and Upgrading

Shaolong Wan[1], Xiaohan Zhang[1], Trung Pham[1], Roberto Galiasso[1], Lance Lobban[1], Daniel Resasco[1] and **Richard G. Mallinson**[2], (1)University of Oklahoma, Norman, OK, (2)School of Chemical, Biological and Materials Engineering, University of Oklahoma, Norman, OK

The operation and product distribution from switchgrass pyrolysis in an integrated kg/h plant are presented. Product quenching with aqueous and organic product streams segregate and stabilize the liquid products. An integrated catalytic cascade that converts fractions of the product vapors between quenches, before condensation, will be discussed.

See more of this Session: Biomass Pyrolysis
See more of this Group/Topical: Fuels and Petrochemicals Division

(208a) Production of Jet and Diesel Fuel Range Alkanes From Waste Hemicellulose-Derived Aqueous Solutions

Rong Xing, Chemical Engineering Department, University of Massachusetts Amherst, Amherst, MA, Ayyagari V. Subrahmanyam, Chemical Engineering, University of Massachusetts, Amherst, Hakan Olcay, Department of Chemical Engineering, University of Massachusetts, Amherst, Amherst, MA, Wei Qi, University of Massachusetts Amherst, Amherst, MA, Michael F. Malone, College of Engineering, University of Massachusetts, Amherst, MA, G. Peter van Walsum, Chemical and Biological Engineering, University of Maine, Orono, ME, Hemant P. Pendse, Managing Director, Forest Bioproducts Rsearch Initiative (FBRI), University of Maine, Orono, ME and George W. Huber, Chemical Engineering, University of Massachusetts-Amherst, Amherst, MA

We present a novel four-step process for the production of jet and diesel fuel range straight chain alkanes from hemicellulose-derived aqueous solutions which are byproducts from a Kraft mill located in the Northeastern U.S. The hemicelluloses extract contains xylose oligomers (21.2 g/l xylose after the acid hydrolysis) as well as 0.79 g/l glucose, 1.1 g/l arabinose, 9.77 g/l acetic acid, 0.32 g/l formic acid, and other minor products. The first step in this process is an acid-catalyzed biphasic dehydration to produce furfural in yields up to 87%. The furfural is extracted from the aqueous solution into a tetrahydrofuran (THF) phase which is then fed into an aldol condensation step. The furfural-acetone-furfural (F-Ac-F) dimer is produced in this step by reaction of furfural with acetone with yields of 96% for the F-Ac-F dimer. The F-Ac-F dimer is then subjected to a low temperature hydrogenation to form hydrogenated dimer (H-FAF) at 110-130 °C and 800 psig with a 5 wt% Ru/C catalyst. Finally the H-FAF undergoes hydrodeoxygenation to make large straight chain alkanes primarily C13 and C12 in yields up to 91%. The theoretical yield for this process is 0.61 kg of alkane per kg of xylose (monomer or oligomers) in the hemicelluloses extract. Experimentally we were able to obtain 76% of the theoretical yield for the overall process. We estimate that the jet fuel range alkanes can be produced from between \$2.06/gal to \$4.39/gal depending on the feed xylose concentration, the size of the biorefinery, and the overall yields. Sensitivity analysis shows that the prices of raw materials, the organic to aqueous mass ratio in biphasic dehydration, and the feed xylose concentration in the hemicellulose extract significantly affect the product cost.

See more of this Session: Catalytic Processing of Fossil and Biorenewable Feedstocks: Fuels I
See more of this Group/Topical: Catalysis and Reaction Engineering Division

Catalytic strategies to convert biomass into liquid hydrocarbons

David Martin Alonso, Jesse Q. Bond, Dong Wang, and James A. Dumesic

Department of Chemical and Biological Engineering, University of Wisconsin-Madison

Research on lignocellulosic biomass conversion is receiving increased attention as a means to provide a renewable source of carbon to reduce CO_2 emissions and to replace petroleum as a feedstock for the production of fuels and chemicals. In this presentation, we describe the conversion of cellulose to liquid hydrocarbon fuels using levulinic acid as a platform chemical intermediate, which is subsequently converted to γ-valerolactone (GVL). In particular, we present an integrated strategy in which GVL is converted to butene, which is then coupled by oligomerization to form hydrocarbons appropriate for use as components in gasoline or jet fuels.

The catalytic strategy for processing γ-valerolactone (GVL) to form liquid hydrocarbon fuels is illustrated in Figure 1.

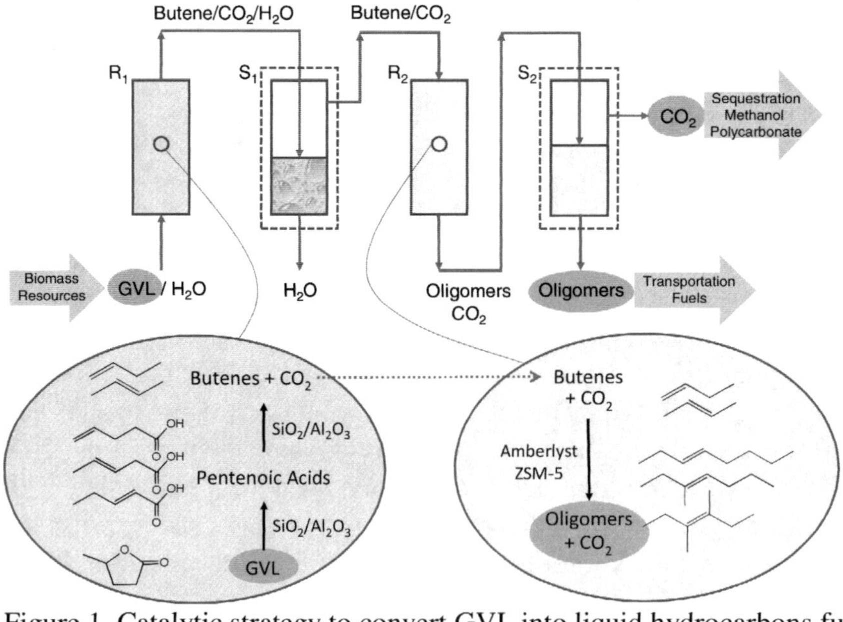

Figure 1. Catalytic strategy to convert GVL into liquid hydrocarbons fuels.

In this cascade process, aqueous solutions of GVL are converted to pentenoic acid by ring opening over SiO_2/Al_2O_3 catalysts, and in the same catalytic bed and at the same conditions, pentenoic acid undergoes decarboxylation to produce CO_2 and butene. In a second catalytic reactor, butene is converted to C_{8+} alkenes by oligomerization over an acid catalyst, such as ZSM-5 or Amberlyst 70. The decarboxylation reaction is favored at lower pressures, but we found that it can be carried out at pressures appropriate for olefin oligomerization (e.g., 500 psi), providing several important advantages. First, we obtain a stream of butene and CO_2 at conditions that favor the downstream oligomerization reaction. Second, working at elevated pressures allows removal of water as a condensed liquid in a simple gas-liquid separator, while butene remains in the gas phase to be fed into the second reactor. And third, we obtain a high purity stream of CO_2 at elevated pressure, suitable for use in other processes, such as reduction to methanol or liquefaction for transportation to sites for sequestration.

Table 1 presents results obtained in the integrated reactor system at different conditions and using HZSM-5 and Amberlyst as oligomerization catalysts[1]. All the reactions and the separation step can be carried out at high yields. Entry 6 shows that 77% of the GVL fed to the reactor system can be converted to C_{8+} alkenes.

Table 1. Performance of integrated catalytic system consisting of two flow reactors in series with an inter-reactor separator. Second reactor operated at 36 bar.

Entry	Reactor 1 (GVL to butene)				Reactor 2 (Butene to alkenes)				GVL to liquid C_8-C_{16} /C_{8+}(%)
	T (K)	GVL conversion (%)	Butene yield (%)	Butene out of 1st separator (%)	Catalyst	T (K)	Butene conversion (%)	Liquid selectivity to C_8-C_{16}/C_{8+} (%)	
1*	648	63	37	75	HZSM-5 (14g)	498	95	63/90	17/24
2†	648	98	91	90	HZSM-5 (14g)	498	44	76/86	28/31
3†	648	99	92	88	Amberlyst (3g)	443	92	74/94	50/62
4‡	648	99	90	89	Amberlyst (4g)	443	94	64/93	48/66
5§	648	99	94	93	Amberlyst (4g)	443	81	79/94	53/63
6‖	648	99	98	95	Amberlyst (12g)	443	90	75/95	60/77

* Reactor 1: 2.7 g SiO_2-Al_2O_3.WHSV = 0.68 h^{-1}. First separator at 373 K.
† Reactor 1: 10g SiO_2-Al_2O_3.WHSV = 0.18 h^{-1}. First separator at 383 K.
‡ Reactor 1: 10 g SiO_2-Al_2O_3.WHSV = 0.18 h^{-1}. First separator at 388 K.
§ Reactor 1: 10 g SiO_2-Al_2O_3.WHSV = 0.22 h^{-1}. First separator at 398 K.
‖ Reactor 1: 8 g SiO_2-Al_2O_3.WHSV = 0.22 h^{-1}. First separator at 398 K.

The approach described here provides a complementary strategy for biomass conversion to liquid fuels with several interesting advantages. First, it offers a route for production of branched, fuel range hydrocarbons, which are requisite in gasoline and jet fuels. Further, owing to extensive hydrogen integration, this processing strategy can be executed with minimal demand for external (petroleum derived) hydrogen, ensuring a greater degree of independence from fossil fuels. Finally, we offer a strategy by which one equivalent of CO_2 can be reclaimed from GVL as a pure, high pressure product appropriate for sequestration or chemical transformation.

[1] J.Q. Bond, D. Martin Alonso, D. Wang, R. M. West, J.A. Dumesic, *Science*, **327**, (2010) 1110-1114

(208c) Acid Functionalized Nanoparticles for Hydrolysis of Lignocellulosic Feedstocks

Myles Anton Ikenberry[1], Leidy Pena[2], Donghai Wang[2] and Keith L. Hohn[1], (1)Chemical Engineering, Kansas State University, Manhattan, KS, (2)Biological and Agricultural Engineering, Kansas State University, Manhattan, KS

Cellulosic ethanol may become an important part of a sustainable energy economy. Such ethanol has advantages over other ethanol sources such as corn ethanol, in that cellulose is overly abundant and the production of cellulosic ethanol does not require specific crops or influence global food prices. However, widespread production of cellulosic ethanol is inhibited by the high cost of converting the cellulose into glucose.

Currently, acids and enzymes are the preferred methods for converting cellulose to glucose. Enzymes cannot be re-used and are somewhat expensive to produce. Acids are cheaper, but also cannot be re-used, and have the additional unwanted effect of degrading some of the glucose produced into waste by-products, and require costly separation or neutralization before disposal into the environment. For cellulosic ethanol to compete in the marketplace with fossil fuels, the cost of converting cellulose to glucose must be minimized. An ideal catalyst would provide fast conversion of cellulose to glucose, minimal degradation of glucose into waste products, and energy efficient, inexpensive recycling and re-use capabilities.

Nanosized particles with superparamagnetic cores and acidic functional ligands possess the desired catalyst traits. Unlike enzymes, the acidic ligands are chemically stable, and resist oxidation. Superparamagnetic nanoparticle cores have diameters that are smaller than the magnetic domains of the material, allowing them to be pulled from solution by the presence of a moderate external magnetic field, and return to a non-magnetic state when the external magnetic field is removed.

Preliminary work has been conducted on the ability of some acid functionalized magnetic nanoparticles to catalyze cellulose hydrolysis.

See more of this Session: Catalytic Processing of Fossil and Biorenewable Feedstocks: Fuels I
See more of this Group/Topical: Catalysis and Reaction Engineering Division

(208d) Heterogeneous Catalysts for the Conversion of Glucose to 5-(hydroxymethyl) Furfural (HMF)

Eranda Nikolla, Manuel Moliner, Yuriy Roman and Mark E. Davis, Chemical Engineering, California Institute of Technology, Pasadena, CA

Growing environmental concerns and dwindling fossil fuel resources have increased the need for a shift toward renewable, carbon neutral resources. Biomass has been identified as a promising renewable resource. A possible avenue toward establishing generalized, biomass-based platforms for the production of fuels and chemicals is based on the catalytic conversion of biomass-based carbohydrates (i. e. fructose and glucose) to compounds that can be used directly as fuels or as gateway chemicals for the production of high value products. One such compound is 5-hydroxymethylfurfural (HMF), which effectively links the chemistry of carbohydrates with that of transportation fuels and chemicals. Our objective is to design environmentally friendly and robust heterogeneous catalysts that enable the conversion of carbohydrates to HMF. We recently reported that a Sn containing silica molecular sieve with the zeolite beta topology (Sn-Beta) can efficiently catalyze the isomerization of glucose to fructose in aqueous media even at low pH (Proc. Natl. Acad. Sci., 107 (2010) 6164). In this molecular sieve, the Sn atoms act as isolated Lewis acid centers. We will discuss our efforts to design a catalytic system, where in addition to the isomerization of glucose to fructose, the catalyst(s) also activate the dehydration of fructose to form HMF. For example, Sn-Beta (1:50 Sn:glucose) can convert a 10 wt% glucose feed to a mixture of glucose (28 wt/wt%), fructose (18) and HMF (11) in 120 minutes at 413K when HCl is added to the reaction mixture to give a pH=1 (without the Sn-Beta there is essentially no reaction). Further catalyst combinations and reaction conditions have been explored and will be reported in this presentation.

See more of this Session: Catalytic Processing of Fossil and Biorenewable Feedstocks: Fuels I
See more of this Group/Topical: Catalysis and Reaction Engineering Division

(208e) Theoretical Study of the Mechanism of Fructose Dehydration to Hydroxymethylfurfural in the Aqueous Phase

Stavros Caratzoulas, Chemical Engineering, Dpt of Chemical Engineering and Center for Catalytic Science and Technology (CCST), University of Delaware, Newark, DE and Dionisios G. Vlachos, Dpt of Chemical Engineering and Center for Catalytic Science and Technology (CCST), University of Delaware, Newark, DE

Hydroxymethylfurfural (HMF) and its furan derivatives can replace key, petroleum-based, building blocks. However, HMF is not yet a high-volume chemical, due to difficulties in cost-effective production. HMF synthesis by dehydration of D-fructose is a problem that illustrates the selectivity challenges involved in the processing of highly functionalized carbohydrates, because the reaction intermediates and HMF itself degrade through various processes. Effective process design is currently impeded by our limited knowledge of the reaction pathways. In this talk, we focus on fructose dehydration and employ theoretical methods in order to elucidate the multi-step mechanism and its energetics in aqueous solutions with acid (proton) catalyst. To that end, we propose a mechanism for the dehydration reaction of fructose to HMF and study the energetics of each elementary step by performing free energy calculations in the aqueous phase; we calculate reaction and activation free energies and thus obtain much needed kinetic constants. Our methodology involves hybrid QM/MM (Quantum Mechanics/ Molecular Mechanics) Molecular Dynamics simulations in the canonical ensemble at 363 K, with explicit solvent molecules. To account for reactive events, namely, bond cleaving and forming, the fructose molecule and all reaction intermediates are treated quantum mechanically at the PM3 theory level. Explicit water molecules are treated at the Molecular Mechanics level using the SPC/E parameterization. Furthermore, to account for rare events in reaction steps involving high energy barriers, we employ umbrella sampling and the weighted-histogram analysis method. Entropic contributions to the calculated free energies are also obtained, revealing the importance of the solvent in the stabilization of the reaction intermediates. For example, we find that upon protonation of the anomeric carbon (C2) hydroxyl group, water removal leads to the formation of a cyclic species whose quantum mechanical energy is ~23 kcal/mol higher than that of the protonated fructose, whereas its free energy is of ~2.3 kcal/mol lower, revealing the importance of solvation in the stabilization of the reaction intermediates. We uncover similar phenomena in the subsequent elementary steps of the dehydration reaction.

See more of this Session: Catalytic Processing of Fossil and Biorenewable Feedstocks: Fuels I
See more of this Group/Topical: Catalysis and Reaction Engineering Division

(208f) Kinetics and Mechanism for Acetone Hydrogenation by Ru/Carbon

Mohit Bhatia, Adriaan van Heiningen, G. Peter van Walsum and M. Clayton Wheeler, Chemical and Biological Engineering, University of Maine, Orono, ME

Biomass as a source of energy and chemicals has gained importance due to the decreasing oil reserves, fluctuating oil prices and environmental concerns. Ethanol is now used as a blend in conventional transportation fuels, however Ethanol energy density (29.84 MJ/Kg) is quite low compared to conventional gasoline (46.53 MJ/Kg). This reduces overall energy density of ethanol blend gasoline. Alcohols such as Propanol (33.6 MJ/Kg) and Butanol (37.3 MJ/Kg) have higher energy densities and can be used as bends in conventional transportation fuels. Production of these higher alcohol liquid fuels via acidogenic digestion and chemical upgrading of industrial biomass streams offers many advantages over sterile fermentation processes such as lower capital cost, no need for sterility or genetically modified organisms, and ability to produce a variety of chemicals such as carboxylic acids, ketones, esters, and alcohols. Carboxylate salts formed during acidogenic digestion of biomass can be thermally-converted to ketones. The ketones can then be hydrogenated to form higher mixed alcohols. In the current work, the activity of ruthenium supported on activated carbon for hydrogenating ketones is explored using acetone as a model compound. A simple kinetic model which involves the effects of temperature, H_2 pressure, and acetone concentration will be presented. Also, results will be discussed which show that the reaction pathway involves an enol intermediate and qualitatively explains an enhancement in the reaction rate by water as well as the product isopropanol.

See more of this Session: Catalytic Processing of Fossil and Biorenewable Feedstocks: Fuels I
See more of this Group/Topical: Catalysis and Reaction Engineering Division

(208g) Gas Phase Butanal Self-Condensation Over Supported MgO and CeZrOx: A Searching for Acid-Tolerant Solid Base Catalysts

Wenqin Shen, Goeff A. Tompsett, W. Curtis Conner Jr. and George W. Huber, Chemical Engineering, University of Massachusetts, Amherst, MA

Abstract:

The sustainable biomass economy requires development of different integral processes to lower the cost of producing fuels and chemicals form ligno-cellulosic biomass. Accordingly, the new concept for the catalytic conversion of the biomass to monofunctional hydrocarbons and targeted liquid fuels have been provided, where the sugars and polyols were first converted into small oxygenates over a Pt-Re/C catalyst. The development of the limited steps for the upgrading those mono-functional oxygenates are necessary in order to achieve value added chemicals or fuels. We conducted the gas phase butanal self-condensation reactions as a model study for the carbon-carbon forming reactions. The unsupported MgO and $CeZrO_x$ solid solution showed less active for the reaction. The silica supported MgO, SrO and a mixture of MgO/SrO catalyst showed improved activity. However, the catalysts deactivated very quickly due to the self-poisoning by butanoic acid, a compound generated through side reactions. $CeZrO_x/SiO_2$ showed a promising performance on butanal self-condensation with about 40% conversion of butanal at 300 ºC and good stability. The effects of the reaction temperatures, addition of butanoic acid and supports were studied and compared over the MgO/SiO_2 and $CeZrO_x/SiO_2$ catalysts. The $CeZrO_x/SiO_2$ catalyst is much more acid-tolerant compared with MgO/SiO_2 and lowered the reaction temperature approximate 100 ºC with the same butanal conversion. The reaction chemistry was studied. The predominate reaction pathway is the aldol condensation of butanal with about 85% selectivity to first addition product and 6 % selectivity to second addition products. The side reactions include Tishchenko esterification reactions, hydrolysis, ketonizations, isomerizations and hydrogenation/dehydration reactions. The nature of the active sizes and the reaction mechanism were also studied by a combination of TPD and in-situ FTIR studies.

Keywords: Butanal self-condensation; $CeZrO_x$ solid solution; Silica supported $CeZrO_x$, supported and unsupported alkaline metal oxide; CO_2-TPD; NH_3-TPD; in-situ FTIR

See more of this Session: Catalytic Processing of Fossil and Biorenewable Feedstocks: Fuels I
See more of this Group/Topical: Catalysis and Reaction Engineering Division

(244a) Latest Advances in Drilling CBM(Coal Bed Methane) Wells

Govindaswamy Balachandran, Natural Resources engineering, Al-jabal-Al-Gharbi-university,Gharyan, Libya, Gharyan, Libya

ABSTRACT

The aim of this paper is to study the challenges and technologies in Drilling the CBM (Coal Bed Methane) wells.Conventional gas will be found after drilling depths upto 7000 m But CBM wells gas will be found at depths of 400m-1000m.therefore these shallower depths make it possible to use smaller more mobile trucks,mounted Rigs in comparison with those used for conventional gas wells.

The majority of the wells project use vertical drilling to develop fields.On the otherhand key advantage of horizontal well is that fewer wells are required relative to vertical wells.Horizontal wells are able to track Coal seams and can access more coal and produce more gas.the main drawbacks with horizontal wells are:technically more difficult to drill and are more expensive than vertical wells.

The other processes which will be discussed on this paper are:hydraulic fracturing of CBM wells,Well completion and gas gathering. As CBM becomes more widespread in USA and Australia,a number of issues are impacting the oil and gas industry.these issues include water disposals,land access,licenses and methane loss to atmosphere.

CBM is an attractive industry for upstream oil companies and a number of opportunities exists for other countries wishing to invest in this sector.

Key words: Coal Bed Methane(CBM), VD-Vertical drilling HW-Hoizontal wells

1 –Asst.Professor ,Department of chemical Engineering, Sohar university, Oman. * - Professor (Dr.) Department of Natural Resources Engineering, Gharyan, Libya E-mail:; govindaswamy. balachandran@gmail.com thangam_32@yahoo.co.in

See more of this Session: Recent Advances: Drilling Under Extreme Conditions and Methane Hydrate Modeling
See more of this Group/Topical: Fuels and Petrochemicals Division

(244b) NETL Extreme Drilling Laboratory: Perspectives

Wu Zhang[1], Jamie L. Brown[1] and William J. Ayers[2], (1)U.S Department of Energy, National Energy Technology Laboratory, Morgantown, WV, (2)U.S Department of Energy, National Energy Technology Laboratory, WV, Morgantown, WV

We have integrated experimental and modeling efforts to attempt to address challenges in drilling at ultra-deep depth where high pressures and temperatures prevail. This presentation gives an overview of this effort.

See more of this Session: Recent Advances: Drilling Under Extreme Conditions and Methane Hydrate Modeling
See more of this Group/Topical: Fuels and Petrochemicals Division

(244c) Filter Cake Formation in a Horizontal Well: CFD Modeling and Simulation

Mohd A. Kabir[1], Isaac K. Gamwo[2], Jamie L. Brown[3] and Jimmy D. Thornton[3], (1)ORISE, U.S Department of Energy, National Energy Technology Laboratory, Pittsburgh, PA, (2)National Energy Technology Laboratory, U.S. Department of Energy, Pittsburgh, PA, (3)U.S Department of Energy, National Energy Technology Laboratory, Morgantown, WV

Horizontal wells offer greater contact area with the reservoir rocks over vertical wells and generally increase the well productivity. Horizontal drilling often enables successful reservoir development for wells which would have been considered uneconomic. During the drilling process, the multiphase drilling fluid is pumped down into the drilling zone through the drilling pipe. The fluid is expected to perform many functions simultaneously. Some of the most important functions are maintain a favorable pressure difference between the wellbore and the rock formation, transport the drilled particles to the surface, reduce friction and wear on the drilling assembly, and significantly reduce large losses of drilling fluid to the surrounding formation. This is achieved by forming a filter cake at the well wall to seal the wall. The aim of this investigation is to increase the understanding of the deposition process and factors that affect the buildup of the filter cake.

This paper reports computational fluid dynamics (CFD) investigation of filter cake formation in horizontal well at both shallow (30oC; 2000 psi) and ultra deep (170 oC; 25,500 psi) drilling processes. Filter cake formation in both processes are compared. The effect of the drilling fluid particle sizes on the buildup of the filter cake will be presented.

See more of this Session: Recent Advances: Drilling Under Extreme Conditions and Methane Hydrate Modeling
See more of this Group/Topical: Fuels and Petrochemicals Division

(244d) Simulation of Methane Hydrate Dissociation and Multiphase Flow with Non-Darcy's Law in Porous Media

Yong Liu, National Energy Technology Laboratory, U.S. Department of Energy, Pittsburgh, PA and Isaac K. Gamwo, National Energy Technology Laboratory, US Department of Energy, Pittsburgh, PA

Methane hydrate, a potential future energy resource, is known to naturally occur in vast quantities beneath the ocean floor and in permafrost regions. HydrateResSim is a free open-source numerical reservoir simulator software available from the National Energy Technology Laboratory (NETL). HydrateResSim, the NETL methane production simulator for hydrate-containing reservoirs, originally developed for NETL by Lawrence Berkley National Laboratory (LBNL), incorporates the multiphase Darcy's law for gas and liquid into both the mass component balances and the energy conservation equations. However, for the multiphase flow in porous media with fractures or close to the production wellbore, the Darcy's law is not valid. We extended the capability of the HydrateResSim by implementing the non-Darcy's law with two sub-models: a kinetic model and pure thermodynamic model. We present the comparison between results with different formula for the non-Darcy's law and Darcy's law. The importance of the non-Darcy's term is analyzed.

See more of this Session: Recent Advances: Drilling Under Extreme Conditions and Methane Hydrate Modeling
See more of this Group/Topical: Fuels and Petrochemicals Division

(244e) Equilibrium of Gas Hydrates Using Molecular Dynamics Simulations

Shaunak Potdar, Chemical & Natural Gas Eng, Texas A&M University-Kingsville, Kingsville, TX, Krishnadeo Jatkar, Chemical & Natural Gas Eng., Texas A&M University-Kingsville, Kingsville, TX, Jae W. Lee, Department of Chemical Engineering, The City College of New York, New York, NY and **Sangyong Lee**, Texas A&M University-Kingsville, Kingsville, TX

Gas hydrates are non-stoichiometric crystalline compounds consisting of water and gas molecules. In gas hydrates, water molecules are connected with each other by hydrogen bonds to form a specific crystal structure named as 'cavity' and gas molecules are captured in cavities. There is no chemical bonding between gas molecules and gas molecules. Gas hydrates are stabilized at low temperature and high pressure conditions, and cause a plugging problem in gas/oil transportation line while natural gas hydrates has a potential to be used as a future energy resources. Equilibrium conditions of various gas hydrates (methane, ethane, propane, ..., etc) have been calculated using Molecular Dynamic Simulation (MD) for a unit cell structure of gas hydrate. In the simulation, we used 'link cell' to calculate short range forces and Ewald sum technique to calculate long range electrostatic forces. The Gaussian thermostat was used to maintain a temperature. For water molecules Jorgansen's TIP4P model was used. With the MD simulation, we calculated the chemical potential for theoretical empty cavities of various gas hydrates based on the distortion theory that assumes the distortion of the hydrate structure due to size of gas molecules captured in water cavity structure. 0.01 to 0.1 ps time step is used during simulation. As an attempt to apply MD simulation to calculate the reference state of gas hydrate, we demonstrate lattice distortion of structure I and II gas hydrates. The chemical potentials for the theoretical empty cavity of various gas hydrates have been calculated and the capacity expansion due to temperature has been also calculated.

See more of this Session: Recent Advances: Drilling Under Extreme Conditions and Methane Hydrate Modeling
See more of this Group/Topical: Fuels and Petrochemicals Division

(248a) Region Specific Economic Modeling of Integrated Biorefining Using Process Simulation and Supply Chain Optimization

Sumesh Sukumara, Chemical and Materials Engineering, University of Kentucky, Lexington, KY, Jeffrey Seay, Chemical and Materials Engineering, University of Kentucky, Paducah, KY and Fazleena Badurdeen, Mechanical Engineering, University of Kentucky, Lexington, KY

Energy dependence and pollution due to the increasing use of fossil fuels for the production of energy has become an area of concern in recent years. Most of the present research is focused on developing processes that rely on renewable resources and using existing processes in an effective and environmentally friendly manner. Feed used for these processes should be available in ample amount and should be economic compared to other fossil fuels used for energy production. Integrated biorefining is one such development which combines all the possible processes for the production of electricity, fuels and other useful products. Biomass, poultry litter and municipal waste are some of the potential feeds for biorefining. However, the density and quality of these feedstocks vary from one region to another. As such, it is necessary to develop a region specific economic model for integrated biorefining. Gasification in an integrated biorefinery is a proven method for the production of synthesis gas from coal, biomass and a combination of both. Gasification of agricultural and municipal waste is not only a good means of disposing unwanted products but can also be a source of additional revenue. Once synthesis gas is produced by gasification, selection of next process or combination of different processes depends on the local demand for a specific product. One of the concerns with gasification is that the ash content coming out of the gasifier will change with any change in the feed used. Hence, the analysis of ash content and proper disposal of it is important to make this process a clean technology. In this presentation we will show a case study on the Jackson Purchase region in Western Kentucky. The Jackson Purchase region not only has good reserves of coal but is also rich in woody biomass, crop residue and has numerous poultry farms. Process simulation models are developed for gasification of coal, biomass and chicken litter in this region. Due to local demand for fuel, gasification is followed by Fischer-Tropsch process for the production of methane, ethane and other alkanes. Ash coming out of the gasifier for different combinations of feed is analyzed and corresponding disposal techniques are suggested. The results obtained show that the operating cost of the process varies significantly with varying combinations of feedstock chosen. Geographic information system modeling is used to locate the actual distance of feedstock from the proposed plant location. Transportation of raw materials to the plant is an important criterion to propose the location of an integrated biorefinery. Most of the poultry farms present in the Jackson Purchase region were located using this technique and a site for biorefining was proposed. An economic framework is developed taking into consideration some key factors such as transportation cost, installation cost and sales revenue along with operating cost. Based on this economic framework, process models and supply chains were optimized which led to the development of a model library for conversion of biomass and other agricultural and poultry waste into fuels in the Jackson

Purchase region. The library includes all possible processes and pathways corresponding to any potential feed available in that region. In a broader perspective, this approach can be applied to connect all the processes for energy and fuel production utilizing local feedstock available in a specific region.

See more of this Session: Sustainable Fuel From Renewable Resources
See more of this Group/Topical: Sustainable Engineering Forum

(248b) The Scope of Using Concentrated Solar Energy to Produce Liquid Fuels From Biomass

Dharik Mallapragada, W. Nicholas Delgass, Fabio H. Ribeiro and Rakesh Agrawal, Chemical engineering, Purdue University, West Lafayette, IN

In a fossil-fuel deprived world, liquid fuels from biomass, or biofuels, can play a major role in satisfying the energy needs of the transportation sector in a sustainable manner. The specific extent of biofuels usage would depend on, among other things, the associated life-cycle CO_2 balance and the land area requirements for the feedstock. Both of these issues can be addressed by maximizing the quantity of biofuel produced from the available biomass. This presentation discusses the scope of using concentrated solar energy as the heat source, in combination with biomass for efficient biofuel production. Details of a synergistic thermochemical process envisioned, involving high temperature biomass steam gasification followed by Fischer-Tropsch (FT) processing to biofuels, are presented. In addition to an increase in the estimated biofuel yield compared to conventional methods, the process offers a means of storing additional solar energy in an easy-to-use form, by upgrading the calorific value of the biomass feedstock. Enhancements to the above process presented, include several scenarios of supplemental use of hydrogen produced from solar energy, in order to convert the residual CO_2 to additional biofuel in the least solar energy intensive manner possible. Overall, the integrated use of hydrogen and concentrated solar energy (heat), results in a maximum threefold increase in the biofuel produced compared to conventional gasification processes.

See more of this Session: Sustainable Fuel From Renewable Resources
See more of this Group/Topical: Sustainable Engineering Forum

(248c) Comparison of Gasification/Synthesis Processes for Converting Wood Into Alcohols or Gasoline

Steven D. Phillips, Joan Tarud, Abhijit Dutta and Mary Biddy, National Bioenergy Center, National Renewable Energy Laboratory, Golden, CO

Process simulations for converting wood to liquid fuels have been developed at the National Renewable Energy Laboratory over the past four years based on a core process model for indirectly-heated gasification, catalytic tar and hydrocarbon reforming,and gas conditioning. The resulting synthesis gas can be catalytically converted into several possible liquid fuels including methanol, ethanol, mixed alcohols, or gasoline. Different product pathways can have significantly different requirements for the syngas conditions that give optimum yields. The degree of thermal integration for these processes can result in large differences in plant gate prices and product yields. A consistent set of technical and economic assumptions between the different processes permits meaningful comparisons to be made. An overview of the wood to alcohols and gasoline processes will be presented including estimated plant gate prices, product yields, and potential external factors that impact each product's fungibility with the existing fuel infrastructure and end users.

See more of this Session: Sustainable Fuel From Renewable Resources
See more of this Group/Topical: Sustainable Engineering Forum

(248d) Alternative Energy through Optimization of Methane Formation by Anaerobic Digesters in Municipal Wastewater Treatment Plants

Stephen M. Ritchie and Ryan Beam, University of Alabama, Tuscaloosa, AL

The operating parameters of anaerobic digesters were optimized to maximize biogas production at the Hilliard Fletcher Wastewater Treatment Plant in Tuscaloosa, Alabama. This process makes use of a renewable resource, municipal waste and wastewater, to produce relatively clean biogas. Currently, most of the produced biogas is burned to heat the anaerobic digesters. However, optimization of this process could result in additional production of biogas which can be converted to electricity and increasing the overall energy efficiency of this and similar water treatment facilities.

Current research focuses on the four distinct steps of the batch anaerobic digestion process as well as determining the affect of operating parameters on gas production. Parameters of interest include pH, temperature, alkalinity, and volatile fatty acid concentration. Our observations show that pH plays a very important role in determining the quality of biogas produced. Contrary to the literature values of 6.5 – 7.5, the optimum pH range during batch trials has been 5.6 – 6.4. During this range, biogas composed of 64 mol % methane was formed, exceeding the highest methane content seen in plant gas samples. Additional results will be given for batch sensitivity trials and implementation of recommended operating parameters in a continuous laboratory scale system. Comparisons will be made with operation at the full scale facility.

See more of this Session: Sustainable Fuel From Renewable Resources
See more of this Group/Topical: Sustainable Engineering Forum

(248e) Optimization of Medium by Response Surface Methodology for Reducing Ethanol Production Cost

Aderonke Adewusi, **Ramalingam Subramaniam**, Cunwen Wang, Mark Zappi, Stephen Dufreche and Rakesh K. Bajpai, Chemical Engineering, University of Louisiana at Lafayette, Lafayette, LA

In this study, Z. mobilis had been used in the fermentation of sweet potato starch to produce fuel ethanol. The yeast extract had significant effect on the production cost of ethanol. Response surface methodology central composite design was applied for medium optimization. Both experimental and model values confirm the influence of yeast extract on lipid synthesis.

See more of this Session: Sustainable Fuel From Renewable Resources
See more of this Group/Topical: Sustainable Engineering Forum

Economic analysis of an annexed ethanol distillery in Africa

B. Amigun[1,*], H. von Blottnitz[2]

[1] *Sustainable Energy Futures, Council for Scientific and Industrial research (CSIR), Stellenbosch, South Africa, *bamigun@csir.co.za*

[2] *Process Engineering Department, Stellenbosch University, Private Bag X1 7602, Matieland, South Africa*

Abstract

Research on improving ethanol production as an alternative to petroleum based fuel has been accelerating for both ecological and economical reasons. Specifically bioethanol has been defined as a gasoline enhancer to reduce greenhouse gases, gasoline imports, and to boost the rural economy. This paper presents the operating cost analysis of an annexed distillery operating in a landlocked and rural African location. The study specifically attempts to analyse the impacts that some of the critical factors influencing bioethanol production, will have on its viability and also present an analysis of the breakdown of the operating costs of the distillery, in an equation format to enables easier and more rapid use of the data in numerical and economic models, and in the preliminary design and optimisation of bioethanol plants. The cost of ethanol production was estimated at $19.57/hl with cost of feedstock (molasses "type C") constituting about one third of the production cost. The study also identified other major, intermediate and minor input factors which can provide insights to both the possible barriers to implementation that should be overcome, and on the technological improvement options that should be stimulated by research and development in ethanol industry.

Keywords: annexed distillery; operating cost; landlocked

1.0 Introduction

In addition to the progressive depletion of energy resources, mostly based on non-renewable fuels, the world faces climate changes and atmospheric pollution as consequences of the combustion of oil-derived fuels. These problems are more serious in large cities where vehicular exhaust emissions rank among the main sources of polluting gases (Quintero et al., 2008). Large increases in sustainable energy demand, reflected in governmental strategy and regulation aimed at harnessing the potential of modern, clean biomass-derived fuels in African countries, is leading to the development of large-scale renewable energy supply which is strategically important for local, regional and global environmental sustainability. Ethanol produced from bioenergy crops and wastes has the potential to become a renewable transportation fuel in place of gasoline. Its use as a gasoline oxygenate increases the oxygen content, allowing better oxidation of hydrocarbons and reducing the amounts of aromatic compounds and carbon monoxide released to the atmosphere. Moreover, CO_2 emissions generated by the combustion of this biofuel are compensated by the CO_2 absorption during the growth of the crops from which ethanol is produced, avoiding a net emission of this gas. Specifically, the expanded use of bioethanol would have significant health benefits in replacing lead as an octane enhancer in most African countries where leaded fuel is still widely used [1]. Africa represents the largest leaded fuel user in the world. Of a total of 49 countries in Sub-Saharan Africa, 22 countries use leaded fuel only, 14 dual system and 13 unleaded.

The study of economic parameters involved in the functioning of an ethanol plant has rarely been carried out from an engineering point of view, and there are no publications in this regard for ethanol plants in Africa. A recent study with the focus on fixed capital investment (FCI) of annexed ethanol distillery in rural African countries was carried out by Amigun and von Blottnitz, (2009). They developed a simplified procedure for rapid ball-park cost estimate that can be used as a research tool by energy policy makers for targeting area of cost reduction in a project, for comparing alternative design and for reviewing achieved costs on completed projects. They further presented analysis of the breakdown of the fixed capital investment (FCI) cost, based on the principle of factorial method of capital cost estimation using Lang factor (f_L) analysis [6]. Although this is not directly relevant to the present study, the paper provide an insight into the economics of ethanol production in Africa (ethanol production has two cost components, capital costs and variable costs, the latter including net feedstock costs and cash operating expenses)

A wide variety of factors affect the economics of ethanol production. These factors include feedstock and energy costs, capital and debt financing costs, the value of products produced and plant design and efficiency. These factors will vary, depending on plant size and location of the facility, technology base (conversion pathway), feedstock and co-product [4]. Most of the available literatures on operating cost analysis are also less transparent due to assumptions that are embedded in the estimate such as, feedstock cost assumptions, process yields, co-product values, etc. This makes cross comparison between studies problematic. The effects of these several factors on variable cost will be very different in a developed country where the existing cost estimation models are concentrated, as compared with a developing African country. It is therefore important that an indigenous theory of cost prediction, central to economic feasibility studies, be developed. It appears that there is no such theory, and not even a good collection of relevant data in most African countries. The existing methods have limited applicability and on many occasions disappointing accuracy because of the inappropriate nature of the estimating techniques

In this study, operating cost analysis of an annexed distillery operating in a landlocked African country is carried out. Specifically, the study attempts to analyse the impacts some of the critical factors influencing ethanol production, will have on its viability. A unique attempt is presented aimed at developing robust and indigenous operating cost model for bioethanol process industry in Africa. This is critical to increase bioethanol development and deployment in Africa.

2. Process model description

A simplified flow diagram of the process is illustrated in Figure 1. The actual process contains more than 200 pieces of equipment and unit operations. The plant is rurally located, with no access to electricity or to landline telephone.

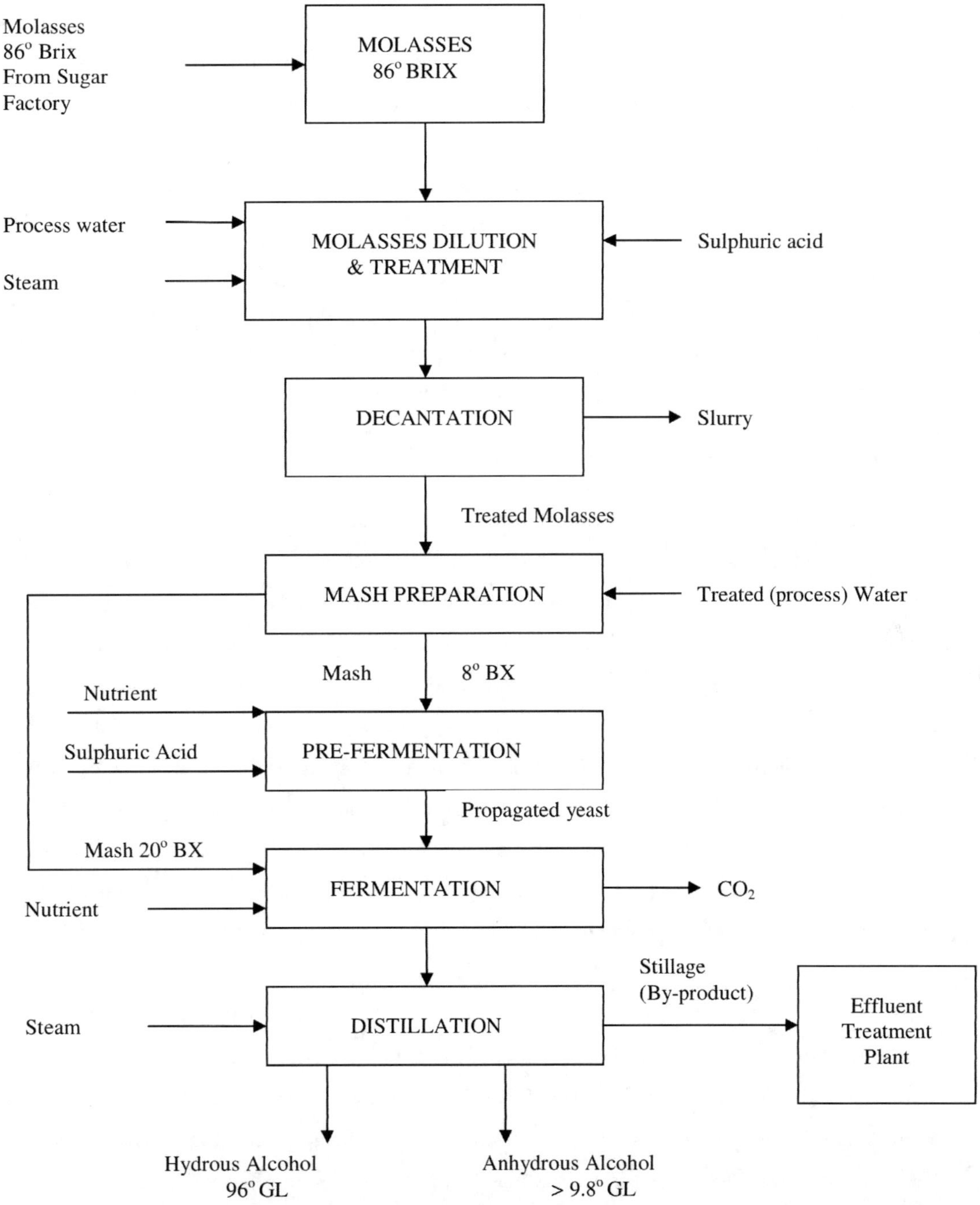

Figure 1: Overview of molasses to ethanol process

The ethanol plant which is annexed to a sugar factory produces a power alcohol (99.5%) and has a design capacity of 45,000 Liters/day with 90% capacity utilization. The ethanol plant makes use of the by-product molasses from the adjoining sugar factory which has a daily crushing capacity of 40,000 quintals (40 million kg) of cane to produce 850,000 quintals (850 million kg) of high quality commercial sugar per annum. Therefore there is no cost incurred on the raw material and its transportation. The ethanol produced can be blended with gasoline (90%) to be used as a motor fuel (power ethanol) without any engine modification and also blended with kerosene (50%) to be used as household cooking fuel. The distillery is in good operating condition, as would be expected from the limited amount of time it has operated. The plant operates 24h/day, 7 months in a year, with the remaining 5 months (down time allowance) due to seasonality of the harvest, equipment repair and maintenance. Therefore, a basis of 212 days per year (4464 hours) operating time was used in all cost analyses and the nominal capacity of the plant is approximately 18.75 Hl/hr (450 Hl/day). On the downside, the plant is difficult to manage because of the limited number of control equipment and more particularly of flow meters (the plant is semi-automated). Energy consumption is fairly high. The distillery was not designed for energy efficiency, since excess bagasse and excess steam are readily available from the sugar factory.

The molasses (86° Brix*) is received from the factory molasses storage tank and pumped by means of molasses pumps in to a scale receiving tank with an overflow facility to the molasses tank. The molasses flows by gravity from the molasses scale in to a weight molasses tank. The next stage is the heating of the molasses. However, for this to be more efficient, it is necessary to reduce the molasses viscosity. For this purpose, pre-dilution should be done to a concentration close to 50° Brix. Pre-dilution is done with process water and condensate water, which under normal working conditions, comes from vaporizer (L) of the Dehydrating column, the molasses heater themselves as well as the molasses pre-dilutors (Figure 2). The advantage of using condensate water in pre-dilution is that there will be a pre-heating of the diluted molasses to about 50°C, therefore diminishing its viscosity and reducing steam consumption for final heating. Before entering the pre-dilutors, the molasses (86° Brix) goes through an instant measuring and totalizing instrument for process and yield control purposes.

After measuring, the concentrated juice goes to a pre-dilutor bottle which has an internal centrifugal type to facilitate dilution. The pH of molasses is adjusted to ensure precipitation by the dilution of the sulphuric acid. The diluted, acidified molasses is then heated to 95°C and held at this temperature for approximately twenty minutes in order to sterilize it before being pumped into molasses settling tanks, where it is allowed to settle for approximately 8 hours. The clarified molasses is decanted into a

* Brix is a unit of measure which is related to the sugar (sucrose) content of a sugar solution.

clarified molasses tank and the sludge from settling tank flows into sludge mixing and dilution tank, where it is diluted using cooling water.

From the mixing and dilution tank, the diluted sludge is transmitted to sludge settling tank where after a settling period, the sweet water is decanted in to sweet water tank while the sludge flows to drain. The decanted diluted juice (50-52° Brix) at a temperature of between 90 and 95°C is diluted and cooled to about 18 - 20 m^3/hr of cold water is further consumed for a final concentration to 20-22° Brix, this operation is done in the pre-fermentation section. From this moment onward, this raw material is called Mash, the temperature of which is from 55 to 60°C. Mash is cooled to 30-32°C in a plate heat exchanger, and its water consumption is around 60-65m^3/hr. The final volume of mash (20-22° Brix) is 31 - 33 m^3/hr adequate for obtaining a good fermentation. Thus it is free of a large part of the impurities, blockage of the centrifugal nozzles and eventual scaling at the distillation stage is avoided. It is not possible to obtain absolute alcohol[†] by simple fractional distillation, because a mixture containing around 95.6% alcohol and 4.4% water becomes a constant boiling mixture (an azeotropic mixture). To obtain absolute alcohol, a small quantity of benzene is added to rectified spirit and the mixture is then distilled. Absolute alcohol is obtained in the third fraction that distils over at 78.2 °C (351.3 K). Because small quantity of the benzene used still remains in the solution, absolute alcohol produced by this method is not suitable for consumption as benzene is carcinogenic.

3. Operating cost data

Operating costs are costs associated with the day– to–day operation of an ethanol distillery. This cost must be estimated before the economic feasibility of a process can be assessed. Important cost factors affecting the manufacturing cost of the distillery are therefore introduced. The data were obtained from process information provided on the process flow diagram (PFD), and the information on the number of operators required to run the distillery.

- **Raw material**

The primary raw material used in the distillery is molasses. The molasses is being supplied by the adjoining sugar plant with a capacity average of 4000 tonnes of sugarcane per day; hence the transportation cost (which is US$ 0.053/km) is not included. Generally, the composition of molasses varies depending on the variety of cane, composition of soil, climatic conditions, harvesting practices, sugar manufacturing process and method of handling and storage. Tropical climatic conditions may also

[†] Absolute or anhydrous alcohol generally refers to purified ethanol, containing no more than one percent water.

influence the technical aspects of molasses to ethanol fermentation. The nominal composition of molasses used for the ethanol production is illustrated in Table 2.

Table 2: Characteristics of Distillery Molasses type "C"

Item	Composition of molasses (%)	
	Range	Mean
Sucrose	30 – 35	33
Reducing Sugar	15 – 20	18
Unfermentable sugar	3.5 – 4.5	4
Sulphited Ash	12 – 13	13
CaO	1 – 2	2
$CaSO_4$	2.9 – 3	3
Potassium Ash (K_2O)	3.5 – 5	4
Nitrogenous compounds	1.2 – 1.5	1.4
pH	5.2 – 6.0	5.6

Other inputs are yeast (*Saccharomyces cerevisiae*[‡]) to convert the sucrose in the molasses to ethanol and carbon dioxide, and a small amount of Nutrient (DAP) for yeast propagation, sulphuric acid (H_2SO_4) for sterilization of molasses to avoid contamination. The volumes of materials utilized and their costs have been incorporated into the model.

- **Product value**

Three products are produced in the conversion of "C "molasses (or "final" molasses) for ethanol production. They are ethanol, carbon dioxide (CO_2 gas) and vinasses (spent washes). There were no local markets for the blending of fuel ethanol produced with gasoline. The denatured ethanol is exported to Europe (Italy and Germany). This is despite the fact that the government imports fuel by truck from neighboring countries. This constitutes a huge drain on foreign exchange of the country which is rated as one of the poorest in the world. The major local use for the ethanol produced is for household cooking purpose by mixing it with kerosene (50-50 ratio). There was no end use for the vinasses produced, but efforts are in place to utilise it as a source of fertilizer by spreading it in the cane plantation. The carbon dioxide is often vented to the atmosphere since the cost of purifying and transportation to the end user, often out-weighs any economic gain from selling it. Nevertheless, if the distillery was planning to install a calco-carbonic epuration (installation of a back end refinery), the

[‡] a specie of budding yeast

pure CO_2 can be used for carbonation (in usual installations, CO_2 gas comes from the boiler exhaust). The mass of CO_2 produced is close to the mass of the alcohol produced simultaneously.

- **Utility costs**

Costs for utilities are based on the utilities required in the production process which includes electricity, steam, process water and cooling water. The distillery was not designed according to energy saving principle, since excess bagasse and excess steam were available from the adjoining sugar factory. This is the main drawback of the ethanol distillery. The energy consumption is about 600 kg of steam per hectoliter of pure alcohol (PA). These utilities are treated as purchased utilities in this study and as such, the capital costs associated with their generation are not included. The distillery is heated by direct injection of steam in the distillation column A, rectification B, dehydrating column C and by heating surface on regeneration column P (Figure 2).

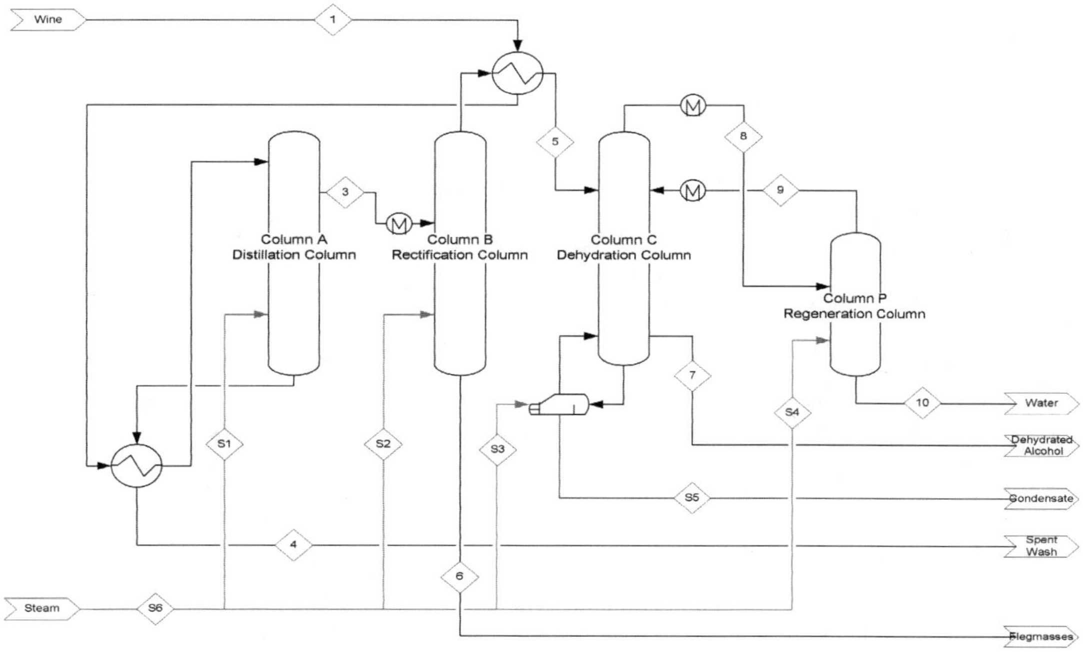

Figure 2: Distillation plant layout of the distillery

1: wine feed; 3: Distilled alcohol; 4: Spent Wash; 5: Rectified Alcohol; 6 Fleg-masses; 7: Pure alcohol; .8: Wet Benzene; 9: Dry Benzene; 10: Removed Water; S1: Column A Steam; S2: Column B Steam, S3: Column C Steam; S4: Column P Steam; S5: Column C Condensate; S6: Steam Header.

The consumption of the steam is as follows:

- 175kg of steam per hectoliter of pure alcohol for Distillation column A
- 85kg of steam per hectoliter of pure alcohol for Lutter column (rectification) B ;

- 150 kg of steam per hectoliter of pure alcohol for the Dehydrating column C;
- 30kg of steam per hectoliter of pure alcohol for benzene Recovery column P, regeneration of the heavy phase.
- 160 kg of steam for the must preparation.

The distillery water utilisation is illustrated in below

- Process water 193,236 m^3/annum
- Cooling water (fermentation) 34,119 m^3/annum
- Cooling water (distillation) 46,526 m^3/annum

The electric consumption of the distillery is reported to be 246 kWh; the motor list received from the factory is illustrated in Table 3.

Table 3: Electricity consumption at the distillery

Item No.	Description	Installed power		Consumed power
	Distillery	Working **HP**	Standby **HP**	Power **KW**
1.	Diluter for nutrient	0.25		0.2
2	Sulphuric acid pump	2	2	1.3
3	Benzene motor pump	2	2	1.3
4	Alcohol loading pump	1	1	0.6
5	Motor pump for slurry	3	3	1.9
6	Diluter continuous	3		1.9
7	Batch slurry pump	3		1.9
8	Motor pump for 1st grade alcohol	3	3	1.9
9	Motor pump for 2nd grade alcohol	3	3	1.9
10	Condensate water pump	3	3	1.9
11	Prefermenter	5.5		3.4
12	Prefermenter	5.5		3.4
13	Prefermenter	5.5		3.4
14	Continuous diluter for molasses	7.5		4.7
15	Continuous diluter for molasses	7.5		4.7
16	Motor pump for treated molasses	7.5		4.7
17	Motor pump for diluted molasses	7.5	7.5	4.7
18	Motor pump for treated molasses	7.5	7.5	4.7
19	Mash/beer pump	7.5		4.7

20	Mash/beer pump	7.5		4.7
21	Mash/beer pump	7.5		4.7
22	Mash/beer pump	7.5		4.7
23	Mash/beer pump	7.5		4.7
24	Mash/beer pump	7.5		4.7
25	Mash/beer pump	7.5		4.7
26	Stillage/exhaust flegma motor pump	7.5	7.5	4.7
27	C/P solution motor pump	7.5		4.7
28	Motor pump for molasses	7.5	7.5	4.7
29	Mash pump	10	10	6.3
30	Beer to distillation pump	12.5	12.5	7.8
31	Beer to distillation pump	15	15	9.4
32	Centrifugal separator	50	50	31.3
	Sub total	239.75	134.5	150.0

Item No.	Description	Installed power		Consumed power
		Working	Standby	Power
	Tank farm area	**HP**	**HP**	**KW**
1	Molasses transfer pump	15	15	9.4
2	Fuel oil transfer pump	15		9.4
3	Technical alcohol delivery pump	3		1.9
4	Gasoline unloading pump	3		1.9
5	Gasoline blend pump	1		0.6
6	Gasoline blend pump	1	1	0.6
7	Denatured ethanol delivery pump	2		1.3
8	Denatured ethanol delivery pump	2	2	1.3
	Sub Total	42	18	26.3

Item No.	Description	Installed power		Consumed power
		Working	Standby	Power
	Effluent treatment	**HP**	**HP**	**KW**
1	Effluent recycle pump	40	40	25
2	Effluent recycle pump	40	40	25
	Sub Total	80	80	50

Item No.	Description	Installed power		Consumed power
		Working	Standby	Power
	Lighting consumption	**HP**	**HP**	**KW**
1	Lighting fixture	24.6		12.9
2	Laboratory electrical equipment	13		6.8

Sub Total	37.6	0	19.7

- **Capital recovery**

The operating cost estimate should contain some mechanism for capital recovery. Depreciation is the method commonly used for conventional construction projects. Exact data, amounting to 4% of the distillery capital investment cost were obtained from the plant for the capital recovery which is a function of the interest rate and the distillery's expected life.

- **Maintenance**

Maintenance and upkeep of the facility components is also considered in the operating cost estimate. This cost includes costs for labour, materials, and supervision for facility equipment and vehicles used at the facility, buildings and their infrastructures at the facility, including phone lines and power generators and regular housekeeping service. A value corresponding to 4% of total per year was obtained. The cost of maintenance is very low as the distillery is in good condition, as would be expected from the limited amount of time it has operated.

- **Downtime allowance**

Facilities will not operate 100 percent of the time. Weather-related shutdowns, equipment repair and maintenance, emergencies, and employee work schedules will affect the number of hours operated per year. Some facilities may be prevented from operating pending acceptance of the facility work plan or other permit. Proper allowance for downtime is important if the operating costs were estimated on a percentage of hours operated basis. For conventional construction facilities, this is not a significant factor unless the facility's operating budget is dependent upon the number of hours operated. Maintenance and repair costs may also be a function of the number of hours of facility operation. The total number of downtime is 5 months (between June and October). This is due to heavy rainfall experience around time which makes transportation of sugarcane from the field to the adjoined Sugar factor difficult.

- **Labour Costs**

The labour requirements are a function of plant size. The distillery employs various types of labour, including operations, technical, administrative, and clerical labour. This cost includes provisions for salaries and labour burden, including medical benefits, vacation and holidays, and other employee compensation items. Labour overhead will consist of administrative costs for scheduling, payroll, etc., as well as costs for employee workspace maintenance. Labour overhead will be present regardless of the facility operating schedule, but labour costs may be a function of the facility's operating schedule, especially if shift work is involved.

- **Distillery human resources management**

The figures provided by the distillery for the total labour costs for one year are distributed as shown below while the breakdown of other staffs is illustrated in Table 4.

- Management: 25,714 US$
- Other staffs: 42,857 US$
- Total labour cost: 68,571 US$

Table 4: Breakdown of labour cost per month at the distillery

Labour description	Personnel cost (US$)
Ethanol Technologist	427
Fermentation Technologist	365
Shift Chemist	309
Foremen	233
Chief Analyst	187
Laboratory Analyst	147
Fermentation Operators	147
Distillation Operators	128
Fermentation and Distillation Helpers	95
Labourers	28

4. Method of analysis

The method of analysis employed is the factorial method of operating cost estimation (using the multiplication factors proposed by Turton *et al.* (1998). This approach is similar to that presented in other chemical engineering design texts (Peters and Timmerhaus, 1990; Ulrich, 1984; Valle-Riestra, 1983). The process information provided on the PFD, the actual costs for the operating personnel in addition to that of the actual fixed capital investment were appropriately analysed, and correlated using the multiplication or factored approach for operating cost model development.. This approach has in common the factoring of process labour and fixed capital investment costs in addition to the costs of raw materials and utilities.

5. Results and Discussion

Figure 3 represents the breakdown of the operating costs of this distillery. The total ex-distillery cost for the produced fuel ethanol includes a significant cost component for feedstock supply. Utilizing sugar molasses "type C" which is the case in this analysis as feed at $ 0.12/kg, and a 450 hl distillery (45,000 liters), operating 7 months in a year (212 days), the cost of ethanol production has been estimated at $19.57/hl. This production cost is a little higher than the net production cost of $16.37/hl for 550,000hl bioethanol plant in Brazil (Henniges and Zeddies, 2004). However, ethanol market value depends on end use. The market value of ethanol as a replacement fuel would generally be measured relative to gasoline prices using the Basic Fuel Price (BFP) approach. BFP linked or tied the domestic retail price of ethanol to international crude oil prices. Because of the lower energy content of ethanol compared to petroleum, the fuel grade ethanol is valued at 75% of the monetary value of petroleum. The market value of ethanol when blended with gasoline may be higher that gasoline because of the increased octane value of ethanol/gasoline blend.

The major economic factor to consider for input costs of ethanol production are feedstock cost which constitute one third of the cost of production, depreciation of capital cost, which is represented by cost on an annual basis and energy related expenses. The type, availability, and price of molasses all factor into the profitability of producing ethanol. The energy costs include that of steam and electricity, with costs being critical for profitability. Energy expenses are one variable in location selection that can affect the success of the ethanol plant. Hence, ethanol plants located near existing manufacturing plant that produces excess steam or have some type of cogeneration potentials, as is common in the sugar industry will result into a better economics. Intermediate economic factors include the cost of labour, both in operating expenses and administrative costs. These costs are expected to be directly related to economies of size for ethanol plants. Other minor economic input factors also include the cost of enzymes, water and denaturant. These costs have little to do with economies of scale. Minor economic concerns have little impact on either the location of a plant or the economies of size associated with large plants.

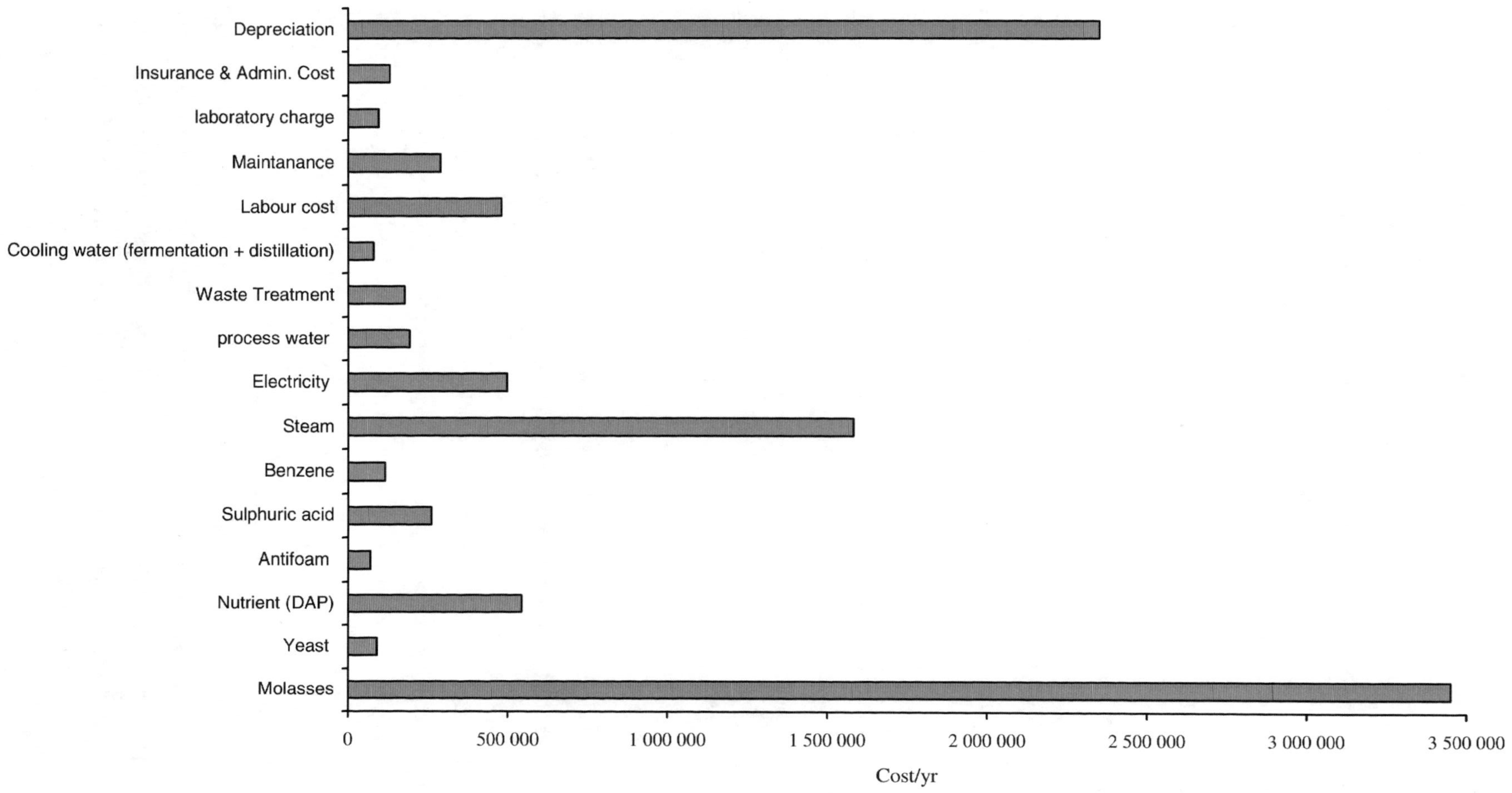

Figure 3: Distillery operating expenses (without by-product value)

Hence, ethanol plants located near an existing manufacturing plant that produces excess steam or that have some type of cogeneration potentials, as is common in the sugar industry will result into a better economics. Intermediate economic factors include the cost of labour, both in operating expenses and administrative costs. These costs are expected to be directly related to economies of size for ethanol plants. Other minor economic input factors include the cost of enzymes, water, denaturant and waste treatment. According to the distillery, about ten volume of effluent will be generated for each volume of fuel ethanol produced. A portion of the effluent is been recycled for use in the distillery. This quantity is however unknown. Factors that influence the volume of wastewater include the quality of the water, the treatment process and the amount of water recycling. The cost of waste treatment depends on the environmental regulation condition in the location. The cost of effluent treatment is shared by the sugar mill and the ethanol distillery. Minor economic input factors have little to do with economies of scale. Minor economic concerns have little impact on either the location of a plant or the economies of size associated with large plants.

Other factors, aside from ethanol production costs and the market value of ethanol, may also be significant to the economic analysis. Displacement of imported petroleum with domestically produced renewable fuel may improve balance-of-payment deficits and may be economically advantageous despite relatively higher ethanol costs. Opportunities for rural employment, alternative markets for agricultural commodities, and energy independence may provide significant economic advantages in addition to a direct accounting of plant profitability.

Important factors although not covered in this dissertation, that need to be considered when determining the profitability of ethanol plant are the competition threat from future technologies such as converting woody products, herbaceous plant like switchgrass, municipal waste materials, and agricultural resources , government subsidies, which is a driving force in ethanol production in some countries, technology of production and the availability of water as production from sugar cane crops uses significant amount of water in the agricultural and industrial processing phases.

A sensitivity analysis was performed by calculating the ethanol price required for molasses cost ranging from 10% increase in the original cost ($0.12/kg) to 200% increase, keeping all other parameters constant. The result of the analysis is illustrated in Figure 4. A 10% raise in the molasses cost increase the cost of ethanol produced from $0.19/liter to $0.20/liter; while a 200% increase in the molasses cost will increase the cost of the ethanol produced by $0.13/liter. It is therefore evident that the price paid for molasses over the life time of a plant can easily "make" or "break" the financial success of an ethanol project.

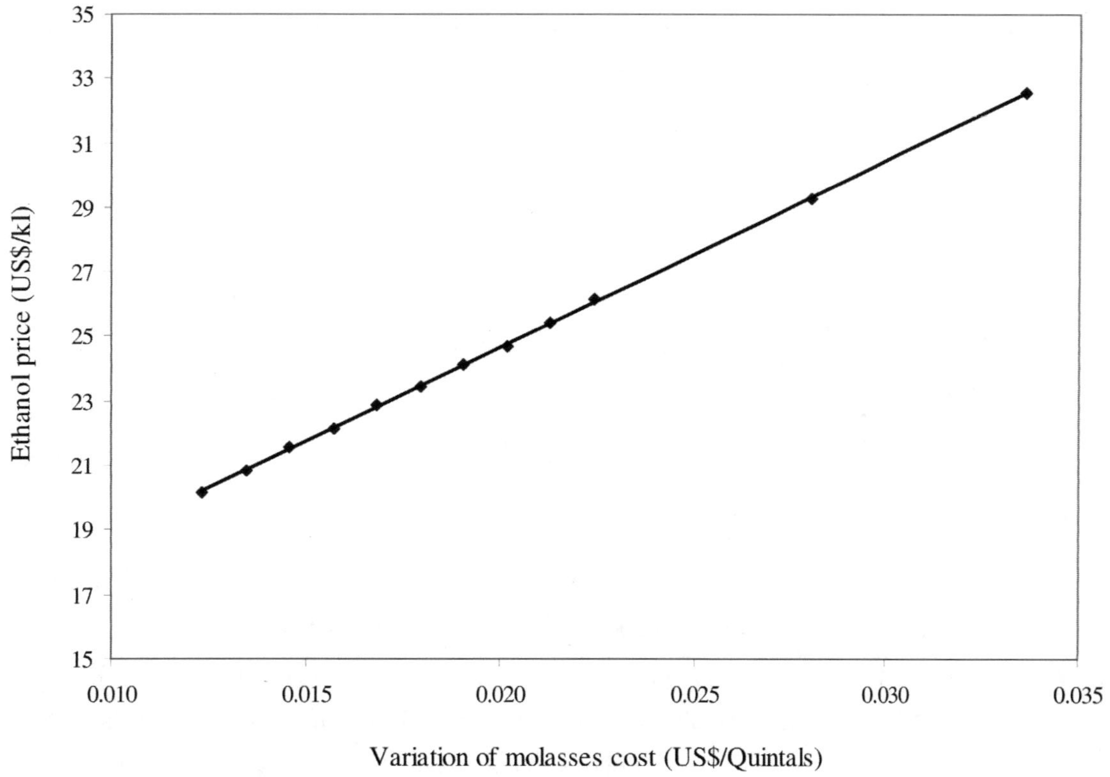

Figure 4: Effect of changing feedstock cost on fuel ethanol

The predominant energy requirement of an ethanol production plant is the steam required for the distillation process. The distillery was not designed according to energy saving principle, since excess steam was available from the sugar factory. The total distillery steam consumption per hectolitre (hl) of pure alcohol is 600kg, which translates to 7.x kg of steam per kg of ethanol, more than double the theoretical value (Mouris estimate) of 3.5kg of steam required per kg of ethanol (Enecon Pty Ltd, 2002).

Fixed operating costs are generally incurred whether or not the plant is producing at full capacity. These costs include labour and various overhead items. The total production period per annum at the distillery is 212 days. The operating labour of 480,000 constituting about 5% of total cost of production is low due to low labour cost at the plant location as well as possible sharing of labour (managerial) with the adjoining sugar mill.

The theoretical estimation of operating labour requirement was calculated to ascertain if the method of estimating operating labour requirement proposed by Ulrich and reported in Turton *et al.*, 1998 will work for the distillery. This method assigns number to different types of equipment categorised as auxiliary and process equipment as illustrated in Table 5:

Table 5: Operator requirements for the ethanol plant

Unit Operations	Number	operator	Total number of operator
Tower	14	0.35	4.9
Reactors	8	0.5	4
dryer (vaporizer)	1	0.5	0.5
heat exchanger	17	0.1	1.7
Total	40	1.45	11.1

This analysis would indicate a requirement of 12 operators employed in the ethanol plant per shift. This does agree with the total number of staff working on a shift in this distillery, but it is worth pointing out that only 8 of the shift positions correspond to those expected from the analysis in Table 5, with the remainder being supervisory. It therefore appears that the use of Ullrich's method would overestimate the labour requirement – an error that is more likely to be specific to the nature of this industry than the African location (where rather, one would expect a first world estimating tool to result in an under-estimate). Annual maintenance materials were calculated as 1.22% of the total equipment cost. Possible explanation for this could be due to the limited amount of operating time. Additionally, insurance and taxes were estimated at 1.3% of the total cost.

Table 6 presents the multiplying factors calculated in the dissertation and the corresponding factors reported by Turton *et al.*, 1998. The cost items for each of the three categories from the breakdown of the distillery cost data were added together to provide the total cost for each category as follows:

$$DC = C_{RM} + C_{UT} + C_{WT} + 1.83C_{OL} + 0.005FCI \quad (1)$$

$$FC = 0.05C_{OL} + 0.043\ FCI \quad (2)$$

$$GE = 18C_{OL} + 0.0006FCI + 0.014COM \quad (3)$$

Table 6: Multiplication factor for estimating manufacturing cost determined for the rural African distillery and the corresponding factor from Turton *et al.*, 1998.

1	**Direct Manufacturing Costs**	Typical rage of multiplying factors (Distillery analysis)	Typical range of multiplying factors (Turton *et al.*)
A	Raw material	C_{RM}	C_{RM}
B	Utilities	C_{UT}	C_{UT}
C	Operating Labour	C_{OL}	C_{OL}
D	Direct supervisory labour	$0.63C_{OL}$	$(0.1\text{-}0.25)C_{OL}$
E	Maintenance and repair	0.005FCI	(0.02-0.1)FCI
F	Laboratory Charge	$0.2C_{OL}$	$(0.1\text{-}0.2)C_{OL}$
G	Patent and Royalties	0	(0-0.06)COM
H	Waste Treatment	C_{WT}	C_{WT}
2	**Fixed Manufacturing cost**		
I	Depreciation	0.041FCI	0.1FCI
J	Local Taxes and Insurance	0.002FCI	(0.014-0.05)FCI
K	plant overhead (distillery analysis)	0.03 (Line 1C + Line 1D + Line 1E)	
	plant overhead (Turton *et al.*)	(0.5-0.7) (Line 1C + Line 1D + Line 1E)	
3	**General Manufacturing Expenses**		
L	Ethanol departmental cost (distillery analysis)	0.11(Line 1C + Line 1D + Line 1E)	
	Administration costs (Turton *et al.*)	0.15(Line 1C + Line 1D + Line 1E)	
M	Distribution and selling cost	0.014COM	(0.02-0.2)COM
N	Research and Development	-	0.5COM

We can then obtain the total manufacturing cost by adding these three cost categories together and solving for the total manufacturing cost (COM).

Cost of manufacturing (COM) =

$$COM = C_{RM} + C_{UT} + C_{WT} + 2.06C_{OL} + 0.044FCI + 0.014COM - C_{BP} \quad (4)$$

The equation for the manufacturing cost categories proposed by Turton et al, 1998 are illustrated below:

$$DC = C_{RM} + C_{UT} + C_{WT} + 1.33C_{OL} + 0.069FCI + 0,03COM \quad (5)$$

$$FC = 0.708C_{OL} + 0.0168FCI \quad (6)$$

$$GE = 0.177C_{OL} + 0.009FCI + 0.16COM \quad (7)$$

$$\text{Cost of manufacturing (COM)} = C_{RM} + C_{UT} + C_{WT} + 2.215C_{OL} + 0.246FCI + 0.190COM$$

It appears that there is a remarkable difference in the supervisory labour, which is high for the distillery compared with the Turton *et al.* prediction. The higher supervisory labour attests to annexed nature of distillery. The remarked difference observed between the manufacturing cost model obtained from the dissertation and that proposed by Turton et al, (1998) could be due to the sharing of utilities and support systems in the annexed distillery plant. However, variables like laboratory charge fit very well with the Turton *et al.,* prediction. This variable is assumed to be independent of the plant type. The low maintenance and repair cost could also be associated with the high distillery downtime. The local tax and insurance factor for the distillery is also very low compared to the literature value. Non cash expenses such as depreciation depend on first cost and whatever depreciation methodology employed. This analysis used a straight line method which is often the case in preliminary cost estimation. A value of 4% of capital investment cost was obtained in the analysis compared to a value of 10% reported by Turton *et al.*

6. Conclusion

From the operating cost analysis of the distillery, it emerged that the factorial approach to cost estimation is principally a sound one, with no indication of untypical cost items. Some of the typical cost items do however display ratios to the base cost that are outside of previously reported limits. The major economic factor to consider for input costs of ethanol production are feedstock cost, depreciation of capital cost (or the equivalent annualised capital financing cost) and energy related expenses. The type, availability, and price of molasses all factor into the profitability of producing ethanol. The energy costs include those of steam and electricity, with these costs being more critical for profitability. Energy expenses are one variable in location selection that can affect the success of the ethanol plant. Hence, ethanol plants located near existing manufacturing plants that produce excess steam or have some type of cogeneration potentials such as commonly found in the sugar industry will result in more favourable economics. Other cost savings are via labour, warehousing and wastewater treatment facilities.

Intermediate economic factors include the cost of labour, both in operating expenses and administrative costs. In this regard it is worth noting that the process flow sheet approach to estimating operating labour requirement (as proposed in standard texts such as [16]) has been shown to be appropriate, but that the use of dated factors can lead to over-estimation, even for African installations. Other minor economic input factors include the cost of enzymes, water and denaturant. These costs have little to do with economies of scale.

The ability to present an analysis of the breakdown of the operating costs of a rural African distillery, in an equation format enables easier and more rapid use of the data in numerical and economic models, and in the preliminary design and optimisation of bioethanol process plants. The identification of the economic input factor of production also provides insights to both the possible barriers to implementation that should be overcome, and on the technological improvement options that should be stimulated by research and development.

References

Amigun, B; Sigamoney, R; von Blottnitz, H. (2008). Commercialisation of biofuel industry in Africa: A review. Renewable and Sustainable Energy Reviews, 12(3), 690-711.

Da Silva, E.J; Ratledge, S.A, (eds.) (1992). Biotechnology: Economic and Social Aspects - Issues for Developing Countries, published by Cambridge University Press (UK) in association with UNESCO; Cambridge (UK).

Enecon Pty Ltd, (2002). Ethanol scooping study. Final report prepared for Johnstone Shire Council, Queensland, Australia by Enecon Pty Ltd and Stephen Schuck & Associates Pty Ltd.

Ethanol Producer magazine, (2004). Mauritius sends ethanol to EU. Retrieved 16 August 2007 from http://www.ethanolproducer.com/article

Henniges, O; Zeddies, J. (2004). Fuel ethanol production in the USA and Germany-A cost comparison. F.O.Litcht's World Ethanol and Biofuels Report,1(11),204-208.

Kartha, S, Leach, G., Rajan, C., 2004. Advancing bioenergy for sustainable development: roles for policymakers and entrepreneurs. Report prepared by Stockholm Environment Institute (SEI) for the World Bank Energy management Assistance Programme (ESMAP)

Peters, M. S; Timmerhaus, K. D. (1980). Plant Design and Economics for Chemical Engineers, McGraw-Hill Book Co, New York, USA.

Shell, Southern Africa, (2007). Shell fuel price online. http://www.shell.co.za/vpower/pprice.htm (accessed September 2007)

Thomas, V; Kwong, A. (2001). Ethanol as lead replacement: phasing out leaded gasoline in Africa. Energy policies, 29, 1113-1143.

Turton, R; Baillie, R. C; Whiting, W. B; Shaewitz, J. A. (1989). Analysis, Synthesis and Design of Chemical Process. Parentic Hall, OTR, New Jersey.

Ulrich, G. D. (1984). A guide to Chemical engineering process design and economics. John Wiley and sons, Inc, Canada, 265-317.

United State Department of Agriculture (USDA), (2006).The economic feasibility of ethanol production from sugar in the United State. Online available at http://www.usda.gov/oce/EthanolsugarfeasibilityReport3.pdf (accessed December 2007)

Valle-Riestra, J.F., (1983). Project evaluation in the chemical process industries, McGraw-Hill, New York

Wanambwa, L., (2005). Ethanol fuel production and use in Kenya for sustainable development. M.Sc. thesis in Environmental Science, Lund University, Sweden.

(265a) Catalytic Hydroprocessing of Bio-Oil Derived From Catalytic Fast Pyrolysis of Biomass to Produce Hydrocarbon Products

Alan H. Zacher, Doug Elliott and Daniel Santosa, Pacific Northwest National Laboratory, Richland, WA

Catalytic hydroprocessing has been used to produce fuels from bio-oils generated from both catalytic and non-catalytic fast pyrolysis of pine wood. The intent of the research was to evaluate the effect of processing mild catalytic pyrolysis derived oils by catalytic hydroprocessing and compare it to bio-oil produced without catalyst present in the pyrolysis reaction in the same bench-scale continuous-flow fixed-bed reactor system. This paper details the experimentation and product analysis undertaken at PNNL. A series of oils were produced under varying catalyst to biomass ratios on a harmonized pine sawdust feedstock in the bench-scale fluidized bed reactor, which were then subjected to hydroprocessing over a typical catalyst under a limited set of operating conditions. Detailed analyses of the products were performed including elemental analysis, mass and carbon balances, and limited chemical functional content. In summation, the paper provides an initial understanding of the potential benefits and consequences on hydroprocessing bio-oils that have been produced through mild catalytic pyrolysis.

See more of this Session: Catalytic Processing of Fossil and Biorenewable Feedstocks: Fuels II
See more of this Group/Topical: Catalysis and Reaction Engineering Division

(265b) Bio-Oil Upgrading through Hydrodeoxygenation

Weihua Deng, Zheng Li and Robert C. Brown, Center for Sustainable Environmental Technologies, Iowa State University, Ames, IA

The development of alternative energy, especially from biorenewable resources, such as biomass, has attracted more research interest because of depletion of fossil oil resources, global climate change and national security concerns. Compared to the other alternative energy choices, the production of energy and chemicals from biomass is more attractive due to the vast potential availability of biomass materials and the availability of existing technologies that potentially enable the production of energy and chemicals in commercial scales in shorter time period for near future. Researchers at Center for Sustainable Environmental Technologies (CSET) have been working on developing thermo-catalytic techniques to produce biofuels and other value added chemicals by upgrading bio-oil through hydrodeoxygenation reactions. In the present work, heterogeneous catalysts including traditional Co/Mo (Ni/Mo) HDO catalysts, noble metal catalyst that are supported on either metal oxide or carbon materials have been synthesized, characterized and tested for their performances using high-pressure reactor systems. Whole bio-oil and fractionated bio-oil produced from fast pyrolysis reactor systems at CSET are upgraded using these catalysts under typical HDO reaction conditions. Model compounds such as guaiacol or D-glucose are used to reduce the complexity of HDO reactions to obtain more insight into reaction mechanisms. The technical goal is to establish a comprehensive upgrading strategy for whole (and/or fractionated) bio-oil through a bottom up methodology.

68.20.52 on 5-3-2010-->

See more of this Session: Catalytic Processing of Fossil and Biorenewable Feedstocks: Fuels II

See more of this Group/Topical: Catalysis and Reaction Engineering Division

Effect of sulfur on $CoMo/Al_2O_3$ catalyst for the coal pyrolysis

Hao Qinglan [1], Wang Chang[2], Li Guiju[2]

1. College of Material Science & Chemical Engineering, Tianjin University of Science & Technology, Tianjin 300457, China; 2. College of Marine Science & Chemical Engineering, Tianjin University of Science & Technology, Tianjin 300457, China

Abstract: A series of $CoMo/Al_2O_3$ catalysts incorporated with sulfur were prepared by the impregnant method. The tests of the coal catalytic pyrolysis were carried out in a power particle fluidized-bed (PPFB). It is found that sulfur has the effect on the catalyst properties. The catalyst activity decreases with the increase of sulfur content. There was no obvious differences in the total catalyst yields when the catalyst CoMo/A incorporated with a few of sulfur. CH_4 was the main product during the coal pyrolysis using series of CoMo/B catalyst. The modified catalyst with sulfur, CH_4 selectivity with time on stream was obviously changed.

1. Introduction

Pyrolysis is a common stage of many coal utilization processes, such as gasification, combustion, or liquefaction. Pyrolysis can also be considered as a possible pre-treating method to obtain clean feedstock and fuels from raw coal. However, during pyrolysis chemical forms and the state of trace elements may also change, which may affect the final fate of the elements and subsequently results in different environmental effect [1-5].

The presence of sulfur in coal is generally undesirable and presents problems in the utilization of coal in various processes, so the transformation of sulfur and desulfurization of coal during thermal processing have long been a topic of great importance. Sulfur transformation during pyrolysis directly influence and control the sulfur distribution in gas, tar and char for the whole process. Many studies have reported the effect of sulfur on the coal pyrolysis [5,6]. However, few studies were focused on the effect of sulfur on $CoMo/Al_2O_3$ catalyst for the coal pyrolysis.

This paper will discuss the sulfur content on the catalytic performance of the $CoMo/Al_2O_3$ catalyst. Catalyst samples with different sulfur contents for the coal pyrolysis were characterized by BET and AAS. The results present in this study are aimed at providing a better understanding of the relationship between the catalyst modification with sulfur and the catalyst performances of $CoMo/Al_2O_3$ catalysts used in a power particle fluidized-bed (PPFB) during the coal pyrolysis process.

2. Experimental

2.1 Coal sample and Catalyst preparation

The ultimate and proximate analysis of the coal was shown in Table 1. The coal was ground to fine particles of 44-74 μm in diameter. The sample was dried in vacuum at 380 K for 1 h before use.

Table 1 Ultimate and Proximate analysis of coal sample

Sample	Ultimate analysis (wt%,daf)					Proximate analysis (wt%,daf)			
	C	H	N	S	O	Mois	V.M.	Ash	F.C.
Coal	77.9	6.4	1.3	0.3	14.1	3.0	46.8	11.5	38.7

The CoMo catalyst precursors, the weight composition is $CoO:MoO_3:Al_2O_3$=4:13:83, used in this study were commercial product. The commercial $CoMo/Al_2O_3$ catalyst was labeled as CoMo/A. And the other four catalysts were prepared by the impregnant method. In brief, 2 g CoMo/A catalyst was impregnanted in different concentrate H_2SO_4 solutions by vigorous stirring for about 5 min at the room temperature. The catalyst samples were dried in air at 98 ^{o}C for 10 h. And then the dried catalysts were calcined in air at 600 ^{o}C for 2 h. The treated catalysts are CoMo/B with different sulfur, which were labeled as CoMo/B1, CoMo/B2, CoMo/B3, and CoMo/B4 with the increasing sulfur content. The catalysts were ground and sieved to 250-500 μm before using. The sulfur contents of the catalysts are listed in Table 2.

Table 2 Sulfur contents of the catalyst samples

catalyst	S/Cat. (wt%) (Calculated value)	S/Cat. (wt%) (Measured value)
CoMo/A	0/100	0/100
CoMo/B1	0.5/100	0.38/100
CoMo/B2	1.0/100	0.84/100
CoMo/B3	2.0/100	1.53/100
CoMo/B4	3.0/100	2.26/100

2.2 Reactor system

The experiments were carried out in a power particle fluidized-bed (PPFB). Details of the reactor system can be found in previous works [7-9]. Briefly, the experiment apparatus consisted of a continuous flow powder-particle fluidized bed reactor installed in an electric furnace, a temperature controller with a thermocouple inserted in the fluidized bed, a micro-feeder for the continuous supplying of fine chip powder, and a product collecting system including a cyclone separator, two cold-traps and a gas-bag. The distributor in the reactor is made by a sintered quartz filter with a quartz tube of 25 mm i.d.. The distance between the distributor and the gas exit is 300 mm. Fine woody powders are fed into the bed where large catalyst particles are fluidized.

A given weight of fine coal powder was packed into the micro-feeder, and the catalyst particles were loaded into the reactor. After evacuation, the particles in the reactor were fluidized at a desired flow rate in an inert N_2 atmosphere. The reactor system was heated from room temperature to the desired reaction temperature with the same N_2 flow. After steady conditions were established, dried coal particles were continuously fed at about 5 g/h into the fluidized bed through the micro-feeder and kept on stream for 40 min. The process was operated under essentially atmospheric pressure.

2.3 Product analysis

During the pyrolysis, all the products were collected. The cyclone wrapped by a ribbon heater for separating the char from gases. The gasbag is used to collect the gas product. And the cold-trap is for determining the liquid product. The gas, liquid and solid products were collected to be analyzed quantitatively. H_2, CO, CO_2, CH_4 (IOG: H_2, CO, CO_2) and C_1-C_3 hydrocarbons (HCG) in tail gas were analyzed by Shimadzu gas chromatograph 2014 (GC-2014) with a MS-13X column (N_2 carrier) and a P-Q column (H_2 carrier) equipped with a TCD and a FID. Light aromtic liquid (HCL), benzene, toluene, xylene and naphthalene (BTXN) were analyzed using the same GC with a 2 m activated carbon column (FID, N_2 carrier).

2.4 Catalyst characterization

BET surface area of the catalysts was measured by N_2 physisorption at -196 ^{o}C using a Micromeritics ASAP 2500 instruments. The samples were degassed under vacuum at 120 ^{o}C for 6 h prior to BET measurement.

The composition of the catalyst and coal was determined by atomic absorption spectroscopy (AAS) on a EA1108-ELEMENTAL ANALYZER.

3. Results and discussion

3.1 The sulfur loss

To study the effect of sulfur content on the catalyst performances, a series of pyrolysis tests under the same reaction conditions are carried out. The sulfur loss during pyrolysis was shown in Fig. 1. It is found that sulfur loss in the catalysts during pyrolysis initial decreases notable with the extending pyrolysis duration and then gradually decreases to the stable values. Meanwhile, the sulfur loss increases with the increase of the sulfur content in the fresh catalysts. The sulfur amount is high in the catalyst, the combination between the catalyst and sulfur is not fastness. And a part of sulfur lost with time on stream (TOS). From the beginning of TOS 25 min, the amount of sulfur in the catalysts reaches a stable value.

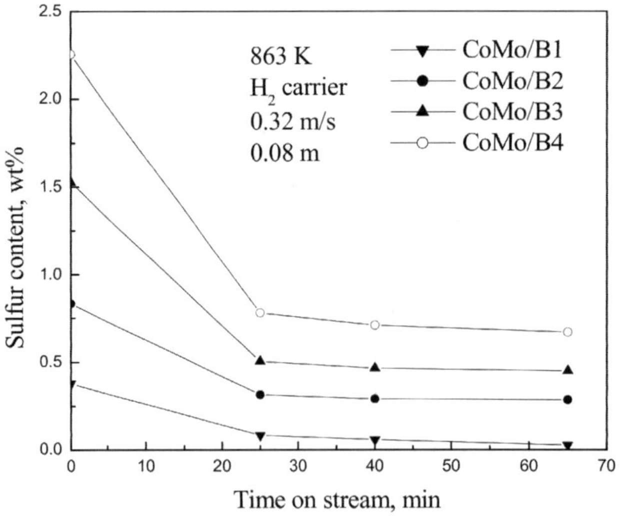

Fig. 1 The variation of the sulfur content with time on stream

3.2 The catalyst activity and selectivity

A series of pyrolysis tests under the same reaction conditions are carried out to systematically investigate the effect of sulfur content on CoMo catalyst activity and selectivity. The time on stream is typically for 40 min. The catalyst activity measured by the product yields is shown in Fig. 2. During the pyrolysis, the catalysts with different sulfur content exhibit significant differences in activity and selectivity. The catalyst activity decreases with the increase of sulfur content, from 35.02% (CoMo/B1) to 19.10% (CoMo/B4). The CoMo/A catalyst activity is 33.73%. The sulfur covered with the activity site results in the catalyst activity decreasing. However, the activity of the CoMo/B1 and CoMo/A is very resemble. This result shows there is no obvious differences when the catalyst CoMo/A incorporated with a few of sulfur.

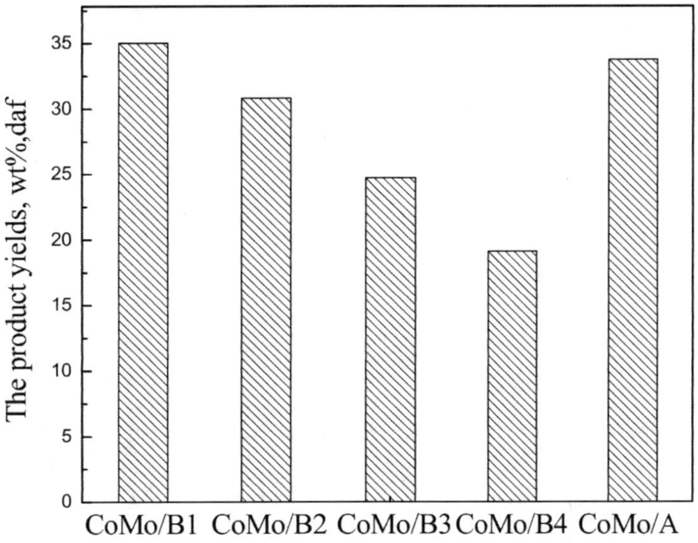

Fig. 2 Effect of sulfur content on the catalyst activities

The effect of sulfur content on the product distributions is shown in Fig. 3 and Fig. 4, respectively. The product distributions display considerable differences with the different catalyst sulfur content during pyrolysis. CoMo/B2 has the highest BTXN (benzene, toluene, xylene, and naphthalene) selectivity, however the lowest BTXN selectivity is obtained by CoMo/B1. The modified catalyst with sulfur, CH_4 selectivity with time on stream was obviously changed. CH_4 is the main product during the coal pyrolysis using series of CoMo/B catalyst.

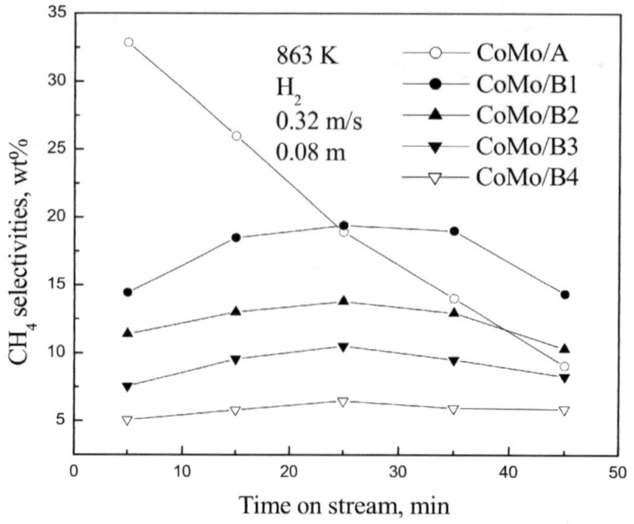

Fig. 3 CH_4 selectivities of the different catalysts

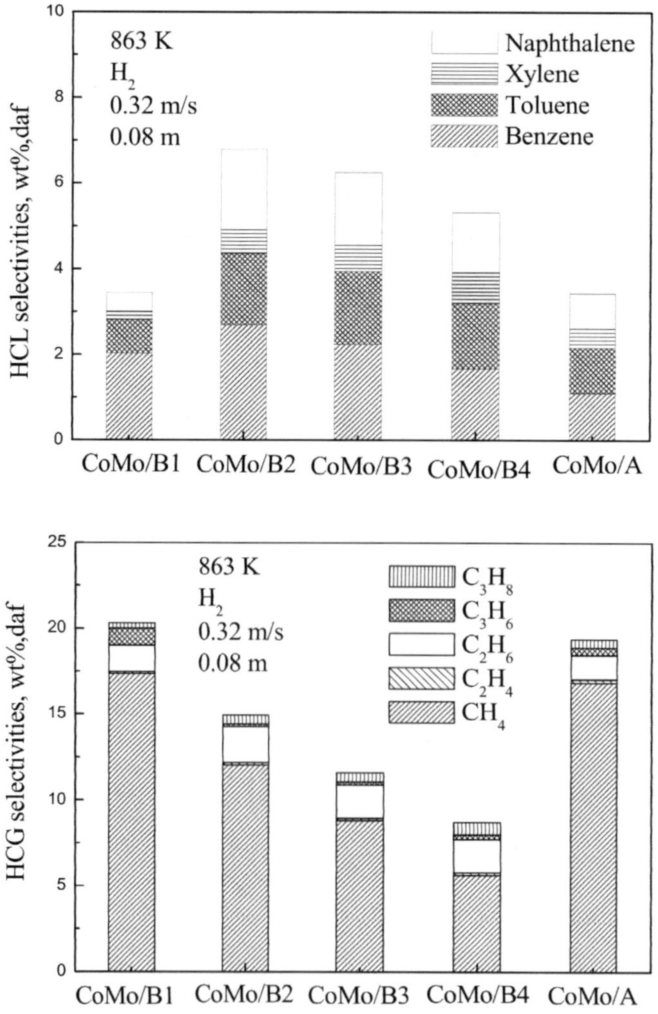

Fig. 4 Effect of sulfur content on the catalyst product distributions

4. Conclusions

The tests of the coal pyrolysis using CoMo catalyst were carried out in a power particle fluidized-bed (PPFB). It is found that the catalyst CoMo/A incorporated with a mount of sulfur could reform the catalyst activity. The catalyst activity decreases with the increase of sulfur content. There was no obvious differences in the total catalyst yields when the catalyst CoMo/A incorporated with a few of sulfur. CH_4 was the main product during the coal pyrolysis using series of CoMo/B catalyst.

Acknowledgement

We thank the National Natural Science Foundation of China for financial support under the contact numbers 20976132.

References

(1) Guo R. X.; Yang J. L.; Liu D. Y.; Liu Z. Y. *Am. Chem. Soc., Div. Fuel Chem.*, **2002**, *47*(1), 201.

(2) Attar A. *Fuel*, **1978**, *57*(4), 201.

(3) Labiano F. G.; Hampartsoumian E.; Willians A. *Fuel*, **1995**, *74*(7), 1072-1079

(4) Ibarra J. V.; Bonet A. J.; Moliner R. *Fuel*, **1994**, *73*(6), 933.

(5) Xu L.; Yang J. L.; Li Y. M.; Liu Z. Y. *Am. Chem. Soc., Div. Fuel Chem.*, **2002**, *47*(1), 198.

(6) Sugawara T.; Sugawara K.; Nishiyama Y. et al. *Fuel*, **1991**, *70*(9), 1091.

(7) Wang C, Hao QL, Lu DQ, Jia QZ, Li G J, Xu B. Chinese J Catal **2008**, *29*(9): 907.

(8) Wang C, Jia QZ, Nakagawa N, Kato K. J Chem Industry and Engineering (China) **2004**, *55*(8):1341.

(9) Wang C, Cui YY. CN20071114, **2007**

(265d) Aqueous-Phase Hydrogenation of Acetic Acid On Monometallic Catalysts: A Combined Experimental and Theoretical Study

Ye Xu[1], Hakan Olcay[2] and George W. Huber[2], (1)Center for Nanophase Materials Sciences, Oak Ridge National Laboratory, Oak Ridge, TN, (2)Department of Chemical Engineering, University of Massachusetts, Amherst, Amherst, MA

Hydrogenation of organic acids is an important step in the conversion and upgrading of biomass-derived feedstock for the production of valuable liquid fuels and chemical intermediates. To develop a fundamental understanding for this class of reaction, we are investigating the aqueous-phase hydrogenation of acetic acid, one of the simplest carboxylic acids, to ethanol, by a combined experimental and theoretical approach. The catalytic activity of monometallic Ni, Cu, Ru, Rh, Pd, Ir, and Pt catalysts have been measured in a fixed bed reactor and found to be distinctly different. In agreement with previous studies of the hydrogenation of organic acids,[1-4] Ru is by far the most active and selective catalyst, with the rest of the metals being at least an order of magnitude less active. For additional mechanistic insight, periodic density functional theory calculations have been performed to investigate the activation and subsequent hydrogenation of acetic acid on the seven metals. Our results suggest that the different activity is dictated largely by the first C-O bond scission step of acetic acid forming acetyl (CH_3CO)[5-7] across the metals. A simple reaction model allows the rate of acetic acid conversion to be estimated based on readily calculated reactivity parameters. Further investigation suggests that the selectivity for ethanol on Ru is controlled by reaction conditions. Our study provides insight for the challenging task of developing new catalysts for efficient biomass feedstock processing.

Work at UMass was supported with a grant from the ACS Petroleum Research Fund. The Center for Nanophase Materials Sciences is sponsored at Oak Ridge National Laboratory by the Division of Scientific User Facilities, U.S. Department of Energy.

(1) Varadarajan, S.; Miller, D. J. *Biotechnol. Prog.* **1999**, *15*, 845.

(2) Chen, Y. Q.; Miller, D. J.; Jackson, J. E. *Ind. Eng. Chem. Res.* **2007**, *46*, 3334.

(3) Dalavoy, T. S.; Jackson, J. E.; Swain, G. M.; Miller, D. J.; Li, J.; Lipkowski, J. *J. Catal.* **2007**, *246*, 15.

(4) Garcia, A. R.; da Silva, J. L.; Ilharco, L. M. *Surf. Sci.* **1998**, *415*, 183.

(5) Santiago, M. A. N.; Sánchez-Castillo, M. A.; Cortright, R. D.; Dumesic, J. A. *J.*

Catal. **2000**, *193*, 16.

(6) Pallassana, V.; Neurock, M. *J. Catal.* **2002**, *209*, 289.

(7) Alcala, R.; Shabaker, J. W.; Huber, G. W.; Sanchez-Castillo, M. A.; Dumesic, J. A. *J. Phys. Chem. B* **2005**, *109*, 2074.

See more of this Session: Catalytic Processing of Fossil and Biorenewable Feedstocks: Fuels II
See more of this Group/Topical: Catalysis and Reaction Engineering Division

Layered Solid Acid Catalysts for the Production of Biofuels From Algal Lipids

Griffin W. Roberts, University of Kansas, Lawrence, KS
Susan M. Stagg-Williams, University of Kansas, Lawrence, KS

Introduction

There exists a significant need to develop transportation fuels from sources other than fossil fuels; the forefront of the problems in the U.S. is the dependence upon foreign oil and the need to reduce greenhouse gas emissions. As a nation we must solve this problem without creating new ones, such as the food vs. fuel debate with some biomass sources. Algae can be a renewable fuel feedstock that does not compete in any existing market (such as corn or soybeans) but instead lays the foundation to create new markets in an already distraught economy. The high lipid content of algae allows for a substantial fuel production per acre of biomass (up to 3000 gal./acre) 100-fold increase to traditional renewable feedstocks (corn and soybeans)[1]. This paper reports on the synthesis of a layered zeolite-like solid acid catalyst or better known as pillared intercalated clays (PILC) to convert algal oils to biofuels. Initial studies have focused on the simultaneous esterification/ transesterification reaction of pure canola oil mixed with oleic acid to produce biodiesel. It is well known that base catalysts are better suited for the transesterification reaction, however when dealing with used cooking oil or algal oils there are significant amounts of free fatty acids (FFA). When FFAs are introduced with base catalysts they undergo saponification producing soap, essentially wasting feedstock. Thus, acid catalysts become the primary way to utilize the FFAs toward fuel conversion. The PILC studied herein uses the base or parent clay from a natural occurring clay mineral, Montmorillonite, and is pillared with an aluminum/ gallium polyoxycation. The catalyst has been characterized by NH_3- temperature programmed desorption to determine the density and strength of acid sites, N_2-adsorption at 77K for study of BET surface area, and SEM imaging to see the layered material and estimate the interlayer spacing. Further characterization will be implemented by used of x-ray diffraction. The intent is to determine how strength and number of acid sites, surface area, and pore size affect catalytic activity. The effects of coking, thermal stability, and reusability of the catalyst will also be determined. The introduction of a scalable pillaring vessel will be discussed to produce pillared clays in large quantities.

Background

In the early years of hydrocarbon processing clays were the standard catalyst used for creating fuels from crude oil until Mobil introduced their "Mobil Olefin toward Gasoline and Distillate" (MOGD)[2] process which uses today's standard, ZSM-5 microporous zeolite. The same ZSM-5 catalyst cannot be used for renewable feedstocks because of the much larger oxygenated compounds compared to the C_2-C_5 olefins used in MOGD. The small micropores don't allow the reactants to reach the catalytic active sites and high acid strength of ZSM-5 create too much coke and gaseous products. Therefore, a larger mesoporous catalyst with mild-to-moderate acid strength is needed to crack and oligomerize the lipids of algal oils to fuels. Pillared clays have the potential to suit the chemistry needed to create fuels from these algal oils.

Studies with our collaborating research groups have shown that algal oils contain fatty acids which are similar to those found in common seed oils such as canola. These are C_{16} and C_{18} fatty acids with varying degrees of unsaturation and concentration comprised mainly of stearic, linolenic, palmitic, oleic, and linoleic acids. Oxygenates within a feed stream to a cracking and oligomerization processes can have unwanted effects on the catalyst in terms of life cycle and turn over rate. If the oxygenates prove to be a problem for the pillared clay it has been shown that amorphous silica-alumina can be used to pretreat oxygenate containing feeds to significantly reduce the amount of oxygen[3].

As is known from the MOGD process, zeolites are efficient cracking catalysts because of their thermal stability, acid strength, and well-defined pore structure. Pillared clays are closely related to zeolites with a highly ordered framework allowing for the incorporation of metals to achieve desired activity. The synthesis of zeolites involves expensive chemicals and energy intensive processes to create; where clays are very cheap common materials found all over the world, Montmorillonite being among one of the more common clay minerals and easily modified to effective solid acid catalyst[4-6]. Synthesis of PILCs however has not been made effectively at high volumes due to the extensive amount of water that is needed during the pillaring process. In recent times there have been many advances in concentrating the clay while pillaring using such techniques as dialysis and sonication[7] assisted pillaring. There has been only one report to my knowledge on producing pillared clays in large quantities (>1 kg)[8] however even then no "one-pot" method or system has been developed to produce pillared clays in large quantities while in concentrated media. The following sections will introduce a pillaring vessel that will be able to synthesize pillared clays in concentrated media while utilizing minimal amounts of water during the entire pillaring process.

Montmorillonite is a smectite mineral comprised of phyllosilicates or layer silicates. The layers are arranged in a 2:1 fashion where two tetrahedral silica oxide sheets sandwich an octahedral alumina/magnesium oxide sheet, cations exist inside the interlayer spacing between each 2:1 layer. For a review of smectites and pillared clays refer to Kloprogge[9] along with the Handbook of Clay Science. The cations in the interlayer spacing can be exchanged with polyoxycations, which upon calcination leaves an irreversible metal oxide pillar between the smectite layers. These pillars create a well-defined pore structure in the mesoporous range. The type of pillar can have significant effects on the catalytic performance by influencing acid type and strength, pore size, surface area, and thermal stability. The following synthesis method uses a $GaAl_{12}$ based polyoxycation to create their corresponding pillars; these polyoxycations are well characterized by NMR studies[10].

As stated earlier the catalytic activity test used herein is to produce biodiesel; the conventional method uses a base catalyzed reaction which suffers from loss of catalyst and expensive feedstocks. Solid acid catalysts can help solve this problem by allowing high FFA feeds (less expensive) to be used as well as minimizing catalyst loss during processing. It has been shown that solid acids can catalyze the simultaneous esterification/ transesterification reactions, however at slightly higher temperatures than traditional homogenous base catalyzed transesterification reaction[11,12].

Experimental

Catalyst Preparation

Although pillared clay catalysts and their polyoxycations have been synthesized and characterized since the mid 70's[13,14], understanding their performances and applications are still of great interest. We have developed our process primarily from the work of Gonzalez et al.[4] and from the understanding of the polyoxycations from Bradley et al.[10]. The clay source uses is a Na-bentonite (Alfa Aesar) after performing a <2μm fraction using the centrifugation technique. The aluminum source for the polyoxycations is a ReagentPlus® aluminum chloride hexahydrate 99% and the gallium source is gallium(III) chloride 99.999% anhydrous beads 10- mesh both obtained from Sigma-Aldrich.

The first step in the pillaring process is creating the polyoxycation solutions by hydrolyzing the mixed metal complex. Aqueous solutions of the combined metal complexes where mixed with sodium hydroxide, dropwise with constant stirring, to achieve OH^-: M^{3+} in a ratio of 2.0. The hydrolyzed solution where then aged at 80°C overnight for roughly 24 hours. Mixed polyoxycation solutions were at an Al: Ga ratio of 12:1. Montmorillonite fraction was swelled for 24 hours in deionized water examining weight percents ranging from 7-17% clay. The dry clay was first placed in a dialysis sack, water was added, and then the sack was placed into a reservoir of deionized water to swell the interlayers of the clay. The water reservoir was switch to the polyoxycation solution at 60 meq M^{3+}/ g of dry clay with a contact time of 3 days while in a sonicator bath. The dialysate was then changed back to water to remove excess chloride ions and exchanged daily for three days or until there is no chloride ions present determined by precipitation with silver nitrate. The clay was then removed from the dialysis sack and freeze dried. The dry pillared clay was ground to a fine powder and calcined at 700°C for 2-24 hours under nitrogen flow of 100mL/min. N_2-adsorption at 77K was implemented with a Micromeritics Gemini. NH_3-TPD was used with a Micromeritics Autochem 2910. SEM imaging was taken of the samples using a Carl Zeiss Leo 1550 FESEM.

Catalytic Activity

Simultaneous esterification/ transesterification was carried out using 10% oleic acid purchased from Fisher Scientific and 90% pure canola oil purchased from a local supermarket (Lawrence, KS). Approximately 100g of oil was used with ~1wt% (after sieving with 120 mesh) of catalyst and a molar ratio of 10:1 methanol to oil. The oil and methanol was placed in a 450mL reactor (model 4560, Parr) with catalyst placed in a catalyst addition device installed in the reactor. The reaction ran at 140°C and 200 psi for 5 hours with a stir rate at 350 rpm. Once the liquid reactants were at temperature the catalyst was introduced in the reaction media and reaction time started. After the reaction time expired the heater was released and the outside of the reactor was cooled quickly with a wet sponge to quench the reaction as soon as possible. The excess methanol was boiled off and the products were analyzed by GC-MS (Agilent 6890 with a 5973N mass spectrometer).

Scalable Pillaring Vessel

The proposed vessel is still under testing stages. This vessel implements concentrated clay suspension as well as concentrated polyoxycation solution to eliminate the most amount of water used during the pillaring process. The same techniques used in the lab scale preparation are used with the larger scale pillaring vessel. The same reagents and same ratio of reagent are used as well. The difference is the vessel itself; it allows for larger quantities of clay to be used during the pillaring process and can be easily scaled. The vessel is made from Teflon and orientated into segments separated by dialysis sheets and sealed with an o-ring assisted by clamps. There are ports on each segment to allow addition and removal of liquids during the washing stages. The vessel is oriented as shown below.

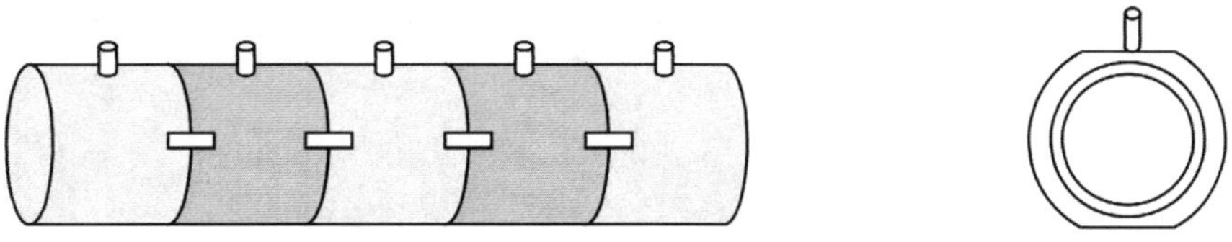

Figure 1 A) Shows the segmented body of the pillaring vessel where blue is the polyoxycation solution (or water if during the washing stage) and orange is the clay. The clamps are shown on the sides with the inlet/ outlet ports on top. B) Shows a cross section of the vessel where the center is the area where the pillaring or ion exchange between segments occurs. The o-ring can be seen in between the center and outer edge; this is where the dialysis sheet would be placed between each segment.

The vessel that has been made fits only a 10cm dialysis sheet however this can be scaled to the size of dialysis sheet available. Scaling of this vessel can also be done by simply adding more segments of alternating clay and polyoxycation solution. Once the vessel is full it can then be immersed in a water bath with ultrasonication to promote ion exchange between the polyoxycation solution and clay suspension.

Results

The catalyst obtained a surface area of 240 m^2/g after calcining for 24 hours at 700°C as compared to the parent clay Montmorillonite 31 m^2/g (air dried) and 40 m^2/g (freeze dried); which shows how the drying procedure clearly effects to layer aggregation; Kloprogge[9] reports this as the lamellar aggregation (air-dried) and house-of-cards aggregation (freeze dried). This retention of high surface area after such extensive calcinations shows that the AlGa-PILC has significant thermal stability. The acidic density of the fresh catalysts was 2.12 mmol NH_3 /g. The NH_3-TPD plot is shown in the following figure. It can easily be seen that most of the acid site are located in the weak regime while there are still some moderate to strong sites available. The TPD experiment was performed up to 700°C because this was the calcination temperature. The acid strengths are generally located from 100°C- 200°C for weak sites, 250°C- 350°C for moderate, and 350°C- 500°C for strong acid sites. The noise associated with the experiment was from the slow ramp rate (5°C/ min) and the high rate of the sweep gas.

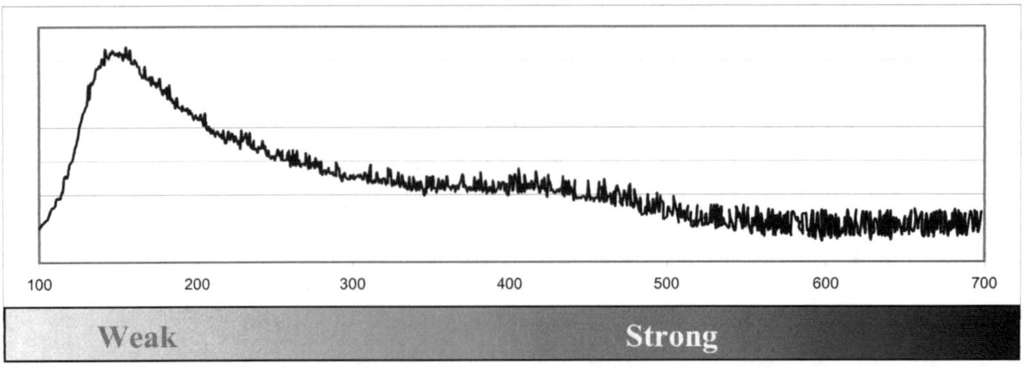

Figure 2 NH_3-TPD using Micromeritics Autochem 2910. X-axis represents the temperature and the y-axis is the TCD signal.

The SEM image below shows the striations indicating the layered structure of the catalysts also showing the ~2nm interlayer spacing.

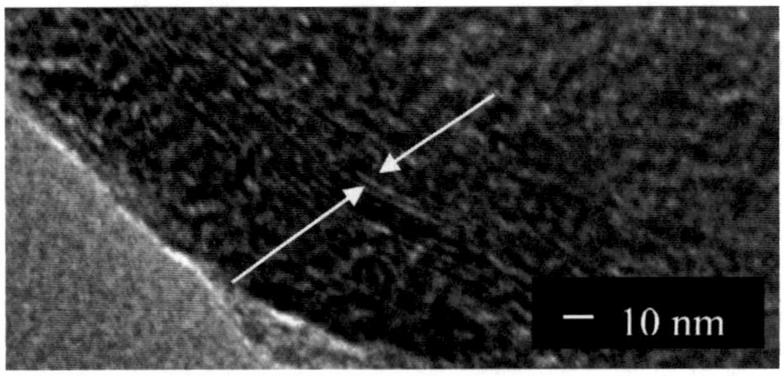

Figure 3 SEM image of AlGa-PILC. Yellow arrows indicate the ~2nm interlayer spacing.

The catalytic activity test (as described in the experimental section) showed that all of the FFAs was converted to fatty acid methyl esters (FAME) but only very little of the triglycerides were converted (~7%). This is congruent with reports of Shu et al.[15], which showed similar results at the temperature of 140°C; however when Shu and co-workers increased the temperature to ~220°C they seen the remaining triglycerides were converted. Further testing of our catalyst at the elevated temperature of 220°C is also expected to convert the remaining triglycerides.

Acknowledgements

We would like to acknowledge the Center for Environmentally Beneficial Catalysis (CEBC) for providing laboratory space and equipment, the Transportation Research Institute for providing funding resources, the "Feedstock to Tailpipe©" Initiative which is a multidisciplinary collaborative effort to develop biofuels from algae, and the Madison and Lila Self Graduate Fellowship for providing personal financing; all based at the University of Kansas.

References

1. Hu, Q., Sommerfeld, M., Jarvis, E., Ghirardi, M., Posewitz, M., Darzins, A., (2008) "Microalgal triacylglycerols as feedstocks for biofuel production: perspectives and advances", *The Plant Journal*, 54, pp. 621-639.
2. Quann, R. J., Green, L. A., Tabak, S. A., Krambeck, F. J., (1988) "Chemistry of Olefin Oligomerization over ZSM-5 Catalyst", *Ind. Eng. Chem. Res,* 27, pp. 565-570.
3. DeKlerK, Arno, (2007) "Effects of Oxygenates on the Oligomerization of Fischer-Tropsch Olefins over Amorphous Silica-Alumina", *Energy & Fuels*, 21, pp. 625-632.
4. Gonzalez, F., Pesquera, C., Benito, I., Herrero, E., Poncio, C., Casuscelli, S., (1999) Pillared clays: catalytic evaluation in heavy oil cracking using microactivity test, *Applies Catalysis A: General*, 181, pp. 71-76.
5. Casagrande, M., Storaro, L., Lenarda, M., Rossini, S., (2007) "Solid acid catalysts from clays: Oligomerization of 1-pentene on Al-pillared smectites", *Catalysis Communications*, 6, pp. 568-572.
6. Liu, Y., Hanaoka, T., Murata, K., Okabe, K., Inaba, M., (2007) "Hydrocracking of Fischer-Tropsch Wax to Diesel-range Hydrocarbons over Bifunctional Catalysts Containing Pt and Polyoxycation-pillared Montmorillonite", *Chemistry Letters*, 36, pp.1470-1471.
7. Sanabria, N. R., Molina, R., Moreno, S., (2009) "Effect of Ultrasound on the Structural and Textural Properties of Al-Fe Pillared Clays in Concentrated Medium", *Catayst Letters*, 130, pp.664-671.
8. Moreno, S., Gutierrez, E., Alvarez, A., Papayannakos, N. G., Poncelet, G., (1997) "Al-pillared clays: from lab synthesis to pilot scal production Characterization and catalytic properties", *Applied Catalysis A: General*, 165, pp.103-114.
9. Kloprogge, J. T., (1998) "Synthesis of Smectites and Porous Pillared Clay Catalysts: A Review", *Journal of Porous Materials*, 5, pp. 5-41.
10. Bradley, S. M., Kydd, R. A., and Yamdagni, R., (1990) "Study of the Hydrolysis of Combined Al^{3+} and Ga^{3+} Aqueous Solutions: Formation of an Extremley Stable $GaO_4Al_{12}(OH)_{24}(H_2O)_{12}{}^{7+}$ Polyoxycation", *Magnetic Resonance in Chemistry*, 28, pp. 746-750.
11. Kulkarni, M. G., Gopinath, R., Meher, L. C., and Dalai, A. K., (2006) "Solid acid catalyzed biodiesel production by simultaneous esterification and transesterification", *Green Chemistry*, 8, pp 1056-1062.
12. Lien, Y. S., Hsieh, L. S., and Wu, J. C. S., (2010) "Biodiesel Synthesis by Simultaneous Esterification and Transesterification Using Oleophilic Acid Catalyst", *Ind. Eng. Che., Res,* 49, pp. 2118-2121.
13. Shabtai, J., Frydman, N., and Lazar, R., (1976) "Synthesis and catalytic properties of a 1,4-Diazabicyclo (2,2,2) octane-Montmorillonite system- A novel type of molecular sieve", *Proc.* 6th *Int. Congr. Catal.,* B5, pp. 135-145.

14. Lahav, N, Shani, U., and Shabtai, J., (1978) "Cross-linked Smectite. I. Synthesis and Properties of Hydroxy-Aluminum-Montmorillonite", *Clays and Clay Minerals*, 26, pp. 107-115.
15. Shu, Q., Nawaz, Z., Gao, J., Lio, Y., Zhang, Q., Wang, D., Wang, J., (2010) "Synthesis of Biodiesel from a model oil feedstock using a carbon-based solid acid catalyst: Reaction and separation", B*ioresource Technology*, 101, pp5374-5384.

(265f) Simultaneous Esterification and Transesterification for Biodiesel Production

Shuli Yan[1], Craig DiMaggio[2], Siddharth Mohan[1], Manhoe Kim[3], Huali Wang[4], Steven Salley[5] and Simon Ng[5], (1)Chem. Eng., Wayne State University, Detroit, MI, (2)National Biofuel Energy Laboratory, Detroit, MI, (3)Chemical Engineering, Wayne State University, Detroit, MI, (4)Wayne State University, Detroit, MI, (5)Chemical Engineering and Materials Science, Wayne State University, Detroit, MI

Biodiesel, a mixture of fatty acid esters, is obtained from natural oil resources. As it is nontoxic, renewable and beneficial for environmental protection, it is considered as an alternative fuel to traditional petroleum-based fuels. A major barrier to the commercialization of biodiesel is its high feedstock cost, which is primarily related with using refined vegetable oils as raw material during the production process. Using waste oils is one of the economical sources for biodiesel production. However waste oils generally contain some content of free fatty acids (FFA), the current catalytic approach is to use a two-step process to handle oils with high FFA content. In this study, we developed a one-step process based on Pr2O3/ZnO catalysts, which effectively converted crude soybean oil, crude coconut oil, crude palm oil, crude corn oil from DDGs, crude algae oil, waste cooking oil brown grease, crude algae oil and waste fishing oil into FAME. It was found that the acidic groups on catalyst surface played an important role in esterification reaction; and the base groups catalyze transesterification reaction. Catalyst structure was characterized by XRD, XPS, BET, FTIR, SEM, EDS. The bulk phase of Pr2O3/ZnO catalyst was ZnO. Pr is well disbursed in ZnO crystal. In comparision with other ZnO catalysts, Pr increases the surface acid and base sites which promote the conversion of acidic oils into FAME.

See more of this Session: Catalytic Processing of Fossil and Biorenewable Feedstocks: Fuels II
See more of this Group/Topical: Catalysis and Reaction Engineering Division

(265g) Understanding Deoxygenation Reactions to Produce Fungible Transportation Fuels On An Optimized Supported Pt Catalyst

Kyle W. Elam, Phuong T. Do, Min Shen, **Rolf Jentoft**, L.L. Lobban and Daniel E. Resasco, Chemical, Biological, and Materials Engineering, University of Oklahoma, Norman, OK

The removal of oxygen from biomass derived molecules is of fundamental interest for the production of fungible transportation fuels. Vegetable oils conversion to "green diesel" by catalytic deoxygenation using a solid catalyst produces a fungible fuel that has a higher energy density than fatty acid methyl esters produced through transesterification. This work focuses on the deoxygenation of model esters, triglycerides and of natural lauric oils over supported platinum catalysts. A series of catalysts containing Pt, Sn and K supported on SiO2, in combination or separately, were used in liquid phase or gas phase reactions. For liquid phase tests a 300 ml Parr reactor was used at 300 - 350 psi and 300 - 330 oC in batch mode without hydrogen, or in semi-batch mode with hydrogen flow. In the gas phase, methyl hexanoate was converted in a fixed bed tubular reactor in He flow at atmospheric pressure and 380 oC. Products were analyzed by GC, GCMS and GCxGC MS. The catalysts were characterized using CO chemisorption, TPR, TPO, and XPS. For the reactions of triglycerides, the concentration profiles for the batch conversion of dilute solutions of the individual triglycerides, trilaurin and trimyristin, without hydrogen, were similar. A power law fit to the data resulted in the rate equation: . Products include primarily hydrocarbons, C11 or C13, as well as oxygenated intermediates and condensation reaction products. The conversion of mixtures of individual triglycerides or of pure palm kernel oil (PKO) was complicated by transesterification reactions. The conversion of triglycerides in pure PKO, in the presence of hydrogen, was found to fit the equation: . The products included hydrocarbons formed through both decarbonylation/decarboxylation and hydrodeoxygenation as well as fatty acids, ketones and esters. For the conversion of methyl hexanoate, the primary products are mainly hexanoic acid and condensation products (C11 symmetric ketones). The main secondary products are C5 unsaturated hydrocarbons, indicating that in the absence of hydrogen in the feed the main reactions are decarbonylation/decarboxylation of condensation products. The reactions of methyl hexanoate help to elucidate the bifunctional nature of the Pt-Sn-K/SiO2 catalyst. The oxide surface of the catalyst functions to form coupling products that then undergo decarbonylation/decarboxylation reactions to form alkenes on the metal surface. The addition of K to the Pt-Sn catalyst plays a key roll in the distribution and nature of the Sn in the reduced catalyst. In the absence of co-fed hydrogen, the optimal catalyst has a balance of a Pt metal function stabilized by the addition of Sn and an oxide surface with enhanced activity associated with the addition of Sn and K.

See more of this Session: Catalytic Processing of Fossil and Biorenewable Feedstocks: Fuels II
See more of this Group/Topical: Catalysis and Reaction Engineering Division

Abstract ID#: 201966

Password: 589438

Title: State-of-Health of Lithium-Ion Battery Estimation Method Using Reduced Equivalent Circuit Model(ECM).

Session Selection: 16D01 Lithium Ion Battery Technology

Preferred Presentation Format: Oral

Lead Presenter's Email: masterkg81@gmail.com

Keywords Data Analysis, Li-ion Battery, State-of-health

State-of-Health(SOH) indicates a remained capacity and the performance of Lithium-Ion Battery. There are a lot of factors decrease the performance of the battery such as normal ageing, high temperature operation, rapid current charging and others. However, only limited variables are available for measuring that make difficult to direct estimation of SOH. The SOH estimation method based on reduced electrochemical cell model is developed to on-line application. This model consists of semi-empirical model and equivalent circuit model. Moving-horizon parameter estimation method was adopted to minimize error of parameter for semi-empirical model. The results of SOH estimation by model are compared to experiment results under various driving patterns and conditions of lithium-ion battery for HEV and PHEV. Consequently, the accuracy is high enough for applied to battery management system in HEV and PHEV as real time on-line SOH estimation.

(286b) Diffusion Induced Stress in a Phase Transforming Spherical and Cylindrical Insertion Electrodes for Lithium Ion Batteries

Rutooj D. Deshpande, Chemical and Materials Engineering, University Of Kentucky, Lexington, KY, Yang-Tse Cheng, Chemical and material engineering, University Of Kentucky, Lexington, KY, Mark W. Verbrugge, Materials and Processes Laboratory, General Motors Research and Development Center, Warren, MI and Adam Timmons, Electrochemical Energy Research Laboratory, General Motors Global Research and Development Center, Warren, MI

In lithium ion batteries, insertion and de-insertion of lithium in and out of the battery electrodes result in diffusion induced stresses (DISs). Diffusion induced stress may be one of the main reasons for mechanical degradation of electrode . Recently, several mathematical models have been proposed that attempt to understand DISs in battery electrodes. It is well known that some of the electrode materials undergo formation of two or more phases during lithium insertion and de-insertion. This phase transformation during lithiation or de-lithiation in an electrode particle may be assumed to form a core shell structure with a moving phase boundary. In this work, we mathematically model such a system to study DISs in phase transforming electrodes. We find that the concentration jump at phase boundaries can result in stress discontinuities at interfaces. We believe that this stress discontinuity is one of causes for electrode cracking. We are able to quantify the effect of concentration jump at the interface on the stress discontinuity. We can also investigate the effects of mechanical properties of the two phases on stress behavior and stress discontinuity. Our model can be used to help select electrode material properties to minimize the stress discontinuity at the interface to possibly enhance battery life.

See more of this Session: Lithium Battery Technology and Materials
See more of this Group/Topical: Fuels and Petrochemicals Division

(286c) Sulfur Encapsulated in Carbon Hollow Structures as Positive Electrodes for Lithium-Sulfur Batteries

Juchen Guo and Chunsheng Wang, Department of Chemical and Biomolecular Engineering, University of Maryland, College Park, MD

Rechargeable batteries with high energy density are essential for emerging demands such as electric vehicles. Lithium-sulfur batteries hold great promise to achieving high energy density due to the high specific capacity of sulfur (1675 mAh g-1) as a cathode material. However, there have been two major challenges for the realization of Li-S batteries: one is the low electrical conductivity of the sulfur, and the other is the formation of polysulfides through charge/discharge cycles. The low conductivity leads to poor rate performance and low utilization of active materials. The formed polysulfides are dissolvable in the liquid electrolyte thus resulting in very fast complete loss of capacity. In the present study, we attempt to solve these two problems by encapsulation sulfur into hollow carbon structures including carbon tubes synthesized with anodic aluminum oxide (AAO) template method, hollow carbon nanofibers from electrospin technique, and hollow carbon spherical structures using emulsion polymerization with various templates.

See more of this Session: Lithium Battery Technology and Materials
See more of this Group/Topical: Fuels and Petrochemicals Division

(286d) Phase Transformation Dynamics in Lithium Iron Phosphate Nanoparticles

Martin Z. Bazant[1], Peng Bai[2], Dan Cogswell[2], Matthew Pinson[3], Gerbrand Ceder[4] and Katsuyo Thornton[5], (1)Chemical Engineering and Mathematics, Massachusetts Institute of Technology, Cambridge, MA, (2)Chemical Engineering, Massachusetts Institute of Technology, Cambridge, MA, (3)Physics, Massachusetts Institute of Technology, Cambridge, MA, (4)Materials Science and Engineering, Massachusetts Institute of Technology, Cambridge, MA, (5)Materials Science and Engineering, University of Michigan, Ann Arbor, MI

Lithium iron phosphate is a promising material for high-rate Li-ion battery cathodes, but its strong tendency for phase separation (into Li-rich and Li-poor regions) poses a significant challenge for mathematical modeling. Battery models focus on diffusion and reactions in dilute solutions, while phase transformation models usually consider closed bulk systems, without coupling to (Faradaic) reaction kinetics at surfaces. We have developed a unified approach to phase transformation dynamics driven by Faradaic insertion/extraction reactions, based on the Cahn-Hilliard equation with (modified) Butler-Volmer boundary conditions. In the limit of strong crystal anisotropy, suitable for LixFePO4 nanoparticles, the model reduces to a nonlinear generalization of the Allen-Cahn equation for phase transformations in an open system, driven by the local overpotential. The theory predicts new phenomena, such as layer-by-layer phase-transformation waves ("domino-cascade"), current-dependent overshoot of the voltage plateau, and nonlinear interactions between phase transforming nanoparticles in composite electrodes, consistent with experiments. The latter effect allows us to fit recent data for ultrafast battery materials with only one parameter (the "wave resistance"), given the observed particle size distribution. This suggests a new means to enhance rate capability, by optimizing not only the average structure of the porous electrode (porosity, mean pore size, etc.) but also the geometrical fluctuations.

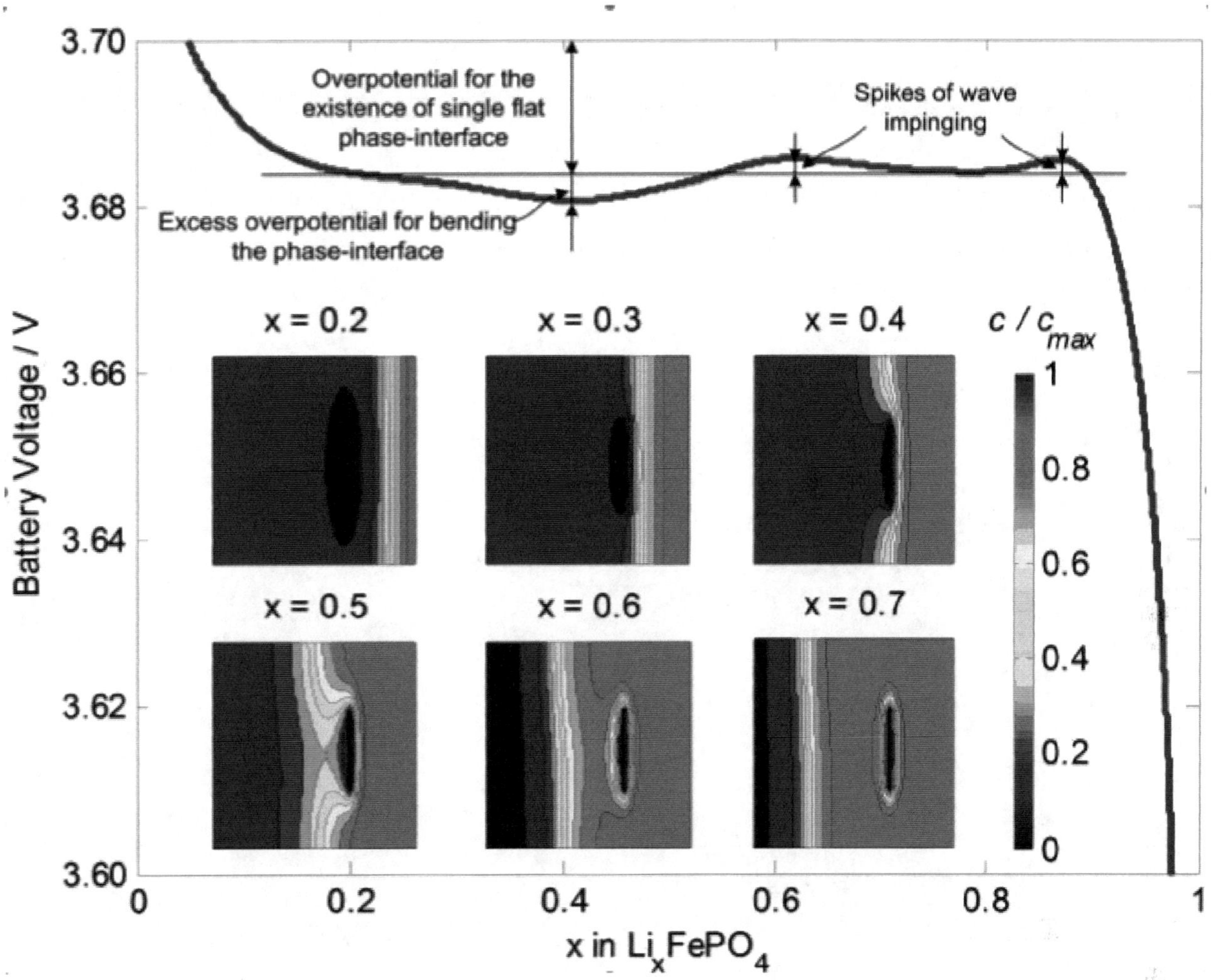

See more of this Session: Lithium Battery Technology and Materials
See more of this Group/Topical: Fuels and Petrochemicals Division

(286e) Lithium Flow Battery Technology

Galen J. Suppes[1], Michael J. Gordon[2], Bryan D. Sawyer[2], Benjamin J. Roseburrough[2] and David Guinta[2], (1)University of Missouri-Columbia, Columbia, MO, (2)Chemical Engineering, University of Missouri-Columbia, Columbia, MO

Flow battery technology is being applied to lithium battery chemistry to increase energy density and eliminate dendrite failure modes. This presentation is on the initial results and comparison to the performance of zinc-alkali flow batteries that use similar packed-bed electrodes.

See more of this Session: Lithium Battery Technology and Materials
See more of this Group/Topical: Fuels and Petrochemicals Division

(305a) Production of High Quality Biodiesel Fuel From Waste Cooking Oil Using Anion-Exchange Resin Catalyst

Naomi Shibasaki-Kitakawa, Takahiro Tsuji, Masaki Kubo and Toshikuni Yonemoto, Department of Chemical Engineering, Tohoku University, Sendai, Japan

1. Introduction

The anion-exchange resin has a high catalytic activity for the transesterification of triglycerides to produce fatty acid esters, used as biodiesel fuel[1]. The principal advantage of the production process of biodiesel fuel using the anion-exchange resin catalyst can be summarized as follows: 1) The alkali soap, contaminant for the fuel, is never formed by saponification[2]. 2) The impurities contained in the raw oils, such as free fatty acid (FFA) and dark pigment, are removed by adsorption on the resin[2,3]. 3) The complete conversion can be achieved even at the stoichiometric molar ratio of 3:1[2].

By combining the catalytic activity and adsorption ability of the anion-exchange resin, the high quality biodiesel fuel would be produced without the pretreatment of the raw oils and the purification of the products.

2. Experimental methods

By supplying the mixed-solution of waste cooking oil and methanol to the expanded-bed reactor packed with the anion-exchange resin, the continuous production experiment of the fatty acid methyl esters (FAME) was performed. The effluent from the reactor was analyzed to check the quality of the biodiesel fuel.

3. Results and discussion

Figure 1 shows the time courses of the triglyceride residual ratio and the FAME concentration in the effluent from the column. The starting time to supply the reaction solution to the column was set to zero. The open and filled symbols were the results using the original resin and the reused one after regeneration, respectively. Both symbols almost overlapped each other and the resin could be repeatedly used without any loss in the catalytic activity. The FAME concentration increased after 3 h and became constant after 10 h, when the steady state was attained. The triglyceride residual ratio was almost zero up to 25 h. so that the transesterification was found to completely proceed.

The result to check the quality of the biodiesel fuel is listed in Table 1 compared with standards of Europe (EN14214) and USA (ASTM D6751). The component properties are picked up. The contents of impurities (FFA and water), unreacted reactants (methanol and glycerides), and the byproduct (glycerin) in the effluent from the reactor fully satisfied with both standards. Only the FAME content was a little lower than that in EN14214, but it was not limited in ASTM D6751. From these results, water and glycerin as well as FFA and pigment were found to be removed by the adsorption on the resin. The distillation process to remove the residual methanol from the product can be omitted in this system.

4. Reference

1) N.Shibasaki-Kitakawa *et al., Bioresour. Technol.,* **98, 416 (2007)**

2) T.Tsuji *et al., Energy Fuels,* **23**, 6163 (2009)

3) N.Shibasaki-Kitakawa *et al., Energy Fuels,* (2010) in printing

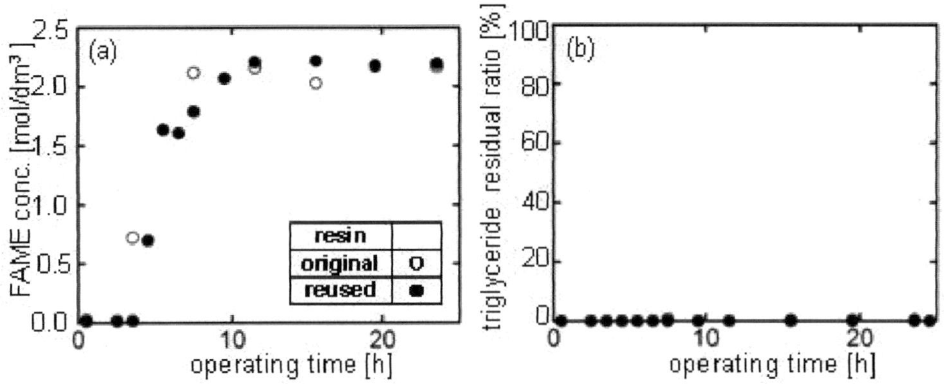

Fig. 1 Time courses of FAME concentration (a) and triglyceride residual ratio (b) in effluent from reactor

Table 1 Component properties of effluent from reactor compared with standards of biodiesel fuel

property	units	effluent	standards	
			EN14214	ASTM D6751
acid value (FFA)	[mg-KOH/g]	0.08	<0.5	<0.8
water	[mg/kg]	190	<500	-
methanol	[wt%]	0.03	<0.20	-
monoglyceride	[wt%]	<0.25	<0.80	-
diglyceride	[wt%]	<0.05	<0.20	-
triglyceride	[wt%]	<0.05	<0.20	-
free glycerin	[wt%]	<0.005	<0.02	<0.02
total glycerin	[wt%]	<0.10	<0.25	<0.24
FAME	[wt%]	91.5	>96.5	-

See more of this Session: Sustainable Diesel Fuel From Renewable Resources
See more of this Group/Topical: Sustainable Engineering Forum

(305b) Electrochemical Conversion of Fatty Acids to Produce Green Diesel

Charles J. Coronella, Chemical Engineering, University of Nevada, Reno, Reno, NV, Cody Wagner, University of Nevada, Reno, Reno, NV and Rainer Busch, Pacific Propellants, Reno, NV

Electrochemical conversion of fatty acids to produce green diesel (aka renewable diesel)

We are developing an electrochemical process to convert fatty acids to green diesel. Under appropriate conditions of temperature, current density, solvent mixtures etc., free fatty acid salts can be decarboxylated in a specialized electrolysis cell. The resulting alkyl fragments form hydrocarbons with a carbon chain shortened by one atom, producing $C_{15}$and C_{17} hydrocarbons from typical, lipid-based fatty acid feedstocks.

The chemistry is relatively straightforward. An unsaturated fatty acid, e.g. oleic acid, subjected to this process is transformed into unsaturated C17-hydrocarbons:

$$CH_3(CH_2)_7CH{=}CH(CH_2)_7COO^- \rightarrow CH_3(CH_2)_7CH{=}CH(CH_2)_5CH{=}CH_2 + CO_2 + H^+ + 2e^-$$

Traditional refining and upgrading processes, such as “hydrotreating”, are commonly being run at 300-500 Celsius as well as thousands of psi of pressure. The process under development permits the stereospecificity (i.e. cis-double bonds) in oleic and other unsaturated acids to be retained. This contributes to the excellent low-temperature performance of the hydrocarbons so produced. Thus, this fuel has properties that easily meets or even exceeds those of fossil-derived diesel fuel.

As shown in the equation above, the process also produces molecular hydrogen as well as carbon dioxide as byproducts. Hydrogen is a valuable byproduct, useful in other areas of a biorefinery, or directly as a fuel or salable commodity. The carbon dioxide can be easily separated and used as is, or fed back to an algae colony or other CO_2-consumer.

The talk presents results from experiments with a prototypical reactor, that has processed pure oleic acid, palmitic acid and commercially available mixtures to produce high quality hydrocarbon fuel products. The effects of pH, temperature, and current density are evaluated.

See more of this Session: Sustainable Diesel Fuel From Renewable Resources
See more of this Group/Topical: Sustainable Engineering Forum

Parametric Study of Biodiesel Quality and Yield Using a Bench-Top Processor

Rebecca Wilson and Ihab H. Farag, Chemical Engineering Department, University of New Hampshire (UNH), Durham NH 03824-3591

Abstract

A 1.5 liter bench-top biodiesel processor was constructed to study the effect of feedstock and operating conditions on the conversion and the biodiesel quality. Both the one-stage base process and the base-base two-stage production process were tested. The two-stage process exhibits greater conversion, lower cost and improved biodiesel quality.

1- Background

1.1 Introduction

Energy is consumed constantly, and is often taken for granted. There are finite resources accessible to us, and future generations will have to deal with the consequences when petro fuels are no longer available. Most sources of energy (such as petro fuels, wood, oil, coal and natural gases) release undesirable emissions, such as sulfur dioxide, carbon dioxide, hydrocarbons, and others that affect air quality. This has provided an incentive to research renewable, ecofriendly fuel sources such as biodiesel.

1.2 Biodiesel

Biodiesel is a plant-derived fuel that can be used to replace petro diesel (Wikipedia). Biodiesel is essentially CO_2 neutral, biodegradable and much less toxic than petro diesel. Figure 1 (Farag 2008) shows the environmental benefits of biodiesel relative to petro diesel. Significant emission reductions are observed in CO, polycyclic aromatic hydrocarbons (PAH) and nitro PAH (nPAH) (both PAH and nPAH are potential cancer causing substances), unburned hydrocarbons (largely responsible for the formation of smog), particulate matter (a health hazard when inhaled) and sulfates (a major component of acid rain). Clearly biodiesel is an environmentally friendly alternative to petro diesel.

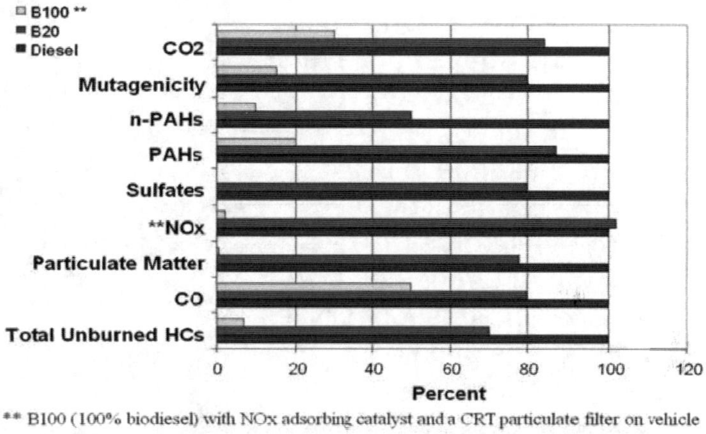

Figure 1: Emission Reduction of Biodiesel (pure or B100, and a mixture of 20% Biodiesel and 80% diesel) relative to petro diesel (Basis petro diesel emission of each component is 100).

1.3 Biodiesel Production

Biodiesel is made by the transesterification reaction of renewable resources like a vegetable oil or animal fat with an alcohol (methanol or ethanol,) in the presence of a catalyst (sodium hydroxide, NaOH, or potassium hydroxide, KOH), which also neutralizes the free fatty acid in the oil. The oils used in the present work include virgin vegetable oil (VVO), canola oil and waste vegetable oil (WVO). Canola oil is also known as Rapeseed oil. WVO is essentially free, ecofriendly and an excellent resource when making biodiesel. VVO is

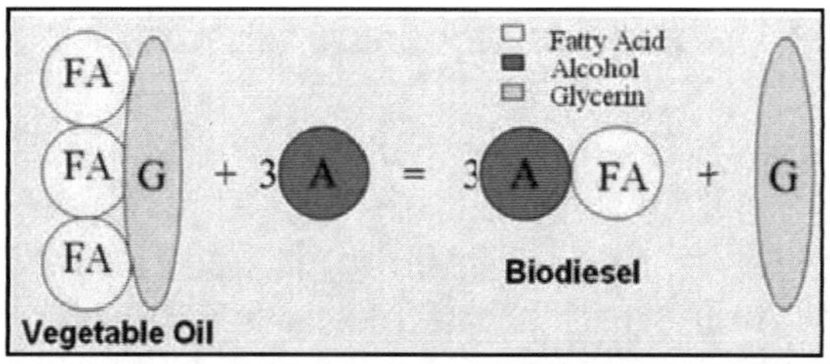

an arduous process to produce, starting with growing the crop, harvesting it, extracting the oil, and then transporting the oil to a facility. This is energy intensive and costly. WVO can both be transported to a biodiesel facility and then cleaned, or can be sent to a cleaning company then transported to the biodiesel facility. The alcohols investigated were methanol and ethanol. Both have pros and cons (Mulumba and Farag 2010). Methanol was used because it is less expensive and is a more efficient reactant. Important parameters that affect the transesterification are ratio of reactants, reaction temperature and reaction time (Briggs, Pearson and Farag 2004, Math 2007). One major problem is soap formation due to the reaction of the base catalyst (KOH in the present study) and the free fatty acids (FFA) which are present mainly in WVO and some VVO, i.e.,

Free Fatty Acids in Oil + KOH = Soap (undesirable)

Residual soap in Biodiesel may plug the fuel filters. Also, if soap is burned in a diesel engine it forms ash residues which result in higher levels of soot out the tailpipe.

1.4 Bench-Top processor

A major constraint in the wide scale use of biodiesel is the production cost. Optimizing biodiesel production is very important in reducing cost and increasing production efficiency. This involves studying and testing different parameters and feedstocks used to make biodiesel. One option that has been the subject of several studies is the use of low cost feedstocks such as WVOs. However, the challenge in using WVO is the high content of free fatty acids (FFA). In order to experiment with different feedstocks and process parameters, our UNH Biodiesel group designed and constructed the bench-top processor shown in Figure 2. It is a scaled down partially transparent processor (1.5 L) that allows for testing runs that would be more difficult and expensive on a full-sized industrial processor.

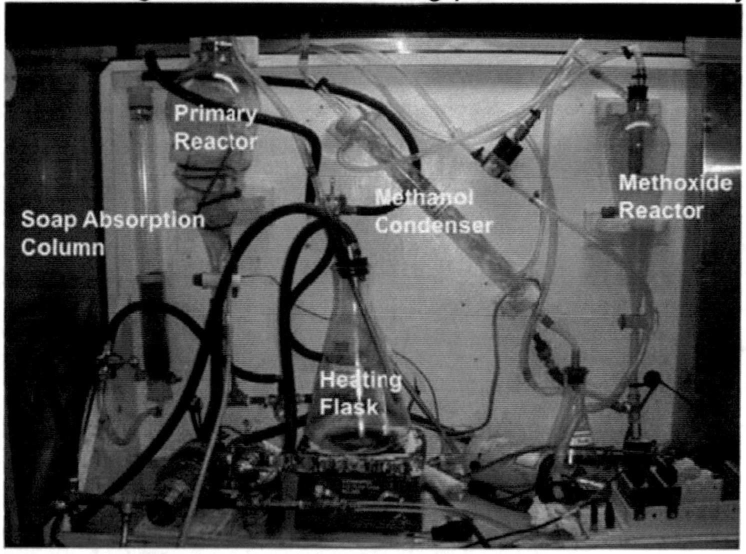

Figure 2: Bench-top Biodiesel Processor

The bench-top biodiesel processor includes 5 major parts (Wilson and Farag, 2010): 1- Methoxide Reactor: KOH and methanol are mixed to form methoxide, 2- Primary Reactor: the oil and methoxide are fed and the transesterification reaction takes place, 3- Soap adsorption column where Amberlite™ BD10DRY™ is used to adsorb soap from the biodiesel product of the reactor, 4- Heating flask: the excess methanol is removed from the biodiesel, and 5- Methanol condenser where the evaporated methanol is recovered for possible reuse.

1.5 Base-Base Two-Stage processing

The simplest method to produce biodiesel is the one-stage base processing. This involves adding all the methoxide to the oil in the reactor, run the reaction at about 135-140° F for 2 hours and let the biodiesel and glycerin products separate for at least four hours. After draining the glycerin, the biodiesel goes through soap adsorption and methanol recovery.

In the base-base two-stage process only some of the methoxide (for example, 80%) is mixed with oil in the primary reactor. The reaction proceeds for 1.5 hours, and the glycerin is drained. The rest of the methoxide (20%) is then added and the reaction proceeds for another 1 hour. After separating the glycerin the biodiesel is cleaned of soap and methanol. The base-base two-stage process studied most in depth was one in which first 80% methoxide was used, and then 20% (this is denoted an 80/20 process).

The base-base two-stage process is intended to increase the conversion of the oil to biodiesel. Removing the glycerin at the end of the first stage will reduce the likelihood of a reverse transesterification reaction. The reaction will proceed forward after more methoxide has been added. It is also possible to reduce the amount of methanol used, to achieve the same conversion as the one-stage process. The base-base two-stage process also reduces the amount of soap formed because the oil is in contact with less KOH in each stage.

2. Goal/Objectives

2.1 Goal

The overall goal of this project is to develop a bench-top biodiesel processor and use it to investigate different feedstocks and operating conditions that result in the highest yield of good quality biodiesel.

2.2 Objectives

The specific objectives of this research project are; 1- construction and flow testing of the bench-top biodiesel processor unit in the UNH Biodiesel Lab. 2- Establish that the processor unit is working properly before employing it with vegetable oil and other chemicals. 3- Run a base case for the production of biodiesel in the bench-top processor to establish operating parameters, 4- Do a parametric investigation to study the effect different variables/operating conditions have on oil conversion and biodiesel quality, 5- Compare the one-stage process with the base-base two-stage process.

3. Methodology/Approach

3.1 Approach

Our first objective to accomplish is to complete the construction and testing of our bench-top biodiesel processor. Operating conditions investigated are reaction temperature, ratio of reactants and the processing time. The quality of biodiesel is assessed by measuring properties such as viscosity, density, the heating value and the soap impurities. The metrics to evaluate optimal biodiesel yield are the oil conversion to biodiesel and the production cost.

3.2 Transesterification

The oil feedstocks used are VVO, canola oil and WVO. The alcohol used was methanol (MeOH), and the catalyst used was KOH in pellet form. Excess methanol was used to increase conversion because the transesterification reaction is reversible. Additional KOH was needed to neutralize the FFAs in the oil. The FFA content in the oil was determined by titration with KOH solution.

3.3 Soap Removal

The Amberlite™ BD10DRY™ purification ion exchange was used for soap adsorption. There are other techniques, such as the water washing or using a synthetic magnesium silicate called Magnesol XL. Water washing is not cost effective because it generates wastewater and is energy intensive. Magnesol XL can be used once, and needs a paper filter to be used. Amberlite™ BD10DRY™ can be used until saturated with soap, then cleaned with methanol, and reused, typically 5-6 times. Its lifespan is about 6 months.

4. Metrics and Measurement Methods

In order to determine the quality of the biodiesel being produced, the percent conversion, the viscosity, the heat of combustion, and the leftover soaps were measured.

4.1 Percent Oil Conversion: 25/225 Test

The percent conversion is measured by dissolving 25 mL of biodiesel in 225 mL of methanol. Biodiesel is soluble in methanol, while vegetable oil is not. If there is 100% conversion, then the biodiesel will be completely dissolved. If conversion is less than 100% than there is some vegetable oil in the biodiesel that will settle at the bottom. For every 1mL of vegetable oil settled there was 4% oil that was not converted.

4.2 Viscosity

Viscosity is a viable measure of quality biodiesel because it is an indicator of the resistance of biodiesel to flow. When the fuel is injected into a combustion chamber, it will atomize and combine with air to combust. If the ignition temperature is low and the fuel is too viscous, then it may not completely combust, and it will form engine deposits. This is why vegetable oil is not typically used in a diesel engine. Vegetable oil has a much higher kinematic viscosity between 36 mm^2/s and 48 mm^2/s (or centistokes, cSt), increasing the probability of engine deposits. Biodiesel typically has a kinematic viscosity between 4 mm^2/s and 6 mm^2/s at 40°C. Diesel viscosity range at 40°C is 1.9 mm^2/s to 4.1 mm^2/s (Knothe, Van Gerpen and Krahl, 2005). In the present study, the kinematic viscosity was measured using a Parr Ball capillary viscometer, which required the density of the biodiesel or oil being tested.

4.3 Heat of Combustion/Heating Value

The heating value is a measure of the potential power obtained from combustion. It was measured using a Parr bomb calorimeter with benzoic acid pellets as the standard.

4.4 Residual Soap: 50/50 Test

Residual soap can be detected by mixing 50 mL of water with 50 mL of biodiesel. If there is clear separation between the water and the biodiesel, then there is no soap left in the biodiesel. If the separation is not clear then the soap made during the reaction was not completely removed by the Amberlite™ BD10DRY™ ion exchange resin. This may indicate that there was probably excess potassium hydroxide (KOH) in the reactants.

5. Results

5.1 Percent Conversion

Figure 3 shows the percent conversion of the biodiesel samples, using the 25/225 test. The results show that the percent conversion increased with the 80/20 process, compared to the one-stage process. The WVO runs show that the percent conversion increased with the 80/20 process, as well as with increased KOH catalyst. WVO contains more FFAs than VVOs so more potassium hydroxide is needed to neutralize the free fatty acids. The conversion results of the vegetable oil show that using less methanol decreases the conversion of the vegetable oil to biodiesel. It also shows that using the 80/20 two-stage process increased the conversion using the same reactants as the one-stage process, and using less methanol the 80/20 process resulted in the same conversion as the one-stage process. This supports one of the benefits of the two-stage process that less alcohol is needed in the two-stage process to achieve the same conversion as the one-stage process.

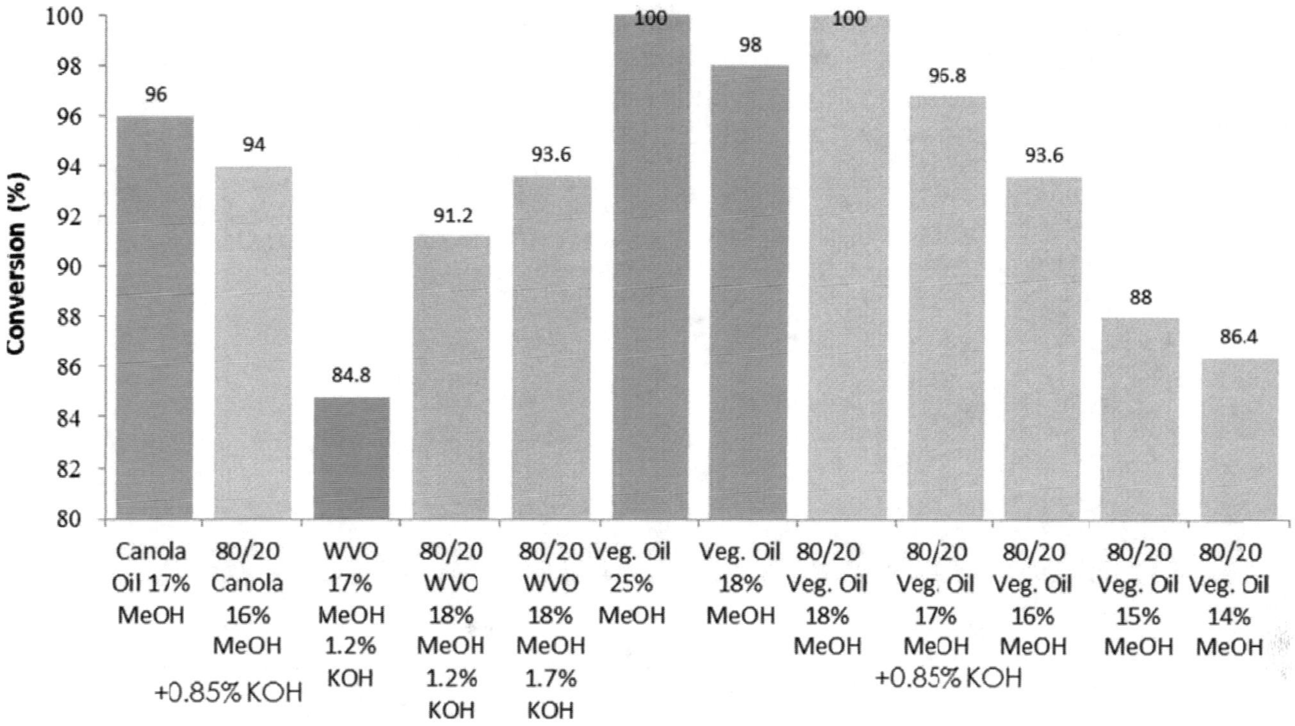

Figure 3. Percent conversion of oil, labeled with the reactants used to make the biodiesel and percent by weight relative to oil, e.g., WVO 17% MeOH 1.2% KOH has 100 g WVO + 17 g MeOH + 1.2 g of KOH pellets. All canola oil and vegetable oil batches were mixed with 0.85% KOH:oil ratio by weight of oil, i.e., 100 g oil + 0.85 g KOH.

5.2 Viscosity

Figure 4 shows that the kinematic viscosity of all biodiesel samples produced in our bench-top processor was in the usual range of 4 mm^2/s to 6 mm^2/s. The WVO and VVO feedstocks have a much greater viscosity (36 – 48 mm^2/s). The biodiesel produced in the 80/20 two-stage process typically had a lower viscosity than the one-stage process. The lower viscosity in the 80/20 two- stage process biodiesel is due to the higher conversion and reduced vegetable oil in the biodiesel, which will raise the viscosity.

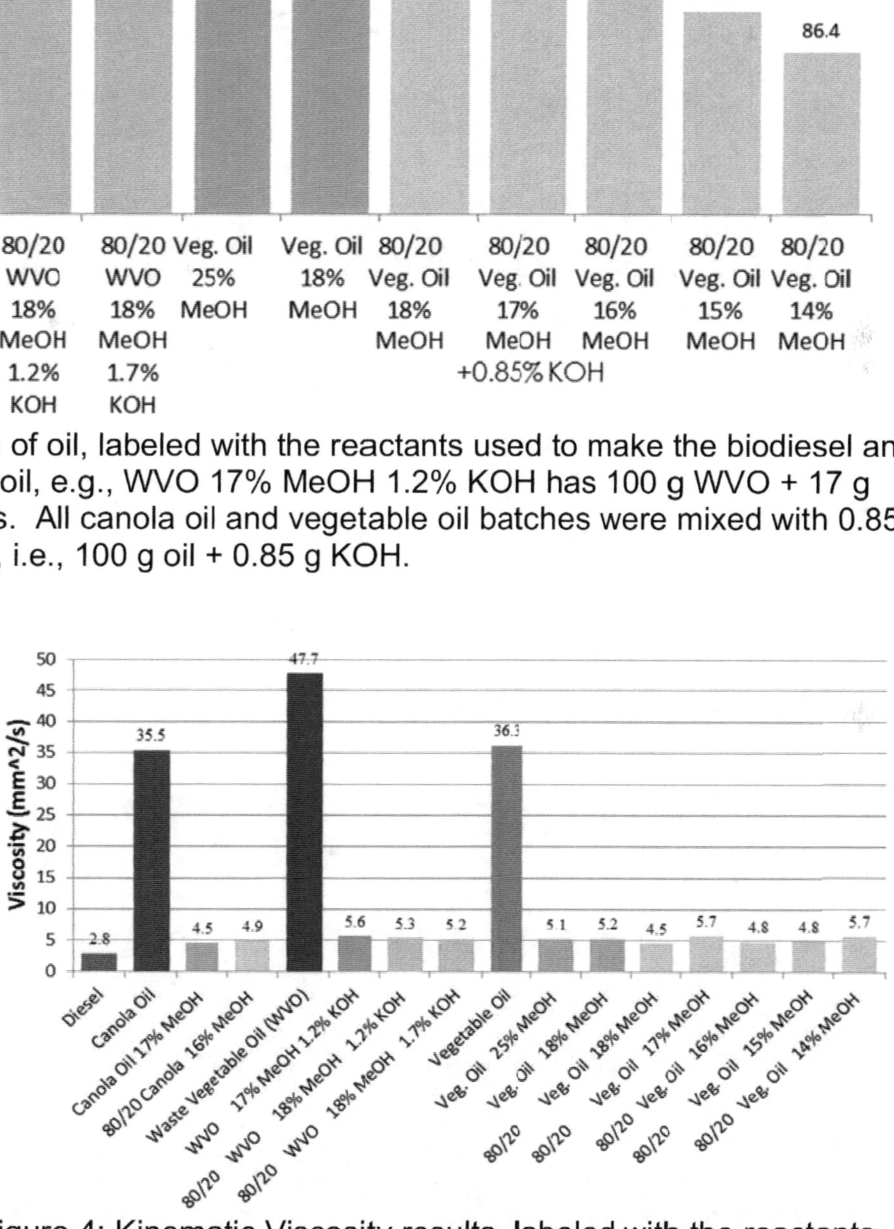

Figure 4: Kinematic Viscosity results, labeled with the reactants used to make the biodiesel and percent by weight. All canola oil and vegetable oil batches were mixed with 0.85% KOH by weight.

5.3 Heat of Combustion

Figure 5 shows that the heat of combustion (HOC) of diesel fuel is greater than that of biodiesel or feedstock oil. The HOC of pure vegetable oil is lower than the biodiesel made from the respective oil. Virgin vegetable oil (VVO) and waste vegetable oil (WVO) show upward trend in the HOC from the one-stage to the 80/20 two-stage process.

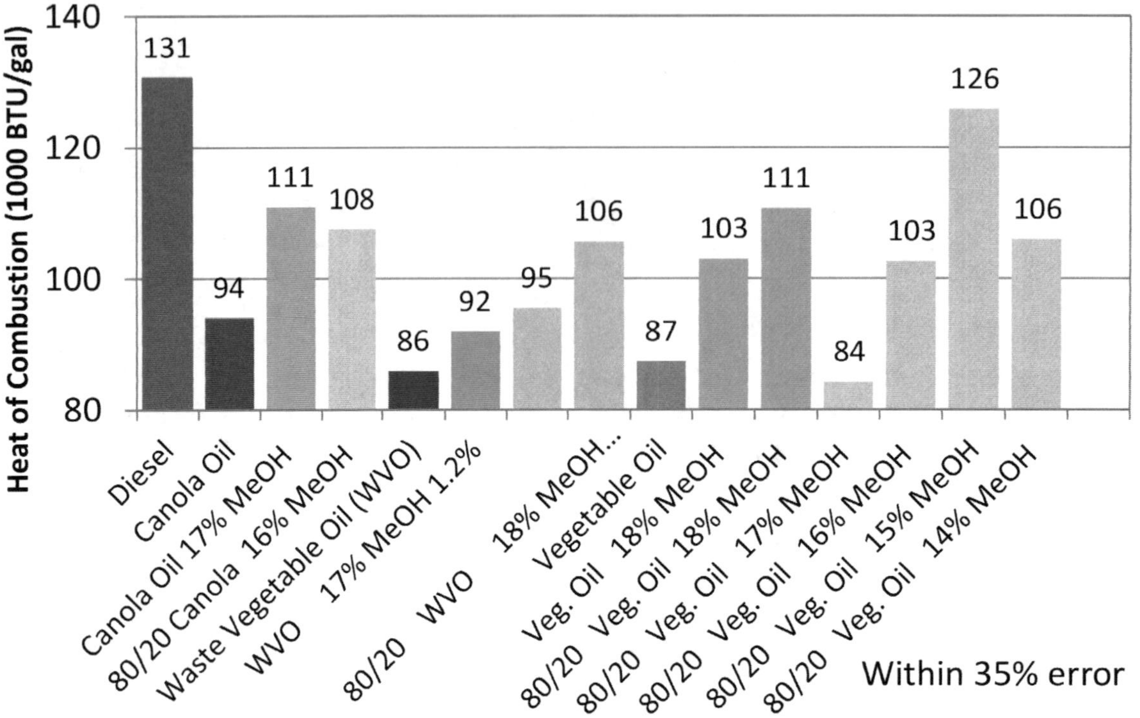

Figure 5. Heat of Combustion of Biodiesel (within 35% error), labeled with the reactants used to make the biodiesel and percent by weight. All canola oil and vegetable oil batches were mixed with 0.85% KOH by weight relative to oil.

5.4 Residual Soap

Figure 6 displays pictures from the soap tests. The 80/20 two-stage processes show less residual soap than the one-stage processes with the same reactants (up to 17% MeOH). The WVO 80/20 process using 0.5% more KOH formed more soap due to the increase in KOH. Also, the vegetable oil one-stage process using 25% MeOH had comparable separation as the 80/20 two-stage process using 18% MeOH/oil ratio.

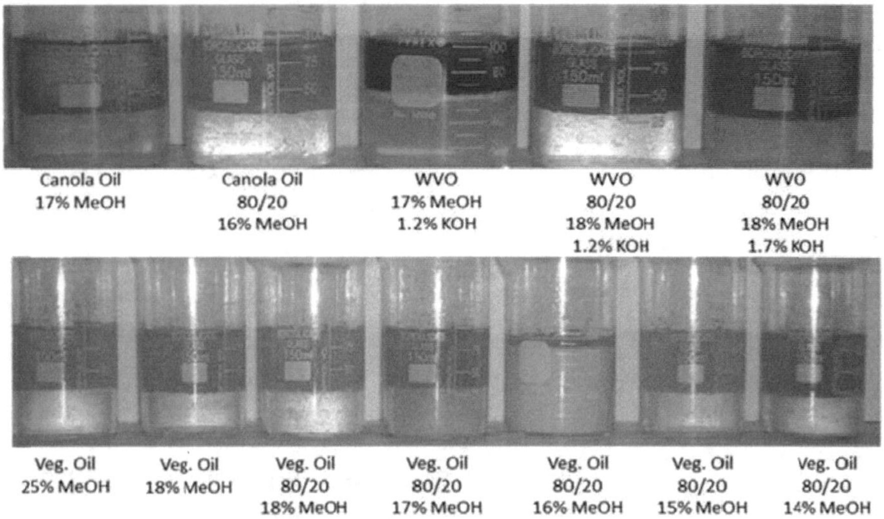

Figure 6: Leftover soap test, labeled with the reactants used to make the biodiesel and percent by weight. Canola oil and vegetable oil batches were mixed with 0.85% KOH by weight.

5.4 Effect of Methanol/oil Ratio

The effects of the methanol/oil ratio on the oil conversion, the viscosity, the heat of combustion and the residual soaps have been discussed in the previous sections. The specific results of the methanol/oil ratio are presented in this section. Figure 7 shows the effect in one-stage operation using vegetable oil. Figure 8 shows the 50/50 residual soaps test results for the same samples in Figure 7. Figures 7 and 8 show a drop in conversion with a decrease in the MeOH/Oil ratio from 25/100 to 18/100, and a slight increase in residual soap. The same effects are found in the base-base 80/20 two-stage process using vegetable oil, as shown in Figures 9 and 10. These results show that using a methanol/oil ratio of more than 18/100 in a two-stage process of vegetable oil is uneconomical because it will not improve the conversion.

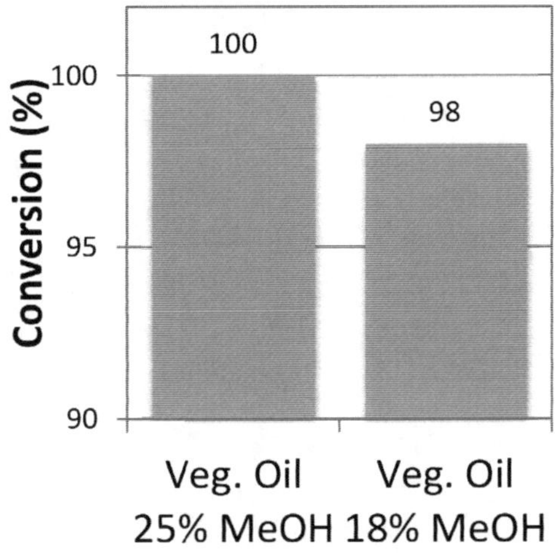

Figure 7. Effect of Methanol/Oil ratio on conversion of vegetable oil, one-stage process

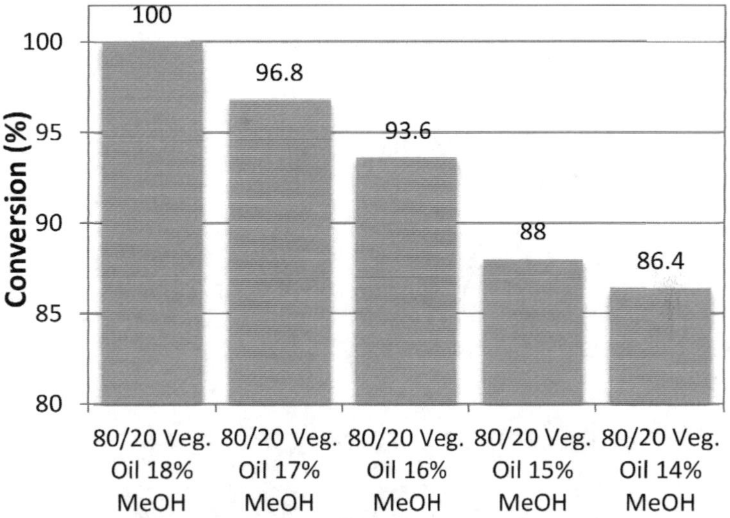

Figure 9. Effect of Methanol/Oil ratio on conversion of vegetable oil, base-base two-stage process

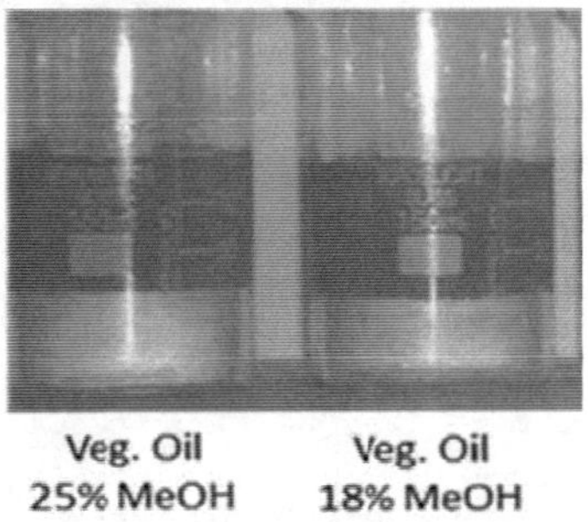

Figure 8. Effect of Methanol/Oil ratio on residual vegetable oil, one-stage process

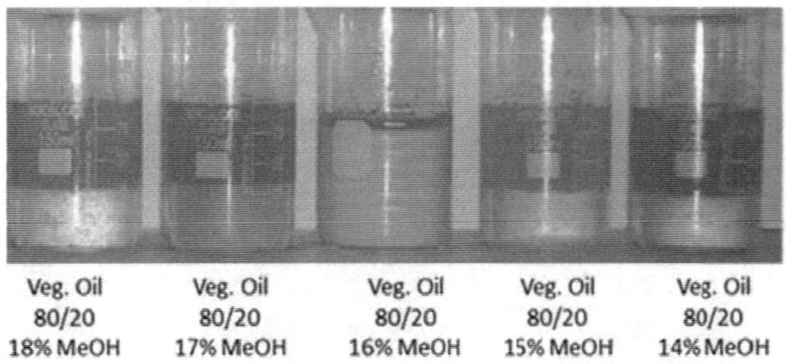

Figure 10. Effect of Methanol/Oil ratio on the residual vegetable oil, base-base two-stage process

5.5 Effect of Catalyst (KOH)/Oil Ratio

Figure 11 shows the effect of the amount of catalyst (KOH in the present study) on the conversion of waste vegetable oil (WVO). The data was obtained from the case of an 80/20 base-base two-stage run. Increasing the catalyst (KOH)/oil ratio from 1.2:100 to 1.7:100 caused an increase in the oil conversion. It also caused an increase in the heat of combustion, as shown in Figure 12. However, increasing the KOH also increased the residual soap, shown by Figure 13.Further experiments are needed to fully investigate the effect of KOH. Based on our results we recommend the use of WVO in a two-stage process using 18% MeOH/oil ratio and 1.2% KOH/oil ratio because of the lower residual soap relative to the case of 1.7% KOH/oil ratio even though this compromises the oil conversion and the heat of combustion.

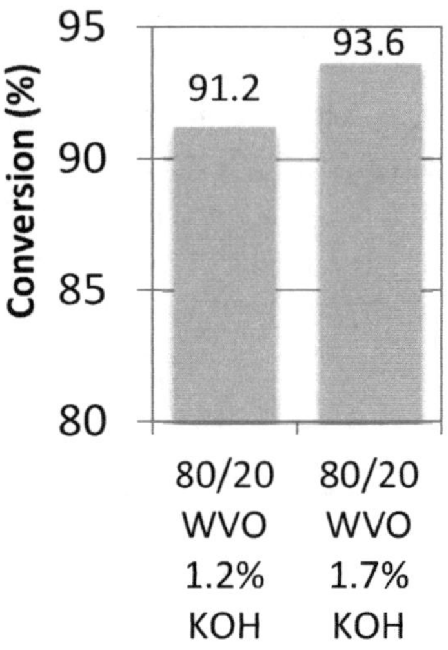

Figure 11. Effect of KOH/oil ratio on conversion of WVO base-base two-stage process

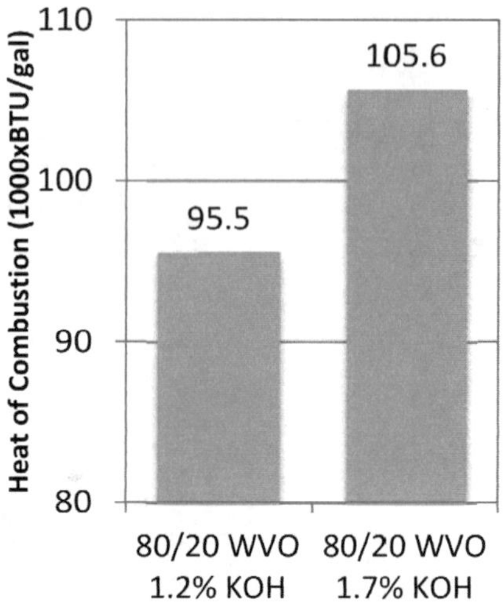

Figure 12. Effect of KOH/oil ratio on heat of combustion of biodiesel made from WVO base-base two-stage process

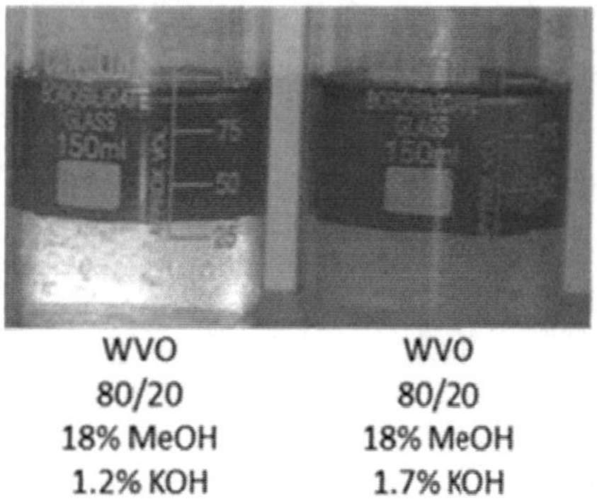

Figure 13. Effect of KOH/oil ratio on leftover soap in biodiesel made from WVO base-base two-stage process

5.5 One-Stage versus Two-Stage Operation

Figure 14 shows the layers of biodiesel and water after mixing. The one-stage process (left beaker) evidently has more residual soap than the two-stage process (right beaker). This demonstrates that the base-base two-stage process produces less soap than the one-stage process. Figure 15 indicates that for the same oil/ methanol ratio the two-stage process results in a higher oil conversion than the one-stage process.

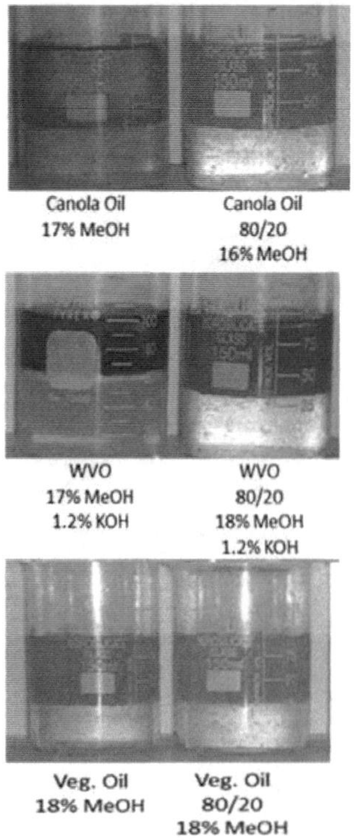

Figure 14. Effect of one-stage versus two-stage process on residual soap in biodiesel.

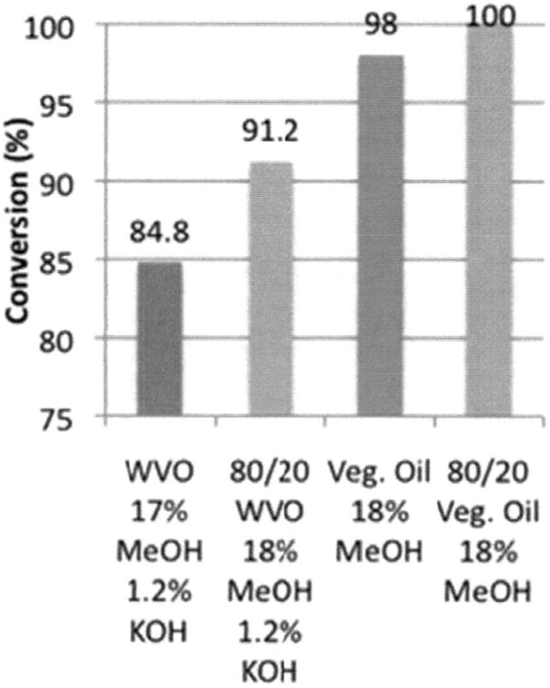

Figure 15. Effect of one-stage versus two-stage process on percent conversion of biodiesel made from WVO and Veg. Oil.

25% MeOH/oil ratio gave the same conversion (100%) as using 18% MeOH/oil ratio in an 80/20 batch. This is shown in Figure 16. It is evident from these results that the base-base two-stage process achieves the same oil feedstock conversion as the one-stage process, but with smaller methanol/oil ratio. This makes the base-base two-stage process more economically viable, as well as environmentally friendly.

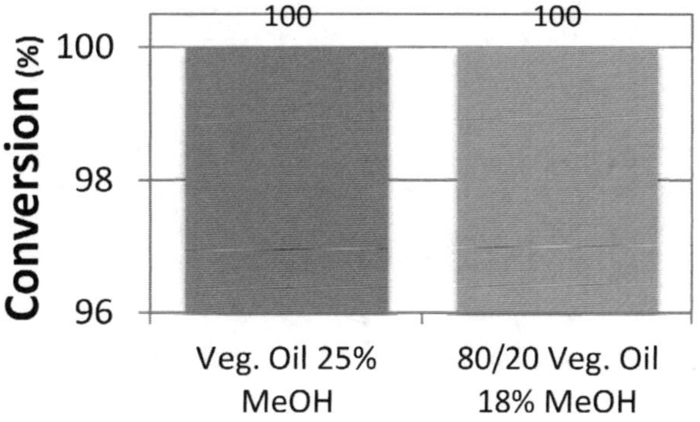

Figure 16. Effect of one-stage versus two-stage process on percent conversion of biodiesel made from Virgin Vegetable Oil (VVO).

6-Conclusions

It is evident from our study that an 80/20 base-base two-stage process is more economically viable because it uses less methanol while producing the same quality biodiesel as a one-stage process. Our results also show that using waste vegetable oil (WVO) is comparable to virgin vegetable oil. It is evident that an 80/20 base-base two-stage process using 18% MeOH/oil ratio and 1.2% KOH/oil ratio is the optimal choice for producing biodiesel. This is the least expensive option, and it had a 91% oil conversion and the biodiesel produced had a viscosity of 5.2 mm^2/s, a reasonable heat of combustion (92,000 BTU/gal) and no residua soap. While better quality biodiesel was produced using a virgin vegetable oil (such as the 80/20 vegetable oil batch using 18% MeOH/oil ratio), the waste vegetable oil is much less expensive and overall better for the environment, so we feel it is the optimal choice.

Acknowledgment

The authors would like to acknowledge the support of the UNH Hamel Center for Undergraduate Research, the UNH Dairy Research Center, Nancy Whitehouse and Sue Chase for use of their equipment.

References

- http://en.wikipedia.org/wiki/Biodiesel last Modified: February 16, 2010.
- Briggs, M., Pearson, J., Farag, I.H., (2004) "Biodiesel Processing," UNH Biodiesel Group. http://www.unh.edu/p2/biodiesel
- Farag, I.H. (2008), "Biodiesel: Challenges and new Fronts", Keynote Paper, International Biodiesel Workshop, National Research Center, March 23-25, 2008, Cairo, Egypt
- Knothe, Gerhard; Van Gerpen, Jon; Krahl, Jurgen (2005), ""The Biodiesel Handbook," AOCS Press, Champaign, Ill.
- Math, M.C. and G. Irfan (2007), "Optimization of Restaurant Waste Oil Methyl Ester Yield," Journal of Scientific and Industrial Research, 66, 772-776.
- Mulumba, N and I.H. Farag (2010), "Production of Biodiesel from Microalgae," M.S. Thesis, Chemical Engineering Dept., University of New Hampshire (UNH), Sep. 2010.
- Wilson, R. Grohs, J and I. H. Farag (2010) "Optimal Biodiesel Production Through Bench-top Processor Testing," Northeast Biomass Conference and Expo, August 4-6, 2010, Westin Copley Plaza, Boston.

(305d) Disruption of Microalgal Cells for Biodiesel Production

Ronald Halim, Michael Danquah and Paul Webley, Chemical Engineering, Monash University, Melbourne, Australia

The production of biodiesel from microalgae involves lengthy processing steps. Cell disruption is an integral part of this downstream protocol as it facilitates the release of the intracellular lipids that will be used for biodiesel production. This study examines different mechanical methods for the disruption of microalgae cells, including High Pressure Homogenization or HPH, ultrasonication and bead beating. The performance of cell disruption method was evaluated in terms of two key indicators: reduction in the intact cell counts per unit volume (C/C0) and reduction in the average diameters of microalgal colonies (D/D0). The most effective disruption was obtained using HPH (optimum C/C0 = 0.06, D/D0 = 0.35) followed by bead beating (optimum C/C0 = 0.55, D/D0 = 0.25). Even though ultrasonication failed to disrupt the microalgal cells (optimum C/C0 = 0.91), it still managed to disintegrate cellular agglomerations (optimum D/D0 = 0.23). It is hoped that the finding from this study can be used for future microalgal biofuel research.

See more of this Session: Sustainable Diesel Fuel From Renewable Resources
See more of this Group/Topical: Sustainable Engineering Forum

(305e) Production of Biodiesel From Alternative Feedstocks – a Bubble Reactor for Converting Oils Containing Free-Fatty-Acids to Biodiesel

Richard A. Cairncross[1], Cory Melick[2] and Minhazuddin Mohammed[2], (1)Chemical and Biological Engineering, Drexel University, Philadelphia, PA, (2)Chemical & Biological Engineering, Drexel University, Philadelphia, PA

Biodiesel is a diesel substitute that is typically produced by transesterification of refined vegetable oils (soybean oil in USA) with methanol; this chemistry is simple and effective with base catalysts and refined oils. However, crude vegetable oils and many alternative oil feedstocks contain significant (> 1%) concentrations of free-fatty-acids (FFA); for example waste fryer oil typically is 10-15% FFA and trap grease is >85% FFA. Free-fatty-acids cause low conversion and difficult separation with the traditional chemistry because they form soaps with base catalysts. There are several alternative techniques for producing biodiesel from high-FFA oils including two-step reactions, high pressure reactions, and supercritical reactions. In this research a two-phase bubble column reactor is used to esterify FFA at moderate temperatures (100-150°C) and atmospheric pressure. Methanol added to the reactor boils at these temperatures and produces bubbles that rise through the reactor. Despite having lower methanol concentrations in the reacting liquid phase, acceptable reaction rates and high conversions are achieved with this configuration. Our research has also shown that bubble reactors are effective at overcoming equilibrium limitations of the esterification reaction. Because the bubbles remove by-product water, the reactor is much more robust for feedstocks containing water, and conversions approaching 100% are achieved with up to 10% water in methanol. Similar results have been obtained for reactions with ethanol/water mixtures showing that it is not necessary to break the ethanol-water azeotrope when purifying ethanol for this reactor. A model of the reactor has been developed that accurately predicts trends of the experimental data and can be used to design improved reactors. Feasibility studies of full-scale production of biodiesel from trap-grease have been performed using the bubble reactor and two other reaction schemes – supercritical reactors and medium pressure CSTR's in series. The bubble reactor process has the highest return on investment and lowest greenhouse gas emissions.

See more of this Session: Sustainable Diesel Fuel From Renewable Resources
See more of this Group/Topical: Sustainable Engineering Forum

High-Pressure Viscosity of Bio-diesel from Soybean, Canola, and Coconut Oil

Andrew M. Duncan[1], Azita Ahosseini[1], Reece McHenry[1], Christopher D. Depcik[2,4], Susan M. Stagg-Williams[1-3,] and Aaron M. Scurto[1-3,*]*

[1] *Department of Chemical & Petroleum Engineering*
[2] *DOT-KS Transportation Research Institute (TRI)*
[3] *Center for Environmentally Beneficial Catalysis*
[4] *Department of Mechanical Engineering*

1. Introduction

Biodiesel, as an alternative fuel, has undergone significant research, development, and large-scale distribution over the past years. Biodiesel is a mixture of fatty acid esters and is commonly used as an alternative to, or blended with, No. 2 petroleum diesel. The utilization of biodiesel as an alternative fuel has been driven by the fact that, compared to petroleum based diesel, biodiesel is: domestically produced, renewable, has low sulfur and aromatic content, has low particulate emissions, has high energy content, and can be used in conventional Compression-ignition (diesel) engines with relatively little or no modification.[1]

Typical biodiesel is produced via the transesterification of fats and oils with an alcohol such as methanol. The composition of the fat or oil feedstock determines the fatty acid methyl ester profile and ultimately the chemical and physical properties of the resulting biodiesel. For example, biodiesel produced from soybean oil typical contains methyl ester profiles rich in oleate (C18:1) and linoleate (C18:2), while biodiesel from coconut oil is rich in laurate (C12:0) and myristerate (C14:0). These differences in compositions can affect properties such as cloud point, viscosity, density, oxidative stability, etc.[2, 3]

Current fuel injection systems for compression ignition (diesel) engines reach pressures approximately 1600 and 1400 bar for passenger car and commercial vehicle systems respectively.[4] Future systems are targeting even higher pressures in excess of 2000 bar while also expanding the possible number of injections per cycle from five to nine.[5] Under these pressures, the liquid viscosity of the fuel can increase up to approximately ten times atmospheric levels.[6] Upon injection of the fuel in the cylinder, the large de-pressurization of the fuel results in a significant gradient of the viscous properties of the fluid. A proper understanding of these values is therefore required for modeling the discharge coefficient and flow rate through the injector. These parameters are key in ensuring the proper rate shaping of the fuel injection spray in order to optimize performance while minimizing emissions.[7]

Generally speaking, biodiesel and biodiesel blends from a range of feedstocks are considered to have lower emissions of carbon monoxide (CO), total hydrocarbons (THC) and particulate matter (PM) than standard diesel fuel, but equal or slightly higher levels of nitrogen oxides (NO_x).[8] In some cases, recalibration of engines based on blending of biodiesel with other fuels has resulted in decreasing NO_x concentrations.[9, 10] Part of the change in NO_x relates to the inadvertent advancing of injection timing caused by the rapid transfer of the pressure wave from the fuel injection pump to the fuel injector causing it to open earlier. This pressure wave is a function of the properties of the fuel. Choi *et al.* attributed the change in fuel injection timing to the higher viscosity of biodiesel [11]; however, others conclude that viscosity does not affect timing and the bulk modulus, which effects the speed of sound as the main factor.[12-15] In any case, viscosity does affect the pressure dynamics within the fuel lines and during the spray break-up process inside the cylinder. This is a concern for engine manufacturers while designing the fuel injection systems for these engines under the range of potential pressure and temperature conditions (up to 3500 bar and 273.15 K to 373.15 K).[16, 17]

2. Experimental High-pressure and ambient viscosity measurement

A Cambridge Applied Systems (currently Cambridge Viscosity, Inc.) high pressure viscometer was utilized for these measurements (ViscoPro 2000 System 4- SPL-440 with Viscolab software). The apparatus has been described in high detail in Ahosseini and Scurto[18], but an overview will be given here. The apparatus consists of a temperature-controlled oven (±0.1 K) which houses the high-pressure viscometer sensor, a precision pressure transducer (PT), and a resistance temperature detector (RTD) (±0.05 K). The viscometer is connected to a manual high-pressure syringe pump purchased from High Pressure Equipment Company (HIP) (Model No. 50-575-30; 30,000 psi, capacity of 10 cm^3 with PolyPak) to pressurize the diesel samples. The viscometer utilizes the principles of annular flow around an axially oscillating piston[19] and has recently been given an ASTM certification.[20] The sensor is capable of measurements from 0.2 to 10,000 cP (mPa·s), at a maximum pressure of 137.9 MPa and in a temperature range of 233.15 K to 463.15 K. The diameter of the inner chamber is 0.314 inches. Various size pistons are used according the nominal viscosity range suggested by the manufacturer: 0.25-5 cP (0.3085 inch diameter), 1-20 cP (0.3055 inch diameter) and 2.5-50 cP (0.3025 inch diameter).

The apparatus was used to find the viscosity of soybean, Vistive (Monsanto's low-linlolenic soybean), canola, used canola, coconut-based biodiesels, and #2 ultra low sulfur diesel fuel. The viscosities were measured over a temperature range of 283.15 K to 373.15 K and a pressure range of 0.1 MPa to 130 MPa.

2. Correlation and Results

Various theoretical models and empirical expressions can be found in the literature that can be used to represent the viscosity of liquids (η) as functions of pressure and temperature. Litovitz[21,22] has suggested an empirical equation, which has been used at a single pressure over wide temperature ranges

$$\eta = A\exp(B / RT^3)$$

where R is the gas constant and A and B are fitted parameters.

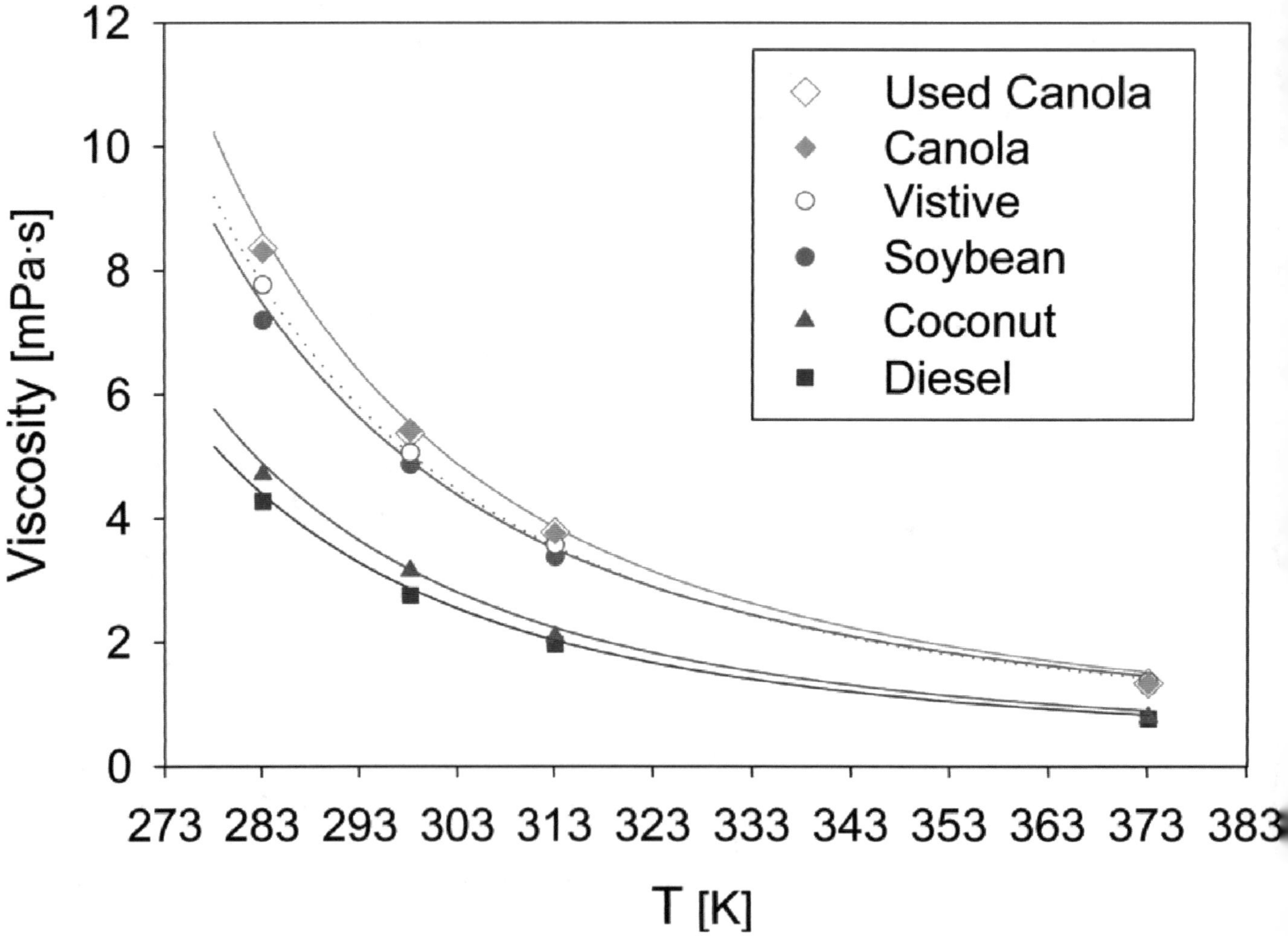

Figure 1. Ambient-pressure viscosity with the temperature for the number 2 diesel and biodiesel samples. The lines represent the correlation from the Litovitz equation from parameters regressed from all data.

Figure 1 above shows the results of the regression of the ambient pressure viscosities of each fuel source tested in this investigation to the Litovitz parameters. The data, correlate very well to the Litovitz equation—represented by the solid or dashed lines. The soybean and canola fuels had very similar behavior, with a very sharp increase in viscosity with decreasing temperature. The coconut biodiesel had very similar viscous behavior to the petroleum-based fuel.

After the ambient pressure values were fit to a Litovitz equation, the high-pressure data were fit to a Tait equation. The Tait equation is well-known to represent the pressure volume-temperature relationship of many liquids.[23,24]

$$\ln\left(\frac{\eta_P}{\eta_o}\right) = E\ln\left(\frac{D+P}{D+0.1}\right)$$

where η_P and η_0 are the viscosities at a pressure P and at 0.1 MPa, respectively. This equation contains only two fitted parameters and has yielded good correlation with experimental data as shown by Kashiwagi and Makita 4 for aromatic hydrocarbons and cyclohexane up to 110 MPa.

This investigation will utilize a hybrid Tait-Litovitz equation at elevated pressures for the viscosity data for a series of biodiesels:

$$\eta_P = A\exp(B'/T^3)((D+P)/(D+0.1))^E$$
$$B' = B/R$$

The Tait-Litovitz hybrid equation was applied and regressed for each of the six fuels tested. The results of this hybrid equation can be seen by the solid lines in Figures 2 and 3. The accuracy of the correlation in these figures was typical of all fuel samples.

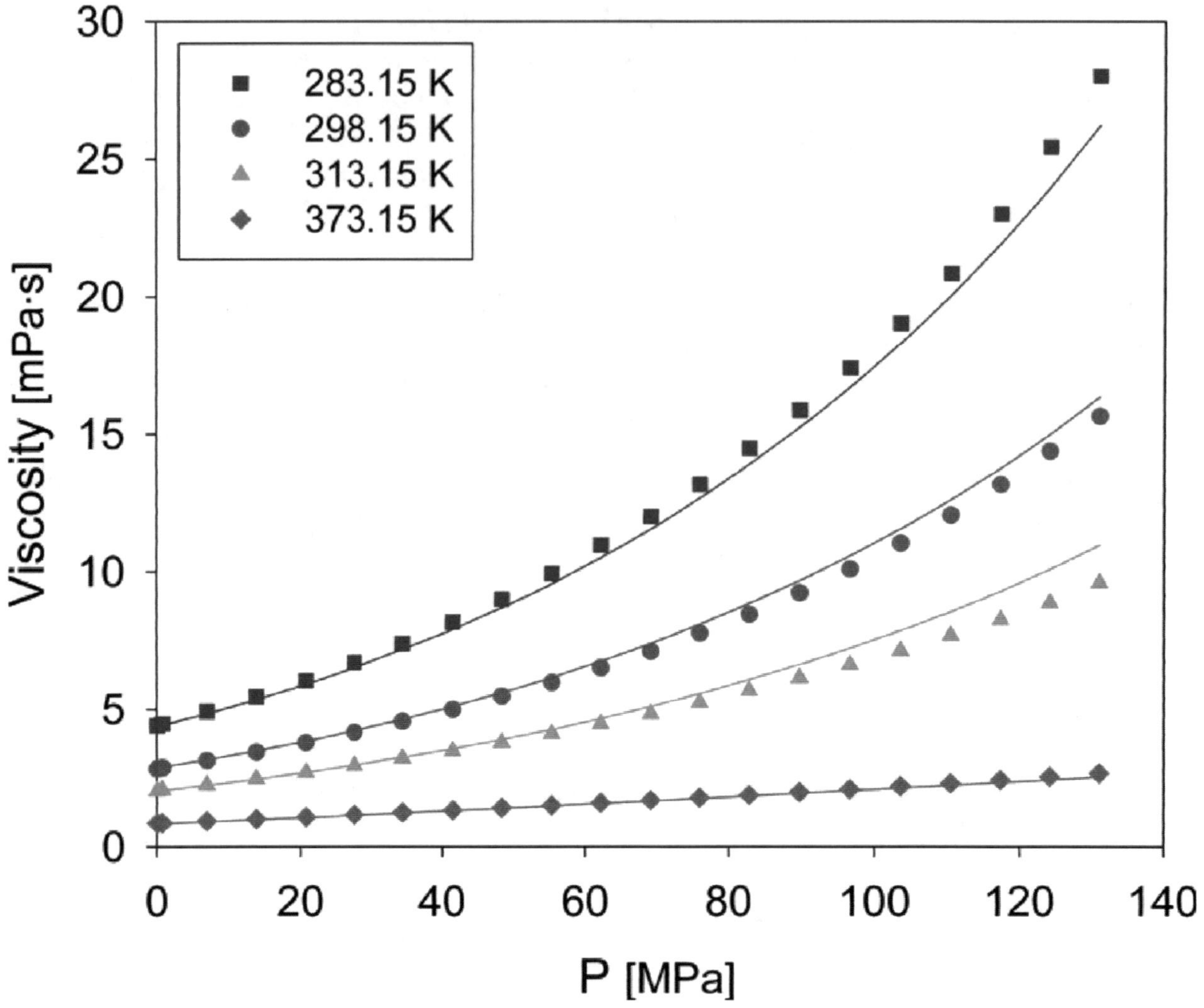

Figure 2. Viscosity with pressure for petroleum-based diesel (number 2) at four isotherms: 283.15, 298.15, 313.15, and 373.15 K. Lines represent the Tait-Litovitz equation from temperature dependent parameters regressed from all four isotherms.

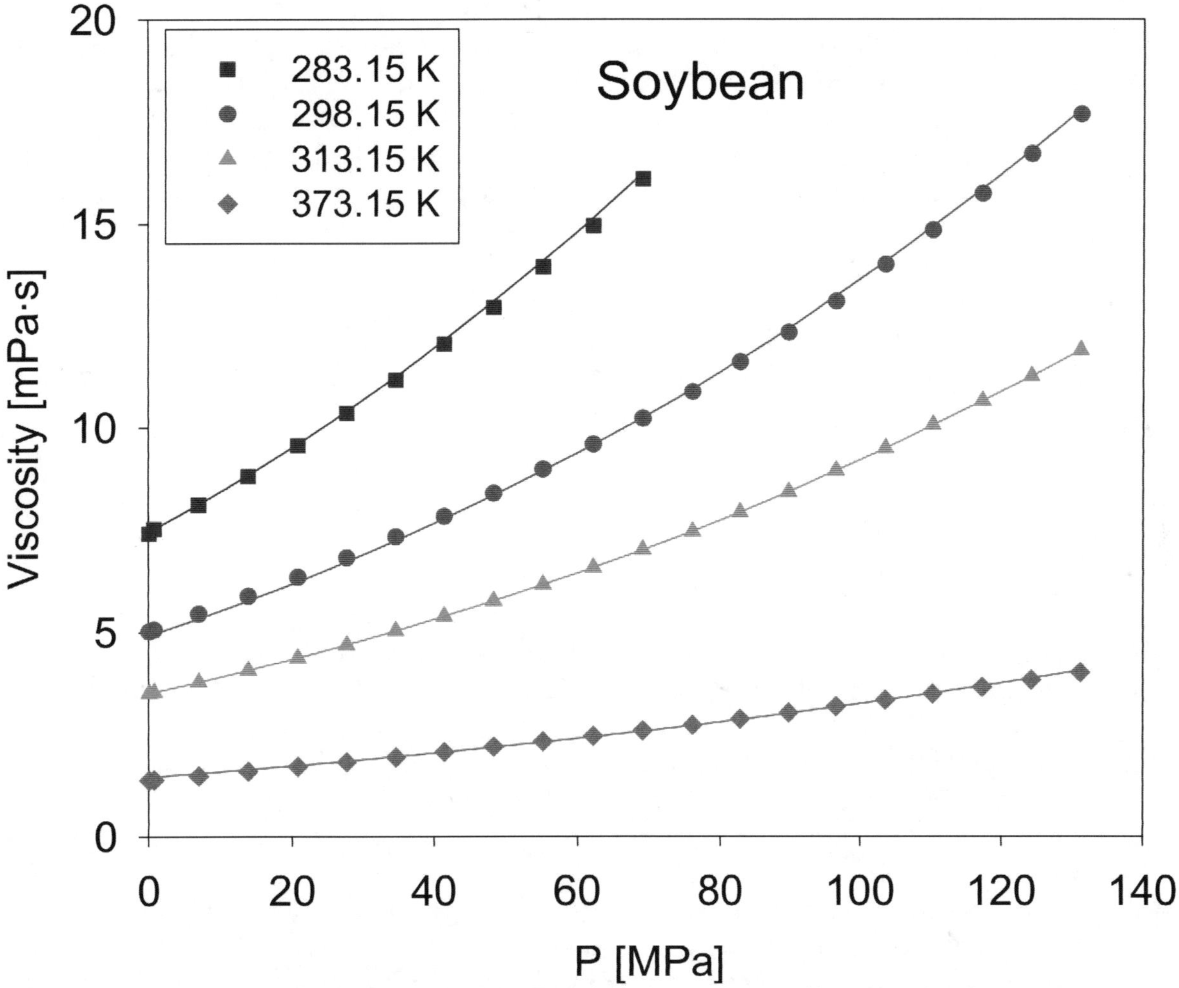

Figure 3. Viscosity with pressure for biodiesel from soybean oil at four isotherms: 283.15, 298.15, 313.15, and 373.15 K.

While the viscosities of the soybean biodiesel shown in Figure 3 is higher compared to the diesel isotherms shown in Figure 2, the percentage change for each respective isotherm was found to be greater for the diesel fuel. This percentage change was typical for the biodiesels compared to the petroleum-based fuel.

Conclusions

The viscosity of a series of biodiesels from various biorenewable sources was measured to determine the effect of high pressures (up to 130 MPa) and temperatures from 283.15 to 373.15 K. These data are the first to explore comprehensively the effect of temperature, pressure, and vegetal source on the viscosity of biodiesel. The viscosity change that a biodiesel would experience in a fuel injector would be up to several hundred percent greater than the ambient pressure viscosity. While differences in ambient pressure viscosity exist among the different biodiesels, the effect of pressure is relatively similar. The viscosity data have been correlated very well with a hybrid Tait-Litovitz equation with average absolute relative deviation percentages (AARD%) from 0.9% for coconut derived biodiesel to 2.1% for canola derived biodiesel. The correlation could be used for a wide range of simulation and design.

References:

1. Graboski, M.; McCormick, R., Combustion of fat and vegetable oil derived fuels in diesel engines. *Prog. Energy Comb. Sci.* **1998,** 24, (2), 125-164.
2. Rodrigues, J.; Cardoso, F.; Lachter, E.; Estevão, L.; Lima, E.; Nascimento, R., Correlating chemical structure and physical properties of vegetable oil esters. *J. Am. Oil. Chem. Soc.* **2006,** 83, (4), 353-357.
3. Ramos, M.; Fernandez, C.; Casas, A.; Rodriguez, L.; Perez, A., Influence of fatty acid composition of raw materials on biodiesel properties. *Bioresource Tech.* **2009,** 100, (1), 261-268.
4. Bosch, *Bosch Automotive Handbook.* 7th ed.; Robert Bosch GmbH: 2007.
5. Tomishima, H.; Matsumoto, T.; Okl, M.; Nagata, K., The Advanced Diesel Common Rail System for Achieving a Good Balance Between Ecology and Economy. *SAE Paper 2008-28-0017* **2008**.
6. Giannadakis, E.; Gavaises, M.; Theodorakakos, A., The Influence of Variable Fuel Properties in High-Pressure Diesel Injectors. *SAE Paper 2009-01-0832* **2009**.
7. Schmidt, D. P.; Corradini, M. L., The internal flow of diesel fuel injector nozzles: a review. *Int. J. Engine Res.* **2001,** 2, (1), 1-22.
8. Agency, E. P. *A Comprehensive Analysis of Biodiesel Impacts on Exhaust Emissions; Draft Technical Report*; EPA420-P-02-001; Office of Transportation and Air Quality: Washington, DC, 2002.
9. Kegl, B.; Pehan, S., Influence of biodiesel on injection, fuel spray, and engine characteristics. *Therm. Sci.* **2008,** 12, (2), 171-182.
10. Lu, X. C.; Ma, J. J.; Ji, L. B.; Huang, Z., Simultaneous reduction of NOx emission and smoke opacity of biodiesel-fueled engines by port injection of ethanol. *Fuel* **2008,** 87, (7), 1289-1296.
11. Choi, C. Y.; Bower, G. R.; Reitz, R. D., Effect of Biodiesel Blended Fuels and Multiple Injections on D.I. Diesel Engines. *SAE Paper 970218* **1997**.

12. Arcoumanis, C.; Gavaises, M.; Yamanishi, M.; Oiwa, J., Application of a FIE computer model to an in-line pump-based injection system for diesel engines. *SAE Paper 970348* **1997**.
13. Rakopoulos, C. D.; Hountalas, D. T., A simulation analysis of a DI diesel engine fuel injection system fitted with a constant pressure valve. *Energy Conv. Managm.* **1996,** 37, (2), 135-150.
14. Szybist, J. P.; Boehman, A. L., Behavior of a Diesel Injection System with Biodiesel Fuel. *SAE Paper 2003-01-1039* **2003**.
15. Szybist, J. P.; McFarlane, J.; Bunting, B. G., Comparison of Simulated and Experimental Combustion of Biodiesel Blends in a Single Cylinder Diesel HCCI Engine. *SAE Paper 2007-01-4010* **2007**.
16. Gault, R. In *Biodiesel Outlook - An Engine Manufacturer's Perspective*, Diesel Engine-Efficiency and Emissions Research (DEER), Dearborn, MI, August 7, 2008; Dearborn, MI, 2008.
17. Beyer, T. The Measurement of Diesel Fuel Properties. Georgia Institute of Technology, Atlanta, GA, 1998.
18. Ahosseini, A.; Scurto, A. M., Viscosity of Imidazolium-Based Ionic Liquids at Elevated Pressures: Cation and Anion Effects. *Int. J. Thermophys.* **2008,** 29, 1222-1243.
19. Bird, R. B.; Stewart, W. E.; Lightfoot, E. N., *Transport Phenomena*. Wiley: New York, 1960.
20. *ASTM D7483 - 08 Determination of Dynamic and Kinematic Viscosity of Liquids by Oscillating Piston Viscometer.*; 2008.
21. Litovitz, T. A., Temperature Dependence of the Viscosity of Associated Liquids. *J. Chem. Phys.* **1952,** 20, 1088-1089.
22. Litovitz, T. A., Temperature Dependence of the Viscosity of Associated Liquids. *J. Chem. Phys.* **1952,** 20, 1980.
23. Dymond, J. H.; Malhotra, R., The Tait equation: 100 years on. *Int. J. Thermophys.* **1988,** 9, (6), 941-951.
24. Kashiwagi, H.; Makita, T., Viscosity of Twelve Hydrocarbon Liquids in the Temperature Range 298-348 K at Pressures up to 110 MPa. *Int. J. Thermophys.* **1982,** 3, (4), 289-305.

(305g) Activated Sludge Oil: Identification and Characterization of Components

Patrisha J. Pham, Rafael Hernandez, Emmanuel D. Revellame and William T. French, Dave C. Swalm School of Chemical Engineering, Mississippi State University, Mississippi State, MS

The demand for alternative sources to petroleum is growing. Recently, sewage sludge as a raw material for biodiesel production has been demonstrated in our group[1-2]. In this present study, activated sludge oil (ASO) extracted from activated sludge was characterized to provide information for activated sludge oil refinement. The refinement process aims for activated sludge to be considered as a feedstock for biodiesel production and as a possible source of value-added products. Fractionation of the ASO extract with solvents of varying polarities gave five ASO solvent fractions. The % composition of the ASO hexane (HX), chloroform ($CHCl_3$), acetone (ACE), methanol (MeOH) and methanol/ammonium hydroxide (HPC) fractions were 7.0, 26.0, 32.0, 31.0 and 4 %, respectively. The ASO HX fraction was found to contain a homologous series of n-alkanes and squalene. The ASO $CHCl_3$ fraction was found to contain free fatty acids (C14:0-C20:0), tocopherols, monoglyceride, free sterols, diglyceride, wax esters, steryl esters and triglycerides. The ASO ACE fraction contained two unknown glycolipids. Approximately 1.3% of the ASO ACE fraction is carbohydrates with glucose (57%) as the main glycosyl moiety present. Majority of the ASO ACE fraction are sterols or steryl derivatives. The ASO MeOH fraction was 35 % phospholipids based on ^{31}P NMR. Phosphatidylethanolamine at 18% was the major phospholipid component detected. Conversion into fatty acid methyl esters was at 38, 18 and 15 % for ASO $CHCl_3$, ACE and MeOH fractions, respectively. The fatty acid profiles show C14:0, C16:0, C16:1, C18:0 and C18:1 as the dominant fatty acids in all three major ASO solvent fractions. However, longer and unsaturated FAMEs were observed in the ASO MeOH fraction. ASO extract (13%) of dry AS is a source of biodiesel but simultaneously can be refined to obtain value-added products such as squalene and tocopherols. Although the utilization of activated sludge oil can reduce the cost for feedstocks, the actual process requires additional expense for refinement but may be compensated by the separation and purification of high value-added compounds of nutraceutical or biologically importance.

[1]Dufreche S, Hernandez R, French T, Sparks D, Zappi M, Alley E. 2007 *J Amer Oil Chem Soc* 84, 181-187.

[2]Mondala A, Liang K, Toghiani H, Hernandez R, French T. 2009 *Biores Technol* 100, 1203-1210.

See more of this Session: Sustainable Diesel Fuel From Renewable Resources
See more of this Group/Topical: Sustainable Engineering Forum

(317a) Effect of Inorganics in Biomass On Rhodium Catalysts for Millisecond Autothermal Reforming

Reetam Chakrabarti and Lanny D Schmidt, Chemical Engineering and Materials Science, University of Minnesota, Minneapolis, MN

Biomass can be upgraded to fuels and chemicals by gasification, which produces synthesis gas (commonly referred to as syngas, a mixture of carbon monoxide and hydrogen). Syngas can be converted to fuels and chemicals by the Fischer Tropsch process. Recently, Schmidt et al. have demonstrated that lignocellulosic biomass can be gasified autothermally by a technique called 'Reactive Flash Volatilization' to a product stream containing equilibrium selectivities to syngas.(1-3) This process takes place on Rhodium-based catalysts at high temperatures (~600-1000° C) with a residence time on the order of milliseconds. Biomass contains inorganic constituents such as Ca, Na, K, Mg, S, Si, Cl, P which may be present in different forms such as oxides, sulfates, carbonates, sulfates, silicates, phosphates etc.(4) Reactive Flash Volatilization has been demonstrated using feeds which contain low or negligible concentration of inorganics. These inorganics can affect the catalyst activity by acting as promoters or poisons, hence understanding their effect on Rhodium catalysts for catalytic gasification is necessary to design appropriate biomass processing strategies. These inorganics were added to the Rhodium catalyst through precursors and the effect on autothermal reforming was studied by studying the change in reactor performance, product distribution and catalyst microstructure. Inorganic constituents such as Si, P, S acted as strong poisons for the Rh-based catalyst. SEM, XRD, EDS and chemisorption studies were carried out to evaluate the change in catalyst microstrucutre.

REFERENCES

1)Dauenhauer PJ, Dreyer BJ, Degenstein NJ, Schmidt LD. Millisecond Reforming of Solid Biomass for Sustainable Fuels. Angewandte Chemie International Edition. 2007;46:5864-5867. 2)Colby JL, Dauenhauer PJ, Schmidt LD. Millisecond Autothermal Steam Reforming of Cellulose for Synthetic Biofuels by Reactive Flash Volatilization. Green Chemistry. 2008;10:773-783. 3)Salge JR, Dreyer BJ, Dauenhauer PJ, Schmidt LD. Renewable Hydrogen from Nonvolatile Fuels by Reactive Flash Volatilization. Science. 2006;314:801-804. 4)Raveendran K, Ganesh A, Khilar KC. Influence of mineral matter on biomass pyrolysis characteristics. Fuel. 1995;74:1812-1822.

See more of this Session: Catalytic Processing of Fossil and Biorenewable Feedstocks: Fuels III
See more of this Group/Topical: Catalysis and Reaction Engineering Division

(317b) Reverse Water Gas Shift Chemistry and Catalytic Oxidative Reforming of Volatile Organics Derived From Boiling Intermediate Cellulose

Paul J. Dauenhauer, Department of Chemical Engineering, University of Massachusetts, Amherst, Amherst, MA

Advanced catalytic process technologies developed for direct conversion of biomass to synthetic chemicals and fuels will require innovative techniques to integrate fast pyrolysis chemistry with high temperature inorganic catalysts. This combination will permit tunable chemistry that minimizes char formation and directs pyrolysis to specific chemical compounds favorable for the production of fuels. Using catalytic exothermic oxidation integrated with endothermic pyrolysis, cellulose-derived volatile oxygenates can be reformed to synthesis gas in a single, fixed-bed reactor. High speed photography reveals the thermal mechanism of cellulose decomposition through an intermediate liquid state that exhibits violent boiling on high temperature catalytic surfaces. The intermediate liquid is observed to wet the surface maximizing thermal and chemical interaction between the pyrolyzing biomass and the oxidation catalyst. Additionally, droplets are observed to form volatile organics through rapid boiling. Co-feeding of carbon dioxide and methane are demonstrated to simultaneously favor reverse water gas shift chemistry and hydrogen generation thereby producing complete conversion of biomass to carbon monoxide.

See more of this Session: Catalytic Processing of Fossil and Biorenewable Feedstocks: Fuels III
See more of this Group/Topical: Catalysis and Reaction Engineering Division

(317c) Biomass Conversion to Hydrogen-Rich Product Gas Via Alkaline Hydrothermal Treatment

Thomas Ferguson, Earth and Environmental Engineering and Chemical Engineering, Columbia University, New York, NY and Ah-Hyung Alissa Park, Earth and Environmental Engineering, Columbia University, New York, NY

Due to issues of environmental sustainability associated with anthropogenic carbon emission and energy security, there is strong interest to develop a new generation of energy conversion technologies that utilize domestic energy sources. As a feedstock, biomass represents a major potential for the sustainable generation of energy worldwide because it is a widespread and carbon neutral resource. Gasification and pyrolysis are the two major technologies that have been widely investigated for energy production from biomass. However, biomass has a much lower energy density as compared to fossil fuels. For this reason, in order for energy generation schemes involving biomass to be viable, they must be implemented within close proximity to the feedstock. Therefore, a distributed energy generation system would be an ideal addition to the biomass-to-energy technologies. In light of this, the present study explores a method to produce high-purity hydrogen from biomass for a fuel cell stack (e.g. PEM fuel cell). As biomass is reacted with a hydroxide (e.g., NaOH), the majority of carbon is separated as solid carbonate while the product gas is hydrogen rich. One of the main benefits of this technology is that chemical conversion can be achieved at temperatures (250 – 300 °C) and pressures (ambient pressure) significantly lower than gasification or pyrolysis. These moderate reaction conditions would make the design of a compact reactor for a distributed energy generation system feasible. Glucose and cellulose are chosen as model biomass compounds. Each is mixed with solid NaOH prior to each experimental study. The gaseous and solid products are analyzed using analytical tools such as GC-MS and NMR, respectively. TGA/DTA data are also obtained and compared to thermodynamic calculations done on these systems. Parameters including reactant ratios of biomass to sodium hydroxide, reactor temperature, heating rate in the reactor, and vapor conditions for the cellulose system are evaluated for their effects on the conversion and selectivity for hydrogen production.

See more of this Session: Catalytic Processing of Fossil and Biorenewable Feedstocks: Fuels III
See more of this Group/Topical: Catalysis and Reaction Engineering Division

(317d) Microkinetic Modeling of Ethylene Glycol Decomposition On Pt and Ni-Pt Bimetallic Catalysts

Michael Salciccioli, Chemical Engineering, University of Delaware, Newark, DE and Dion Vlachos, Director of Center for Catalytic Science and Technology (CCST), University of Delaware, Newark, DE

There has been increasing interest in the utilization of biomass for renewable energy and chemical production. Biomass, which is made up of oxygenated hydrocarbon building blocks, can be more efficiently utilized for chemical and energy production through the use of small-scale catalytic reforming technologies. Small-scale reactors can increase localization of synthesis gas production and increase reaction rates over current large-scale and enzymatic technologies. Distributed synthesis gas will allow for local production of methanol, ammonia, liquid fuels (Fischer-Tropsch process) and hydrogen for fuel cell applications. Recent technological successes in catalytic reforming of biomass derived oxygenates have shown high reaction rates and high product selectivity towards hydrogen for smaller oxygenates such as methanol, ethylene glycol and glycerol. Future improvements and extensions of this technology to larger oxygenates will require rational design for reactor and catalyst optimization.

Microkinetic models provide necessary insights into surface chemistry that are useful for reactor design and catalyst optimization. The focus of this study is the development of a thermodynamically consistent microkinetic model that describes the detailed mechanisms of thermal decomposition and reforming of ethylene glycol on a Pt(111) surface. A hierarchical methodology was implemented for refinement of sensitive kinetic parameters via DFT and allows for the inclusion of adsorbate-adsorbate interactions for abundant surface species. The surface mechanism describing oxygenate chemistry on Pt(111) includes over 100 reversible elementary reactions of the following classifications: adsorption/desorption, hydrogen extraction, carbon-carbon bond cleavage, hydrogen oxidation, carbon monoxide oxidation, hydrogen and carbon monoxide coupling reactions via the carboxyl intermediate, and oxidative dehydrogenation. Sensitivity analysis on rate constants shows that early dehydrogenation reactions are kinetically important.

Additionally, previous experimental research has shown that Ni/Pt bimetallic catalysts have increased activity toward oxygenate reforming compared to that of the parent metals. This work also explores fundamental descriptors for catalytic activity which explain the increased activity of the Ni/Pt bimetallic catalyst, and offers insight into future catalyst design for oxygenate reforming.

This presentation will focus on key model results and interpretation to reactor design applications and extensions of the model to other chemistry sets. A key focus will be on the mechanistic differences of the reforming of ethylene glycol on platinum compared to the reforming of ethylene glycol on the Ni/Pt bimetallic catalyst.

(317e) Yield of Dimethyl Ether and Gasoline as a Function of Size of ZSM-5 Crystals

Ali A. Rownaghi, Caroline Häggström, Olov Öhrman and Jonas Hedlund, Chemical Engineering, Luleå University of Technology, Luleå, Sweden

In this work we have studied different acidic zeolitic solid catalysts ZSM-5 with different crystal size, measuring their physical-chemical properties and correlating this with the activity, selectivity at different process condition during the conversion of methanol to dimethyl ether (DME) and gasoline boiling range hydrocarbons. The catalytic properties of these catalysts depended mainly on the crystal size and surface properties. Methanol feed in the DME and gasoline process is derived from black liquor-based syngas. The methanol to DME and gasoline process involves the conversion of black liquor syngas to crude methanol and methanol to DME and gasoline.

See more of this Session: Catalytic Processing of Fossil and Biorenewable Feedstocks: Fuels III
See more of this Group/Topical: Catalysis and Reaction Engineering Division

(317f) Catalytic Conversion of Biogas to Transportation Fuels and Chemicals

Sagar B. Gadewar[1], Zachary Komon[1], Eric W. McFarland[2], Peter Stoimenov[1] and Aihua Zhang[1], (1) GRT, Inc., Santa Barbara, CA, (2)Chemical Engineering, University of California, Santa Barbara, Santa Barbara, CA

Efficient transformation of methane to more valuable chemical and fuel products continues to be an important intellectual challenge with strong economic rationale. A big driver is the abundance of methane both from fossil resources and renewable biological sources. Selective partial oxidation of methane and other alkanes can be performed effectively using bromine for C-H bond activation. The overall conversion occurs in a three-step process involving, i) bromination of methane, ii) catalytic reaction of alkyl bromide intermediates to form hydrocarbon products, and iii) recovery and recycle of bromine. The process does not utilize a synthesis gas intermediate and is tolerant to common methane contaminants including high concentrations of carbon dioxide present in biogas.

Activation of methane using bromine generates methyl bromide as the key intermediate. It is critical to develop catalysts for converting methyl bromide into higher hydrocarbons that can be used in gasoline and jet fuel. There is a remarkable similarity between methyl bromide and methanol in terms of reactivity and conversion to higher hydrocarbons over suitable catalysts. Conversion of methanol to olefins and also to gasoline was developed and commercialized by Mobil. A big advantage of the GRT technology is the ease and cost effectiveness of producing methyl bromide compared to methanol production. High-throughput catalyst screening was employed to identify catalyst leads to produce range of products including olefins, gasoline, and jet fuel components. Catalysts screened were based on ZSM-5, ferrierites, mordenites, MAPO, ALPO, SAPO and other zeolites. ZSM-5 based catalysts were the top performers in the production of aromatics and gasoline range components, while SAPO based catalysts were the top performers for olefins production. Novel methods for catalyst modification were used to improve product yields and reduce coke formation. Operating conditions such as temperature, pressure and residence time, have a big impact on the product distribution and coke formation.

This talk will provide an overview of the process to convert methane derived from biomass or fossil sources to chemicals and transportation fuels. The focus of the talk will be on catalyst screening for converting the methyl bromide intermediate into hydrocarbon products.

References:

1. I. Lorkovic, M. Noy, M. Weiss, J. Sherman, E.W. McFarland, G.D. Stucky, P.C. Ford, Chem. Commun. (2004) 566.

2. I. Lorkovic, M.L. Noy,W.A. Schenck, C. Belon, M.Weiss, S. Sun, J.H. Sherman, E.W. McFarland, G. D. Stucky, P.C. Ford, Catal. Today 98 (2004) 589.

3. A. Breed, M.F. Doherty, S. Gadewar, P. Grosso, I.M. Lorkovic, E.W. McFarland, M.J. Weiss, Catal. Today 106 (2005) 301.

See more of this Session: Catalytic Processing of Fossil and Biorenewable Feedstocks: Fuels III
See more of this Group/Topical: Catalysis and Reaction Engineering Division

(317g) Potassium Promoted CoMo Sulfide as a Catalyst for the Synthesis of Higher Alcohols From Syngas

Juan J. Bravo Suarez, Bala Subramaniam and R. V. Chaudhari, Center for Environmentally Beneficial Catalysis, Chemical & Petroleum Engineering Department, University of Kansas, Lawrence, KS

Cellulosic biofuels (e.g., ethanol) are receiving tremendous academic and industrial attention due to potential economic rewards associated with an almost exponential increase in renewable biofuels demand for the next decade, as mandated by the Energy Independence and Security Act of 2007. Catalytic conversion of biomass derived syngas to higher alcohols is one attractive alternative for production of biofuels; however, active and selective catalysts such as promoted metal sulfides and rhodium based catalysts still lack appropriate performance (CO conversion, alcohols STY, etc.) for a commercial process. Better understanding of reaction mechanism and of the influence of process variables should assist in the design of improved catalysts. In this work, unsupported K-$CoMoS_x$ was investigated as a catalyst for the synthesis of ethanol and higher alcohols from syngas. The catalyst was prepared by coprecipitation methods and tested in a fixed bed reactor with online analysis of products which mainly consisted of C_1-C_5 hydrocarbons and linear alcohols. Carbon balances closed to 100 ± 5%. Under typical reaction conditions 330 °C, 82 bar, 4000 scc g_{cat}^{-1} h^{-1}, and $H_2/CO=1$ alcohols space time yield was about 200 g kg_{cat}^{-1} h^{-1}. The major byproduct is CO_2 which is mainly formed via water gas shift reaction ($CO + H_2O = CO_2 + H_2$). Results suggest that water gas shift reaction is quasi-equilibrated under the studied reaction conditions and acts as an internal chemical control for oxygen rejection as CO_2. We have also observed the presence of oscillations in the rates of formation of the products which appears to be directly linked to an oscillatory behavior of the water gas shift reaction on the K-$CoMoS_x$ catalyst.

See more of this Session: Catalytic Processing of Fossil and Biorenewable Feedstocks: Fuels III
See more of this Group/Topical: Catalysis and Reaction Engineering Division

(327a) Optimization of Biodiesel Production Under Uncertainty

Sheraz Abbasi, Bioengineering, University of Illinois at Chicago, Chicago, IL and Urmila D. Diwekar, Vishwamitra Research Institute, Clarendon Hills, IL

Optimization of Biodiesel Production under Uncertainty

This work focuses on the optimization of the biodiesel production process under uncertainty using ASPEN Plus and CAPE-OPEN compliant stochastic simulation and the novel Better Optimization of Nonlinear Uncertain Systems (BONUS) algorithm.

Biodiesel is an alternative to conventional diesel. Biodiesel production involves a trans-esterification reaction where long chain triglycerides are converted to methyl esters and glycerol. Out of available biodiesel process technologies, transesterification followed by separation either in batch or continuous process is the most widely used commercial process. The glycerol and esters are separated out in a settling vessel. The alcohol is separated from the glycerol and ester by either evaporation or flashing. The methyl esters are then neutralized by washing with slightly acidic water and stored. The uncreated triglycerides are recycled and the biodiesel (consisting of methyl esters and some free fatty acids) is either sold as is or blended with diesel for use as B20 fuel. [1][4]

The production of biodiesel suffers from inherent uncertainties because the biodiesel feedstock is derived from plant and animal matter having a range of components and different distributions. Due to limited availability of feedstock in an area as a result of geographical and weather constraints, biodiesel processing plants must be flexible enough to process feedstock from a variety of sources and locations, which further increases the uncertainties in feedstock triglyceride composition. The common uncertainties in the biodiesel production process are: feedstock compositions, reactor operating conditions, and methanol to oil ratio of feed to the transesterification reactor. These uncertainties can have a significant impact on the process efficiency and product quality. The recent sharp swings in the price of biodiesel and biodiesel feedstocks have demonstrated that for biodiesel to compete with conventional diesel, the biodiesel process must be efficient and cost effective. In addition, quality of fuel must meet the specifications for diesel so that biodiesel can be used in internal combustion engines of cars without the need for blending with diesel. Thus maximization of process efficiency requires optimal operating parameters while adhering to the constraints of meeting diesel fuel specification in terms of viscosity, 90% liquid boiling point and ocetane number.

Studies for the optimization of biodiesel production processes have focused largely on process integration and evaluated the feasibility of different process pathways using a constant feedstock composition and sensitivity analysis [1][4].

In their current unpublished work, the authors of this abstract have quantified the uncertainties in the biodiesel production process by representing them in the from a probability distribution curve and applied stochastic modeling to published optimized flow sheets to evaluate its impact on process efficiency and product specifications. Results have indicated that when compared to the deterministic models, stochastic modeling shows that process efficiency varies widely with the feedstock composition and operating parameters. In addition, the biodiesel fuel produced has difficulty meeting product specifications. It is evident that the optimization of biodiesel production through process integration is not a truly optimized process because it does not consider uncertainties in the biodiesel feedstock and operating parameters. Optimization under uncertainty will capture the uncertainties in feedstock components and operating parameters and incorporate them in the optimization scheme. The objective of the optimization scheme will be to maximize process efficiency under tighter product specification constraints.

Application of optimization in the field of process systems often involves dealing with non linearity and uncertainty necessitating the solution of stochastic non linear programming (SNLP) problems. But current algorithms in this area suffer from various limitations. Diwekar and Shastri [2] have applied L-shaped BONUS algorithm, which is the integration of BONUS (Better Optimization of Nonlinear Uniform Systems) algorithms and sampling based L-shaped method to cases studies and their results confirm the computation efficiency of the L-shaped BONUS algorithm. The sampling based optimization algorithm structure is shown below.

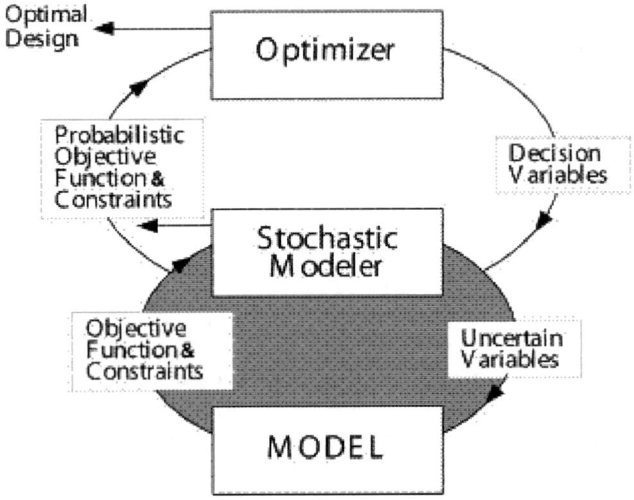

Figure 1 Sampling-based optimization algorithm structure [3]

Diwekar and Salazar [3] have applied the algorithm to minimization of water consumption in coal fired power plants. Optimization under uncertainty for such large scale complex process cannot be solved with conventional stochastic programming because of the large computational expense. However, the novel optimization algorithm BONUS when implemented in ASPEN Plus dramatically decreased the computation requirements of stochastic optimization with results which led to minimization of water consumption in coal fired plants.

Biodiesel production process is very complex and the use of ASPEN Plus and BONUS will dramatically decrease the computational requirements of the stochastic optimization. This study will comprise of two main steps: sensitivity analysis to determine the major decision variables and optimization of the process to maximize process efficiency. The characterization of uncertainties in the biodiesel production process has already been accomplished by the authors in their unpublished work.

The result of the study will be the maximization of process efficiency while using product quality specifications as a constraint. The results of this study can be used for the design of a plant which is flexible to accept any feedstock and the finished product does not require additional blending of fuel to meet diesel fuel specifications.

References:

1. Myint, L.L. and El-Halwagi, M.M (2008), 'Process Analysis and Optimization of Biodiesel Production from Soybean Oil', Clean Technologies and Environmental Policy, pp the 85-110, 2008.

2. Shastri Y. and U. Diwekar., L-Shaped BONUS algorithm with application to water pollutant trading, (2008), In Industrial and Engineering Chemistry Research, 47, 9417.

3. Salazar J. and U. Diwekar, Minimization of Fresh Water Consumption for Particulate Carbon (PC) Power Plants, Foundations of Computer Aided Process Design 2009:Design for Energy and Environment, Halwagi M. and A. Linninger Eds., 649.

4. Ai-Fu Chang and Y. A. Liu, Integrated Process Modeling and Product Design of Biodiesel Manufacturing, Ind. Eng. Chem. Res. 2010, 49, 1197–1213

See more of this Session: Energy Systems Design and Alternative Energy Sources
See more of this Group/Topical: Computing and Systems Technology Division

(327b) New Strategies for Polygeneration: Hybrid Natural Gas Reforming and Coal Gasification Techniques for Production of Methanol, Electricity, and Fischer-Tropsch Fuels

Thomas A. Adams II, Chemical Engineering, Massachusetts Institute of Technology, Cambridge, MA and Paul I. Barton, Department of Chemical Engineering, Massachusetts Institute of Technology, Cambridge, MA

Polygeneration is the conversion of fuels such as coal, natural gas, and biomass into multiple products such as synthetic natural gas, hydrogen, methanol, diesel, gasoline, dimethyl ether, and electricity. Polygeneration systems can be constructed with some flexibility, such that the product mix can be altered in response to changing market conditions over the lifetime of the plant. This provides financial security and reduced risk in the face of rapid price fluctuations and an uncertain future. Additionally, the incorporation of multiple inputs or outputs provides many opportunities for energy integration which can lead to overall efficiency improvements.

In this work, novel polygeneration processes which convert coal and natural gas to methanol, Fischer-Tropsch products (i.e., diesel, naphtha) and electricity with CO_2 capture capability are presented. To create the liquid products, syngas must be generated with a H_2/CO ratio of about 2.0. The key characteristic of each process lies in how this ratio is achieved. For example, coal gasification is commonly considered, which can be used to generate syngas with a H_2/CO ratio of about 0.7-1.3 and then upgraded through the addition of H2 gas from an external source (such as solar-powered electrolysis of water) or through the water-gas-shift reaction. However, the exothermic, high-temperature gasification step (~1300°C) is inherently inefficient since the high-temperature waste heat is underutilized for steam generation at best and discarded at worst.

Instead, we explore the combination of natural gas reforming with coal gasification. Natural gas can be reformed by adding steam at high temperatures to create syngas with a H_2/CO ratio of greater than 3.0. This is then mixed with coal-derived syngas to obtain the desired ratio, avoiding the need for H_2 generation or the water-gas-shift reaction. However, because the reforming step is both high-temperature and highly endothermic, the reformer can be heat integrated with the gasifier, such that the high-temperature generated by the gasification reaction can be used to provide for the heat needs of the reformer directly. This synergistic configuration creates efficiency improvements throughout the plant and has interesting consequences when considering environmental issues such as CO_2 capture and water consumption. Using Aspen Plus simulations and nonlinear optimization techniques, a techno-economic analysis is presented which compares several design options to highlight these improvements. The effects on the optimal design and profitability caused by changes in the prices of the fuels and products are discussed, as well as the impact of a potential carbon tax.

See more of this Session: Energy Systems Design and Alternative Energy Sources
See more of this Group/Topical: Computing and Systems Technology Division

(327c) An Integrated Multi-Scale Spatiotemporal Modelling Framework for Optimal Design of CO_2 Capture, Transport and Storage Networks

Ahmed Alhajaj[1], Niall Mac Dowell[1], Murthy Konda[2] and Nilay Shah[2], (1)Chemical Engineering, Imperial College London, London, United Kingdom, (2)Imperial College London, London, United Kingdom

In general, when the carbon capture, transport and storage (CCTS) problem is considered, each component of the CCTS network is analysed independently. However, in order to realise a cost-optimal system design, the extensive inter-connectedness of these networks must be accounted for. In this contribution, we propose a whole-systems approach to the design of CCTS networks. To achieve this goal, we integrate validated CO_2 capture process models[1,2] with a spatio-temporally explicit approach for network modelling. Using the process models, we determine a cost- optimal degree-of-capture (DOC) for a given source. A typical CCTS network comprises a number of CO_2 sources and a number of potential CO_2 storage sites. Further, the source-sink relationships may be extremely complex in terms of the proximity of a given source to an appropriate sink. Finally, it is important to account for the evolution of CCTS networks by accounting for the gradual incorporation of additional sources and sinks into the network in such a way as to minimise the total lifetime cost whilst complying with CO_2 emission mitigation goals. Thus a comprehensive framework that is both spatially and temporally explicit is required for the design of these systems. The dynamic aspect of the proposed models is of paramount importance, especially within the context of future energy-related networks design. For instance, a recent study[3] has revealed the long-term prospects of hydrogen in the transport sector (and mapped out the optimal roadmaps), using a detailed spatio-temporal framework, which could not have been identified with static models. Since the future emission reduction targets, sink availability and capacity, and injection rates are uncertain, a detailed stochastic analysis is an integral part of our framework. We then demonstrate the applicability and usefulness of our integrated multi-scale approach with a real case study by applying it to design CCTS networks for the UAE. As the UAE has the second largest CO2 emissions-per-capita in the world, the development of cost-optimal CCS networks is of special relevance to the UAE. The potential CO_2 emission sources considered include 17 power plants, 5 cement plants and 2 refineries – altogether these contribute up to 30 million tons of CO_2 annually. The potential sinks considered include 5 geological sites based on enhanced oil recovery (EOR). CO_2 transportation is based on pipelines and economies of scale for the pipeline transport are explicitly accounted for (using piecewise linearization). The detailed framework and the resulting CCTS networks will be presented during the conference. Finally, it is worthwhile to mention that the framework is generic and is equally applicable to any other geographical region.

1.Mac Dowell, N. et al., Ind. Eng. Chem. Res., 49(4), 1883-1899, s2010 2.Mac Dowell, N. et al., ESCAPE 20, Italy, June 2010 3.Konda, N. V. S. N. M. et al., PSE Asia 2010, Singapore, July 2010

See more of this Session: Energy Systems Design and Alternative Energy Sources
See more of this Group/Topical: Computing and Systems Technology Division

(327d) Optimization Studies of An Integrated Gasification Combined Cycle (IGCC) Plant with CO2 Capture

Debangsu Bhattacharyya, Department of Chemical Engg., West Virginia University, Morgantown, WV, Richard Turton, Chemical Engineering, West Virginia University, Morgantown, WV and Stephen E. Zitney, Collaboratory for Process and Dynamic Systems Research, National Energy Technology Laboratory, U.S. Department of Energy, Morgantown, WV

Integrated Gasification Combined Cycle (IGCC) plants with CO2 capture have a strong potential to be the leading technology in view of its numerous advantages over other commercially available technologies. However, its energy efficiency should be improved to make these plants commercially viable. To improve its energy efficiency, a systematic, top-down, design approach is taken considering commercially available technologies. In this three stage approach, first important global design decisions are made based on optimization studies. This is followed by taking optimum local design decisions. At the end, the operating conditions are optimized. The operability of the plant is taken into account before taking final design decisions. The IGCC reference plant includes an entrained, downflow, General Electric Energy (GEE) gasifier with a radiant syngas cooler (RSC), a two-stage water gas shift (WGS) conversion process, and two advanced "F" class combustion turbines partially integrated with an elevated-pressure air separation unit (ASU). A subcritical steam cycle is considered for heat recovery steam generation. Syngas is selectively cleaned by a SELEXOL acid gas removal (AGR) process. Sulfur is recovered using a two-train Claus unit with tail gas recycle to the AGR. A multistage intercooled compressor is used for compressing CO2 to the pressure required for sequestration. A number of important global design decisions are identified that have strong impact on system performance. As an example of a global design decision, the optimum combination of CO conversion in the WGS reactors and CO2 capture in the SELEXOL unit is considered. An optimal design decision results in about 1.2% increase in the net plant efficiency (HHV basis) compared to a previously published study by NETL1. Further improvement in the plant efficiency is achieved by optimizing local design decisions and operating conditions.

1 Cost and Performance Baseline for Fossil Energy Power Plants Study, Volume 1: Bituminous Coal and Natural Gas to Electricity," National Energy Technology Laboratory, www.netl.doe.gov, August 2007

See more of this Session: Energy Systems Design and Alternative Energy Sources
See more of this Group/Topical: Computing and Systems Technology Division

(327e) Computer-Aided Selection and Design of Working Fluids for Organic Rankine Cycles

Patrick Linke[1], Mirko Z. Stijepovic[1] and Athanasios I. Papadopoulos[2], (1)Chemical Engineering, Texas A&M University at Qatar, Doha, Qatar, (2)Chemical Process Engineering Research Institute, Center for Research and Technology Hellas, Thermi-Thessaloniki, Greece

Organic Rankine Cycles (ORCs) are receiving increasing attention from the scientific community due to their potential in renewable energy applications. A very large number of working fluids have been proposed for ORCs over the past years. However, the identification of the best working fluids for given heat source and sink temperature profiles remains an open research challenge. A systematic method is needed to guide the identification of high-performance ORC systems for a given application. We will present a systematic methodology to the selection and design of optimal working fluids for Organic Rankine Cycles (ORCs). The approach is based on thermodynamic analysis to identify property-performance relationships, targeted property database searches, computer aided molecular design (CAMD) and ORC optimization techniques. Through thermodynamic analysis we will derive physical property objectives to target searches for those molecules that offer optimum ORC performance with respect to economic, operating, safety and environmental objectives. The methodology enables the quick identification of existing molecules from databases that offer maximum performance in the ORC in terms of electricity cost. The method further enables the identification of novel working fluids for which no experimental physical property data are available. Here, we utilize group contribution methods in combination with multi-objective optimization technology for the generation of optimum working fluid candidates.

For the identified optimum ORC working fluids, the ORCs are optimized to identify the cost-optimal optimal performance of the integrated system with acceptable safety and environmental characteristics. The proposed approach is illustrated with case studies in the design of working fluids for solar and geothermal ORC applications. The methodology systematically and quickly identifies both novel and conventional molecular structures that enable optimum ORC process performances.

See more of this Session: Energy Systems Design and Alternative Energy Sources
See more of this Group/Topical: Computing and Systems Technology Division

(327f) Estimation of State of Charge of a Lithium-Ion Battery Pack

Venkatasailanathan Ramadesigan[1], Ravi N. Methekar[1], Sumitava De[1], Richard D. Braatz[2] and Venkat Subramanian[1], (1)Energy, Environmental and Chemical Engineering, Washington University in St. Louis, St. Louis, MO, (2)Department of Chemical Engineering, Massachusetts Institute of Technology, Cambridge, MA

The high energy density of lithium-ion batteries make them the preferred candidate for portable and mobile applications. Many manufacturers of electric vehicles combine lithium-ion cells into a battery pack to fulfill the requirement of large energy in a small space. When many cells are integrated into a single pack, it poses a challenge of maintenance and safety. The battery pack may face failure due to overcharging, increasing internal pressure, a short circuit, or a drop in the state of charge (SOC) of an individual cell in the pack. To maintain the battery pack in a good state of health, it is important to maintain the SOC of each cell in the pack, so the prediction of the SOCs is essential.

The energy stored in the battery pack is governed by the cell having highest percentage of SOC, as this cell will reach the cutoff voltage earlier than the other cells in the pack. If charging of the pack is continued then the cell with higher percentage of SOC will overcharge and lead to thermal runaway and cause a failure of the battery pack. If any cell of the battery pack enters into thermal runaway, then the internal pressure will build up, leading the battery pack even to explode. In fact, there are many cases of Li-ion batteries going into thermal runaway in laptops leading to recalls by major brands and manufacturers.

In a battery pack, direct measurement of SOC is not possible and the only measurement is the terminal voltage of the battery pack. It is evident that the voltage obtained from the battery pack essentially depends on the SOC of the cells of the pack. During the charging and discharging cycles, each cell in the battery pack undergoes different forms of capacity fade and aging mechanisms due to different profiles across the pack and hence significantly differ from each other in their SOC even though they were started at the same initial conditions. This makes the estimation of SOC with cycle number a challenging task. Several techniques have been proposed for SOC estimation, such as model-based observers and black-box methods [1-3]. The accuracy reached is about 2% [4]. The probable reason for this accuracy is the use of a simple model for prediction of SOC. This motivates the use of a reformulated model [5] for the lithium-ion battery that reduces the computational burden while keeping accurately describing the battery dynamics.

In this work we estimate the SOC of the battery pack consisting of 16 cells in series as needed for a battery pack produced by Tesla [6]. The technique involves predicting the terminal voltage of each cell in the battery pack with initial guesses for the values of SOC for each cell and summing the cell voltages to obtain a predicted terminal voltage of the battery pack to be compared with the measured value in real time. Minimizing a suitable norm of the error between the measured and predicted terminal cell voltage using numerical optimization methods [7, 8] enables the fast and accurate estimation of the SOC of each cell. The mathematical representation of the objective function is

$$\underset{\theta}{\operatorname{argmin}}\ E(V_{pack}, V_{pack,estimated}, \theta)$$

$$\theta = \{SOC_i,\ i = 1.....n\}$$

$$V_{pack,estimated} = V_1 + V_2 + \cdots + V_n$$
$$\text{where } V_i = f(SOC_i),\quad i = 1,\cdots,n$$

Figure 1 is an example result comparing the measured and estimated terminal voltage of two cells in a battery pack. These two cells are maintained as 0% and 50% SOC. Figure 1 shows that the presented approach predicts the SOC of each cell in the battery pack reasonably well. Table 1 show that the SOC of each cell can be estimated within ±1% error with a small computational time. The computation time for estimation of SOC for two cells is very much less than the charging time of the pack. The presentation will show results for 4, 8, and 16 cells per pack and for varying disturbances and nonidealities in the model.

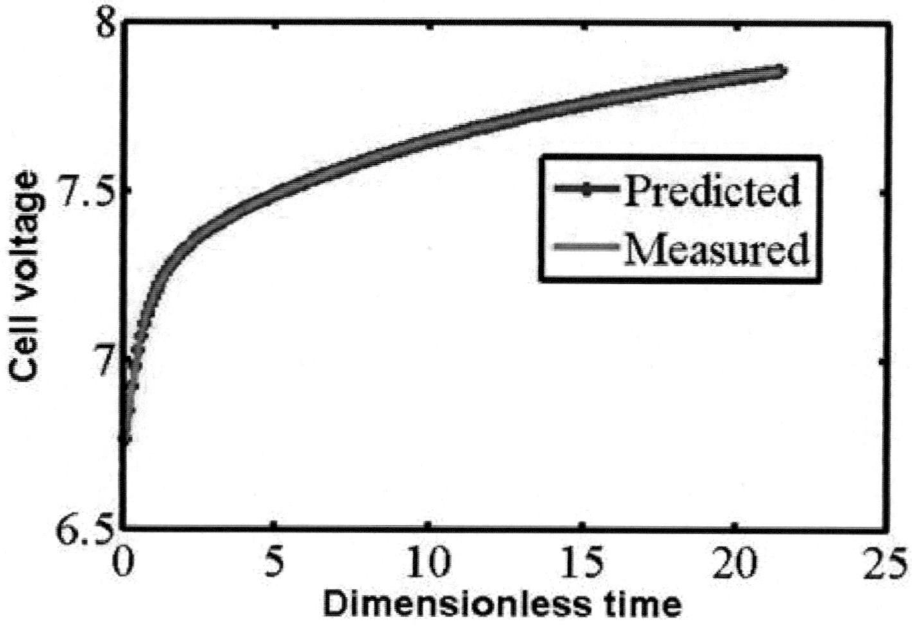

Figure 1: Comparison of predicted discharge curve with measured discharge curve.

Table 1: Percentage error and CPU time for the SOC estimation

SOC	Actual	Predicted	% Error	CPU time (s)
SOC_1	0.9456	0.945599	1E-4	526

SOC_2	0.7725	0.772499	1E-4	

Acknowledgements

The authors acknowledge partial financial support by the National Science Foundation under contract numbers CBET-1008692, CBET- and CBET-0828123, I-CARES (WUSTL) and the United States Government.

References

[1] O. Barbarisi, F. Vasca, and L. Glielmo, Control Engineering Practice, **14**, 267-275, 2006.

[2] D. Di Domenico, G. Fiengo, and A. Stefanopoulou, Proceedings of 2008 IEEE Conference on Control Applications, **1**, 702-707, 2008.

[3] A.J. Salkind, C. Fennie, P. Singh, T. Atwater, and D.E. Reisner, Journal of Power Sources, **80**, 293-300, 1999.

[4] V. Pop, H. J. Bergveld, J. Veld, P. Regtien, D. Danilov, and P. Notten, Journal of the Electrochemical Society, **153**, A2013-A2022, 2006.

[5] V.R. Subramanian, V. Boovaragavan, V. Ramadesigan, and M. Arabandi, Journal of the Electrochemical Society, **156**, A260-A271, 2009.

[6] http://www.teslamotors.com/display_data/TeslaRoadsterBatterySystem.pdf accessed on 05/03/2010.

[7] L.T. Biegler, Computers & Chemical Engineering, **8**(3), 243-247, 1984.

[8] L.R. Petzold, "A Description of DASSL: A Differential/Algebraic System Solver," Scientific Computing, eds. R.S. Stepleman et al., North-Holland, Amsterdam, 65-68, 1983.

See more of this Session: Energy Systems Design and Alternative Energy Sources
See more of this Group/Topical: Computing and Systems Technology Division

(327g) Process Development and Techno-Economic Analysis of a Novel Process for MeOH Production From CO2 Using Solar-Thermal Energy

Jiyong Kim[1], Carlos Henao[1], Christos Maravelias[1], James E. Miller[2], Daniel E. Dedrick[3] and **Terry A. Johnson**[3], (1)University of Wisconsin, Madison, WI, (2)Ceramic Processing and Inorganic Materials, Sandia National Labs, Albuquerque, NM, (3)Thermal Fluids Science and Engineering, Sandia National Labs, Livermore, CA

Mitigating and overcoming environmental problems brought about by the current worldwide fossil fuel-based energy infrastructure requires the creation of innovative alternatives. In particular, such alternatives must actively contribute to the reduction of carbon emissions via carbon recycling and a shift to the use of renewable sources of energy. Carbon neutral transformation of biomass to liquid fuels is one of such alternatives, but it is limited by the inherently low energy efficiency of photosynthesis with regard to the net production of biomass. Researchers have thus been looking for alternative, energy-efficient chemical routes inspired in the biological transformation of solar power, CO2 and H2O into useful chemicals; specifically, liquid fuels.

Methanol has been the focus of a fair number of publications for its versatility as a fuel, and its use as an intermediate chemical in the synthesis of many compounds. In some of these studies, (e.g. Joo et al., (2004), Mignard and Pritchard (2006), Galindo and Badr (2007)) CO2 and renewable H2 (e.g. electrolytic H2) are considered as the raw materials for the production of methanol and other liquid fuels. Several basic PFD diagrams have been proposed. One of the most promising is the so called CAMERE process (Joo et al., 1999). In this process, carbon dioxide and renewable hydrogen are fed to a first reactor and transformed according to:

H2 + CO2 <=> H2O + CO Reverse Water Gas Shift (RWGS)

After eliminating the produced water the resulting H2/CO2/CO mixture is then feed to a second reactor where it is converted to methanol according to:

CO2 + 3.H2 <=> CH3OH + H2O Methanol Synthesis (MS) CO + H2O <=> CO2 + H2 Water Gas Shift (WGS)

The approach here is to produce enough CO to eliminate, via WGS, the water produced by MS. This is beneficial since water has been proven to block active sites in the MS catalyst.

In this work a different process alternative is presented: One that combines the CO_2 recycling of the CAMERE process and the use of solar energy implicit in some of the biomass-based process, but in this case with the potential high energy efficiency of thermo-chemical transformations.

PROCESS DESCRIPTION The proposed process is composed of 5 different subsystems:

1. Dish-CR5 solar array: This system takes the main raw material CO_2 and partially converts it into CO, finally generating a CO_2-CO mixture. It consists of a parallel arrangement of many identical parabolic dishes. Each one concentrates solar power on a small thermo-chemical splitting reactor (called CR5) where CO_2 is transformed into CO and O_2. From the thermodynamic point of view, the array uses solar energy to transform the low energy molecule CO_2 into the higher energy molecule: CO. In fact, it is the energy stored in the resulting CO that is later used to drive the rest of the chemical transformations in this process.

2. WGS reaction system: This system takes the CO_2-CO mixture produced by the solar array (later adjusted for CO_2 content in the CO_2 absorption system) along with H_2O to generate H_2 and CO_2 via WGS. This system is responsible for supplying the H_2 required by the MS reaction system. The reaction is carried out on a commercial $Cu/ZnO/Al_2O_3$ catalyst used as well in the MS reaction system, but at conditions that favor WGS specifically. Only a fraction of the feed CO is used to produce the H_2 leaving enough to control water poisoning in the MS reaction system.

3. MS reaction system: This system takes the CO-CO_2-H_2 mixture produced in the WGS reaction system (later adjusted for CO_2 content in the CO_2 absorption system) to generate MeOH and H_2O. This H_2O tends to poison the catalyst but it is continuously removed from the surface via WGS reaction with CO, producing extra H_2 and CO_2 which end up as additional MeOH. As mentioned before, this reactor uses a commercial $Cu/ZnO/Al_2O_3$ catalyst, while the reaction conditions are adjusted to favor both the WGS and the MS reaction.

4. Amine based CO_2 absorption system: Both the methanol synthesis and the water gas shift reaction systems are fed with gas mixtures having significant content of CO_2 and CO. However, each one of their reactors achieves an optimum performance at specific feed CO_2/CO molar ratios. To adjust these ratios, an amine based CO_2 absorption system is included. The system uses an aqueous solution of monoethanolamine (MEA) to absorb CO_2 in two different columns connected in a loop with a solvent regeneration column. The first column absorbs part of the CO_2 in the stream leaving the Dish-CR5 array, producing a rich CO mixture that is later fed to the WGSR system. The objective here is to drive the WGS equilibrium to the production of H_2. The second column absorbs part of the CO_2 generated in the WGS system, producing a mixture with only the required amount of CO_2 to optimally drive the methanol synthesis reaction. The objective here is to drive the WGS equilibrium to counter the water poisoning while still favoring the MS equilibrium as to obtain an optimum carbon to methanol yield.

5. Methanol purification system: This system is responsible for the separation incondensable materials (i. e. CO_2, CO, H_2) and water from the main product MeOH. The system includes a simple flash vessel to

eliminate the light materials and a standard distillation column to perform MeOH-H2O separation.

ECONOMIC ANALYSIS This work presents an assessment on the current economic feasibility of the process for a plant capacity of 83000 MT/y (i.e. 319 kmol/hr) of MeOH. Nominal prices of CO2, H2O and MeOH were used according to recent technology analysis and market trends. Detailed Net Present Value (NPV) sensitivity analysis studies were also performed. The main conclusions of this analysis are:

- The main economic bottlenecks appear to be the high utility consumption for the CO/CO2 separation and the high capital investment associated primarily with the Dish-CR5 array. - In terms of capital cost, the Dish-CR5 array is the dominating factor, accounting for more than 90% of the capital expenditure. - In terms of operating costs, the low pressure steam consumption of the amine CO2 absorption system is the higher contributor followed by the compression electricity consumption for the CR5 system. - The sensitivity analysis indicate that the NPV sensitivity to the CO2 price is much lower that than its sensitivity to the price of MeOH; in fact, modifying the CO2 price alone it is not possible to reach a breakeven point. However, it is also important to remember that if emission regulations or emission trading schemes are introduced CO2 consumers will get credits, which means an improvement in the process economics.

Our analysis indicates that the process has potential, and points towards future research efforts. Particularly, it is clear that there is much room for improvement in the development of a less expensive and more efficient CR5 system, work that is currently under way at Sandia National Labs. In addition, the use of advance process synthesis techniques can bring much improvement in terms of energy efficiency to the preliminary process design proposed here.

"This work was supported by the Laboratory Directed Research and Development program at Sandia National Laboratories, in the form of a Grand Challenge project entitled Reimagining Liquid Transportation Fuels: Sunshine to Petrol. Sandia is a multiprogram laboratory operated by Sandia Corporation, a Lockheed Martin Company, for the United States Department of Energy's National Nuclear Security Administration under Contract DE-AC04-94AL85000".

See more of this Session: Energy Systems Design and Alternative Energy Sources
See more of this Group/Topical: Computing and Systems Technology Division

(363a) Relationship Between Kinetic and Flow Parameter Representations in Complex in-Situ Reactive Processes

Jacob Bauman, Chemical Engineering, University of Utah, Cottonwood Heights, UT and Milind D. Deo, Chemical Engineering, University of Utah, Salt Lake City, UT

Modeling reactive-transport processes is complex because it is necessary to calculate heat transfer, multicomponent/multiphase fluid flow, thermodynamics, reaction kinetics, and geomechanics, all models which may have continuously changing parameters. Some examples of these processes are in-situ combustion, pyrolysis, or gasification and CO2 sequestration with geochemical reactions. In this study a novel statistical approach has been implemented to understand the interdependences between reaction and flow parameters.

The approach is applied to in-situ oil shale pyrolysis. The appropriate molecular representation of kerogen, the insoluble solid hydrocarbon in oil shale, is highly uncertain due to its complex nature and variableness across resources. Component lumping schemes have different levels of detail, and the appropriate level of detail is debatable at large scales. Due to the complex nature of kerogen the reaction kinetics of conversion to products are uncertain. The appropriate representation of relative permeability is also a major uncertainty, and is often approximated. Sensitivities of simulated results to these parameters and their interactions are compared using fractional and full factorial experimental designs. Simulations were run with STARS, a thermal reservoir simulator created by CMG.

Ranges of activation energies in hydrocarbon cracking reaction schemes were tested, and it was discovered that specific reaction parameter combinations affect numerical behavior of the simulations. The distance between heating wells and producer wells is a crucial parameter for a successful processing strategy. Heat of reaction for kerogen pyrolysis is insignificant compared to the heat supplied. Relative permeability representation, activation energy distribution for kerogen pyrolysis, and activation energies for kerogen conversion and light oil cracking are dominant parameters.

The statistical approach described exposed parameters that dominate this complex process. Efforts for fundamental understanding and successful production strategies can focus on the most dominant parameters. The statistical methods applied to oil shale pyrolysis can be applied to other complex reactive-transport processes.

See more of this Session: Underground Thermal Treatment of Carbonaceous Materials
See more of this Group/Topical: Topical D: Chemical Engineering in Oil and Gas Production and Other Complex Subsurface Processes

(363b) Flow Simulation with the Lattice Boltzmann Method in 3D Porous Structures of Pyrolyzed Oil Shale Cores Using Multiscale X-Ray CT Imaging

C.L. Lin[1], Jan D. Miller[2] and Ching Hao Hsieh[1], (1)Department of Metallurgical Engineering, University of Utah, Salt Lake City, UT, (2)Metallurgical Engineering, University of Utah, Salt Lake City, UT

The pore structure and connectivity of pore space are important features which determine fluid flow in oil shale during pyrolysis in the production of shale oil. Using a combination of Micro and Nano X-ray computed tomography (CT) and 3D image analysis software, 3D pore networks, kerogen phase, crack structure and flow channels of Mahogany oil shale core samples were characterized before and after pyrolysis at a voxel resolution from several microns to 60 nm. The pore and crack network structure deduced from the images was used in Lattice Boltzmann simulations to calculate the permeability in the pore space. The permeabilities of the pyrolyzed samples of the silicate-rich zone were on the order of milli-Darcies, while the permeabilities of the kerogen-rich zone after pyrolysis were very anisotropic and about four orders of magnitude higher

See more of this Session: Underground Thermal Treatment of Carbonaceous Materials
See more of this Group/Topical: Topical D: Chemical Engineering in Oil and Gas Production and Other Complex Subsurface Processes

(363c) "Cold" Extraction and Liquefaction of Fossil Fuels

L. T. Fan and **Shahram Shafie**, Green Source Energy, LLC, Austin, TX

Our process for "cold" extraction and liquefaction of various fossil fuels, facetiously named analogous to the "cold" fusion of hydrogen, has been invented based on our serendipitous observations of unexpected phenomena when an essentially green solvent is mixed with powdered coal. It is worth recalling that Salt Lake City is the birthplace of the "cold" fusion of hydrogen, which has totally fizzled. In contrast, however, our "cold" extraction and liquefaction process for fossil fuels has survived third-party verifications by several reputable testing laboratories with flying colors. Moreover, it is well on the way to sustainable commercialization, environmentally and economically, although the domestic and foreign filings of patents for the process have barely been completed.

Various fossil fuels, such as oil sands, heavy crude oil, coal and oil shale, can be extracted or liquefied by our process under or near ambient conditions with a single green solvent. In addition, our process is effected without the aid of additional reagents, such as transport media (water and diluents), catalysts, and surfactants as well as without external supply of large mechanical forces and uncommon forms of energy, e.g., ultrasonic, microwave or nuclear radiation. Apparently, this is hardly the case for any of the available processes for extraction or liquefaction of fossil fuels. These available processes, therefore, tend to be exceedingly expensive to construct and be environmentally unsustainable.

Our "cold" extraction or liquefaction process for fossil fuels is deployable in the subsurface (in situ) as well as on the surface (ex situ). An independent pilot-scale test performed at a leading reservoir engineering laboratory has recovered more than 90% of the residual heavy crude remaining in samples of Berea sandstone plug core with the injection of only 5 pore volumes of our solvent. A third-party verification test carried out at one of the nation's major contract R&D companies has extracted at least 96% of bitumen from Alberta oil sands with an average tar content of 13% in a semi-pilot-scale facility within 5 minutes; the test conditions were 70°C under ambient pressure (1 atm), and the ratio of sample to solvent was 1:1. The residual bitumen on the sands after extraction was determined by an analytical laboratory of Alberta Research Council. Highly encouraging results have also been obtained with a variety of coal by the same company as well as in our laboratory. The extraction of hydrocarbons from coal, however, required about 30 minutes at temperatures approximately between 90 and 100°C with the pressure remaining at 1 atm.

See more of this Session: Underground Thermal Treatment of Carbonaceous Materials
See more of this Group/Topical: Topical D: Chemical Engineering in Oil and Gas Production and Other Complex Subsurface Processes

Thermal Gravimetric – Mass Spectrometry Analysis of Oil Shale

Pankaj Tiwari and Milind D. Deo, Chemical Engineering, University of Utah, Salt Lake City, UT 84112.

Abstract

The fast depleting conventional oil reserves may be substituted by proven vast recourse of petroleum generated source rock known as oil shale. The petroleum source rock contains organic matter laminated with complex mineral matrix. The organic matter is mainly in kerogen form which posses an undefined high molecular weigh structure and upon heating releases the volatile matters. The generated volatile products are both, lighter gases as well as condensable higher boiling point substances known as shale oil, a prospective alternative for conventional oil. The yield of potential oil and composition distributions of pyrolysis process products depend primarily on time-temperature history. The shale oil production yield and improved quality oil from source rock could be enhanced from the knowledge of evolved products composition, competitive reactions mechanism and good knowledge of kinetics triplet. A Thermo-gravimetric analysis instrument with mass spectrometry is potent technique to capture the basic chemistry of the complex process. The evolved components identification based on molecular weight and their thermal behavior at different heating programs provide the in-depth understanding of the products formation and their reaction mechanisms. In this study we report the overall kinetic distribution of Mahogany oil shale pyrolysis using the TGA. The uncertainty quantification is estimated in the distribution of activation energies over conversion scale and simulated model free isoconversional kinetic model (not includes transport effects) is extrapolated to derive the conversion profiles at slow pyrolysis and flash pyrolysis. Single ion monitoring of mass spectra from combined TGA-MS analysis is used to reveals inflections in TGA thermograms during the continuous pyrolysis at different thermal ramps. Identification, relative quantification and thermal behavior of many hydrocarbon products and mineral gases are studied to understand the competitive pyrolysis mechanism of naturally complex petroleum source rock.

Introduction

Oil shale thermal treatment can be performed in both in-situ and ex-situ modes but the produced hydrocarbons distribution is different and yield depends of on other experimental conditions too, temperature, heating rate, pressure and residence time of products in heated zone, etc. The standard experimental techniques such Fischer assay and Rock-Eval can be used to perform the experiments at the normal surface retorting and under high pressure insitu retorting. TGA (thermal gravimetric analysis) is a standard technique to observe the chemical process progress in terms of weight change. The organic matter is mainly in the form of kerogen which posseses an undefined high molecular weight, structure and upon heating releases volatiles. The products generated are mainly in vapor phase and after condensation generate liquid known as shale oil. The potential yield of oil from raw shale depends on the organic maturity of the material that in other way depends on the origin and geological formation of shale. Oil releasing organic matter in shale is broadly classified as naturally present bitumen (very less, max

1wt % of total shale) and kerogen (10-12 wt % of total shale). Some oil shales contain unbound and bound water. Because of tightly bound matrix of the oil shale, significant organic decomposition occurs at a temperature range of 300-550^{0}C. The decomposition of organic matter may even occur at low temperatures, but the rate of the evolution of products will be relatively slower. The understanding of weight loss and continuous evolution of hydrocarbon products can truly be mimicked using faster heating rate. Comparison of the data from relatively two (or more) wide and different heating rates can explain the dependency of product distribution on thermal program.

The organic matter is mainly in the form of kerogen which posseses an undefined molecular weight, structure and upon heating releases volatiles. The products generated are mainly in vapor phase and after condensation generate liquid known as shale oil. The potential yield of oil from raw shale depends on the organic maturity of the material that in other way depends on the origin and geological formation of shale. Oil releasing organic matter in shale is broadly classified as naturally present bitumen (very less, max 1wt % of total shale) and kerogen (10-12 wt % of total shale). Some oil shales contain unbound and bound water. Because of tightly bound matrix of the oil shale, significant organic decomposition occurs at a temperature range of 300-550^{0}C. The decomposition of organic matter may even occur at low temperatures, but the rate of the evolution of products will be relatively slower. The understanding of weight loss and continuous evolution of hydrocarbon products can truly be mimicked using faster heating rate. Comparison of the data from relatively two (or more) wide and different heating rates can explain the dependency of product distribution on thermal program.

The thermal treatment of oil shale produce, condensable vapor, non-condensable gases and spent shale. The yield and composition distributions of different products pyrolysis process depend on time-temperature history. The potential gaseous products start evolving when petroleum source rock reaches some definitive temperature and continue as the temperature rises. The generated volatile products are both, lighter gases as well as condensable higher boiling point substances known as shale oil. The shale oil is valuable products and considered a prospective alternative for conventional oil. The yield of total shale oil production from source rock may be enhanced from the knowledge of evolved products compositions and competitive reaction mechanism. It has long been recognized that the fraction of organic carbon converted to oil depends on the H/C ratio of the oil shale. Thus, the ultimate analysis (elemental compositions) of the raw oil shale and spent shale is necessary to figure out and support the theoretical understanding with broad sense to the mechanism of the products formation. Because oil shales have differing origins and are found in differing geological environments, and the compositions are different (H/C ratio, organic % and mineral matter's matrix) it is not surprising that they behave differently when subjected to varied pyrolysis conditions.

The major hurdle in the development of commercial production of oil shale is a variability of parental oil shale compositions, both organics and minerals. The rock formation varies, at depth and horizon in terms of composition. The comprehensive understanding of oil shale pyrolysis requires the detail information of the raw material composition, transport of heat, rate of products formation, distribution of products, interaction and alternation of minerals and organics, resistance of mineral matrix during products release, etc. The presence of some minerals may catalyze the hydrocarbon generation and the resulted pore formation may reduce the possibility of secondary reactions, such as coking, cracking, etc. Thus, multiscale thermal treatment of oil shale, experiments with different size of cores may reveal the insight of thermal heavier of rock substance, resistance distribution offered by pore formation and variability in the products rates and compositions during pyrolysis.

The compositional analysis of the products obtained can be estimated in many ways. GC (Gas chromatography) analysis is usually performed on the samples collected after condensation and provides the information on overall products independent of the thermal behavior. Combined GC-MS (Gas-chromatography- mass spectrometry) analysis is a similar and powerful technique, which provides the identification of the components in the sample. DSC (Differential scanning calorimetry) gives the heat value information of the samples. FTIR (Fourier Transformed Infrared spectrometry) provides the methodology about compositions identification and wax appearance temperature. HPLC (High performance liquid chromatography) is also a similar technique, which quantifies the products in groups (SARA- saturates, aromatics, resins and asphaltene). XRD (X-ray diffraction), SEM (Scanning electron microscopy) and EDS (Electron dispersive spectroscopy) are used for the characterization of solid products (qualitative and quantitative analysis of the minerals in the original and reacted assemblages) from the experiments.

A Thermo-gravimetric analysis instrument with mass spectrometry is potent technique to capture the basic chemistry of the complex process. The evolved components identification based on the molecular weight and their thermal behavior at different heating programs provide an in-depth understanding of the possible products and their reaction mechanisms. The single ion monitoring of water vapor, methane and carbon-dioxide in mass spectra over temperature scale reveals inflections in TGA curve during the continuous pyrolysis and could be used appropriately for distinction of organic and mineral decomposition sections. Identification, relative quantification and thermal history of many hydrocarbon products from pyrolysis of mahogany oil shale were obtained from TGA-MS instrument. The identified organic compounds were classified based on the common structure and bonds, like, aromatic, alkane, alkene, alkyne and some PAH (poly aromatic hydrocarbons). The rational comparison of different products is used to understand the pyrolysis mechanism of naturally complex petroleum source rock.

Experimental setup

Elemental Analysis

The elemental analysis is appropriate to classify the oil shale based on the ratio of hydrogen to carbon (H/C) and oxygen to carbon (O/C) before performing the experiments. The raw oil shale samples were obtained from Utah Geological Survey department and were crushed and screened to 100 mesh (1.49 E – 4 meters) size particles. The CHNSO (carbon, hydrogen, nitrogen, sulfur ad oxygen) elemental analyses of the sample were performed using LECO CHNS-932 and VTF-900 units and results are summarized in the Table 1 with standard deviation (10 runs for each). The elemental analysis shows that studied oil shale is Type I on Van Krevelen classification diagram in the term of H/C (1.1) and O/C (0.67) atomic ratios.

Table 1. Elemental analysis of the oil shale sample

Oil Shale Sample	*Average %*	*Stdev*
Carbon	17.45	0.26
Hydrogen	1.6	0.08
Nitrogen	0.53	0.06
Sulfur	0.18	0.04
Oxygen	15.69	0.79
H/C % (molar)	1.1	
O/C % (molar)	0.67	

Thermo Gravimetric Analysis-Mass spectrometry

The crushed 100 mesh-size samples of Mahogany oil shale were dried for four hours to remove any moisture; if present. There was no significant weight loss during drying, thus the samples are used as received after screening to 100 mesh size. To study the thermal behavior of oil shale pyrolysis (N_2 environment) products, a TGA instrument (TA Instruments Q-500) coupled with mass spectrometry was used for the entire temperature range of kerogen decomposition. The particles size used along with other conditions, was specifically selected to allow the products to leave the heated chamber of TGA with no transport effects. Nitrogen was used as sweeping media to perform the experiment in inert environment. Total flow rate of nitrogen was 100 ml/min (90ml/min as purge and 10 ml/min as balance gas). The possibility of secondary reactions during the decomposition and because of significant residence time of products in the reactor (heated chamber of TGA) was taken care using the high sweep gas flow rate. It is assumed that the total sweep gas flow rate is enough to carry over the entire vapor out from the chamber and prevent any vapor phase decomposition and condensation of the products. The combined TGA (thermal) and MS (monitoring of products) were performed at four different heating rates (Table-2). The TGA furnace chamber outlet was connected to MS instrument through heated (150C) capillary column line to prevent any condensation of the products, and it is assumed that products are in original form as they produced. The vacuum is maintained in the MS to reduce the residence time of the sample. The total number of components to be identified by single ion monitoring of mass spectrometry was changed in each experiment while the basic components such as methane, water, carbon-dioxide and light hydrocarbons (aklanes, olifins and aromatics) were fixed in all experiments to analysis the comparative and competitive reaction mechanism of naturally complex oil shale materials. The experiment at 5C/min thermal rate was repeated adding more compounds in ion monitoring catalog.

Results and Discussion

The TGA is a standard technique, which records the weight change data over the time-temperature history. The weight loss of oil shale is mainly because of the thermal treatment of solid material and due to this phenomenon the evolved products are mainly in the vapor phase. The sufficient flow rate of the purge gas carry those vapor out of the heated chamber and it can be assumed that the resident time is not enough to allow the products to participate for secondary reactions. The combined TGA-MS experiments are performed at different temperature programs. Table-2 summarized the experimental condition and information obtained from TGA weight loss curve.

Table 2: Experimental conditions

Heating rate	Final temperature	Total compounds	Weight, mg	%wt loss @250C	%wt loss @500C	%wt loss @600C	T max, C
0.5C/min	650C	43	22.54	0.34	9.38	20.14	401.03
1C/min	600C	56	13.10	0.65	9.6	15.99	413.09
5C/min-1	960C	22	13.34	0.59	10.76	13.2	442.61
5C/min-2	600C	56	13.26	0.46	10.3	11.29	443.41
10C/min	650C	41	19.05	0.63	11.28	12.47	456.93

TGA produces overall weight loss thermograms during the pyrolysis and facilitate with the products in vapor phase as they generates to be indentified in attached mass spectrometry. The evolved product gases were identified in single ion monitoring response of mass spectrometry based on molecular weight. The list of the molecular weight was created based on the GCMS analysis of the shale oil produced in the lab and results published in the literature. That includes some basic key compounds that may be used as sectional division criteria for the TGA curve as well as represent the basic chemistry when molecules change their form and explain the physical phenomena of thermally activated process products, are H_2O, CO_2 and CH_4. For example, evolution monitoring of the methane may be an indication when organic decomposition occurs and presence of CO_2 and water peaks at different section on thermal scale may be indication of organic and mineralogical transformation. The water sign at the early stage may explain the presence of bounded and unbounded water in the sample. The other compounds molecular weight were used are normal hydrocarbons, could be straight chain, branched, olefins, cyclic, aromatic and PAH.

The combined TGA-MS analysis of any process products is a powerful approach to study the reaction mechanism and further kinetics development but on the other side, limited in various ways specially for complex material like oil shale where many unknown compounds may be in the products. There is a lag on the temperature scale in TGA weight loss data and corresponding ion current signal in mass detector. Most of the organic weight decomposition in the TGA data are completed around 500C (first peak in organic decomposition temperature range). This may be because of the time required for product gases to travel from TGA furnace to ion detector through capillary tube under vacuum. Second possible reasons could be that some components were still being formed at higher temperature while the total weight loss was not significant and corresponding not was contributing in the DTG peak. The TGA weight loss and the area of the components in that specific section cannot be compared because of two primary reasons. First, there could be more compounds, which may be present in the sample but were not taken in to consideration and secondly, the ion current response of components is not on a different scale and cannot be quantified and compared to other runs. But the mass spectrometry ion current for a single run allows us to do the relative quantification of the competitive reaction mechanism, such as the thermal behavior and relative amount by computing the area of one component and compare to the others. The maximum rate temperature from TGA is for combined weight loss and cannot be compared to individual ion monitoring maximum signal temperature for a specific component obtained from mass spectrometry. That's why the temperature range was extended till 600C for components relative quantification in this study. Some potential components molecular weight may conflict with other structural compounds. For example nonane and naphthalene (128 amu) and, carbon-dioxide and propane (44 amu) share the same molecular weight and make it impossible to distinguish based on single ion response. The presence of oxygen, nitrogen and sulfur may release the products with oxygenated, nitrogenated and sulfur attached hydrocarbons. In this study the GCMS information is used to assign the name of components. Mostly components are straight statured and unsaturated hydrocarbon and aromatic structures were found in GCMS total ion spectra (TIC) of shale oil sample. The possible products distribution of oil shale pyrolysis and their corresponding molecular weights along with the onset points for some key components are summarized in Table-2. The first experiment was performed at 5C/min (denoted by 5C/min-1) from ambient to 960C and total 22 compounds molecular weights were added for ion monitoring. The results suggested that all of the organic compounds traces are before the 500C and carbon dioxide and H_2S because of mineral decompositions started after 600C. Thus, to focus the analysis in the organic decomposition section, further experiments were design upto 600C with different heating rates. The experiment at 5C/min is repeated (5C/min-2) to insure the instruments reliability for components identification and therograms (Figure-2). The single ion monitoring compounds were added in the list based on the results in the earlier runs. The total 56 compounds were added in 0.5C/min and 5C/min-2 runs while the total limit of the MS instrument is 63 compounds. The MS instrument is also constrained with the limitation on highest molecular weight, 300 amu. Thus, the products of higher molecular weight (more than 300 amu) can not be detected with TGA-MS instrument.

Analysis of TGA Thermograms

The total weight loss of oil shale from ambient to 500C at different heating rates was found to be in the range of 10-12% of total initial weight. The higher heating rates shift the peak to higher temperature and total organic decomposition happens at relatively lower temperature than lower heating rates. This is because of time-temperature history; at low heating rate, the sample spends more time at small temperature range. The derivative of the TGA thermograms is shown in figure-1. The offset has been set differently to see the clear view. All the run don't have the peak in the samples at early stage except 1C/min has tiny peak in the temperature range of 80 to 180C. This could be either because of water loss or molecular rearrangements. The experiments at 5C/min for wide temperature range (960C) have the second large peak after 600C, because of the mineral decomposition.

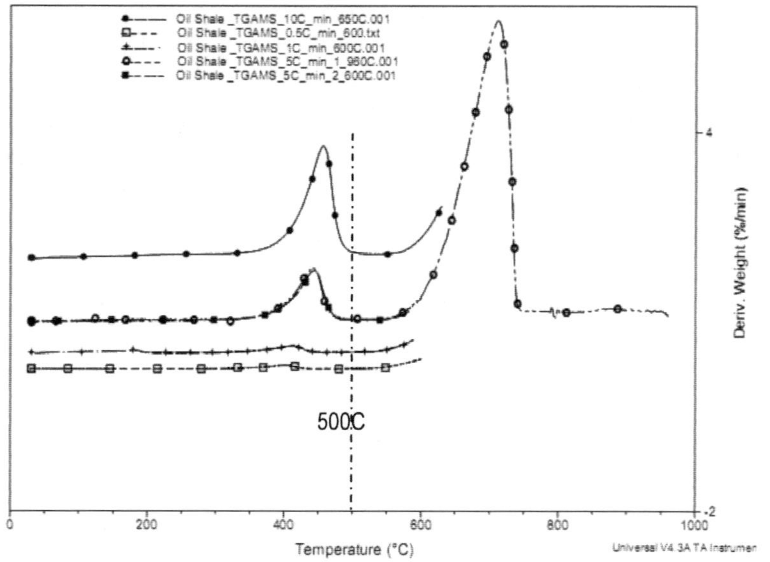

Figure-1: TGA thermograms at different thermal programs

The organic decomposition temperature scale may be divided in the three sections. The weight curve has been divided in three sections, A1-30 to250C, section A2 -200 to 500C and section A3- 500 to 600C. This information can be obtained from table by subtracting the different section weight loss from ambient to that temperature. As can be seen from the tables and also a clear indication in the figure that there is no significant weight loss in the section A1 that indicates a very small or no water presence in the samples, TGAMS curves also shows the water absence below 200C. The weight loss is very less in this section and could be cause of the molecular arrange of the sample as the temperature increases at low heating rate. Section A2 shows the significant weight loss and this weight loss is mainly because of the organic matter decomposition in the samples. The third section (A3) has the significant weight loss information but this could be because of the mineral matter decomposition, at slow rate, better time-temperature relationship this weight loss is more and mostly attributes to mineral decompositions or rearrangement in mineralogy of the sample.

Analysis of ion monitoring – mass spectrometry

Single ion monitoring allows the inflections in the DTG thermograms to be unambiguous identified based on molecular masses. The compounds evolution depends on the heating rates as well as the programs of thermal history. Traces results of the common compounds at different heating were similar except the intensity of signal and thermal behavior. At low heating rates products come out earlier as well as don't have the significant presence in terms of peaks as they spend more time at the low temperatures and continuously evolved. The detector intensity of ion current depends on some factors, first is amount of the sample, second the rate the particular ion in reaching to the detector. In the case of the low heating rates the peaks of identified compounds are present, but not with high rates as for the higher heating rates. The similar trends as we observed on the TGA weight derivative curves were seen in the mass spectra. At low heating rates the peaks are in the same region but the intensity of most of the compounds signals are too low to get the sharp peak onset points. Similar to TGA thermograms, the MS single ion monitoring response for the products generated from raw material at different heating rates like the derivative peaks intensity is higher at higher heating rate (10C/min). Figure-2 shows the single ion response of mass spectrometry for different molecular weight compounds. The data were plotted in the lumped format to get the better view. Alkanes, olefins (alkenes and alkynes), benzene based aromatics and poly aromatic hydrocarbons along with inorganic gases are plotted in lumped format. The Figure-2 reveals that in all hydrocarbons the lighter products get released earlier. At higher heating rates the alkene compounds shows the peaks at low temperature; this could be because at higher temperature these unbounded compounds get released. Higher molecular weight compounds at respective lumped group need more temperature to get formed.

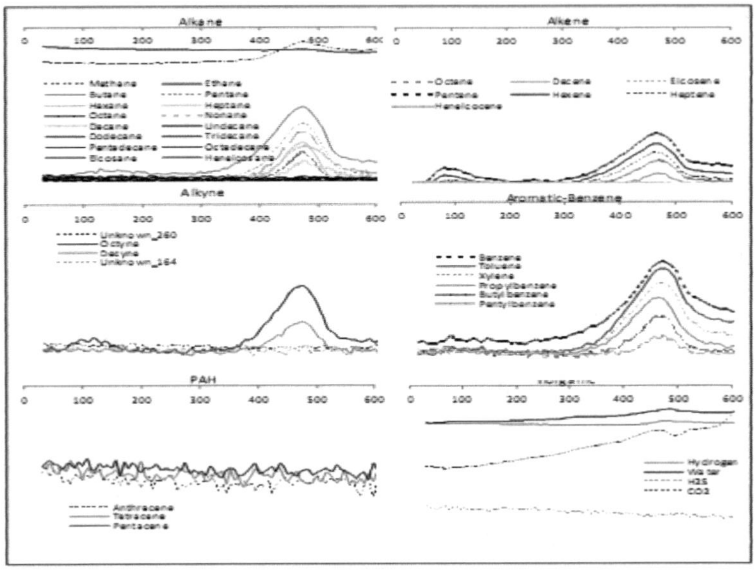

Figure-2: Classified relative components single ion monitoring over thermal scale at 10C/min ramp rate

The PAH compounds have lot of noise and hard to identified are they present in the sample or not and where are the onset points for them. The baseline shift is also different for all the compounds, especially light organic and inorganic gases. Single ion monitoring of the basic key components, water, methane and carbon-dioxide release hypothesis can be explained based on the signal response at different heating rates, shown in figure-3. The reproducibility at 5C/min is good agreement but compare to heating rate (10C/min) the signal are showing the low intensity.

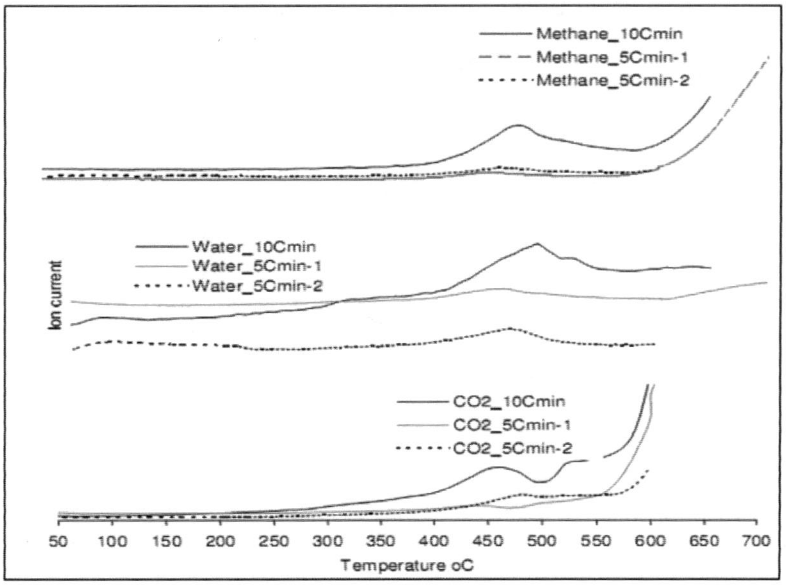

Figure-3: Single ion monitoring signal of methane, water and carbon dioxide at different heating rates

There is a steady stream for all these three compounds before the organic decomposition section (less than 300^0C) that confirms the absence of naturally present water in the sample. All these three compounds have significant signal strength in the organic evolution temperature range, because during thermal loss of organic oxygen as water and through some mineral dehydroxylation, methane as a light hydrocarbon and carbon dioxide as the products of oxygen and carbon contained material pyrolysis. The CO_2 and methane peaks are having the more strength at higher temperature, this may be because the mineral decomposition release CO_2 and decomposition of remain higher molecular weight hydrocarbon goes under cracking and release the light gas, methane.

The comparison of the single ion monitoring for different compounds could be a better illustration of thermal behavior and the same compound could be compared at different heating rates for the thermal response. Figure-4 shows the some alkanes, alkenes and aromatic compounds thermal behavior at two heating rates. The y axis is an ion current response of the mass spectrometry for diffent compounds based on the molecular weight. This cannot be used for quantitative analysis and omitted from all the graphs. Aromatic compounds signal appears at low temperature then respective carbon number hydrocarbons (Benzene vs hexane and decane vs butybezene). The maximum rate temperature for specific compound at different heating rate is almost identical and the traces follow the similar path. Start onset point for olefins of a specific hydrocarbon is at early temperature then the respective normal alkane. The difference on the TGA weight loss curve over temperature and the spectra of some compounds obtained from mass spectrometry show the deviation on the temperature, signals are spread above 500^0C.

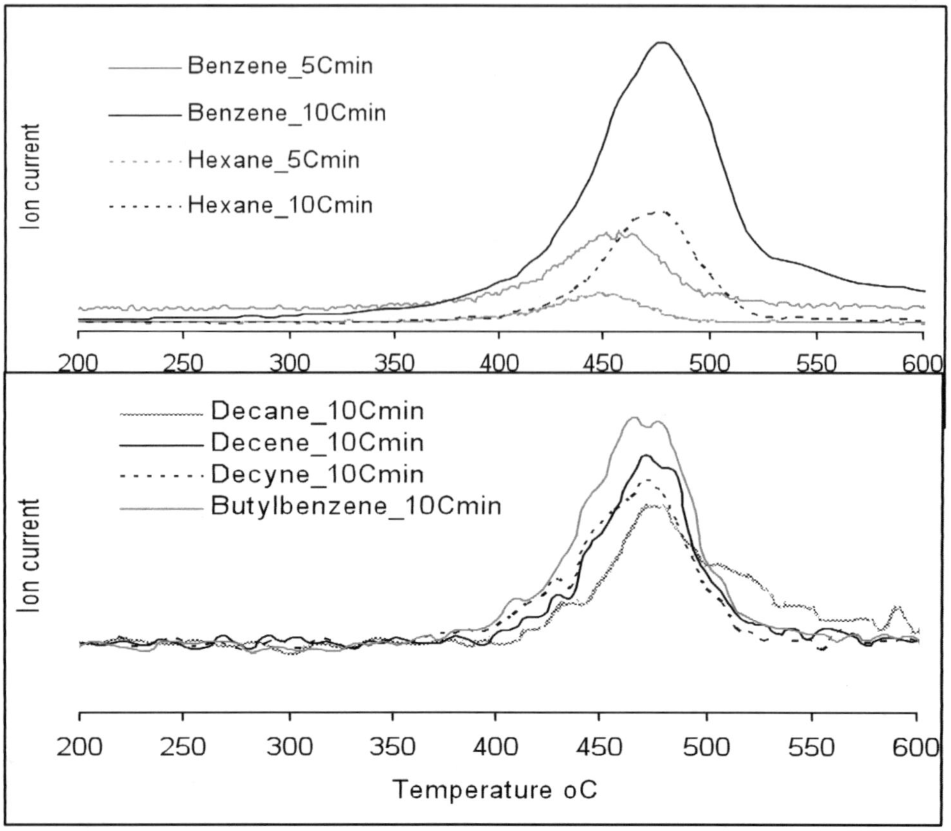

Figure-4: Comparison of the key components on thermal scale at different heating rates

Higher boiling points compounds, can not make an educated judgment just based on their evolution temperature range about their origins but the relative thermal rate dependency with other runs and relative quantitative presence could be obtained from the information obtained at different temperature rate TGAMS runs. The area under the single ion monitoring over temperature range is calculated by trapezoidal method. The correction for shift in the baseline was deducted using the minimum value of the particular component (rectangle of minimum value over temperature range). The area integral is used for the relative analysis of the components evolved. The area under the curves could be a good basis to compare the ratio information for the compounds at particular heating rate and at different heating rates. The area is divided in two section, before the organic decomposition temperature (30^0-250^0C) and organic pyrolysis temperature (250^0-600^0C). Some compounds complete the peak before 600^0C, but it is assumed that the return to base line and does not contribute extra in 600^0C section. Some compounds second peaks, like methane, water and CO_2 started before because of the mineral decomposition, thus were treated separately as not be compared with rest of the hydrocarbon compounds. Another reason is there baseline shift is different then the rest. Figure-5 shows the distribution of some components at 5C/min in two different section. For the hydrocarbons, the section of 30-250C could be ignored or used as another correction of the noise in the signal as they not get released in this section but the area is estimated because of the noise in the signals. At %C/min aromatic (benzene) dominate the other compounds and highest hydrocarbons are mainly present in the products (less than13 carbon number). The small window shoes the absence of key components (or very small percent) as compared to in the second section.

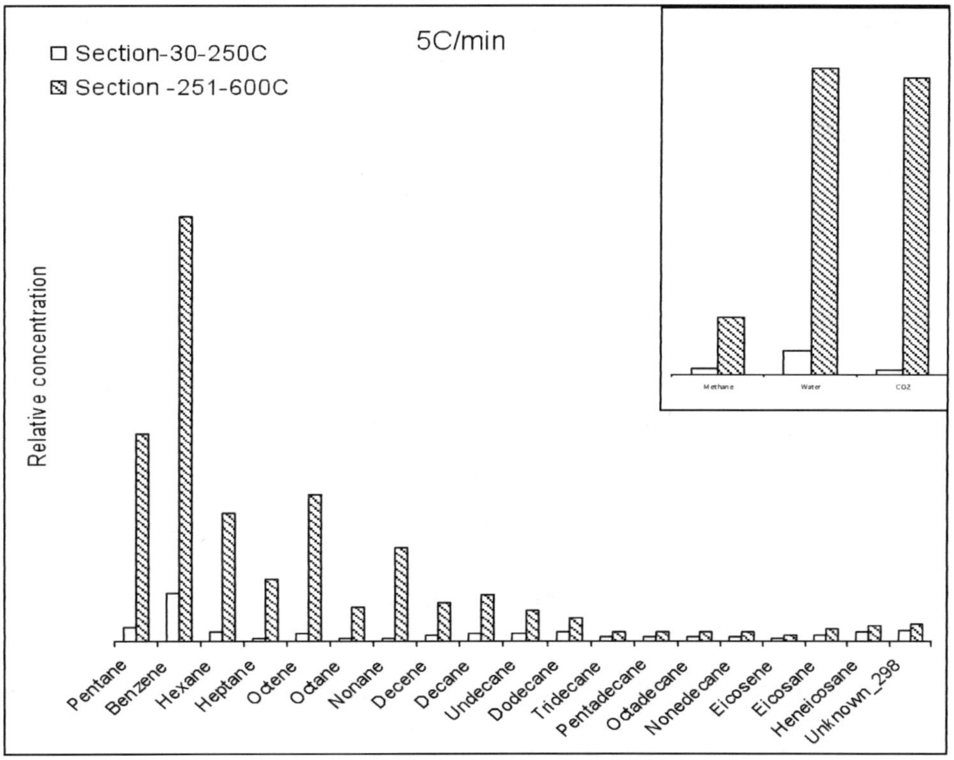

Figure-5: Relative area of ion current response of the components at 5C/min ramp rate

Conclusion

The TGA weight loss information along with the single ion monitoring of mass spectrometry provides the components identification and their thermal behavior at programmed heating rate. The obtained signal from MS identifies specific gas components during thermal analysis. The onset points (start, end and maximum rate temperatures) for a component identified my single ion monitoring reveal not only the presence but also their thermal activities at different time-temperature history. The monitoring of water and CO_2 at significant level and on the temperature scale distinguishes the combined processes such as organic decomposition and mineralogical changes. Different hydrocarbon signals at ramped thermal programs provide the understanding about the products composition, temperature dependency and the relative quantification. The source rock material is very complex and it is not possible to determine the formation route of the products. But, based on the results it could be concluded that the source rock includes a variety of hydrocarbons, like the presence of substantial aromatics and molecules are attached through different light hydrocarbons (saturated molecules and olefins). During thermal treatment, or more specifically when a source rock reaches a definitive temperature at specific heating rate, the bond connection breaks and low molecular weight products formed. At fast heating rates temperature onset points shift to higher scale side and produce more aromatics and olefins. The limitation of TGA-MS data don't allow to perform absolute quantified analysis of the products and relate it to source material. But, this kind of analysis may play a vital role in correlating GC and GCMS analysis of products obtained from pyrolysis of source rock in standard apparatus, such as Fischer-Assay and Rock-Eval. The combined analysis may permit the absolute quantification with temperature distribution of the products and further may be used in the development of a comprehensive kinetic model.

(363e) Reservoir Based Hydrothermal Reforming of Hydrocarbons

Yousef Alshammari, Imperial College London, London, United Kingdom and **Klaus Hellgardt**, Chemical Engineering, Imperial College London, London, United Kingdom

Enhanced Oil Recovery (EOR) operations may offer a potentially clean and economical method for hydrothermal production of syngas or hydrogen as useful product or energy vector. EOR operations usually employ in situ combustion (using oxygen or air), steam injection and/or CO2 injection. In situ combustion has been studied for a number of years, e.g. the THAI process invented in 1993 by M. Greaves. In this process an air injection well is used to combust heavy oil and thus provides local heat to reduce oil viscosity. The oil is then driven towards a producer well (toe to heal). It is suggested that approximately 80% ooip recovery may be accomplished with the additional benefit of some in situ oil upgrading. Air injection results in the release of an equal amount of CO, CO2 and hydrocarbons. This of course begs the question whether the return gas, which would be used to provide the required electricity for the compressors, could be converted into syngas/hydrogen and thus lead to the production of a clean energy vector. Indeed, is it conceivable to employ, for example, redundant production wells as hydrothermal reactor systems with the explicit injection of oil, steam and air/oxygen and thus remove the need for above ground reformers? In the present work a basic parametric equilibrium study has been carried out on the reactions between crude oil and EOR fluids (air, steam, and CO2) using HYSYS and employing the embedded Gibbs Reactor model. This model allows the thermodynamic analysis of the effect of the molar flowrates of injection fluids (ratio of injection fluid to oil), reservoir pressure, reservoir temperature and water cut on the composition of the produced syngas. Only a few similar investigations have been reported in the literature (Ye et al., 2009, Ersoz et al., 2003, Jeon, 2008). These studies do not normally include crude oils containing sulphur compounds, nitrogen compounds and asphaltenes. Adiabatic steam/air reforming has been compared with adiabatic CO2/air reforming and adiabatic partial oxidation in terms of product distribution, temperature, and compression load. The goal was to identify conditions under which hydrogen production is maximised (Figure 1). It was found that steam, containing 20% O2, may potentially provide the highest yield of H2, the lowest yield of CO2, with reasonable compression load compared with air or CO2. The main challenge of air injection is the huge compression load, due to its high N2 content, which can reduce its cost-effectiveness. On the other hand, the main challenge with CO2 reforming is the huge rise in the CO2 compared with the very low H2 yield, which may add additional cost for separation of CO2 above surface. In situ sequestration of CO2 underground before reaching the wellhead might be feasible.

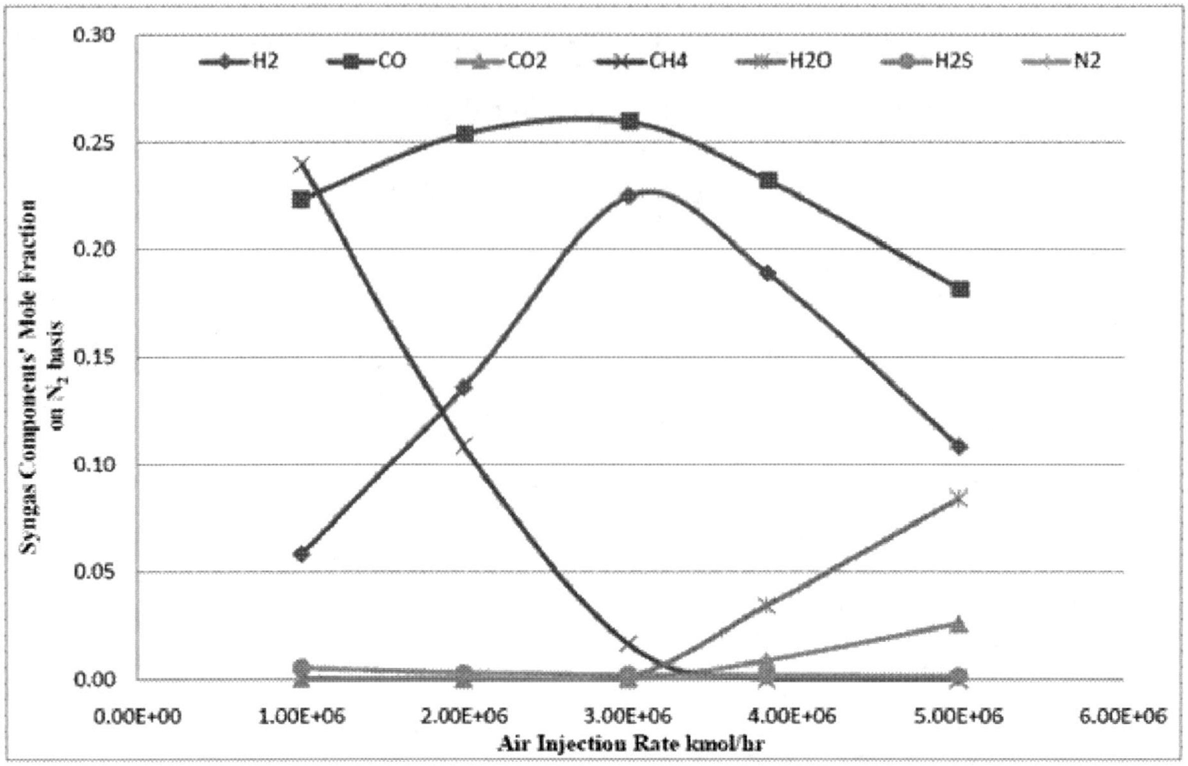

See more of this Session: Underground Thermal Treatment of Carbonaceous Materials
See more of this Group/Topical: Topical D: Chemical Engineering in Oil and Gas Production and Other Complex Subsurface Processes

(363f) Development of CFD-Based Simulation Tools for in-Situ Thermal Processing of Oil Shale/ Sands

Benjamin J. Isaac and Philip J. Smith, Chemical Engineering, University of Utah, Salt Lake City, UT

In-situ technologies are currently being explored because of their potential for reducing the environmental footprint of oil shale/sands development. However, the first generation technologies have proven to be energy-intensive, and many unknowns remain relative to optimal heating strategies, potential contamination of groundwater, and achievable production rates.

Reservoir simulation tools are typically applied to in-situ production processes. However, we are simulating a modified in-situ process where a distribution of rock size and orientation (important due to directional permeability of the rock) exists in the production bed. The rocks are heated by convective currents through the channels of the rubblized bed. Our approach is to apply CFD-based simulation tools to this modified in-situ process. The system consists of a clay-lined volume filled with rubblized oil shale that is heated by pipes fired with natural gas burners. We have employed a suite of commercial software tools, including EDEM, Matlab, Gambit, and Star-CCM+.

We have determined that a simulation with the actual geometry of the pieces of shale will provide more accurate results than treating the rubblized shale volume as a continuum with a defined permeability. DEM Solutions product, EDEM, a discrete element modeling software that is used for modeling particle movement, was used to simulate the filling of an empty bed with pieces of oil shale. This simulation models the interactions between particles and geometry as they fall into the bed. The simulation resulted in a bed full of thousands of pieces of oil shale that can be used as a representation of the actual rubblized bed. The EDEM simulation provides data about the position, size, and rotation of each particle.

Matlab was then used to write a journal file that defines the three-dimensional layout of the particles. The journal file was run in the CAD software Gambit to create the shale geometry. Gambit produces a parasolid transmit file which is a standard geometry file type that can be used in Star-CCM+. Implementation of the correct geometry representation in Star-CCM+ requires being able to represent boundary interfaces between pieces of shale and fluid.

We demonstrate the efficacy of this set of tools for the rubblized shale system, using the tools of validation and uncertainty quantification to perform a consistency analysis between simulation results and experimental data.

See more of this Session: Underground Thermal Treatment of Carbonaceous Materials

(417a) Novel Electrolyte Chemistries for Mg-Ni Rechargeable Batteries

Marie C. Kane, Brenda L. Garcia-Diaz and Ming Au, Savannah River National Laboratory, Aiken, SC

Commercial hybrid electric vehicles (HEV) and battery electric vehicles (BEV) serve as means to reduce the nation's dependence on oil. Current electric vehicles use relatively heavy nickel metal hydride (Ni-MH) rechargeable batteries. Li-ion rechargeable batteries have been developed extensively as the replacement; however, the high cost and safety concerns are still issues to be resolved before large-scale production. In this study, we propose a new highly conductive solid polymer electrolyte for Mg-Ni high electrochemical capacity batteries. The traditional corrosive alkaline aqueous electrolyte (KOH) is replaced with a dry polymer with conductivity on the order of 10^{-2} S/cm, as measured by impedance spectroscopy. Several potential novel polymer and polymer composite candidates are presented with the best-performing electrolyte results for full cell testing and cycling.

See more of this Session: Novel Battery Chemistry and Technology
See more of this Group/Topical: Fuels and Petrochemicals Division

(417b) Ni-Doped Porous Manganese Dioxide as Positive Electrode Material for Fuel Cell/Battery (FCB)System

Gang Zhao, Institution of Industral Science(IIS), The University of Tokyo, Tokyo, Japan, Bokkyu Choi, Institute of Industrial Science, The University of Tokyo, Tokyo, Japan, Chihiro Fushimi, Department of Mechanical and Biofunctional Systems, Institute of Industrial Science, The University of Tokyo, Tokyo153-8505, Japan and Atsushi Tsutsumi, the University of Tokyo, Tokyo, Japan

For a Fuel Cell/Battery (FCB) system, manganese dioxide (MnO2) could serve not only as active positive electrode materials of secondary battery but also as a catalyst for oxygen reduction (ORR) in fuel cells. In the present study, porous MnO2 materials with huge surface areas were prepared using a template method in order to provide enough space for the electrochemical redox reaction and ORR. In addition, the Ni ion was doped on the prepared porous MnO2 to enhance the proton transportation. The obtained materials were characterized by BET, XRD, and SEM. It is found that the porous Ni-MnO2 had the crystal structure like the gamma MnO2. The performance of the positive electrode composed of the porous Ni-MnO2 in FCB system was also investigated via Electrochemical Impedance Spectroscopy (EIS) and galvanostatic discharge curves. The results showed that the electrochemical impedance of this positive electrode obviously decreased while the output voltage increased both in the secondary battery and in the fuel cell, suggesting that the Ni doped MnO2 with porous structure could improve the power density of the FCB system and be suitable for high power output energy generation/storage applications.

See more of this Session: Novel Battery Chemistry and Technology
See more of this Group/Topical: Fuels and Petrochemicals Division

(417c) Non-Aqueous Single-Metal Redox Flow Batteries

Charles W. Monroe, Aaron A. Shinkle, Alice E. S. Sleightholme and Levi T. Thompson, Chemical Engineering, University of Michigan, Ann Arbor, MI

Redox flow batteries (RFBs) promise high-capacity energy storage useful for the leveling of power harvested from stochastic sources like sunlight or wind. Unlike traditional batteries, whose chemical energy is stored in solid electrode materials, RFBs store energy by exchanging charge between chemically inert electrodes and a dissolved component of the liquid phase. The choice of stable, reversible redox-active species is critical.[1] Ideally, the redox couples should be chosen to yield the highest possible potential during discharge. The use of single-metal disproportionation reactions also prevents irreversible side reactions that arise during active-species crossover.[2]

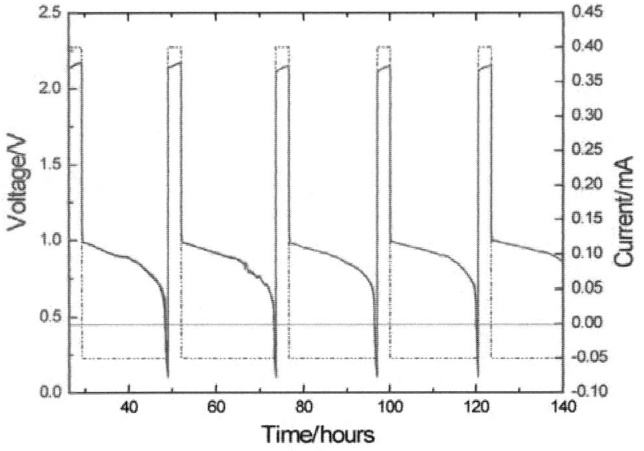

Figure 1. High coulombic efficiency is maintained over extended cycling. Charge-discharge of a nonaqueous vanadium acetylacetonate flow battery, supported by 0.1 M $TEABF_4$ in acetonitrile solvent.

Research in our groups focuses on novel nonaqueous RFB chemistries, which differ from the aqueous systems used commercially. The water-splitting side reaction at 1.23 V (or slightly higher in strong-acid support) limits aqueous RFBs to a relatively narrow voltage range. Cyclic voltammetry indicates that the replacement of water by acetonitrile affords an operating voltage increase–and consequent power-density increase–of more than 50%, while suppressing solvent degradation.

Sparingly soluble active species limit the energy density of the charge-storing liquid phase. Thus, for nonaqueous RFB chemistries, a key technical issue is to increase of active-species concentration. We have shown that this can be achieved by exploiting redox-active organometallic complexes whose ligands enhance solubility.[3] For instance, the inset figure demonstrates how a vanadium(III) acetylacetonate, or $V(acac)_3$, active species can be used to successfully implement a nonaqueous single-metal RFB that achieves coulombic efficiencies of 85-90%, comparable to those achieved when water is the solvent.

But a transition from water poses significant design challenges: it alters the field of options for active species, supporting electrolytes, membrane separator materials, and electrodes. In addition to discussing the new organometallic vanadium single-metal flow battery, we will discuss engineering aspects related to RFB design and optimization.

This presentation will describe our work on the electrochemical analysis of system stability, thermodynamic investigations that elucidate the behavior of cell potential during battery operation, physicochemical characterization to establish the transport properties of liquids and membranes, and kinetic measurements showing how electrode materials catalyze the heterogeneous reactions at the liquid/metal interfaces. Each of these aspects will be discussed in terms of its impact on the charge/discharge response of an RFB.

[1] C. Ponce de Leon et al. *J. Power Sources* **160** (2006) 716–732.

[2] E. Sum and M. Skyllas-Kazacos. *J. Power Sources* **15** (1985) 179–190.

[3] Q. Liu et al. *Electrochem. Comm.* **11** (2009) 2312–2315.

See more of this Session: Novel Battery Chemistry and Technology
See more of this Group/Topical: Fuels and Petrochemicals Division

(417d) Investigations of New Materials for Rechargeable Zinc-Air Batteries

Viktor Hacker[1], Katja Froehlich[1] and Waltraud Taucher-Mautner[2], (1)Institute of Chemical Engineering and Environmental Technology, Graz University of Technology, Graz, Austria, (2)Institute for Chemistry and Technology of Materials, Graz University of Technology, Graz, Austria

The investigations aim to develop and optimise rechargeable zinc-air batteries. Zinc-air cells are characterised by high specific energy density, inexpensive materials and good environmental compatibility. The necessary increase in cycle stability of zinc air cells will be realized by using innovative materials for all components, which are capable to avoid or to minimise the degradation and deactivation.

The characterisation of materials in half-cell tests focuses on the development and optimisation of the electrodes, the investigation of various production procedures and the influence of electrode and electrolyte composition on the rechargeability of the zinc-air cell. The investigated air electrodes consist of the gas diffusion layer, prepared by the calendar rolling technique and the active layer which contains the catalyst for both, oxygen reduction and evolution. The two layers are combined with the current collector made of titanium mesh. Carbon nanofibers are used in both layers of the electrode. Polytetrafluoroethylene (PTFE) is added as hydrophobic component and La0,6Ca0,4CoO3 perovskite is used as catalyst in the active layer. The measurements are performed in half cell tests using 3.2 M KOH as electrolyte.

See more of this Session: Novel Battery Chemistry and Technology
See more of this Group/Topical: Fuels and Petrochemicals Division

(417e) Novel Zinc-Alkali Flow Battery

Galen J. Suppes[1], Bryan D. Sawyer[2], Michael J. Gordon[2] and Michael G. Heidlage[2], (1)University of Missouri-Columbia, Columbia, MO, (2)Chemical Engineering, University of Missouri-Columbia, Columbia, MO

The presentation is on a novel flow battery backed of flow through packed-bed electrodes. Trends in performance substantiate a theoretical foundation for advancing packed-bed flow batteries.

See more of this Session: Novel Battery Chemistry and Technology
See more of this Group/Topical: Fuels and Petrochemicals Division

(443a) Opportunities, Current Status and Challenges of Renewable Transportation Biofuels

Duyen T. Nguyen and Kris Parimi, Energia Technologies, Inc., San Leandro, CA

The energy landscape is rapidly changing as cheap and readily available fossil fuels are becoming scarce. Costs of conventional petroleum based fuels have risen at a phenomenal rate due to increased demand worldwide, uncertainties and insufficiency on supply side, and geopolitical instability. Imagine the possibilities of growing our own transportation fuels from renewable domestic resources without disrupting the food supply chain, making cost effective "designer" biofuels and reducing fossil fuel imports. Alternative transportation biofuels have the potential to mitigate threats to national energy security while providing environmental benefits of being carbon neutral and minimizing harmful emissions. Evaluation of current status of transportation biofuels will be given. Technical challenges and requirements for successful transportation biofuels deployment will be discussed along with innovative approach to production of bio-jet, bio-diesel and bio-gasoline.

See more of this Session: Alternative Fuels and Enabling Technologies
See more of this Group/Topical: Fuels and Petrochemicals Division

Solar-driven gasification of carbonaceous waste feedstocks in a packed-bed reactor – Dynamic modeling and experimental validation

Nicolas Piatkowski [a], Aldo Steinfeld [a,b]*

[a] ETH Zurich, Department of Mechanical and Process Engineering, 8092 Zurich, Switzerland.

[b] Paul Scherrer Institute, Solar Technology Laboratory, 5232 Villigen PSI, Switzerland

Abstract – Thermochemical gasification of various carbonaceous waste feedstocks (specifically: scrap tire powder, industrial sludge, and sewage sludge) for high-quality syngas production is numerically modeled and experimentally demonstrated using concentrated solar process heat. The solar reactor consists of two cavities separated by a radiant emitter plate, with the upper one serving as the solar radiative absorber and the lower one containing the reacting packed bed that shrinks as the reaction progresses. The reactor is modeled by considering combined radiation-conduction-convection heat transfer coupled to the reaction kinetics, with the driving boundary condition being the applied transient solar radiative flux at the reactor aperture. Model validation is accomplished in terms of converted mass, reactor temperatures, efficiency, and solar upgrade based on experiments with an 8 kW reactor prototype subjected to solar flux concentrations up to 2560 suns and packed-bed temperatures up to 1490 K. The transient operation of a 200 kW reactor for gasifying industrial sludge is simulated for a solar day consisting of 8 hours irradiation and 4 hours cooling, yielding a solar-to-fuel energy conversion efficiency of 89%.

1. Introduction

Solar steam-gasification of carbonaceous materials makes use of concentrated solar energy to convert solid feedstocks such as coal, biomass, or carbon-containing wastes into high-quality synthesis gas (syngas) – mainly H_2 and CO – applicable for power generation in combined cycles and fuel cells, or for Fischer-Tropsch processing to liquid fuels. Conventional autothermal gasification requires a portion of the injected feedstock mass to be combusted with pure O_2 to supply high-temperature process heat for the endothermic steam-gasification reaction. For example, 12 MJ/kg are required to steam-gasify a typical bituminous coal of LHV 34 MJ/kg [1]. Thus, about 35% of the injected coal mass must be burned uniquely to power the reaction, which inherently decreases coal utilization and contaminates the product gases. Although coal is a primary feedstock for gasification, carbonaceous waste gasification has also been investigated in the laboratory and modern waste-to-energy systems [2]-[5], where the combustion products were separated from the product gas and discharged to the environment [3].

In contrast, solar-driven gasification is free of pollutants and/or combustion byproducts. The product syngas has a lower CO_2 intensity because its calorific value is solar-upgraded over that of the original feedstock by an amount equal to the enthalpy change of the

* Corresponding author: Tel. +41-44-6327929; Fax. +41-44-6321065; E-mail: aldo.steinfeld@ethz.ch

endothermic reaction. A 2nd-Law (exergy) analysis indicated that Brayton-Rankine combined power cycles running on solar-made syngas from coal can double the specific electric output per unit mass of coal and, consequently, avoid half the specific CO_2 emissions of conventional coal-fired generation plants [1]. Solar gasification is estimated to reduce CO2 emissions by at least 30% vis-a-vis conventional autothermal gasifiers [6]. In the long term, it has the potential of becoming economically competitive even before the application of government subsidies and/or credit for pollution abatement, mainly due to the elimination of the energy-intensive air separation unit as well as significant savings (> 35%) in feedstock consumption [7]. Solar steam-gasification is ultimately a means of chemically storing intermittent solar energy in a dispatchable form.

Solar gasification of petroleum coke and coal was studied in directly irradiated fluidized-bed and vortex-flow solar reactors [6][8]-[12]. Direct irradiation of particle suspensions was found to be an effective means of heat transfer directly to the reaction site, leading to extremely fast heating rates (~1000 K/s) and enhanced kinetics [12]. However, the transparent window needed for optical access of concentrated solar radiation becomes a troublesome, critical component under high pressures and severe atmospheres. The large volume flow rates of inert carrier gas or excess steam necessary for fluidization and for protecting the window displace syngas, decreasing production and energy conversion efficiency [13]. Additionally, effective fluidization requires small particle sizes (< 5mm) and narrow particle size distributions, making particle suspension reactors ill-suited to highly variable carbonaceous feedstocks such as wastes and more costly due to extra feedstock handling operations [13]. Indirectly irradiated reactors eliminate the need for a window at the expense of having less efficient heat transfer – by conduction – through the walls of an opaque absorber. Thus, the disadvantages are linked to the limitations imposed by the materials of the absorber, with regards to maximum operating temperature, inertness to the chemical reaction, thermal conductivity, radiative absorptance, and resistance to thermal shocks. A number of exploratory studies investigated solar gasification of carbonaceous feedstocks such as coal, oil shale, biomass, scrap tire, and polyethylene wastes, and the carbothermal reduction of electric arc furnace dusts in packed-bed reactors [14][15][16]. Recently, solar gasification of carbonaceous feedstocks including wastes was experimentally studied and demonstrated in an indirectly irradiated packed-bed reactor [17]. This reactor concept was found to be robust for these highly variable feedstocks. Similarly, gasification of sewage sludge in a radiation-driven batch reactor was investigated [18]. In this context, solar gasification should be viewed as a viable waste management and fuel recovery option.

Numerical studies performed in the area of packed-bed solar reactors have considered the thermochemical decomposition of CaCO3, carbothermal reduction of ZnO, and the steam-gasification of charcoal [19]-[21]. The packed-beds underwent rapid heating on the irradiated surfaces, but heat transfer through the porous bed was found to be the reaction limiting factor as is typical of ablation regimes [19][21][22].

This paper presents the design, experimentation, and dynamic modeling of a solar packed-bed reactor for the steam gasification of highly heterogeneous waste carbonaceous feedstocks. The unsteady-state governing energy equations and associated boundary conditions are formulated for the model and solved numerically. The model is validated on the basis of reactor temperatures, converted mass, energy conversion efficiency and solar upgrade with experimental results obtained using a 8 kW solar reactor prototype tested in a high-flux solar simulator. Finally, the model is used to assess the performance of a 200 kW solar-driven packed-bed gasifier.

2. Solar Reactor Configuration and Experimental Set-up

The solar reactor is schematically shown in Figure 2. Its design and associated experimental setup have been previously described in detail [17]; only the main features are highlighted here. It is specifically designed for a "beam-down" (Cassegrain) optical configuration of solar towers, which makes use of a hyperbolic reflector at the top of a solar tower to redirect sunlight collected by the heliostat field to a solar receiver located at ground level, facilitating feedstock feeding and off-gas handling [23]. The reactor consists of two cavities in series. The upper cavity functions as the solar absorber and contains a windowed aperture to let in concentrated solar radiation. The lower cavity functions as the reaction chamber and contains the packed bed on top of the steam injector. An emitter plate separates the two cavities and acts as solar absorber and radiant emitter. Its main purpose is to protect the window from deposition of particles or condensable gases. A CPC [24] is incorporated at the reactor aperture and affixed to the metal front to further augment the incident solar flux and minimize re-radiation losses. The reactor is operated in batch mode, with the packed bed shrinking as the gasification reaction progresses. This arrangement enables the processing of a wide range of feedstock's sizes and forms.

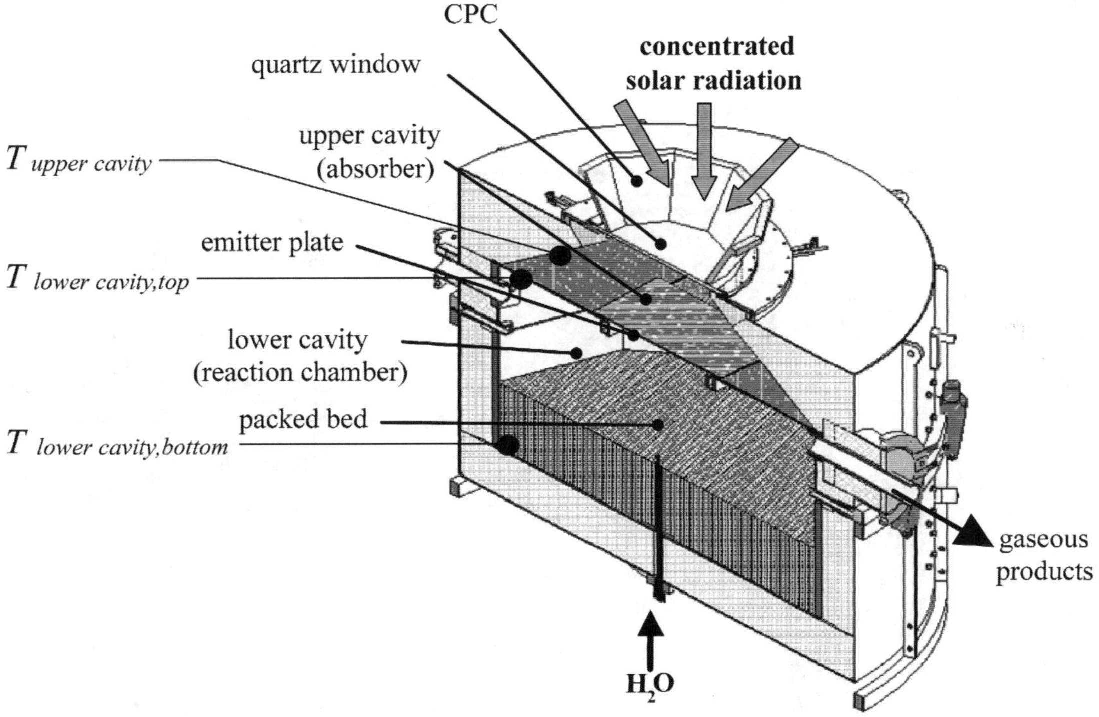

Figure 1 Schematic of the packed-bed solar reactor, featuring two cavities separated by an emitter plate, with the upper one serving as the radiative absorber and the lower one containing the reacting packed bed that shrinks as the reaction progresses.

An 8 kW reactor prototype was fabricated with an upper cavity containing a 6.5 cm-diameter aperture and a 14.3 cm-diameter, 16 cm-height lower cylindrical cavity. The 6-mm thick emitter plate was made of SiC-coated graphite so as to achieve high thermal conductivity and good thermal equalization while maintaining resistance to the highly oxidative atmosphere.

The upper cavity was sealed by a 3 mm-thick fused quartz window located at the aperture. The lower cavity was lined with 6 mm-thick SiC tiles and with 70 mm-thick Al_2O_3-SiO_2-insulation. A steam-argon mixture at 400 K with liquid water flow rates up to 8 ml/min and an Ar flow rate of 2 l_N/min was introduced from below the packed bed. Product gases exited through a lateral outlet port where they underwent initial cooling down to below 500 K and flowed across a wet filter to remove solid matter and accomplish the final cooling to ambient temperature. Gas composition was analyzed by gas chromatography (Agilent High Speed Micro GC G2890A, equipped with molecular sieve 5A and HaySep A capillary columns) with a sampling period of 145 seconds. All product gases were flared. Temperatures were measured at the top $T_{lower\ cavity,top}$ and bottom $T_{lower\ cavity,bottom}$ of the lower cavity with type-K thermocouples, and at the surface of the insulation cone $T_{upper\ cavity}$ with a type-S thermocouple, as indicated in Figure 1. The experiments were carried out at PSI's High-Flux Solar Simulator (HFSS): an array of 10 Xenon arcs, close-coupled to ellipsoidal reflectors, which can simulate the radiative heat transfer characteristics of highly concentrating solar systems [25]. Radiative fluxes incident into the reactor were measured optically with a calibrated CCD camera on a water-cooled Al_2O_3-plasma coated Lambertian target and with a water-cooled calorimeter. The maximum radiative flux at the reactor's aperture was equivalent to a solar concentration ratio of 2560 suns (1 sun = 1 kW/m^2). Figure 2 shows the solar reactor setup with associated peripheral devices as used in the HFSS.

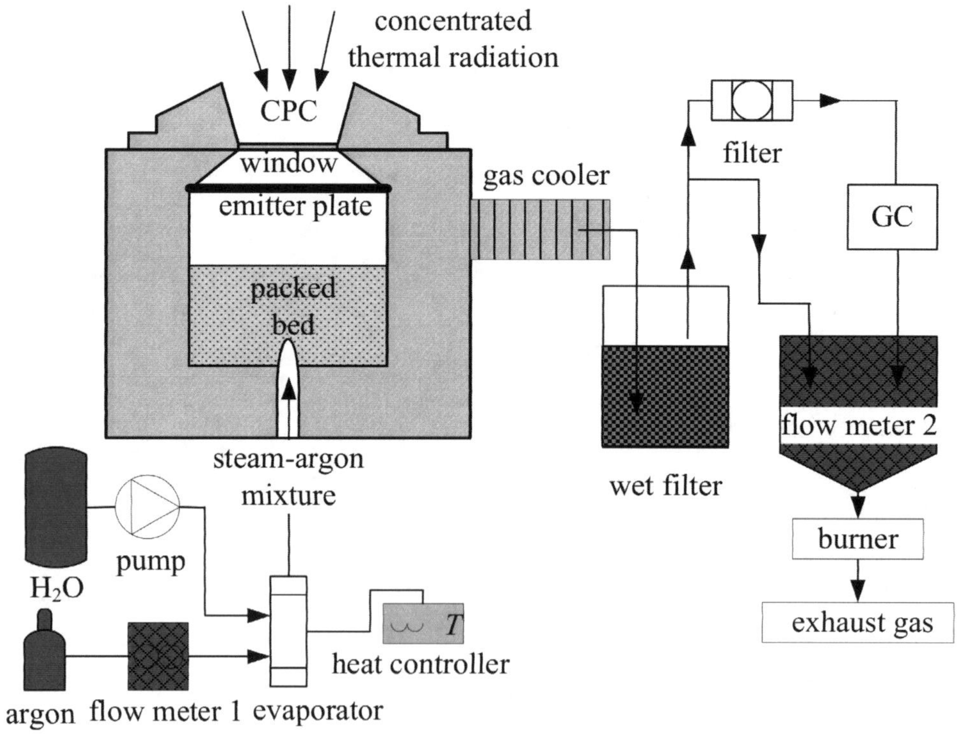

Figure 2 Scheme of solar reactor experimental set-up and associated peripheral devices as used in the High-Flux Solar Simulator.

3. Feedstocks

Three waste feedstocks were loaded to validate the reactor model: scrap tire powder, industrial sludge (a mixture of paints, dyes, inks and solvents), and dewatered sewage sludge. The characteristics of these materials vary greatly in terms of physical and thermal properties as well as proximate and ultimate analyses as shown in Table 1. Of particular note are the large volatile fractions making pyrolytic reactions of principle importance [26]. Variations in particle size, bed porosity, and packed bed effective density were considered for scrap tire powder in the gasification regime and influenced the packed bed dynamics [27]. For the other materials, constant particle sizes, porosities and densities were considered.

Table 1 Ultimate and proximate analyses and packed bed properties of carbonaceous waste feedstocks considered for steam-gasification.

Feedstock	scrap tire powder	industrial sludge	sewage sludge
Ultimate Analysis			
C (wt. %)	82.20	67.48	51.39
H (wt. %)	7.30	8.94	7.59
O (wt. %)	3.60	22.60	32.81
N (wt. %)	0.39	0.91	6.64
S (wt. %)	1.90	0.44	1.57
Proximate Analysis			
Ash (wt %)	4.5	19.9	33.5
Volatile (wt %)	67.5	62.2	51.9
Water (wt %)	1.0	5.5	7.2
C-fix (wt. %)	27.0	12.4	7.4
$LHV_{feedstock}$ (kJ/kg)	35,515	19,305	10,288
Bed Physical Properties			
Particle size range (mm)	0.1-2	10-30	0.1-3
Particle form	regular particles	irregular chunks	regular particles
ρ_{eff} (kg/m^3)	380	440	770
Bed porosity	0.3	0.4	0.3

The gasification process is represented by the simplified net stochiometric reaction:

$$C_1H_xO_yS_uN_v + (1-y)H_2O = \left[\frac{x}{2} + 1 - y - u\right]H_2 + CO + uH_2S + \frac{v}{2}N_2 \qquad (1)$$

The specific reaction enthalpy change Δh_{chem}, shown in Figure 3 for the three feedstocks, was determined from the equilibrium composition of the carbonaceous feedstock-steam system at T_{bed} [1][19]. The reaction enthalpies of pyrolytic reactions were neglected due to the huge uncertainty in their determination and their low value compared to those for gasification [28].

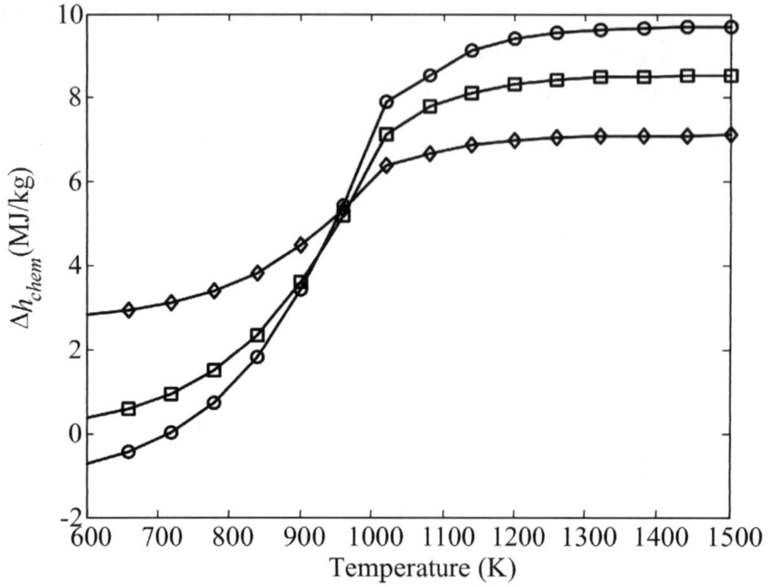

Figure 3 Specific enthalpy change Δh_{chem} of the steam-gasification of the carbonaceous waste material (eq. (1)) as a function of temperature T, with the reactants and products both at temperature T, and the products having the equilibrium composition. Legend: ◊ – sewage sludge, □ – industrial sludge, ○ – scrap tire powder

4. Reactor model

Three regions of interest were modeled: 1) The upper cavity, which receives concentrated solar radiation as the boundary condition, 2) the lower cavity, which contains the reacting packed bed, and 3) the reactor insulation.

Upper cavity model – Figure 4(a) depicts the heat and mass flows affecting the upper cavity of the reactor. Concentrated solar radiative flux q_{solar} arriving at the external surface of the window serves as the driving boundary condition. Due to the CPC placed at the reactor aperture, q_{solar} can be considered diffuse. Diffuse radiative flux $q_{emitter,lower\ cavity}$ is also emitted to the lower cavity from the emitter plate ring elements facing the lower cavity. The insulation cone participates through the conductive flux $q_{inscone,ins}$. An Ar purge gas flow of not more than 2 l_N/min is injected into the upper cavity at the 'cold surface' – a 16 mm water-cooled gap between the insulation cone and the plane of the window – and exits via the space between the insulation cone and the emitter plate.

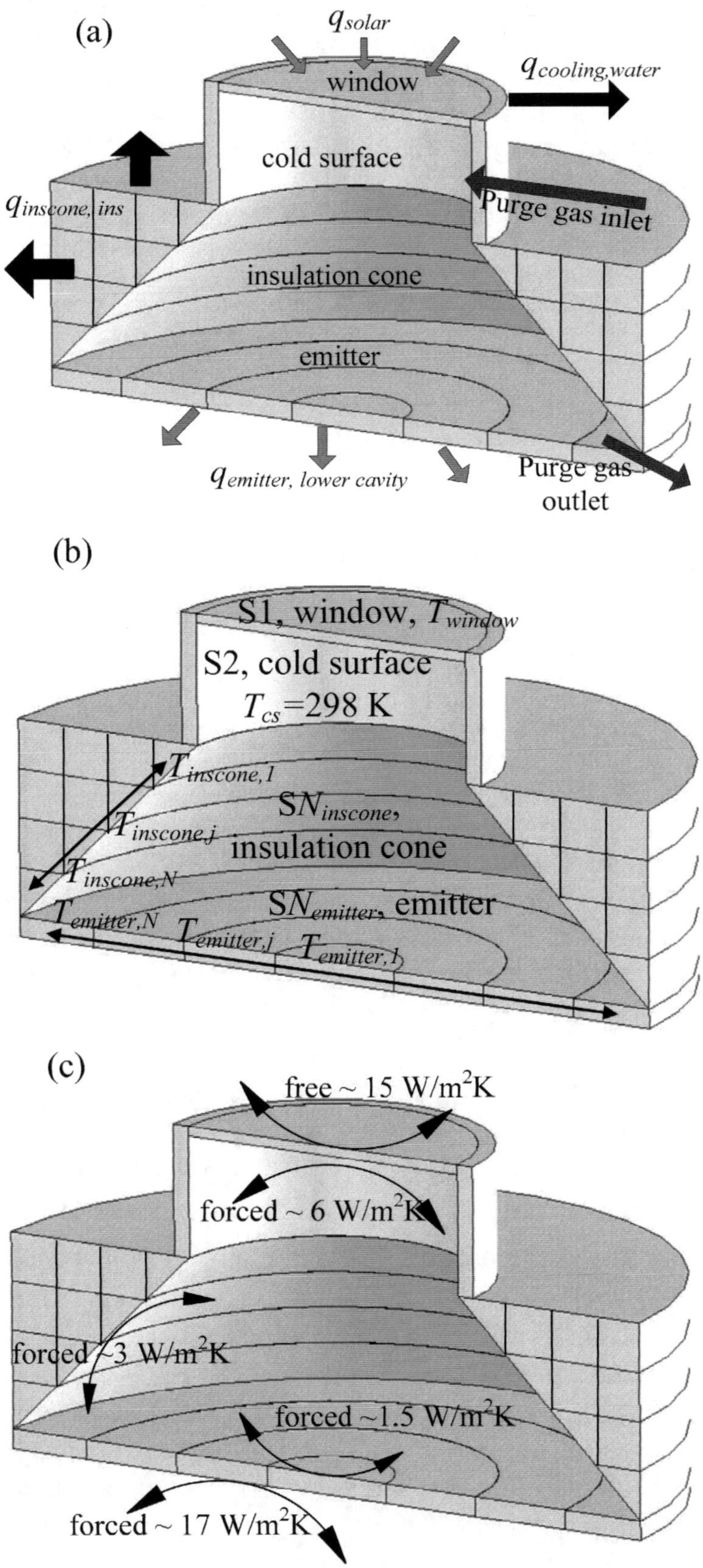

Figure 4 Schematic of the modeled reactor upper cavity: a) heat transfer; b) surfaces considered in radiative exchange; c) convective heat transfer coefficients.

The radiative exchange within the upper cavity was solved by applying the radiosity formulation for an enclosure with N isothermal surfaces, as shown in Figure 4(b): one window, one cold surface, $N_{emitter}$ concentric rings of the emitter plate, and $N_{inscone}$ conical ring surfaces. The spectrally-selective quartz window S1 was modeled as a semi-transparent disk using the semi-gray band approximation over three spectral bands with hemispherical emittance E, transmittance V and reflectance R listed in Table 2 [29][30]. The window increases reflection losses of directly incident solar radiation, but decreases long wave radiation losses compared to a windowless cavity [31].

Table 2 Spectral hemispherical properties of the quartz window

Spectral band	0-0.17 μm	0.17-3.5 μm	3.5-∞ μm
E	0.906	0.028	0.866
V	0.0	0.835	0.0
R	0.094	0.136	0.134

The net radiative flux on each surface k of the upper cavity for each spectral band b is given by [32].

$$q_{k,b} = q_{o,k,b} - q_{i,k,b} + q_{l,k,b} - q_{e,k,b} \tag{2}$$

The first two terms of eq. (2) - radiosity and incident fluxes - are applicable to all surfaces while the last two - externally leaving and applied fluxes - only to semi-transparent surfaces. $q_{o,k,b}$ is the spectral radiosity flux from surface k directed towards the inside of the enclosure and is given by:

$$q_{o,k,b} = FF_{b,\lambda T_k} E_{k,b} \sigma T_k^4 + FF_{b,\lambda T_{solar}} V_{k,b} q_{e,k} + R_{k,b} q_{i,k,b} \tag{3}$$

where $FF_{b,\lambda T_k}$ is the fractional function in spectral band b for radiation originating from a surface at temperature T_k , analytically determined using the series expansion [32]. $q_{e,k}$ is the externally supplied flux to surface k and is equal to q_{solar} for the window S1. The last term of eq. (3) is the reflection of incident flux arriving at surface k from within the enclosure, given by:

$$q_{i,k,b} = \sum_{j=1}^{N} q_{o,j,b} F_{k-j} \tag{4}$$

where $F_{k\text{-}j}$ is the view factor from surface k to surface j and $q_{o,j,b}$ is the radiosity flux from surface j in spectral band b which arrives at surface k. The view factors for all surfaces $F_{k\text{-}j}$ were computed by Monte-Carlo ray tracing. Eq. (3) and (4) are coupled and must be solved iteratively until $\left|1 - q_{i,k,b}^{new} / q_{i,k,b}^{old}\right| \le 10^{-4}$ with superscripts *new* and *old* representing values of successive iteration steps. $q_{l,k,b}$ is zero for opaque surfaces and is the spectral radiative flux leaving surface k, directed towards the environment outside the enclosure,

$$q_{l,k,b} = R_{k,b} q_{e,k} + \underbrace{FF_{b,\lambda T_k} E_{k,b} \sigma T_k^4 + V_{k,b} q_{i,k,b}}_{q_{rerad,b}} \tag{5}$$

with $q_{rerad,b}$ representing the re-radiation losses – self emission of the window and transmission of radiative energy incident on the interior surface of the window – from the upper cavity for spectral band b. The individual spectral band solutions of eq. (2) are summed over all bands to determine the net flux.

$$q_k = \sum_{b=1}^{N_{band}} q_{k,b} \tag{6}$$

The temperature of the window was determined from the unsteady energy balance which considers the net radiation on the window, determined from eq. (2), and the various cooling modes,

$$\frac{dH_{window}}{dt} = \left(q_1 - q_{conv,env} - q_{conv,upper\ cavity} - q_{cooling,water}\right) A_{window} \quad (7)$$

The temperature of the water-cooled 'cold surface' is 298 K, imposing a very strong heat sink in the upper cavity: nearly 50% of the applied radiative power is lost there. The emitter plate ($SN_{emitter}$) couples the upper cavity to the lower cavity. It is modeled as gray-diffuse concentric rings with an total emissivity of 0.88 at a temperature $T_{emitter,j}$ determined for each ring element by:

$$\frac{dH_{emitter,j}}{dt} = (q_{upper\ cavity,emitter,j} + q_{lower\ cavity,emitter,j}) A_{emitter,j} + \sum \dot{Q}_{cond,j} \\ \left(-q_{conv,emitter,upper\ cavity} - q_{conv,emitter,lower\ cavity}\right) A_{emitter,j} \quad (8)$$

where $\sum \dot{Q}_{cond,j}$ represents radial conductive heat transfer and $q_{lower\ cavity,\ emitter,j}$ is the net radiative flux on the surface of the emitter plate facing the lower cavity, obtained from the solution of the radiative exchange of the lower cavity (see eq. (13)). The surfaces of the insulation cone ($SN_{inscone}$) were modeled as gray-diffuse conical rings with an emissivity of 0.6 at $T_{inscone,j}$, determined through the transient energy balance considering 2D conduction and convective heat transfer,

$$\frac{dH_{inscone,j}}{dt} = \left(q_{inscone,j} - q_{conv,inscone}\right) A_{inscone,j} + \sum \dot{Q}_{cond,j} \quad (9)$$

All upper cavity internal surfaces see forced convection heat transfer due to the purge gas flow. The window is additionally cooled by natural convection to the environment. Product gases at a temperature of the bed top surface $T_{bed,top}$ contribute to forced convection on the emitter plate from the lower cavity side. Forced convective heat transfer on the window and emitter plate was calculated with the Nusselt correlation for a flow over a flat plate [33],

$$Nu = \frac{hL_c}{k} = 0.664\, Re_{L_c}^{0.5}\, Pr^{1/3} \quad (10)$$

using window and emitter plate diameters as the characteristic dimensions L_c. Forced convective heat transfer affecting the insulating cone was modeled as developing laminar flow in a duct with mean diameter $\overline{D}$ and length $L_{upper\ cavity}$ [33],

$$Nu = 3.66 + \frac{0.065(\overline{D}/L_{upper\ cavity})\, Re\, Pr}{1 + 0.04\left[(\overline{D}/L_{upper\ cavity})\, Re\, Pr\right]^{2/3}} \quad (11)$$

Heat transfer coefficients were determined in the range 6-4 W/m^2K for the window, 2-1.5 W/m^2K for the insulating ring, and 1.8-0.6 W/m^2K for the emitter plate. Finally, the solar absorption efficiency of the upper cavity is defined as:

$$\eta_{upper\ cavity} = 1 - \frac{q_{l,1}}{q_{solar}} \quad (12)$$

Lower cavity model – The lower cavity contains the reacting packed bed, which is coupled to the upper cavity by means of the common emitter plate ring elements. A detailed quasi-1D heat transfer model was previously formulated [19]. The radiative flux q_j between the emitter plate ring elements, exposed SiC tiles, and the bed surface was solved with the radiosity formulation for a packed bed that shrank:

$$\sum_{j=1}^{N_{lower\ cavity}} \left(\frac{\delta_{kj}}{\varepsilon_j} - F_{k-j} \frac{1-\varepsilon_j}{\varepsilon_j} \right) q_j = \sum_{j=1}^{N_{lower\ cavity}} F_{k-j} \sigma (T_k^4 - T_j^4) \quad for\ k = 1 \ldots N_{lower\ cavity} \tag{13}$$

Effective conductive heat transfer $\dot{Q}_{cond,y}$ in the axial direction within the packed bed made use of an experimentally verified effective thermal conductivity k_{eff} which coupled conductive, convective, and radiative heat transfer modes [19]. Additionally, heat transfer between the SiC tiles and the packed bed was handled by the term $\dot{Q}_{tile,bed}$, considering the temperature at the interface. The 1D unsteady energy conservation equation on the packed bed is:

$$m_{feedstock} Cp_{eff} \frac{dT_{bed}}{dt} = \dot{Q}_{cond,y} - \frac{dm_c}{dt} \Delta h_{chem}(T_{bed}) + \dot{Q}_{tile,bed} \tag{14}$$

Eq. (14) was solved with the fully-implicit finite volume (FV) technique [34]. The reaction rate $\frac{dm_c}{dt}$ was expressed based on the sum of 1st-order reactions of N pseudo-components [26]:

$$\frac{dm_c}{dt} = \sum_{j=1}^{N} m_{c,initial,j} k_j' \left(1 - X_j\right) \tag{15}$$

where the conversion $X_j = x_j(T,t) / x_{t=\infty,j}$ and $x_j(T,t)$ denotes the mass fraction of pseudo-component j having reacted until temperature T and time t, and $x_{t=\infty,j}$ is the total mass fraction of pseudo-component j reacted at time $t = \infty$. k_j' is assumed to obey the Arrhenius law, with apparent activation energies and frequency factors experimentally obtained by thermogravimetric measurements [26]. The bed shrink rate is expressed as:

$$\frac{dy}{dt} = \frac{1}{\rho_{eff} A_{bed}} \frac{dm_c}{dt} \tag{16}$$

Changes in the particle sizes, density and porosity were considered within the gasification regime for scrap tire powder [27]. Due to the relatively high thermal conductivity of the SiC tiles (~25 W/mK), heat conduction through the SiC tiles was an important factor in delivering heat to deeper regions of the packed bed. The SiC tile temperatures were determined using a 1D finite volume grid which considered heat transport by conduction in the SiC tiles, heat exchange with the packed bed, heat exchange with the insulation and the grid size changes associated with the packed bed shrink rate:

$$m_{tile} Cp_{tile} \frac{dT_{tile}}{dt} = \dot{Q}_{cond,y} - \dot{Q}_{tile,bed} - \dot{Q}_{tile,ins} - H_{tile} \frac{dy}{dt} \tag{17}$$

The 1D bed assumption was justified by the observation of a flat bed top surface after each experimental run. This assumption was verified by solving a 2D packed bed thermal model using a fully-implicit FV technique in cylindrical coordinates, resulting in nearly flat radial temperature profiles. CFD simulations in CFX of gas flow through porous media were conducted in order to support the assumption of a constant steam concentration required by the reaction rate expressions used in eq. (14). The pressure drop across the porous packed-bed is given by [35]:

$$\nabla P = -\frac{\mu}{K_{perm}} \overrightarrow{W} + F_s \left| \overrightarrow{W} \right| W \tag{18}$$

with the permeability K_{perm} and Dupuit-Forchheimer coefficient F_s determined from the particle size and bed porosity using the Ergun equation [36],

$$\nabla P = -\underbrace{\left(\frac{150(1-\phi)^2}{d_p^2\phi^3}\right)}_{1/K_{perm}}\mu W - \underbrace{\left(\frac{1.75(1-\phi)\rho}{d_p\phi^3}\right)}_{F_s}W^2 \qquad (19)$$

Water vapor mass fraction and stream lines for a steam-Ar flow through a representative porous packed bed are shown in Figure 5 (a ¼ slice is shown), for a flow of 2 l_N/min Ar and 4 ml/min liquid water in a 14.3 cm-diameter packed bed of representative particles ($d_p \cong 2\text{mm}$, $\phi \cong 0.56$, $\rho \cong 0.6\ \text{kg/m}^3$). The stream lines show a strong radial distribution away from the inlet and are uniformly spaced at the height of the outlet indicating that the strengths of advection to diffusion are comparable (Peclet number ≈ 1). The water vapor distribution is uniform along the radius of the packed bed at regions at and above the inlet which justifies the 1D assumption.

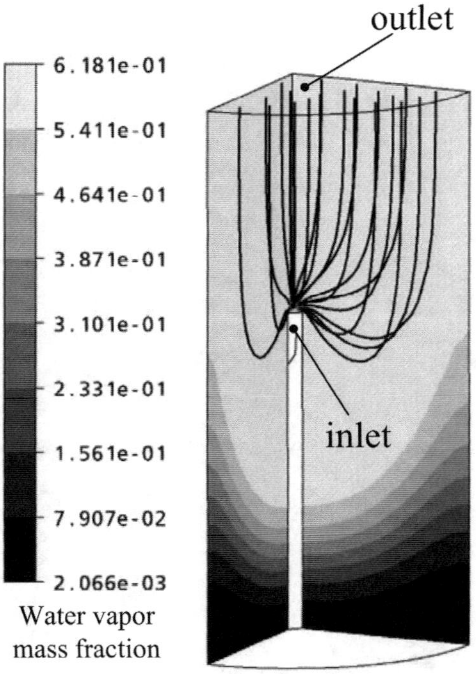

Figure 5 Water vapor mass fraction and stream lines for a steam-Ar flow through the porous packed bed.

Insulation model – The insulation exchanges heat by conduction with the upper and lower cavities and by convection with the environment. The 2D energy equation is given by eq. (20) and solved using a fully-implicit FV technique [34].

$$\frac{dH}{dt} = \frac{1}{r}\frac{\partial}{\partial r}\left(kr\frac{\partial T}{\partial r}\right) + \frac{\partial}{\partial y}\left(k\frac{\partial T}{\partial y}\right) + S \qquad (20)$$

where S is a volumetric source/sink and includes the thermal exchange with the environment, cooling water and/or wall tile components. Natural convection heat transfer to the environment was calculated for local heat transfer coefficients using Nusselt correlations [33] for vertical plate: $$Nu = \left\{0.825 + \frac{0.387 Ra_{L_c}^{1/6}}{\left[1+\left(0.492/Pr\right)^{9/16}\right]^{8/27}}\right\} \quad (21)$$

for horizontal plate (hot surface facing up): $$Nu = 0.54 Ra_{L_c}^{1/4} \quad (22)$$

for horizontal plate (hot surface facing down): $$Nu = 0.27 Ra_{L_c}^{1/4} \quad (23)$$

with L_c being determined for each surface grid element based on the ratio of surface area to perimeter. Heat transfer coefficients were in the 1-2 W/m^{2}K range. Water cooling affecting the reactor front was modeled as an internal forced turbulent flow of water in a duct with an L_c as the hydraulic diameter D_h with the Nusselt correlation [37],

$$Nu = \frac{(f/8)\left(Re_{L_c} - 1000\right) Pr}{1 + 12.7(f/8)^{0.5}\left(Pr^{2/3} - 1\right)} \quad (24)$$

where f is the friction factor with roughness values taken for new stainless steel,

$$\frac{1}{f^{1/2}} \cong -1.8 \log\left[\frac{6.9}{Re_{L_c}} + \left(\frac{\varepsilon_r / D_h}{3.7}\right)^{1.11}\right] \quad (25)$$

5. Model validation and discussion

Figure 6 shows the applied solar radiative flux and the resulting experimentally measured and numerically modeled temperatures for the steam gasification of the three waste feedstocks investigated.

(a)

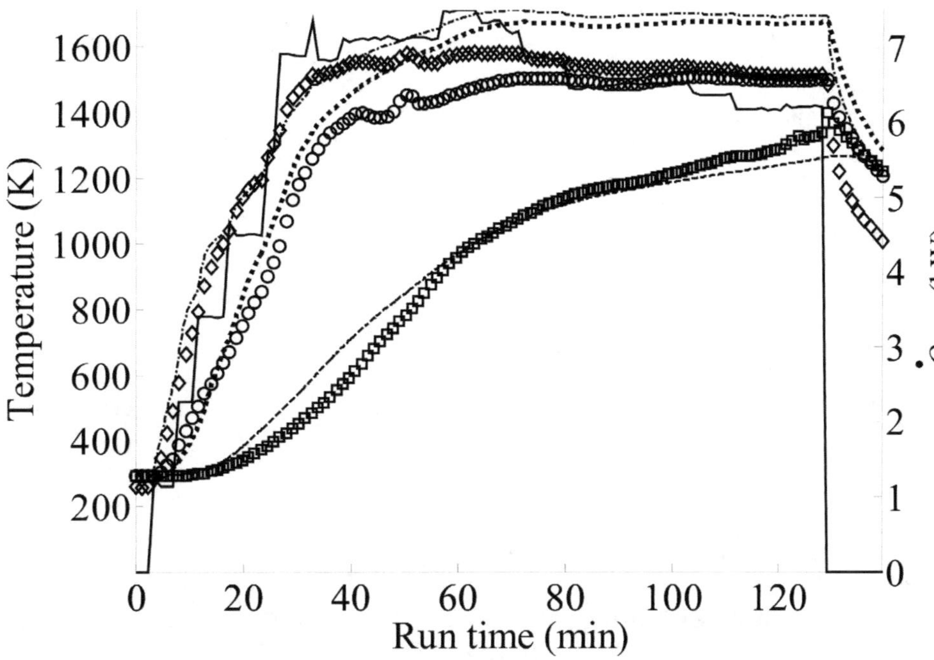

(b)

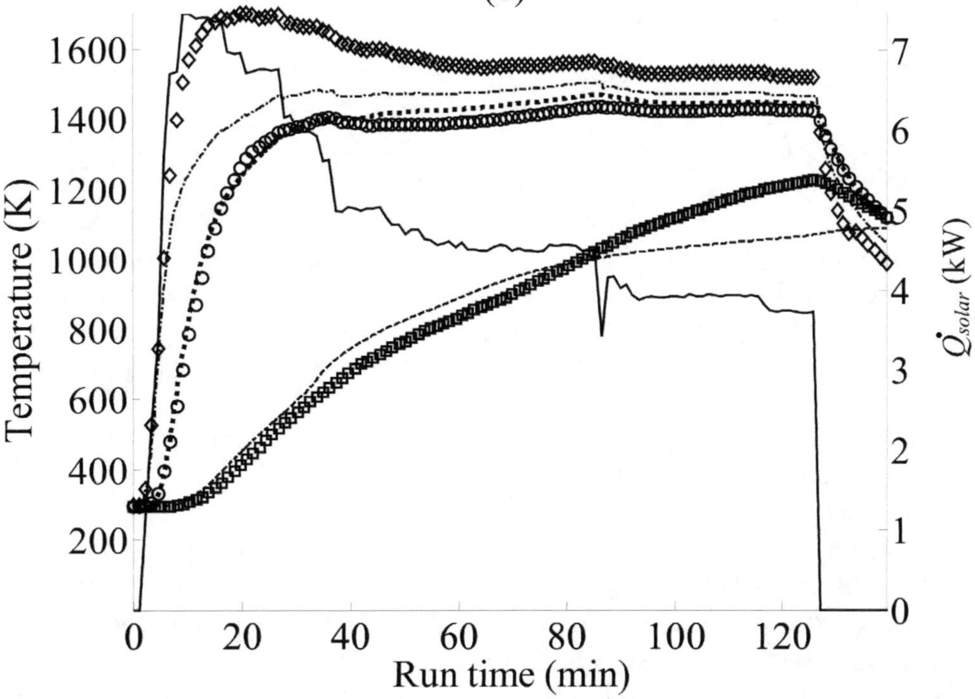

(c)

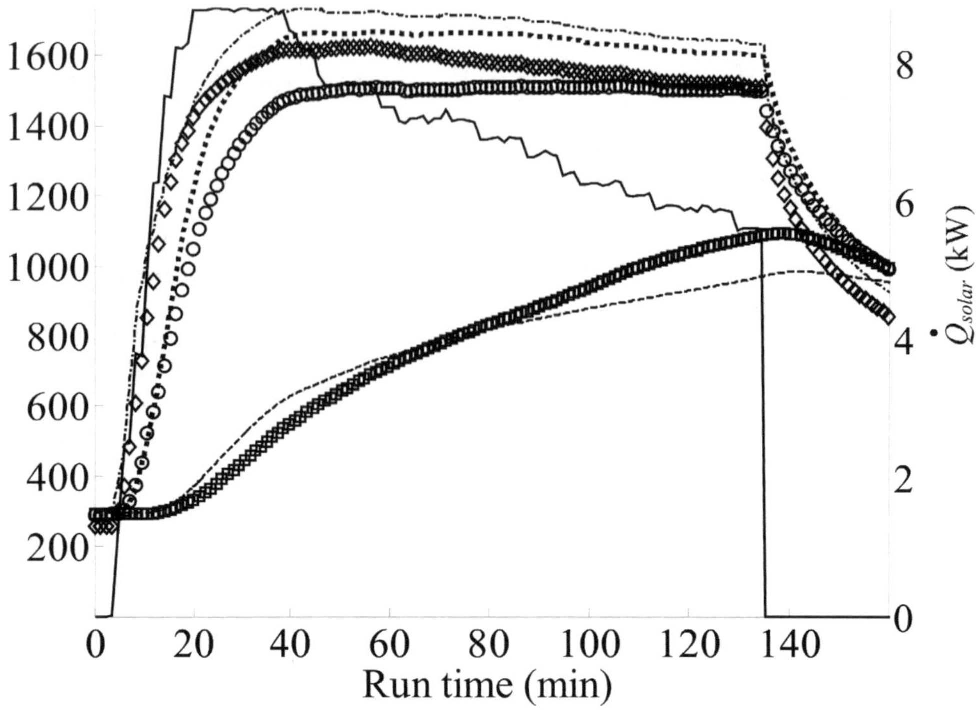

Figure 6 Variation of the experimentally measured and numerically simulated temperatures under variable radiative power inputs during the solar steam-gasification of: (a) scrap tire powder, (b) industrial sludge and (c) sewage sludge. Legends: ◊ $T_{upper\ cavity}$ (exp), ○ $T_{lower\ cavity,\ top}$ (exp), □ $T_{lower\ cavity,\ bottom}$ (exp),

—· $T_{upper\ cavity}$ (model), ·· $T_{lower\ cavity,\ top}$ (model), -- $T_{lower\ cavity,\ bottom}$ (model), – $\dot{Q}_{solar}$

The overall matching of the modeled and measured temperature profiles is good. The dynamics of the measured and modeled $T_{upper\ cavity}$ agree quite well in the heating phase under rather different irradiation conditions. It must be noted that the significant discrepancy between the measured and modeled $T_{upper\ cavity}$ in Figure 6(b) stems presumably from direct irradiation of the thermocouple during the experiment. In Figure 6(c), the lower measured temperatures stem from the melting of ash deposits on the bed top surface which imposed strong heat sinks not considered in the model. Changes in particle size and porosity during the reaction were considered in the gasification regime for scrap tire powder but not for the other two feedstocks. These physical changes, particularly in the strongly represented pyrolytic regimes, may play an important role.

The model appears to replicate well the onset of important endothermic reactions above 1000K for materials with large fixed carbon fractions. This is seen in Figure 6(a) and is characterized by a flattening of the curve of $T_{lower\ cavity,\ bottom}$ for scrap tire with 27.7% fixed carbon. Correspondingly, industrial and sewage sludge Figure 6(b) and (c), with their very low fixed carbon fractions experience a less important endothermic heat sink and lower reaction rates within the gasification temperature range however are affected by slow bed shrink rates and large ash deposits which impose a temperature flattening for the modeled profiles.

Figure 7 shows modeled reaction and bed shrink rates. All three waste feedstocks show high reaction rates associated with volatile release at the beginning of the run below 900K. Above 1000K the gasification of only small quantities of fixed carbon produces a significant reduction in reaction rate. Due to varying density and porosity for scrap tire powder, the bed shrink rate deviates from the reaction rate, as observed in Figure 7(a).

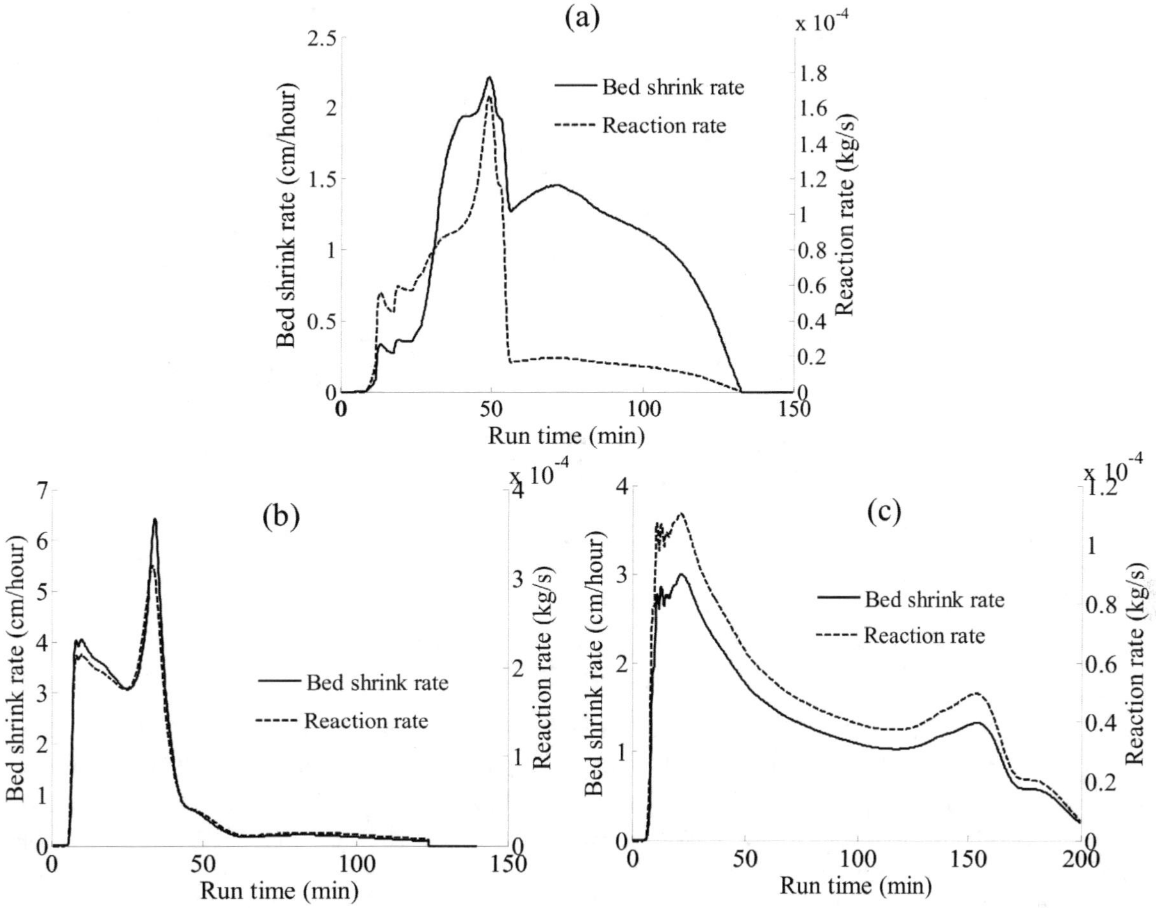

Figure 7 Simulated reaction rates and bed shrink rates for the solar steam-gasification of: (a) scrap tire powder, (b) industrial sludge, and (c) sewage sludge.

Power partitioning for the steam-gasification of sewage sludge is shown in Figure 8 for the irradiation period. The energy going to heat the feedstock and power the reaction is on the order of 15 % of the input solar power. As expected, the predominant heat loss is to the cold surface which indicates the importance of eliminating any such potential heat sinks from high flux regions in the upper cavity.

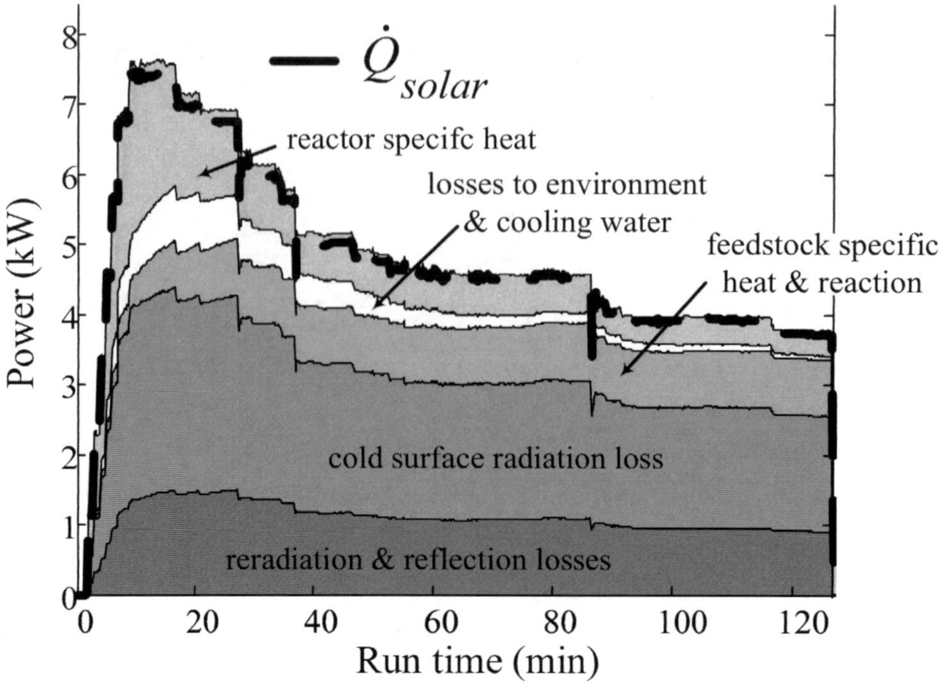

Figure 8 Power partitioning and applied input solar power during the steam-gasification of industrial sludge.

The solar absorption efficiency (eq. (12)) and the temperatures of the upper cavity and window are shown in Figure 9 for the solar steam-gasification of sewage sludge. At the beginning of the run, the upper cavity is strongly influenced by the non-zero reflectance of the window. Thereafter, an increase in re-radiation from the upper cavity and window self-emission lead to a further decrease in retained radiative energy.

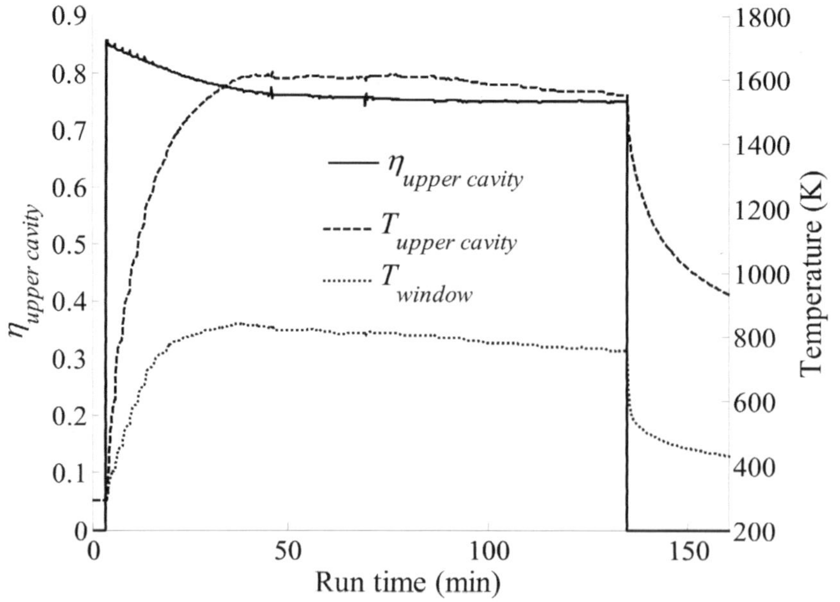

Figure 9 Solar absorption efficiency and the temperatures of the upper cavity and window during for the solar steam-gasification of sewage sludge.

The solar upgrade factor U and the solar-to-fuel energy conversion efficiency $\eta_{solar\text{-}to\text{-}fuel}$ are defined as:

$$U = \frac{m_{syngas} \cdot LHV_{syngas}}{m_{feedstock} \cdot LHV_{feedstock}} \tag{26}$$

$$\eta_{solar-to-fuel} = \frac{m_{syngas} \cdot LHV_{syngas}}{Q_{solar} + m_{feedstock} \cdot LHV_{feedstock}} \tag{27}$$

where Q_{solar} is the total solar energy delivered to the outer surface of the window over the duration of the experimental run, $m_{feedstock}$ is the converted feedstock mass, and m_{syngas} is the evolved syngas mass. For the model, the syngas is assumed to have the equilibrium composition at the temperature of the individual bed volumes integrated over the duration of the experimental run. Experimentally determined and numerically simulated $m_{feedstock}$, $\eta_{solar\text{-}to\text{-}fuel}$ and U are shown in **Table 3**. For scrap tire powder, significant deposits of oils and particles prevented closing the carbon balanced which negatively affected the experimentally determined $\eta_{solar\text{-}to\text{-}fuel}$ and U.

Table 3 Experimental and simulated performance indicators for the solar steam-gasification of carbonaceous waste feedstocks

	Scrap tire powder		Industrial sludge		Sewage sludge	
	Experimental	Modeled	Experimental	Modeled	Experimental	Modeled
$m_{feedstock}$ (kg)	0.263	0.294	0.423	0.507	0.488	0.595
U (-)	0.83	1.25	1.07	1.17	1.16	1.35
$\eta_{solar\text{-}to\text{-}fuel}$ (%)	17.3%	27.4%	28.0%	24.7%	18.0%	17.4%

6. Scale-up to a 200kW solar reactor

The reactor model was applied to analyze the performance of a 200 kW solar gasifier converting industrial sludge under the conditions of a simulated solar day consisting of a solar flux increase of 50 suns/min to 842 suns followed by an 8-hour hold and 4-hour cooling period. The fixed bed reactor parameters are shown in Table 4, with the diameters of the lower and upper cavities as well as the packed bed made equal and varied between 1.5 and 4 m in 0.25m steps.

Table 4 Simulation parameters of the 200kW solar reactor.

Nominal power $\dot{Q}_{solar}$	200 kW
Upper cavity height (m)	0.65
Lower cavity height (m)	0.85
Aperture diameter (m)	0.55
Initial packed bed height (m)	0.5
Radial insulation thickness (m)	0.18
Axial insulation thickness (m)	0.18
Solar flux at aperture (Suns)	842
Cavity/packed bed diameter (m)	1.5 – 4.0

The impact of packed-bed diameter on $\eta_{solar\text{-}to\text{-}fuel}$, $T_{bed,top}$, and the specific syngas power are shown in Figure 10. With increasing bed diameters, the radiation flux decreases in the upper and lower cavities, resulting in a decrease in temperatures from 1850 K at 1.5m diameter to 1300 K at the 4m diameter. The size increase however exposes more material to these temperatures. Furthermore, thanks to the high volatile content (62.2%) of industrial sludge and negligible heat of reaction of devolatilization, 62.2% of the batch is effectively converted without an energy penalty, producing the strong increase in $\eta_{solar\text{-}to\text{-}fuel}$. The specific syngas power, defined as the average lower heating value of the product gas evolved per unit bed area and per unit time, decreases only slightly – from 49 to 39 kW/m^2 - as the rate of the gasification reactions falls with lower reactor temperatures. As the fixed carbon fraction is only 12%, this effect is small. The steam gasification of industrial sludge achieves a peak efficiency of 89.4% at a 4 m diameter lower cavity.

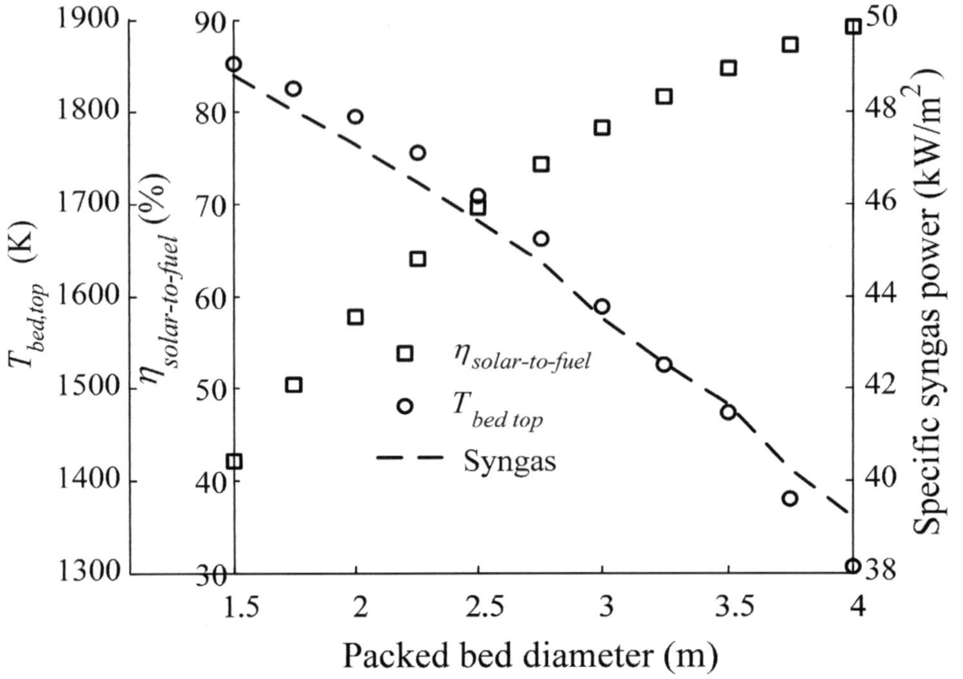

Figure 10 Solar-to-fuel energy conversion efficiency, specific syngas power, and top bed temperature as a function of the packed bed diameter for a simulated 12-hr solar run of the 200kW solar reactor for steam-gasification of industrial sludge.

Figure 11 shows the reaction rates and solar radiative power input for a simulated 12-hr solar run of the 200kW solar reactor with packed-bed diameters 1.5, 2, 2.5, and 3 m. Under the rapidly increasing solar flux, the reaction rate rises thanks to rapid low-temperature volatile release (< 900 K) [26]. As the bed temperatures pass into the gasification range, the reaction rates fall strongly owing to the low fixed carbon fraction. By virtue of larger bed surface areas, larger diameter reactors achieve higher reaction rates for the same radiative power, supporting higher efficiencies. Between the 1.5 and 3.0 m diameters, there is a 67% increase in the peak reaction rate during devolatilization and a doubling in the gasification driven range at the end of the solar day. There is no significant transient penalty in going to larger reactors due to the low temperatures required for devolatilization onset (< 500 K) and the good insulation of the packed bed from the walls by virtue of its own poor thermal conductivity. Notably, during the 4-hour cooling period, the latent heat of the reactor and hot feedstock supports the gasification reaction longer in larger reactors. Gas evolution can continue even at lower temperatures because of the large volatile content. It must be noted that the reaction rates of industrial sludge as shown for this batch configuration are undesirable for an industrial facility as they require sizing equipment to handle large peaks in production rates which occur for only a fraction of the time.

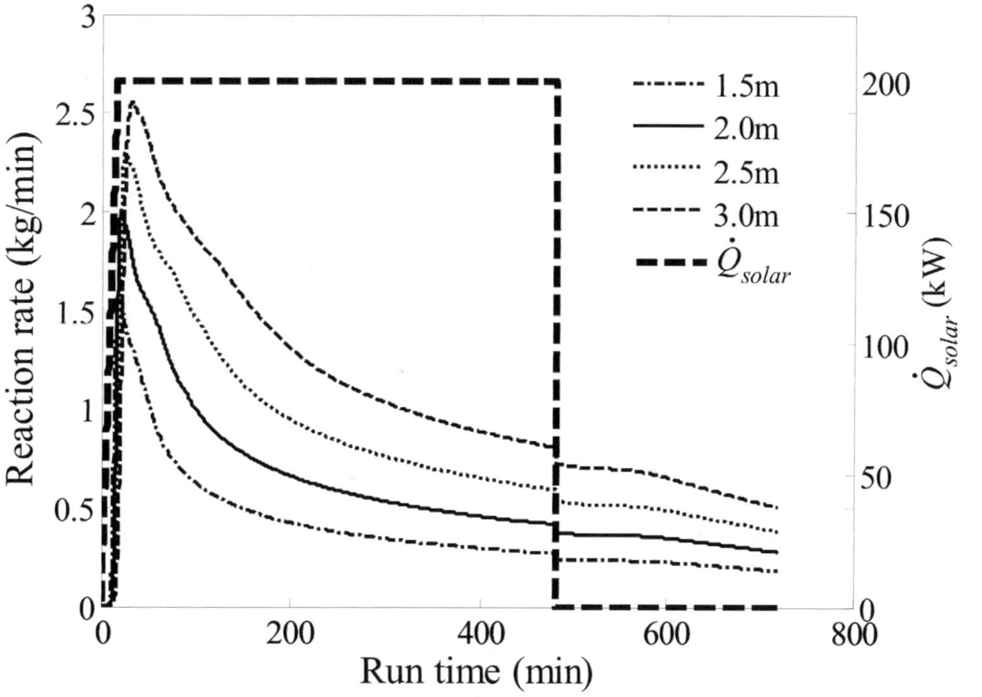

Figure 11 Reaction rates and solar radiative power input during a simulated 12-hr solar run of the 200kW solar reactor for steam-gasification of industrial sludge. The parameter is the packed-bed diameter: 1.5, 2, 2.5, and 3 m.

The power partitioning is shown in Figure 12 for reactors with packed bed diameters 1.5 and 2.5 m during a simulated 12-hour solar run. For the larger packed bed, a larger portion of solar power input is consumed by the sensible heat of the reactor; however, more heat is delivered to the reacting packed bed by virtue of proportionally greater packed-bed surface to reactor surface. Similarly, larger reactors lose less energy by re-radiation due to lower upper cavity temperatures. Notably, the losses to the environment remain nearly constant despite larger external surface areas because of the good insulation and a decrease in reactor inner temperatures. The above effects lead to higher overall $\eta_{solar\text{-}to\text{-}fuel}$.

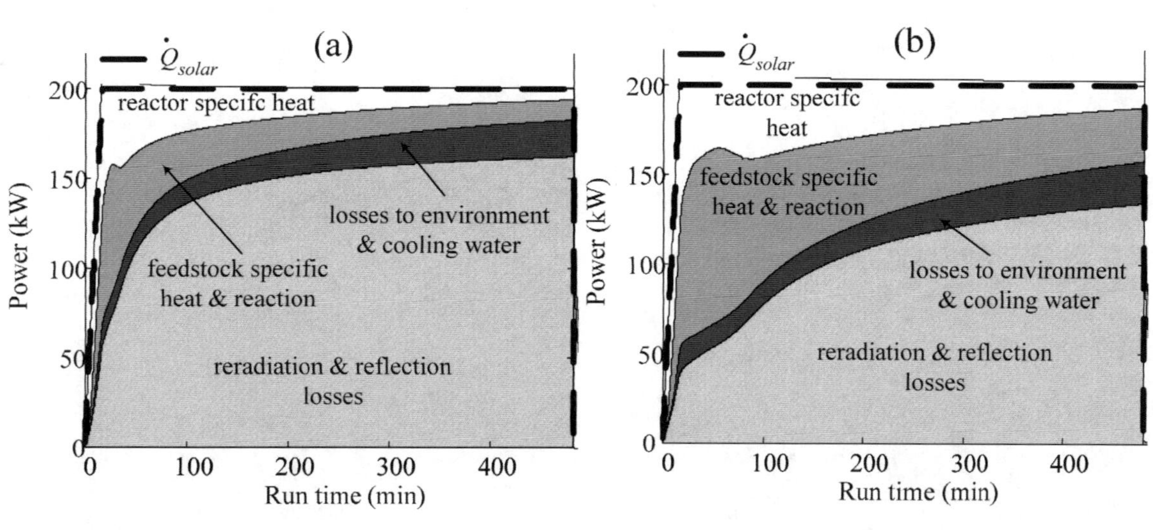

Figure 12 Power partitioning during a simulated 12-hour solar run for the steam-gasification of industrial sludge in a 200kW solar reactor with packed-bed diameters (a) 1.5m and (b) 2.5m.

6. Conclusions

A dynamic model for a solar-driven packed-bed reactor has been developed for the steam-gasification of heterogeneous, high volatile content, carbonaceous waste feedstocks. The boundary condition to the model is the applied solar flux which makes the model suitable for dynamic simulations under transient solar irradiation conditions. Simulation of the upper cavity considered radiative exchange in a windowed enclosure containing the emitter plate which thermally couples the upper and lower cavities as well as heat transfer to the reactor insulation. A quasi-1D model is used for the lower cavity containing the packed bed, which couples the rate of heat transfer to the rate of the gasification reaction. Reaction rates considered pyrolysis and gasification with enthalpies of reaction determined from equilibrium compositions of steam-feedstock systems. A 2D heat transfer model was applied to the reactor insulation which interacts with the environment. The model was validated with experimental data using an 8kW reactor prototype for the steam gasification of scrap tire powder and industrial and sewage sludges in terms of reactor temperatures, converted mass, energy conversion efficiency and upgrade factor. The model was further applied to analyze a 200kW solar gasifier of industrial sludge operating under a simulated solar day. Larger packed-bed diameters were found to convert more material with the same solar input by virtue of lower re-radiation losses and a greater portion of applied heat delivered to the reacting packed bed. Solar-to-fuel energy conversion efficiencies of up to 89.4% were determined for industrial sludge with a reactor having a 4m diameter packed bed.

Nomenclature

δ_{kj}	Kronecker function (□ij = 1 for i = j; □ij = 0 for. i□ j)
ε	surface emissivity
ε_r	relative roughness
ρ	density, kg/m3
σ	Stephan-Bolzmann constant, W/m2K4
μ	viscosity, kg/ms
ϕ	porosity
η	energy conversion efficiency, %
A	surface area, m2
Cp	specific heat capacity, J/kgK
$\overline{D}$	mean diameter, m
D_h	hydraulic diameter, m
d_p	particle diameter, m
E	hemispherical emittance
f	friction factor
F_{k-j}	view factor from surface k to j
$FF_{\lambda T}$	fractional function
F_s	quadratic loss coefficient, kg/m4
Δh_{chem}	specific enthalpy change of reaction, J/kg
H	enthalpy, J
h	convective heat transfer coefficient, W/m2K
k	thermal conductivity, W/mK
K_{perm}	permeability constant, m2
L_c	characteristic dimension, m
LHV	lower heating value, kJ/kg
m	mass, kg
N	number of discrete units
Nu	Nusselt number
Pr	Pradlt number
P	pressure, Pa
$\dot{Q}$	power, W
Q	energy, J
q	flux, W/m2
r	radius, m
R	hemispherical reflectance
Ra	Rayleigh number
Re	Reynolds number
S	volumetric heat source/sink, W/m3
SN	surface number
t	time, s
T	temperature, K
U	upgrade

V	hemispherical transmittance
W	velocity, m/s
$\overline{W}$	mean velocity, m/s
X	conversion
x	mass fraction
y	axial direction

Subscripts

λ	wavelength, μm
b	spectral band index
band	spectral band
bed	packed bed
bottom	bottom
c	carbonaceous
cooling	active cooling
cond	conduction
conv	convection
cs	cold surface
e	external
eff	effective
emitter	emitter plate
env	environment, T = 292K
feedstock	feedstock
free	free convection
forced	forced convection
i	incident
initial	initial condition
j,k	general indices
inscone	insulation cone
ins	bulk insulation
l	leaving
lower cavity	lower cavity
o	radiosity
rerad	re-radiation
syngas	syngas
solar	solar radiation
solar-to-fuel	solar to chemical fuel conversion
tile	tiles
top	top surface
upper cavity	upper cavity
water	water
window	window

Superscripts

new	new iteration value
old	old iteration value

References

[1] Von Zedwitz P, Steinfeld A., The solar thermal gasification of coal – energy conversion efficiency and CO_2 mitigation potential. Energy.2003; 28: 441-456.

[2] Morris M, Waldheim L. Energy recovery from solid waste fuels using advanced gasification technology. Waste Management.1998;18: 557-564.

[3] Filippis P.D, Borgianni C, Paolucci M, Pochetti R. Prediction of syngas quality for two-stage gasification of selected waste feedstocks. Waste Management.2004;24: 633-639.

[4] Senneca O, Chirone R, Masi S, Salatino P. A thermogravimetric study of nonfossil solid fuels. 1. Inert Pyrolysis. Energy & Fuels.2002;16: 653-660.

[5] Andersson B.A, Lundberg M., Heikne B, Josefsson U, Norrköping 75 MW CFB plant and biomass RDF combustion in fluidized –bed boilers. Power Production from waste and biomass IV: Advanced concepts and Technologies, VTT Symposium, 2002;185-210.

[6] T. Kodama, T. Kondoh, T. Tamagawa, A. Funatoh, K-I. Shimizu, Y. Kitayama, Fluidized bed coal gasification with CO_2 under direct irradiation with concentrated visible light. Energy and Fuels.2002;16:1264-1270.

[7] Gregg DW, Aiman WR, Otsuki HH, Thorsness CB. Solar coal gasification. Solar Energy.1980; 24:313-321.

[8] Trommer D, Noembrini F, Fasciana M, Rodriguez D, Morales A, Romero M, Steinfeld A. Hydrogen production by steam-gasification of petroleum coke using concentrated solar power – I. Thermodynamic and kinetic analyses. International Journal of Hydrogen Energy.2005;30:605-618.

[9] Mueller R, Von Zedtwitz P, Wokaun A, Steinfeld A. Kinetic investigation on steam gasification of charcoal under direct high-flux irradiation. Chemical Engineering Science.2003;58:5111-5119.

[10] Z'Graggen A, Haueter P, Trommer D, Romero M, de Jesus JC, Steinfeld A. Hydrogen production by steam-gasification of petroleum coke using concentrated solar power – II Reactor design, testing and modeling. International Journal of Hydrogen Energy.2006;31:797-811.

[11] Von Zedwitz P, Steinfeld A. Steam gasification of coal in a fluidized-bed/packed-bed reactor exposed to concentrated thermal radiation – modeling and experimental validation. Industrial & Engineering Chemistry Research.2005;44:3852-3861.

[12] Z'Graggen A, Haueter P, Maag G, Vidal A, Romero M, Steinfeld A. Hydrogen Production by Steam-Gasification of Petroleum Coke using Concentrated Solar Power - III. Reactor experimentation with slurry feeding. International Journal of Hydrogen Energy.2007;32:992-996.

[13] Van Heek KH. General aspects and engineering principles for technical application of coal gasification,. In: Figuieiredo, JL. Carbon and Coal Gasification, NATO ASI Series, 1986: 399-411.

[14] Gregg DW, Taylor RW, Campbell JH, Taylor JR, Cotton A. Solar gasification of coal, activated carbon, coke and coal and biomass mixtures. Solar Energy.1980;25:353-364.

[15] Flechsenhar M, Sasse C. Solar gasification of biomass using oil shale and coal as candidate materials. Energy.1995;8:803-810.

[16] Schaffner B, Meier A, Wuillemin D, Hoffelner W, Steinfeld A. Recycling of Hazardous Solid Waste Material Using High-Temperature Solar Process Heat – II. Reactor Design and Experimentation. Environmental Science and Technology.2003;37:165-170.

[17] Piatkowski N, Wieckert C, Steinfeld A. Experimental investigation of a packed-bed solar reactor for the steam-gasification of carbonaceous feedstocks. Fuel Processing Technology.2009;90:360-366.

[18] Nipattummakul N, Ahmed I, Kerdsuwan S, Gupta AK. High temperature steam gasification of wastewater sludge. Applied Energy.2010;87:3729-3734.

[19] Piatkowski N, Steinfeld A. Solar driven coal gasification in a thermally irradiated packed bed reactor. Energy and Fuels.2008;22:2043-2052.

[20] Lipinski W, Steinfeld A. Heterogeneous thermochemical decomposition under direct irradiation. International Journal of Heat and Mass Transfer.2004;47:1907-1916.

[21] Osinga T, Olalde G, Steinfeld A. Solar carbothermal reduction of ZnO: Shrinking packed-bed reactor modeling and experimental validation. Industrial & Engineering Chemistry Research.2004:43;7981-7988.

[22] Atkinson B, Merrick D. Mathematical models of the thermal decomposition of coal: 4. Heat transfer and temperature profiles in a coke-oven charge. Fuel.1983;62:553-561.

[23] Yogev A, Kribus A, Epstein M, Kogan A. Solar "tower reflector" systems: A new approach for high-temperature solar plants. International Journal of Hydrogen Energy.1998;23:239-245.

[24] Welford WT, Winston R. High Collection Nonimaging Optics. Academic Press, San Diego, 1989.

[25] Petrasch J, Coray P, Meier A, Brack M, Haeberling P, Wuillemin D, Steinfeld A. A 50-kW 11,000-suns novel high-flux solar simulator based on an array of Xenon arc lamps. Journal of Solar Energy Engineering.2007;129:405-411.

[26] Piatkowski N, Steinfeld A. Reaction kinetics of the combined pyrolysis and steam-gasification of carbonaceous waste materials. Fuel.2010;89:1133–1140.

[27] Haussener S, Lipinski W, Wyss P, Steinfeld A. Tomography-Based Analysis of Radiative Transfer in Reacting Packed Beds Undergoing a Solid-Gas Thermochemical Transformation. ASME Journal of Heat Transfer.2010;132:061201-1/7.

[28] Lopez-Peinado AJ, Tromp PJJ, Moulijn JA. Quantitative heat effects associated with pyrolysis of coals, ranging from anthracite to lignite. Fuel.1989;68:999-1004.

[29] Müller R, Lipinski W, Steinfeld A. Transient heat transfer in a directly-irradiated solar chemical reactor for the thermal dissociation of ZnO Applied Thermal Engineering.2008;28:524-531.

[30] Müller R, Steinfeld A. Band-Approximated Radiative Heat Transfer Analysis of a Solar Chemical Reactor for the Thermal Dissociation of Zinc Oxide. Solar Energy.2007;81:1285-1294.

[31] Maag G, Falter C, Steinfeld A. Temperature of a Quartz/Sapphire Window in a Solar Cavity-Receiver. ASME Journal of Solar Energy Engineering, in press.

[32] Siegel R, Howell JR. Thermal Radiation Heat Transfer, 4th ed., Taylor and Francis-Hemisphere, Washington. 2001.

[33] Çengel YA. Heat Transfer, McGraw Hill, 2003

[34] Patankar SV. Numerical Heat Transfer and Fluid Flow, Taylor and Francis, New York, 1980.

[35] Bejan A, Nield DA. Convection in Porous Media, Springer, New York, 2006.

[36] Ergun S. Fluid flow through packed columns Chemical Engineering Progress.1952;48:89-94.

[37] White FM. Fluid Mechanics, McGraw Hill, 2003.

(443c) Modeling of Fischer-Tropsch Fuels Production From Polygeneration Facilities

Wei Yuan and Mario Richard Eden, Department of Chemical Engineering, Auburn University, Auburn University, AL

The interest in Fischer-Tropsch fuel production is a consequence of stringent environmental regulations, technological developments and changes in fossil energy reserves. It is a potential answer to satisfy many of our future energy needs. The use of FT technology for biomass conversion to synthetic hydrocarbons may offer a promising and carbon neutral alternative to conventional diesel, kerosene and gasoline. FT-liquids are free of sulphur and can be matched directly with conventional fuel markets. Generic process simulation models of gas-phase Fischer-Tropsch Synthesis (FTS) and supercritical phase Fischer-Tropsch Synthesis (SC-FTS) processes were developed based on previous and ongoing work at Auburn University. The models have been developed using data available in the open literature and their performance is currently being validated against the latest experimental data. Synthesis gas production from steam reforming is included in the model. The downstream processing results in two products, i.e. a gasoline and a jet fuel similar to JP-5, including paraffin, olefin and oxygenate products. The product distribution is obtained using the Anderson-Schulz-Flory (ASF) distribution for gas-phase model. In supercritical phase, the product distribution is much more narrow compared to traditional gas-phase FTS due to the vapor-like transport properties and liquid-like thermal properties in the supercritical phase. A hydrocracking reactor is added to convert the heavier compounds back into fuel range products. Using experimental data generated within the Consortium for Fossil Fuel Science (CFFS), models have been developed for analysis of the supercritical water reforming (SCWR) process. The high-pressure hydrogen produced from supercritical water reforming of biomass is intended for injection in the SC-FTS process during the cracking/isomerization step and will be included in the model as well. In addition, the light fraction from gas-phase model is converted to synthesis gas and recycled to the FTS reactor by steam reforming. Economic and environmental analysis was performed through quantifying the level of environmental impact using EPA's WAR algorithm. In this way, experimental and theoretical efforts can supplement each other, providing direction for further experimental efforts.

See more of this Session: Alternative Fuels and Enabling Technologies
See more of this Group/Topical: Fuels and Petrochemicals Division

PROGRESS TOWARDS MODELING OF FISCHER TROPSCH SYNTHESIS IN A SLURRY BUBBLE COLUMN REACTOR

Donna Post Guillen[1], Tami Grimmett[1], Anastasia M. Gribik[1] and Steven P. Antal[2]
[1] Idaho National Laboratory Idaho Falls, Idaho, 83415 USA, Donna.Guillen@inl.gov
[2] Interphase Dynamics, Ballston Lake, NY 12019 USA, antals@rpi.edu

ABSTRACT

This paper presents a computational multiphase fluid dynamics model developed to simulate Fischer Tropsch synthesis in a slurry bubble column reactor. The work is being undertaken under the Modeling and Simulation Platform of the Idaho National Laboratory's Hybrid Energy Systems initiative to develop and demonstrate hybid energy technology. The SBCR model is one element of an integrated suite of research and development activities targeted towards our nation's pressing energy security and climate change challenges. An SBCR can be used to produce alternative fuels and value-added chemicals from syngas derived from secure domestic feedstocks, including biomass, coal, natural gas, or refuse. Although the fundamental technology was developed in the 1920s, details concerning the myriad complex processes occuring in the SBCR during FT synthesis are still not fully understood. Our team has developed a model based on a robust computational multiphase fluid dynamics platform (NPHASE-CMFD) to be used as a numerical tool or testbed for reactor optimization and design, as well as sensitivity analysis. Mechanistic submodels are connected to the main flow solver to incorporate relevant physics, such as bubble breakup/coalescence, two-phase turbulence and interfacial momentum exchange. A methodology has been developed to incorporate the dominant mechanisms of heterogeneous catalysis. Vapor-liquid equilibrium has been assimilated into the product distribution via a simple, robust property method approach. Property data is provided by a chemical process simulation package.

INTRODUCTION

The Hybrid Energy Systems Testing (HYTEST) Laboratory has been established at the Idaho National Laboratory to develop and test hybrid energy systems with the principal objective to safeguard U.S. Energy Security by reducing dependence on foreign petroleum. A central component of the HYTEST laboratory is the slurry bubble column reactor (SBCR) in which the gas-to-liquid reactions will be performed to synthesize transportation fuels using the Fischer Tropsch (FT) process. SBCRs are cylindrical vessels in which gaseous reactants (for example, synthesis gas or syngas) is sparged into a slurry of liquid reaction products and finely dispersed catalyst particles. The syngas bubbles exchange mass and momentum with the liquid phase as they react and travel though the column. The catalyst particles are transported in the slurry by the rising gas bubbles and serve to promote the chemical reaction that converts syngas to a spectrum of longer chain hydrocarbon products, which can be upgraded to gasoline, diesel or jet fuel.

In a commercial process, the gas velocity is high enough that the SBCRs operate in the churn-turbulent flow regime [1, 2] which is characterized by complex hydrodynamics, coupled with reacting flow chemistry and heat transfer, that affect reactor performance. The purpose of this work is to develop a research tool to aid in understanding the physicochemical processes occurring in the SBCR. Our team is developing a robust methodology to couple reaction kinetics and mass transfer into a four-field model (consisting of the bulk liquid, small bubbles, large bubbles and solid catalyst particles) that includes thirteen species – CO reactant, H_2 reactant, hydrocarbon product, and H_2O product in small bubbles, large bubbles, and the bulk fluid plus catalyst. Mechanistic submodels for interfacial momentum transfer in the churn-turbulent flow regime are incorporated. Bubble breakup/coalescence and two-phase turbulence submodels are also incorporated. The absorption and kinetic models, specifically changes in species concentrations, have been incorporated into the mass continuity equation. The reaction rate is determined based on the macrokinetic model for a cobalt catalyst developed by Yates and Satterfield [3]. The model includes heat generation produced by the exothermic chemical reaction, as well as heat removal from a constant temperature heat exchanger. A property method approach is employed to incorporate vapor-liquid equilibrium (VLE) in a robust manner. Physical and thermodynamic properties as functions of changes in both pressure and temperature are obtained from VLE calculations performed external to the CMFD solver. The novelty of this approach is in its simplicity, as well as its accuracy over a specified temperature and pressure range. The focus on this paper is on the property method approach for incorporating VLE into the FT SBCR model. No CMFD results are given in this paper.

DESCRIPTION OF CMFD CODE

The computational platform for this work is the NPHASE-CMFD computer program, an unstructured, finite-volume, multifield, pressure-based computational multiphase fluid dynamics (CMFD) computer code offering both segregated and fully coupled multifield numerical solution methods. To achieve numerical convergence for multiphase flows, simulations are performed using the robust, coupled algorithm, which fully couples the phasic mass and momentum equations to solve for the phasic velocity, pressure, and volume fraction simultaneously. Although, the computer memory requirements are increased, this ability to couple the conservation of mass and momentum equations is more robust than the use of a segregated solver. A full three-dimensional, Eulerian-Eulerian framework is employed as a practical engineering tool, rather than tracking a very large number of individual bubble trajectories. Ensemble-averaged conservation equations for mass and momentum can be solved for a user-specified number of fields or phases. The conservation equations are discretized into a block matrix system, which is solved by an algebraic multigrid solver. A detailed derivation of the ensemble-averaged conservation equations has been given by Drew and Passman [4]. The fully coupled mass/momentum scheme allows any field/phase to interact with any other field/phase within the modeled system. Interfacial mass, momentum and turbulence submodels provide coupling between the fields and phases and are key to making this approach viable.

In this Eulerian-Eulerian formulation, the different phases are subdivided into one or more fields that are treated as interpenetrating continua. The probability of each field at any location is indicated by the field volume fraction. The laws of conservation of mass and momentum are satisfied for both large and small bubbles and the local velocity and volume fraction for each field are calculated at each node. The conservation of mass equation for each field-j of phase-k is expressed as

$$\frac{\partial}{\partial t}\left(\varepsilon_{jk}\rho_k\right)+\nabla\cdot\left(\varepsilon_{jk}\rho_k\underline{v}_{jk}\right)=\Gamma_{jk}'''+m_{jk}''' \tag{1}$$

where Γ_{jk}''' is the volumetric mass transfer rate due to phase change in field-j of phase-k; m_{jk}''' is the mass source of field-j from other fields of phase-k; ε_{jk} is the volume fraction of field-j of phase-k; ρ_k (kg/m^3) is the phasic density; and $\underline{v}_{jk}$ (m/s) is the velocity of field-j of phase-k. The number of fields of each phase is user-defined and allows different flow physics to be included for each field.

The ensemble-averaged, phasic momentum conservation equation can be reduced to the following approximate multifield formulation for field-j in phase-k [5]

$$\frac{\partial}{\partial t}\left(\varepsilon_{jk}\rho_k\underline{v}_{jk}\right)+\nabla\cdot\left(\varepsilon_{jk}\rho_k\underline{v}_{jk}\underline{v}_{jk}\right)=-\varepsilon_{jk}\nabla p+\nabla\cdot\varepsilon_{jk}\underline{\underline{\tau}}_k+\varepsilon_{jk}\rho_k\underline{g}+\underline{M}_{jk}^i+\Gamma_{jk}'''\underline{v_{jk}^i}+m_{jk}'''\underline{v_{jk}} \tag{2}$$

The terms on the right hand side of the equation represent the pressure gradient, shear stress tensor, body force due to gravity, momentum exchange at interfaces and momentum flux due to mass transfer. The volume fractions of the various field-j of phase-k must sum to unity

$$\sum_{k}\sum_{j}\varepsilon_{jk} = 1 \tag{3}$$

The churn-turbulent flow is modeled as a continuous liquid phase with two dispersed bubble size groups. In our four-field, ensemble-averaged CMFD model, field 1 represents the continuous liquid phase, fields 2 and 3 represent the two dispersed bubble size groups, and field 4 is the catalyst. Each bubble group is treated as a separate field to allow appropriate interfacial flow physics for the different bubble size groups to be included via the closure models. The use of mechanistic closure models for bubble forces is discussed in detail in a subsequent subsection.

For churn-turbulent flows an understanding of the expected bubble distribution is needed. Flow structure visualization studies [6] show a clear separation of bubble sizes. In reality, the bubble size distribution is polydisperse as evidenced by the experimental data [7], but since the bubble size distribution is largely bimodal, such flows may be simplified by considering the flow to consist of a small and a large bubble group. Small bubbles are spherical to ellipsoidal in shape, whereas the larger bubbles vary in size with gas flow and exhibit a distorted, spherical cap shape. A minimum of two bubble groups is needed since the interfacial forces are quite different between the groups.

The phasic energy equation implemented in NPHASE-CMFD is written as

$$\frac{\partial}{\partial t}\left(\varepsilon_{jk}\rho_k h_{jk}\right)+\nabla\cdot\left(\varepsilon_{jk}\rho_k \underline{v}_{jk} h_{jk}\right)=\nabla\cdot\left[\varepsilon_{jk}\left(\underline{\underline{q}}'' + \underline{\underline{q}}''^{T}\right)\right]+\varepsilon_{jk}\Delta H_k + UA(T_{liq}-T_w)+\Gamma'''_{jk}e^{i}_{jk}+m'''_{jk}e_{jk} \tag{4}$$

where h_{jk} (J/kg) is ensemble-averaged enthalpy; e_{jk} (W) is the internal energy for field-k; q"(W/m^2) is the interfacial heat flux; ΔH (W/m^3) is the chemical heat of reaction, U (W/m^2-K) is the overall heat transfer coefficient and A (m^2) is the area of the heat exchanger; and T_c and T_w are the temperatures (K) of the continuous liquid phase and the heat exchanger wall, respectively. The superscript "T" indicates the transpose of the tensor. The terms on the right hand side of the equation represent the conductive heat flux, the volumetric heat source due to the chemical reaction, and convective heat transfer to the heat exchanger, respectively. The energy equation is used to predict the change in liquid temperature due to the exothermic chemical reaction and heat exchanger. A simple heat exchanger model has been used which allows heat transfer from the liquid and represented as a constant temperature heat exchanger over the central region of the column. The heat exchanger serves to remove all of the heat of reaction (ΔH=-0.172 MJ/mol$_{CO}$) [1].

The local mass fractions of the chemical species are evaluated using species transport equations

$$\frac{\partial}{\partial t}\left(\varepsilon_{jk}\rho_k Y_s\right)+\nabla\cdot\left(\varepsilon_{jk}\rho_k \underline{v}_{jk} Y_s\right)=\nabla\cdot\varepsilon_{jk}\left(\frac{\mu_k}{Pr_s}+\frac{\mu_k^t}{Pr_s^t}\right)\nabla Y_s+\Gamma'''_{jk}Y_{jk}+m'''_{jk}Y_{jk}+S_s \tag{5}$$

where Y_s is the species mass fraction, μ_k and μ_k^t are the molecular and turbulent viscosity for field-k, Pr_s and Pr_s^t are the laminar and turbulent species Prandtl numbers, and S_s (kg/m^3-s) is the source term containing the reaction kinetics. It is assumed that the species are well mixed at the molecular level, that is, they have the same velocity, pressure and temperature as the carrier field.

NPHASE-CMFD includes three major sub-models to describe the following flow physics [8]: (1) bubble breakup/coalescence, (2) two-phase turbulence, and (3) interfacial momentum exchange. Bubble breakup and coalescence are important phenomena occurring in SBCRs. Small bubbles enter the flow channel and begin coalescing to form new bubbles that are twice the size of the inflow bubbles. As the bubbles rise, the large bubbles can coalesce with either small bubbles or other large bubbles. This coalescence transitions the flow to the churn-turbulent regime. A simple, mechanistically-based breakup and coalescence model is employed [9]. Closure relations are needed to reintroduce the information that was lost as a result of averaging the conservation equations. Turbulence in the bulk fluid was modeled using a standard k-ε model with the two-phase turbulence viscosity given by Sato [10, 11]. Another set of closure models are invoked to account for the interfacial forces, which include drag, lift, wall, turbulence dispersion and virtual mass, that are included in the momentum equation. The coefficients in the interfacial momentum transfer forces can be tuned using a design optimization approach [12].

IMPLEMENTATION OF FT REACTIONS

This section describes the methodology implemented to incorporate the FT reactions into the CMFD code. The Fischer-Tropsch synthesis reaction can be modelled as a chain growth reaction of CO and hydrogen on the surface of a heterogeneous catalyst. The catalytic process converts syngas to higher molecular weight hydrocarbons where the synthetic crude carbon number distribution follows the Anderson-Schulz-Flory (ASF) distribution. This simple statistical model predicts a linear relationship between the logarithm of the molar amount of a paraffin and its carbon number. The growth probability, which fixes the slope of the ASF distribution, is catalyst dependent. The product stream is characterized by a light hydrocarbon fraction ($C_1 - C_4$), a gasoline fraction ($C_5 - C_{11}$), a diesel fraction ($C_{12} - C_{19}$) and a wax fraction (C_{20+}) as its main products, with water and carbon dioxide as by-products [13]. A cobalt catalyst was selected for the initial model since, unlike Fe catalysts, cobalt-based catalysts exhibit very little water-gas-shift (WGS) activity. Due to the minimal WGS activity, the by-product is mostly water.

To simplify the model, the FT product distribution is characterized by a representative liquid hydrocarbon (HC) product, a representative HC vapor product and water/steam. This results in modeling 13 species, which are

- CO reactant in small bubbles, large bubbles, and the bulk fluid
- H_2 reactant in small bubbles, large bubbles, and the bulk fluid
- HC wax product in small bubbles, large bubbles, and the bulk fluid, and
- H_2O product in small bubbles, large bubbles, and the bulk fluid, and
- Catalyst in bulk fluid.

The local mass fractions of the chemical species are tracked using 13 separate species transport equations. The mass fraction is tracked for the species H_2 and CO in the gas phase and HC and H_2O in the liquid and vapor phases. Absorption of gas species from both large and small bubbles into the bulk liquid phase is included. The driving force for the gas across the bubble interface into the bulk liquid will be dependent upon the interfacial species concentration in both small and large bubbles. However, because it is difficult to measure the concentration at the gas-liquid interface, coefficients for convective mass transfer across an interface for the overall driving force between the bulk concentrations in the gas and liquid phases are implemented. The product from the incorporation of absorption is the steady state concentration profile of the absorbed gas species in the bulk liquid phase and the impact of the hydrodynamics on the concentration profile.

The heterogeneous catalysis process has been simplified to model the two dominant steps: (1) absorption from the gas phase reactants into the liquid phase across the bubble interface, and (2) the chemical reaction that occurs at the catalytic sites on the catalyst surface. Other steps, such as film mass transfer, pore diffusion within the catalyst, etc. have been neglected in this model [14]. However, effects due to intraparticle diffusion and mass transfer to the catalyst surface should be investigated. Mass and heat transfer may be altered due to the resistance from the laminar layer formed around each catalyst particle in the turbulent flow field. The resulting temperature and reactant concentrations on the catalyst surface can be much different than that in the bulk liquid.

VLE properties are computed using a chemical process simulation package (Aspen Plus from Aspen Technology, Inc.). Aspen Plus uses properties from DIPPR and other property databanks, which facilitates property validation. Using Aspen Plus, a heater block was created, which allows the temperature and pressure of the hydrocarbon stream exiting the FT reactor to be varied. The hydrocarbon product stream is then sent to a flash drum where it is separated into vapor and liquid streams (Figure 1). A sensitivity operation is used to calculate the various thermodynamic and physical properties for the liquid and vapor streams, as well as the vapor fraction exiting the heater block. The Peng-Robinson thermodynamics package is used. Then, based on the temperature and pressure, the appropriate split between liquid and vapor products can be calculated and the associated properties using the equations regressed from the data. Depending upon the temperature and pressure of the operating column, H_2O may experience a phase change and both liquid and vapor are present. A list of physical and thermodynamic properties needed for the hydrocarbon and water liquid and vapor, and for the hydrogen and carbon dioxide vapor, is given in Table 2.

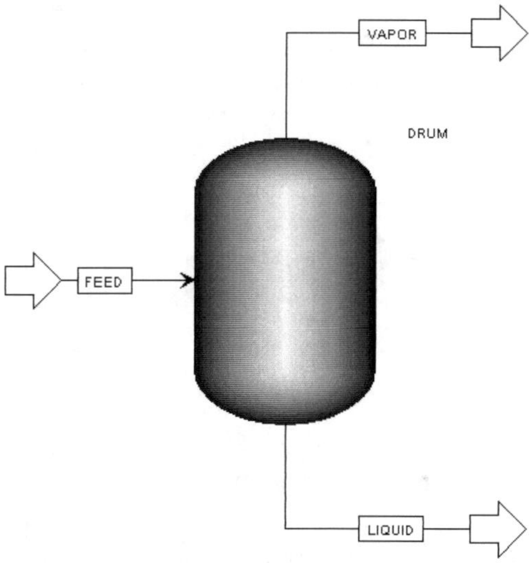

Figure 1. Illustration of flash drum used to separate liquid and vapor components in Aspen Plus.

This methodology provides reactant and product properties that are functions of reactor temperature and pressure. The expressions have the general form

$$\phi(T,P) = \xi_1 TP + \xi_2 P^2 + \xi_3 P + \xi_4 + \xi_5 T + \xi_6 T^2 \tag{6}$$

where $\phi(T,P)$ represents a physical quantity (such as molecular weight, density, viscosity, etc.) listed in the left column of Table 1. All physical properties are calculated based on temperature and pressure of the current computational cell. All properties are in SI units, pressure is in Pa and temperature is in K for all equations listed, unless otherwise noted. The property equations derived by regression from the Aspen Plus simulation results are valid over a temperature range from 423 to 623 K and a pressure range from 2.4 to 3.4 MPa. The regression coefficients are tabulated in the Appendix.

Table 1. Physical and thermodynamic properties specified for the FT reactants and products (L=liquid; v=vapor).

	HC	H_20	H_2	CO
heat capacity, c_p (J/kmol-K)	L,v	L,v	v	v
thermal conductivity, k (W/m-K)	L,v	L,v	v	v
molecular weight , MW (kg/mole)	L,v	L,v	v	v
Prandtl number, Pr (-)	L,v	L,v	v	v
density, ρ (kg/m^3)	L,v	L,v	v	v
viscosity, μ ($N\text{-}s/m^2$)	L,v	L,v	v	v
surface tension, σ (N/m)	L	L	N/A	N/A
diffusivity in HC phase, D (m^2/s)	N/A	N/A	v	v
Henry's constant, H (-)	N/A	N/A	v	v
volume fraction, ε (-)	v	v	N/A	N/A

The trick to making the property method approach work with the CMFD code is to stabilize the solution by referencing the hydrodynamic pressure. If the local system pressure is used for the properties, initial instabilities in the calculation as convergence is approached are amplified causing the CMFD simulation to diverge. For this reason, the hydrodynamic pressure (a function of column height) is used in the equations for the various reactant and product properties. As shown in the energy equation, the CMFD code uses enthalpy, rather than temperature. A subroutine to derive local temperature from enthalpy and specific heat was written, since the property expressions use temperature whereas the CFMD code uses enthalpy. Expressions for the properties are then connected to the main NPHASE-CMFD solver via user-defined subroutines. This approach allows the incorporation of VLE without all the added baggage of doing complex, time-consuming calculations within the CMFD code.

The procedure employed by the property-method approach is described in detail below and depicted in Figure 2. The VLE simulation performed in Aspen is used to obtain equations for the various physical and thermodynamic properties and parameters. The CMFD solution is initialized with a set of initial conditions. The initial temperature and pressure are provided to the routines indicated by "FT reactions." The results of these routines includes: (1) properties, (2) the mass transfer source term, and (3) the species source term, which are used by the CMFD code. The CMFD results are checked for convergence. If the solution has converged within an acceptable residual, the spatial distribution of velocity, temperature, pressure, phase hold-up and species within the SBCR are output. If not, the updated temperature and pressure are used in the "FT reactions" routines and the CMFD code iterates until convergence within a specified tolerance is achieved.

Step 1: Calculate properties for liquid hydrocarbon

First, determine the physical properties of the bulk liquid hydrocarbon using equation coefficients (ξ_1 through ξ_6) listed in Table A1.

Step 2: Calculate properties of gas bubble mixture

Gas mixture properties are calculated by the mixture rule

$$\phi_{mix,k}(T,P) = \sum_s \phi_{s,k}(T,P) Y_{s,k} \tag{7}$$

where the terms are summed over the number of gaseous species, s. The mass fraction, Y_s, is related to the molar concentration c_s (moles/m^3) of species s by

$$Y_s = \frac{MW_s c_s}{\rho_k} \tag{8}$$

where MW_s (kg/mole) is the molecular weight for species, s. The physical properties of the bubbles are determined using Equation 6 with the equation coefficients listed in Tables A2 and A3. For example,

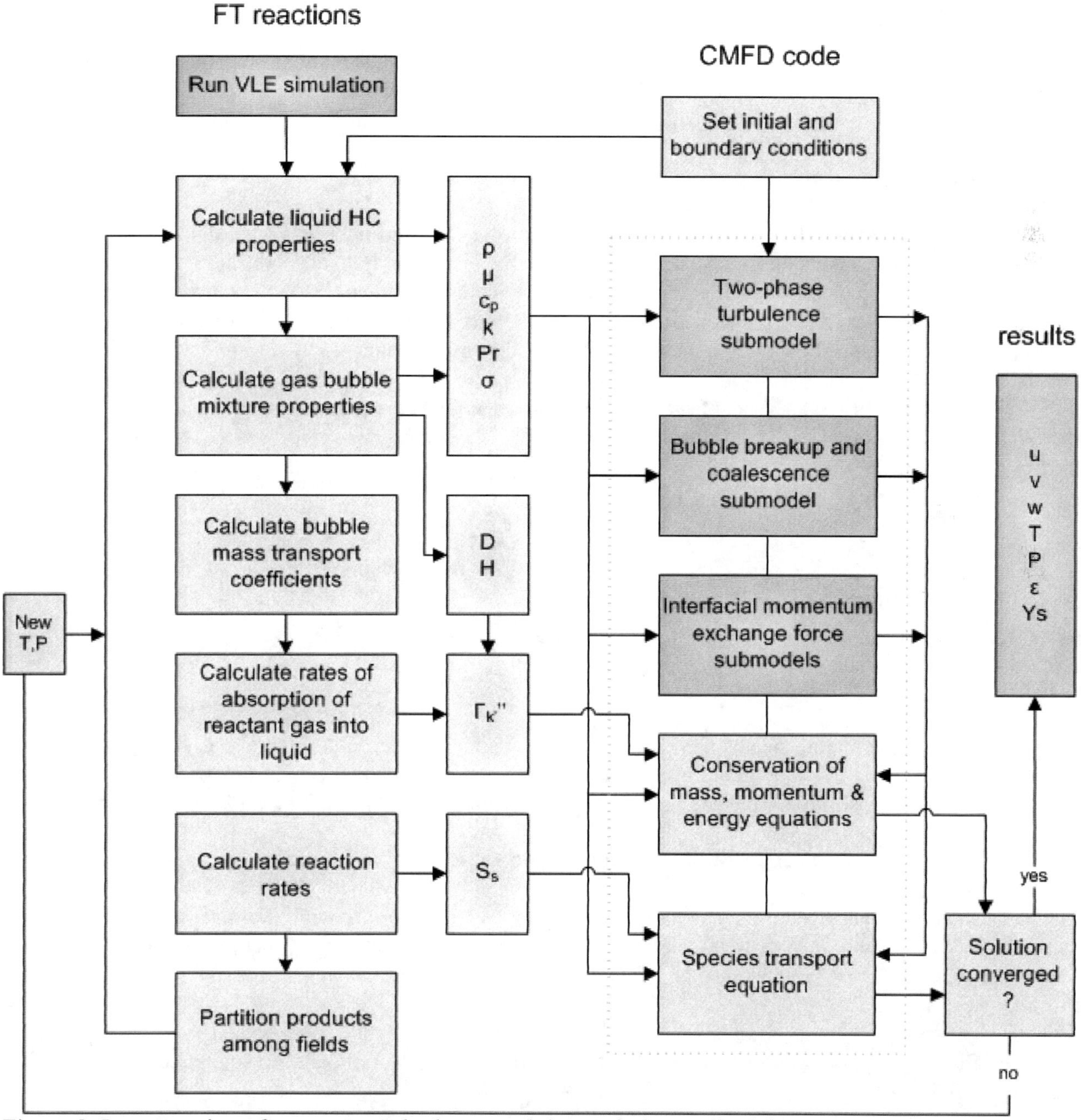

Figure 2. Incorporation of property method approach into solution methodology.

to calculate the initial mixture properties of the gas in the bubbles, multiply the pure CO and H_2 physical properties by their corresponding mass fraction and sum. For a gas mixture comprised of only CO and H_2, the expression for the small bubble (sb) field is

$$\phi_{mix,sb}(T,P) = \phi_{CO,sb}(T,P)Y_{CO,sb} + \phi_{H2,sb}(T,P)Y_{H2,sb} \tag{9}$$

and, similarly, for the large bubble (lb) field

$$\phi_{mix,lb}(T,P) = \phi_{CO,lb}(T,P)Y_{CO,lb} + \phi_{H2,lb}(T,P)Y_{H2,lb} \tag{10}$$

where density, viscosity, specific heat, thermal conductivity and Prandtl number are substituted for the variable ϕ. Once the syngas has reacted and gaseous hydrocarbon products and water vapor are produced, these additional components must be included in the mixture property calculation.

Step 3: Calculate small and large bubble mass transport coefficients

After the initial physical properties are calculated for the wax and bubble mixtures, the mass transport rates and reaction rates are calculated as described in [14]. The mass transfer coefficient for large and small bubbles for each gas species is calculated based on the diffusion coefficients regressed from the Aspen Plus output. First, the diffusion coefficient is calculated for CO and H_2 in the liquid hydrocarbon (wax) product, using the coefficient data for CO and H_2 given in Table A4.

Then, the volumetric mass transport coefficient for CO and H_2 in small and large bubbles is calculated as a function of gas holdup and diffusivity [1]

$$\frac{k_L a_{s,sb}}{\varepsilon_{sb}} = 1.0\sqrt{\frac{D_{L,s}}{D_{L,ref}}} \tag{11}$$

$$\frac{k_L a_{s,lb}}{\varepsilon_{lb}} = 0.5\sqrt{\frac{D_{L,s}}{D_{L,ref}}} \tag{12}$$

where $k_L a_s$ (s^{-1}) is the overall mass transfer coefficient for species s, ε_{lb} and ε_{sb} are the gas holdup for large bubbles and small bubbles (calculated by NPHASE), $D_{L,s}$ (m^2/s) is the diffusivity for species s in the liquid phase, and $D_{L,ref}$ (m^2/s) is the reference diffusion coefficient [1] in the liquid (2×10^{-9} m^2/s).

Step 4: Calculate rates of absorption of reactant gas to bulk liquid phase

Next, the rate of absorption is calculated for CO and H_2 from the large bubbles and small bubbles, using the mass transfer coefficient calculated above and the equilibrium gas concentration in the liquid phase, based on the Henry's law constant. The Henry's law constant, H_s, is calculated from the data listed in Table A5.

The following equation describes the rate of mass transfer ($\Gamma'''_{s,k}$, mol/m^3-s) for the solute (gas) into the solvent (liquid):

$$\Gamma'''_{s,k} = k_L a_{s,k} \left(c_s^* - c_{s,L} \right) \quad (13)$$

where

$$c_s^* = \frac{c_{s,v}}{H_s} \quad (14)$$

1. where c_s^* is the equilibrium concentration of species s in the gas phase with the liquid (mol/m^3), $c_{s,L}$ is the current concentration of species s in the liquid phase (mol/m^3), $c_{s,v}$ is the current concentration of species s in the gas phase (mol/m^3), and H_s is the Henry's law constant for species s, calculated above. Figure 3 depicts the concentration profiles at the gas bubble-bulk liquid interface. The diffusion coefficient controls the rate of mass transfer of gases into the bulk liquid, whereas the solubility determines the amount of gas that will dissolve eventually in the liquid phase. Film mass transfer (bulk diffusion) of the products from the external surface of the solid catalyst particle through the boundary layer has been neglected.

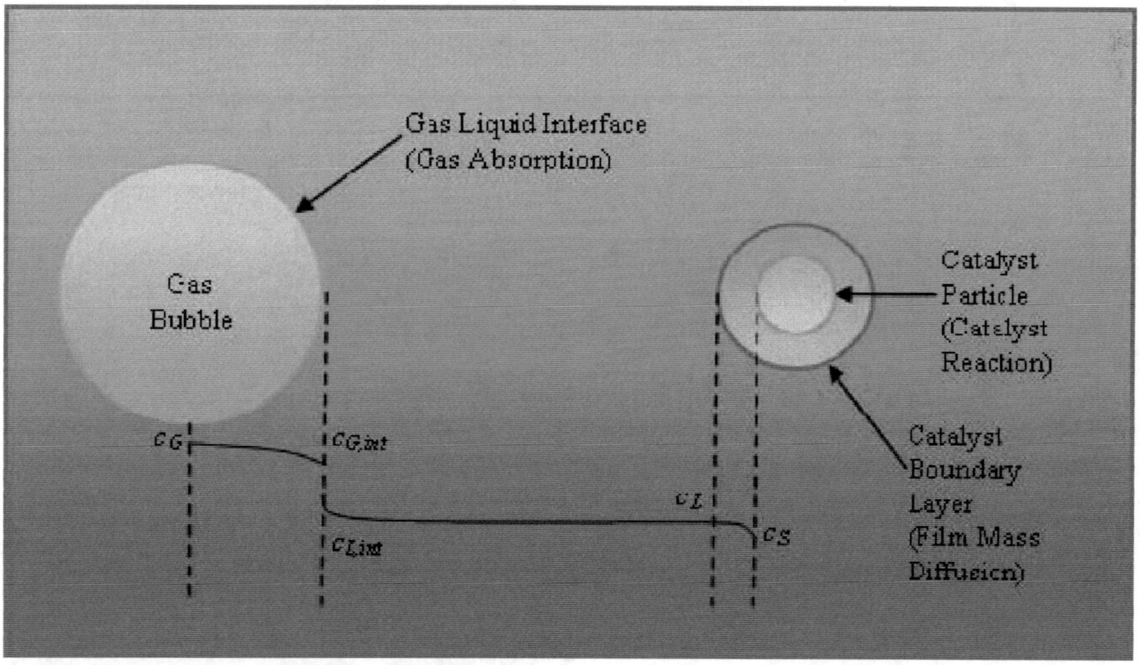

Figure 3. Concentration profile in the SBCR.

Rates of absorption are calculated for CO and H_2 from large and small bubbles, using the above method. Using these calculated rates of absorption, the resulting concentration of CO and H_2 in the bulk liquid is calculated in NPHASE-CMFD. This concentration of CO and H_2 will then be used in the reaction rate expressions given below. By combining the four equations listed above, the source terms for the conservation equations are obtained. The rates of absorption for the small bubbles is expressed as

$$\Gamma'''_{CO,sb} = 1.0 \sqrt{\frac{D_{L,CO}}{D_{L,ref}}} \left(\left(\frac{c_{CO,v}}{H_{CO}} \right) - c_{CO,L} \right) \varepsilon_{sb} \quad (15)$$

$$\Gamma'''_{H2,sb} = 1.0\sqrt{\frac{D_{L,CO}}{D_{L,ref}}}\left(\left(\frac{c_{H2,v}}{H_{H2}}\right) - c_{H2,L}\right)\varepsilon_{sb} \tag{16}$$

and the rates of absorption for the large bubbles is expressed as

$$\Gamma'''_{CO,lb} = 0.5\sqrt{\frac{D_{L,CO}}{D_{L,ref}}}\left(\left(\frac{c_{CO,v}}{H_{CO}}\right) - c_{CO,L}\right)\varepsilon_{lb} \tag{17}$$

$$\Gamma'''_{H2,lb} = 0.5\sqrt{\frac{D_{L,H2}}{D_{L,ref}}}\left(\left(\frac{c_{H2,v}}{H_{H2}}\right) - c_{H2,L}\right)\varepsilon_{lb} \tag{18}$$

In the CMFD code, Γ''' are the source terms for the ensemble-averaged phasic mass transfer rate.

Step 5: Calculate reaction rates

To determine the consumption of CO and H_2 for the FT reaction, it is necessary to calculate the molecules of carbon and hydrogen required to form the product to ensure that enough CO and H_2 have absorbed into the wax phase and reacted on the catalyst surface for the reaction to proceed. This is a limitation of modeling the kinetics macrokinetically. Unfortunately, a microkinetic model of sufficient rigor does not currently exist for FT catalysts, so we are limited to a macrokinetic approach [13]. It is assumed that the catalyst particles are sufficiently small such that external and internal mass and heat transfer are not rate limiting. The intrinsic kinetic expression for the consumption of CO and H_2, which is a Langmuir-Hinshelwood type expression, was based on data collected over a range of industrially relevant conditions. A macrokinetic approach using power rate laws is used, where the heterogeneous catalytic reaction is expressed in terms of the rate of consumption of CO as [3]

$$R_{CO} = \frac{-a c_{CO,L} c_{H_2,L}}{(1 + b c_{CO,L})^2} \tag{19}$$

with the kinetic constant, a $\left(\frac{m^6}{kg_{cat}\, s\, mol}\right)$, and the adsorption coefficient, b $\left(\frac{m^3}{mol}\right)$, defined as follows

$$a = 8.852 \times 10^{-13} \exp\left[4494.41 \frac{1}{K}\left(\frac{1}{493.15} - \frac{1}{T}\right)\right](R_{gas}\, T)^2 \tag{20}$$

$$b = 2.226 \times 10^{-5} \exp\left[-8236.15 \frac{1}{K}\left(\frac{1}{493.15} - \frac{1}{T}\right)\right] R_{gas}\, T \tag{21}$$

where R_{CO}, R_{H2}, R_{H2O}, and R_{HC} (mol/kg$_{cat}$s) are the rates of consumption for CO and H_2 and the rate of production for H_2O and HC, respectively, R_{gas} (J/mol-K) is the gas constant, and T (K) is the temperature of the catalyst.

To determine the number of moles of hydrogen and carbon in the product, it is assumed that the hydrocarbon product is a C_nH_{2n+2} (paraffin) molecule. Based on this calculation, the appropriate rate of hydrogen consumption, hydrocarbon production, and water production can be implemented. For instance, if the current temperature and pressure is 493.15K and 2930271.85 Pa, the volume fraction in the hydrocarbon product is approximately 0.9700. The molecular weight of the liquid is ~280.15 g/mol and the molecular weight of the vapor is 24.47 g/mol. As a result the average molecular weight is

$$MW_{avg} = 0.97 \cdot 24.47 \frac{g}{mol} + (1 - 0.97) \cdot 280.15 \frac{g}{mol} = 32.14 \frac{g}{mol} \tag{22}$$

based on the chemical formula $C_nH_{(2n+2)}$, if the molecular weight of carbon is 12.011 g/mol and the molecular weight of hydrogen is 1.00794 g/mol, n is equal to 2.15

$$32.14 \frac{g}{mol} = 12.011 \cdot n + 1.00794 \cdot (2n + 2) \rightarrow n = 2.15 \tag{23}$$

and if the chemical reaction formula for the production of alkanes is as follows [15]

$$nCO + (2n + 1)H_2 \rightarrow C_nH_{(2n+2)} + nH_2O \tag{24}$$

the rates of consumption and formation of H_2, H_2O, and hydrocarbon product are

$$R_{H_2} = \left(\frac{2n+1}{n}\right)(R_{CO}) \tag{25}$$

$$R_{H_2O} = -(R_{CO}) \tag{26}$$

$$R_{HC} = -\frac{1}{n}(R_{CO}) \tag{27}$$

To obtain reaction rate in units of rate of change of moles per volume $\left(\frac{mol}{m^3 s}\right)$, the reaction rates are multiplied by the density of the catalyst and the corresponding catalyst volume fraction. The species source term due to chemical reactions is summed from the contributions of the reaction rates

$$S_s = MW_s \sum_{n=1}^{N_R} n_{sn} R_n \tag{28}$$

where MW_s is the molecular weight (kg/mol) for species *s*, n_{sn} is the overall stoichiometric coefficient for species *s* in reaction *n* (positive for products and negative for reactants), N_R is the number of reactions and R_n is the reaction rate for the nth reaction. The species source term is fed into the species transport equation.

Step 6: Partition products among fields

After the rates of formation are determined, the products formed must be partitioned to either the liquid or vapor phase. Therefore, the moles of hydrocarbon product formed are partitioned to the liquid wax phase by multiplying the moles formed by one minus the vapor fraction. The moles of vapor product are the vapor fraction times the moles formed. The vapor product is assumed to be equally divided amongst the various bubble sizes. It should be noted that the coefficients of the reaction expressions, i.e. how many moles of CO and H_2 are reacted are dependent upon the number of CO and H_2 molecules formed in the reaction product. To determine this, the volume fraction of the hydrocarbon product is calculated, based on the current temperature and pressure. The molecular weight of the hydrocarbon mixture is then calculated using the corresponding volume fraction of liquid and vapor using the equations for the vapor and liquid molecular weights. The coefficients in the regressed equations for volume fraction, ε, and molecular weights, MW, of the hydrocarbon product are given in Table A6.

$$MW_{HC,mix} = \varepsilon_{HC} MW_{HC,v} + (1 - \varepsilon_{HC}) MW_{HC,L} \qquad (29)$$

The water product formed is then determined to be either a liquid or vapor product. This is accomplished by inputting the current temperature into equation for pressure. The coefficients of this pressure equation are given in Table A7. The regressed equation for water partial pressure is

$$P_{H2O}(T) = \xi_7 T^3 + \xi_8 T^2 + \xi_9 T + \xi_{10} \qquad (30)$$

If the calculated pressure is less than the current system pressure, the water formed is a liquid product and joins the wax phase, if the pressure is greater than or equal to the current pressure the water is in the vapor phase and should be partitioned equally amongst the various bubble sizes.

Step 7: Calculation of gas mixture properties

The resulting properties of the gas phase for all bubble sizes and the liquid phase are calculated using general mixture rules for the next computational node according to steps one and two above. The mixture rule given is used to obtain the properties of the gaseous and liquid mixtures. However, the gas mixture now must include the gaseous properties of the hydrocarbon product formed, and depending on which phase it is present in, either the liquid or gas phases must include the water mixture properties. Coefficients for the property equations are given in Tables A8 through A10.

Step 8: Repeat Steps 3 through 7 above

Steps three through seven described above are repeated with updated temperatures and pressures from the CMFD solution. This procedure continues until convergence of the CMFD simulation has been achieved.

SUMMARY AND FUTURE WORK

A CMFD model developed for simulation of FT reactions in a SBCR has been presented. A four-field, 13-species model was created to predict churn-turbulent flows in a SBCR. The churn-turbulent flow is modeled as a continuous liquid phase with two dispersed bubble size groups using an Eulerian-Eulerian framework. Each bubble group is treated as a separate field to allow appropriate flow physics for the different bubble size groups to be included via the closure models. The use of mechanistic closure models accounts for interfacial momentum transfer between the bubbles and the liquid wax. Bubble breakup/coalescence and two-phase turbulence submodels are incorporated into the NPHASE-CMFD code. A property method approach to incorporating VLE into the calculation is outlined. An Aspen Plus simulation was executed and the resulting data was regressed into a set of equations for the physical and thermodynamic properties that are functions of reactor temperature and pressure. This approach allows the incorporation of VLE without all the complexity of performing time-consuming calculations within the CMFD code.

Physical processes that have been neglected here for the purpose of simplification should be investigated to determine their effect. Specifically, diffusion across the boundary layer near the catalyst surface may affect the heat and mass transfer. Future work to improve on the current model includes obtaining an expression for the ASF product distribution as a function of temperature and pressure. A more detailed model of the heat exchanger tubes within the SBCR, including the pressure drop and other flow effects due to the presence of internals, should be included. Most importantly, validation of the simulation results will be necessary using data from an operating SBCR. Results from the SBCR model will be presented after the validation has been performed.

ACKNOWLEDGMENTS

Funding for this research was supported by the U.S. Department of Energy, Office of Nuclear Energy, under DOE Idaho Operations Office Contract DE-AC07-05ID14517.

REFERENCES

1. Maretto, C. and R. Krishna, *Modelling of a bubble column slurry reactor for Fischer-Tropsch synthesis.* Catalysis Today, 1999. **52**(2-3): p. 279-289.
2. Fernandes, F.A.N., *Modeling and Product Grade Optimization of Fischer-Tropsch Synthesis in a Slurry Reactor.* Ind. Eng. Chem. Res. , 2006. **45**: p. p. 1047-1057.
3. Yates, I.C. and C.N. Satterfield, *Intrinsic Kinetics of the Fischer-Tropsch Synthesis on a Cobalt Catalyst.* Energy & Fuels, 1991. **5**: p. 168-173.
4. Drew, D.A., Passman, S.L.. *Theory of Multicomponent Fluids.* Applied Mathematical Sciences, 1998. **Vol. 135**.
5. Podowski, M.Z., *On the Consistency of Mechanistic Multidimensional Modeling of Gas/Liquid Two-Phase Flows.* Nucl. Eng. Des. , 2009. **doi:10.1016/j.nucengdes.2008.10.022**.
6. Lopez de Bertodano, M., Sun, X., Ishii, M., Ulke, A., *Phase Distribution in the Cap Bubble Regime in a Duct.* Journal of Fluids Engineering, 2006. **Vol. 128**: p. p. 811-818.
7. Lucas, D., Beyer, M., Kussin, J., Schutz, *Benchmark Database on the Evolution of Two-Phase Flows in a Vertical Pipe*, in *XCFD4NRS, OECD/NEA & International Atomic Energy Agency (IAEA) Workshop*. 2008: Grenoble, France.
8. Guillen, D.P., et al., *Development of a Computational Multiphase Flow Model for Fischer Tropsch Synthesis in a Slurry Bubble Column Reactor*, in *XIX International Conference on Chemical Reactors, CHEMREACTOR-19* 2010: Vienna, Austria.
9. Antal, S.P., M.Z. Podowski, R.T. Lahey, Jr., D. Barber, C. Delfino, *Multidimensional Modeling of Developing Two-Phase Flows in a Large Adiabatic Riser Channel*, in *The 11th International Topical Meeting on Nuclear Reactor Thermal-Hydraulic (NURETH-11)*. 2005: Popes Palace Conference Center, Avignon, France.
10. Sato, Y., Sekoguchi, K., *Liquid Velocity Distribution in Two-Phase Bubble Flow.* Int. J. Multiphase Flow, 1975. **Vol 2**: p. 79-95.
11. Sato, Y. and M. Sadatomi, *Momentum and Heat Transfer in Two-Phase Bubble Flow - I.* International Journal of Multiphase Flow, 1981. **7**(2): p. 167-177.
12. Guillen, D.P., et al., *Optimization of a Two-Fluid Hydrodynamic Model of Churn Turbulent Flows*, in *Proceedings of the 17th International Conference on Nuclear Engineering, ICONE17-75113*. 2009: Brussels, Belgium.
13. Bartholomew, C.H., Farrauto, R.J., *Fundamentals of Industrial Catalytic Processes*. Second Edition ed. 2006, Hoboken, NJ: John Wiley & Sons, Inc. p. 402-404.
14. Gribik, A., D.P. Guillen, and D. Ginosar, *Kinetic Modeling of a Fischer-Tropsch Reaction over a Cobalt Catalyst in a Slurry Bubble Column Reactor for Incorporation into a Computational Multiphase Fluid Dynamics Model*, in *International Pittsburgh Coal Conference 2008*. 2008, September 29 – October 2, 2008: Pittsburgh, PA.
15. Steynberg, A.P., and Dry, M.E., *Fischer-Tropsch Technology*. Studies in Surface Science and Catalysis, ed. G. Centi. Vol. 152. 2006: Elsevier.

APPENDIX
Values for regression coefficients in property equations

Table A1. Values for regression coefficients in liquid hydrocarbon property equations.

	ξ_1	ξ_2	ξ_3	ξ_4	ξ_5	ξ_6
MW	-5.26249E-08	2.68795E-12	-1.23341E-05	-9.53522E+01	1.03320E+00	-1.84674E-04
ρ	-7.07429E-10	1.30075E-12	-2.36017E-05	5.96444E+02	6.08341E-01	-9.56955E-04
μ	1.79025E-13	9.57890E-18	-2.03926E-10	1.57893E-03	-2.75838E-06	1.43360E-09
C_p	-2.31585E-04	7.55333E-09	8.51037E-03	-4.86779E+05	2.56158E+03	1.04205E+00
k	-5.64169E-14	2.66274E-16	-3.37194E-09	1.93724E-01	-1.92020E-04	9.18030E-08
Pr	1.71935E-09	1.99551E-13	-3.30858E-06	1.73954E+01	-8.59123E-03	-7.43486E-08
σ	5.96243E-13	1.25675E-17	-6.16461E-10	4.89329E-02	-7.90545E-05	2.40673E-08

Table A2. Values for regression coefficients in carbon dioxide property equations.

	ξ_1	ξ_2	ξ_3	ξ_4	ξ_5	ξ_6
ρ	-1.25390E-08	-2.28081E-14	1.30890E-05	2.00088E+01	-7.73788E-02	7.38009E-05
μ	-2.35447E-16	2.64499E-21	2.09568E-13	3.93160E-06	5.15621E-08	-1.56605E-11
C_p	-7.30591E-07	-2.15127E-12	5.48923E-04	3.11178E+04	-1.02158E+01	1.52000E-02
k	2.10647E-24	-1.33959E-19	7.85074E-13	3.11252E-03	7.96693E-05	-1.90837E-08
Pr	-2.93158E-11	4.41590E-17	2.17002E-08	8.64163E-01	-5.57856E-04	5.35027E-07

Table A3. Values for regression coefficients in hydrogen property equations.

	ξ_1	ξ_2	ξ_3	ξ_4	ξ_5	ξ_6
ρ	-8.75910E-10	-1.39898E-15	9.27675E-07	1.37977E+00	-5.29761E-03	5.01847E-06
μ	-2.48452E-17	1.74648E-22	2.35718E-14	2.71031E-06	2.25144E-08	-5.09891E-12
C_p	-8.62701E-08	-1.61542E-13	6.07591E-05	2.71274E+04	5.45923E+00	-2.65814E-03
k	1.09099E-23	-6.93408E-19	4.06375E-12	3.58612E-02	5.10847E-04	-1.01628E-07
Pr	-4.31587E-12	9.88441E-18	3.17359E-09	7.38622E-01	-1.38315E-04	8.74823E-08

Table A4. Values for regression coefficients in equations for CO and H_2 diffusion coefficients.

$D_{L,s}$	ξ_1	ξ_2	ξ_3	ξ_4	ξ_5	ξ_6
CO	1.18599E-17	-7.99993E-23	-3.27266E-15	2.47081E-08	-1.34701E-10	2.21670E-13
H_2	1.04809E-17	-9.47682E-23	-2.69480E-15	2.16567E-08	-1.28508E-10	2.26763E-13

Table A5. Values for regression coefficients in equation for Henry's law constants.

H_s	ξ_1	ξ_2	ξ_3	ξ_4	ξ_5	ξ_6
CO	4.68681E-04	7.15978E-08	-1.23962E+00	-1.90891E+07	1.45163E+05	-1.22350E+02
H_2	2.61567E-03	1.00931E-07	-2.66928E+00	3.02048E+07	4.36191E+04	-6.82025E+01

Table A6. Values for regression coefficients in equation for molecular weight of hydrocarbon product mixture.

HC	ξ_1	ξ_2	ξ_3	ξ_4	ξ_5	ξ_6
ε	1.43991E-11	2.00076E-16	-1.17690E-08	8.78877E-01	2.81191E-04	-1.46253E-07
MW_l	-5.26249E-08	2.68795E-12	-1.23341E-05	-9.53522E+01	1.03320E+00	-1.84674E-04
MW_v	-9.81403E-10	6.38883E-14	-9.77897E-08	3.14426E+01	-5.14268E-02	8.03856E-05

Table A7. Values for regression coefficients in equation for water partial pressure.

H_2O	ξ_7	ξ_8	ξ_9	ξ_{10}
P	1.40289E+00	-1.74277E+03	7.38805E+05	-1.06411E+08

Table A8. Values for regression coefficients in gaseous hydrocarbon property equations.

	ξ_1	ξ_2	ξ_3	ξ_4	ξ_5	ξ_6
ρ	-6.93966E-09	1.72886E-14	9.51117E-06	2.35494E+01	-9.39044E-02	9.33258E-05
μ	-1.59834E-16	-2.70601E-22	2.69670E-13	8.32169E-07	3.68418E-08	-1.40560E-11
C_p	-5.38051E-06	1.62315E-10	1.85377E-03	6.46160E+04	-1.42580E+02	3.03757E-01
k	1.04817E-12	-5.55914E-17	-1.56222E-11	-1.98474E-02	1.63646E-04	-2.20302E-08
Pr	-6.46457E-11	5.95938E-16	4.40271E-08	8.55932E-01	-5.45348E-04	6.49952E-07

Table A9. Values for regression coefficients in liquid water property equations.

	ξ_1	ξ_2	ξ_3	ξ_4	ξ_5	ξ_6
ρ	1.07411E-23	2.40265E-26	-1.45361E-19	1.09401E+03	4.46546E-02	-1.39892E-03
μ	-1.31567E-28	1.76460E-32	-4.36713E-26	1.76856E-03	-6.19582E-06	5.71432E-09
C_p	-2.10917E-06	9.57962E-13	8.17283E-04	2.14966E+05	-6.91089E+02	9.14277E-01
k	-3.99546E-26	6.42966E-30	-1.85723E-23	-2.55743E-01	4.56380E-03	-5.52945E-06
Pr	-1.28882E-11	9.73843E-18	4.36054E-09	1.28135E+01	-4.68802E-02	4.58569E-05
σ	-1.60679E-26	1.76160E-30	-2.88485E-24	8.14066E-02	-1.71636E-04	6.72944E-08

Table A10. Values for regression coefficients in water vapor property equations.

	ξ_1	ξ_2	ξ_3	ξ_4	ξ_5	ξ_6
ρ	-1.31582E-08	1.61028E-13	1.11162E-05	1.89252E+01	-7.66428E-02	7.65644E-05
μ	-3.66624E-16	6.49154E-21	2.98575E-13	-1.15802E-06	3.44470E-08	5.30844E-12
c_p	-1.20110E-05	1.62851E-10	7.59538E-03	5.60600E+04	-9.42386E+01	1.04107E-01
k	-3.83553E-27	5.74015E-30	-3.19590E-23	-1.09893E-03	4.90209E-05	5.06316E-08
Pr	-3.79074E-10	4.96761E-15	2.38647E-07	1.88419E+00	-3.57598E-03	3.35923E-06

(443e) Critical Evaluation of H_2 Production Technologies: When Do Other Technologies Become Economically Competitive with Steam Methane Reforming?

N.V.S.N. Murthy Konda[1], N. Shah[1], Gert Jan Kramer[2] and Nigel P. Brandon[3], (1)CPSE, Chemical Engineering Department, Imperial College London, London, United Kingdom, (2)Shell Global Solutions International, The Netherlands, Amsterdam, Netherlands, (3)Department of Earth Sciences, Imperial College London, London, United Kingdom

Hydrogen is generally considered as one of the potential fuels of the future transportation. However the high production cost has remained a barrier that is yet to be overcome. There are various H_2 production routes based on thermo-chemical, electrolytic and biological processes. These include steam methane reforming (SMR), coal gasification (CG), biomass gasification (BG) and water electrolysis (WE). Within the context of H_2 production, while SMR, CG and WE are commercially available, BG is still in the development/demonstration stage (Ball and Wietschel, 2009).

While SMR is currently the cheapest production technology, all these technologies are characteristically different in their cost foot-print. For instance, gasification can be twice as much capital intensive as reforming technology (NAE, 2004). On the other hand, the common feed-stocks used in gasification (i. e., coal and biomass) are usually less expensive than the feed-stock (i.e., natural gas, NG) used in SMR. Since the total production cost depends on both the capital investment and the price of raw-materials, the actual production cost in the future depends on the technological advancements and the future feed-stock price. While both the technological advancements and the feed-stock prices in future are uncertain, a recent study by Kramer and co-workers (Schoots, et al., 2008) has concluded, based on an extensive historical data since 1940, that the cost reduction potential on the basis of technology learning for SMR, CG and WE is limited. While they did not rule out the possibility of potential to reduce the electrolytic H_2 production costs due to technological advancements (such as PEM-based electrolysis), the challenges on electrolysis are twofold as it is both capital and energy intensive. Subsequently, the renewable electricity driven H_2 production route will have even more challenges (which is partly due to the competition from electricity sector) to overcome. Thus, in future, the feedstock price can play an even more critical role and it is important to quantitatively understand the circumstances under which the other technologies can be economically competitive compared to SMR. Such analyses in the literature are mostly qualitative - For instance, some studies foresee that CG becoming more economically competitive compared to SMR as NG prices tend to increase more rapidly than that of coal (Stiegel and Ramezan, 2005; Mueller-Langer et al., 2007; Ball and Wietschel, 2009). However, an objective quantitative analysis is lacking.

Further, the relative cost-competitiveness of these technologies can vary significantly with the plant capacity due to the economies of scale (EOS). For instance, at medium production scale (say, 150 ton/day) SMR is more cost-competitive than what it is at large production scale (say, 750 ton/day), compared to CG (NAE, 2004). Further, actual feed-stock prices can vary significantly geographically depending on the local and global (in the case of imported feed-stocks) taxes. Hence, the EOS and geographic considerations need to be explicitly considered when evaluating a technology.

Further, the relative economic merits will also depend on the future CO_2 mitigation policies as these technologies have very different carbon footprint. In a market with high carbon prices, BG may become economically competitive with CG. However, though BG can play a significant role in the H_2 based economy and is one of the first renewable H2 production technologies that is expected to be available at commercial scale, there will be competition for the biomass from other sectors - i.e., to produce liquid biofuels and power/heat generation (Mueller-Langer et al. 2007).

In addition to the large-scale production technologies that are used exclusively for H_2 production (as discussed above), the other production options include co-production (H_2 and electricity) and small-scale forecourt production. Exclusively on the basis of economic and energetic potential, it may rather be more beneficial to consider the co-production of H2 and electricity in a IGCC plant with CCS (Chiesa et al., 2005 and Kreutz et al. 2005). Forecourt production however may become redundant if the by-product (or excess capacity) H2 from the existing production facilities can be streamlined to supply the H2 in the transition period (Konda et al., 2010).

In addition to a single plant level economic analysis, the large-scale H_2 production requires network of several plants optimally located, by exploiting the trade-off between production and transportation costs. Hence network-wide perspective has to be taken to understand the implications of the production technology on the layout of the network. For instance, since biomass is less energy intensive, BG based networks could be relatively less-centralized compared to the network based on, for e.g., CG. In addition, network topology also depends on the H_2 form – i.e., liquid or gaseous (LH_2 or GH_2). As shown by Konda et al., (2009 and 2010), LH_2 based networks are more centralized while the GH_2 based networks are less-centralized.

Hence, in this contribution, based on the historical data and future projections, we try to quantify the future potential (and limitations) of these technologies while explicitly accounting for possible technological advancements, production scale, CO_2 emissions and geographic characteristics (by considering the Netherlands as a case-study). In addition, the presentation will encompass the benefits and limitations of co-production and forecourt production routes together with the importance and the need of optimal topology within the context of network of multiple-plants. Such an objective quantitative analysis helps to understand if these technologies can ever be cost-competitive (compared to SMR) and plan future research direction and strategies for the timely realization of the ultimate potential and market introduction of these technologies.

Disclaimer: Results/comments are of the authors' and do not necessarily represent the views of the associated organizations/institutions.

References:

Ball, M., Wietschel, M., (2009) The future of hydrogen – opportunities and challenges. International Journal of Hydrogen Energy. (34) 615 – 627.

Chiesa, Paolo., Consonni, S., Kreutz, T., Williams, R., (2005) Co-production of hydrogen, electricity and CO_2 from coal with commercially ready technology. Part A: Performance and emissions. International Journal of Hydrogen Energy. (30) 747-767.

Kreutz, T., Williams, R., Consonni, S., Chiesa P., (2005) Co-production of hydrogen, electricity and CO_2 from coal with commercially ready technology. Part B: Economic analysis. International Jornal of Hydrogen Energy. (30) 769 – 784.

Konda, N. V. S. N. M., Shah, N., Kramer, G.J., Brandon, N. P., (2009) Spatio-Temporal Model based Optimization Framework to Design Future Hydrogen Infrastructure Networks. Hydrogen and Fuel Cells Conference (HFC2009). Vancouver, Canada.

Konda, N. V. S. N. M., Shah, N., Kramer, G.J., Brandon, N. P., Dutch Hydrogen Economy: A Multi-billion Dollar Question? Reflections from an Optimization Framework. to be presented at PSE Asia 2010, Singapore, July 2010.

Mueller-Langer, F., Tzimas, E., Kaltschmitt, M., Peteves, S., (2007) Techno-economic assessment of hydrogen production processes for the hydrogen economy in the short and medium term. International Journal of Hydrogen Energy. (32) 3797-3810.

NAE (2004) The Hydrogen Economy: Opportunities, Costs, Barriers and R&D Needs

Schoots, K., Ferioli, F., Kramer, G.J., van der Zwaan, B.C.C., (2008). Learning curves for hydrogen technology: An assessment of observed cost reductions. International Journal of Hydrogen Energy (33) 2630 – 2645.

Stiegel, G.J. and Ramezan, M. (2005) Hydrogen from coal gasification: An economical pathway to a sustainable energy future. International journal of Coal Geology (65) 173 – 190.

See more of this Session: Alternative Fuels and Enabling Technologies
See more of this Group/Topical: Fuels and Petrochemicals Division

(443f) Char Oxidation Characteristics of Ashless Coal Particles Burning in a Drop Tube Furnace

Chunsung Lee[1], Song-gon Kim[1], Byoung-hwa Lee[1], Sihyun Lee[2], Chung-hwan Jeon[3], Young-june Chang[3] and Juhun Song[3], (1)Mechanical Engineering, Pusan National University, Pusan, South Korea, (2)Climate Change Research Division, Korea Institute of Energy Research(KIER), Daejeon, South Korea, (3)mechanical engineering, Pusan Clean Coal Center, Pusan National University, Pusan, South Korea

Recently, it is essential to develop alternative energy as the fuel in the power plant in the world which is faced with energy crisis because of economic constraints. The ashless coal, one of cleanest fuel, is known to contain relatively low ash and high calorific value, which could increase combustion efficiency and reduce carbon dioxide emission. The combustion processes of the ashless coal char were examined to understand how it could influence the performance of a pulverized coal-fired boiler. In this study, an experiment has been conducted on the kinetic analysis for single particle of an ashless coal char burning at different temperatures of drop tube furnace (DTF). Temperature distributions of single particles during the oxidation process were monitored by a two-color pyrometer mounted on the top of the furnace. An amount of char mass reacted during char oxidation is calculated from the flue gas measurement. The char oxidation kinetic parameters(A, E) of the ashless coal with determined char oxidation rate and the reaction rate coefficient was also obtained using an n-th order Arrhenius model. The results show the oxidation reactivity of the ashless coal was comparable with those of the conventional coals which could justify the use of the ashless coal in the real utility boiler.

See more of this Session: Alternative Fuels and Enabling Technologies
See more of this Group/Topical: Fuels and Petrochemicals Division

(452a) Fast Pyrolysis Bio-Oils From High Protein Containing Feedstocks

Charles A. Mullen and Akwasi A. Boateng, Eastern Regional Research Center, Agricultural Research Service, USDA, Wyndmoor, PA

Typically biomass considered for conversion to biofuels is comprised mostly of carbohydrates and lignin, with small amounts of other materials such as protein, lipids, and ash. Fast pyrolysis of such lignocellulosic materials generally produces bio-oils that are acidic, thermally unstable and have approximately half the energy content of petroleum fuels. However, the properties of bio-oils produced from biomass that contains a higher percentage of protein are different. Fast pyrolysis was performed on several high protein feedstocks, including distiller's grains with solubles (DDGS) from grain fermentation to ethanol, extracted oil seeds (pennycress, camelina), and animal rendering by-product. We found that during the pyrolytic breakdown of protein, basic organo-nitrogen compounds are formed and therefore the bio-oils have a higher pH and lower acid number than typical bio-oils. They also tend to have higher energy contents than wood pyrolysis oils. In this presentation, the results of these experiments, including mass and energy balances, carbon conversion efficiencies, and detailed product characterizations will be presented. Bio-oil chemical characterizations include elemental analysis, GC/MS, GPC, NMR and stability assessments. The fuel properties of the bio-oils including heating values and viscosities will also be presented. Finally, the bio-char co-products of high protein feedstocks will be compared with bio-chars from wood-like feedstocks.

See more of this Session: Biorefinery - Thermochemical Conversion of Biomass I
See more of this Group/Topical: Forest and Plant Bioproducts Division

Fast pyrolysis of lignocellulosic biomass to produce biooil and biochar : Parameter optimization and physicochemical analysis of them

Kwang Ho Kim, Seoul National University, Seoul, South Korea
In Yong Eom, Seoul National University, Seoul, South Korea
Sun Joo Moon, Seoul National University, Seoul, South Korea
Tai Seung Kim, Seoul National University, Seoul, South Korea
Soomin Lee, Korea Forest Research Institute, Seoul, South Korea
Hwangmyeong Yeo, Seoul National University, Seoul, South Korea
In Gyu Choi, Seoul National University, Seoul, South Korea
Joon Weon Choi, Seoul National University, Seoul, South Korea

Introduction

Fast pyrolysis, which is characterized by its simple process and short reaction time, has the capability of transforming inhomogeneous and recalcitrant lignocellulosics into a liquid form of fuel, called biooil [1]. This process is one of the most promising thermal technologies in the world. Recently, liquid biooil has been highly regarded both as an alternative energy [2] and green chemical resources. The fast pyrolysis process also has another advantage, as compared to other biofuel production processes, all different types of biomass can be considered as potential raw materials for fast pyrolysis [3]. A wide variety of biomass has been studied for pyrolysis, from agricultural residues to lignocellulosic biomass [4-11].

Fast pyrolysis is operated in the absence of oxygen producing gas, liquid and char in different proportions, according to experimental conditions. During fast pyrolysis, temperature, residence time of volatiles, heating rates, and particle size are well known as crucial parameters affecting the yield and fuel properties of biooil, as well as its chemical composition. Uzun et al. [12] studied the effect of temperature and nitrogen gas flow rates on the yield of biooil using soybean residues after the extraction of biodiesel. Zheng [13] produced biooil from rice husks in different temperature ranges and analyzed the fuel properties of the biooil.

In this study, yellow poplar wood, a representative fast-growing species of woody biomass, was pyrolyzed at various temperatures and residence times, using a fluidized bed-type reactor. The goal of this study was to elucidate the effect of temperature and residence time during fast pyrolysis on the conversion rate of woody biomass to biooil and to analyze the physicochemical properties of the biooil, as well as the biochar, which is a co-product of fast pyrolysis of woody biomass

Materials and methods

Feedstock

Yellow poplar wood *(Liriodendron tulipifera)* was used in this experiment as the raw lignocellulosic biomass. Bark-free stems were chipped and filtered through a 0.5mm aperture screen. The wood meal was air dried to about 8% moisture content.

Pyrolysis

Fast pryolysis of yellow poplar wood meal was performed using a fluidized bed reactor. A schematic diagram of the fluidized bed reactor is shown in figure 1. Before the experiment, the reactor was purged for 30 min with nitrogen to create an oxygen-free atmosphere. A sample weighing 150g was fed into the reactor for 1hr. After the experiment, the produced biooils were kept in the refrigerator at 4°C for further analysis.

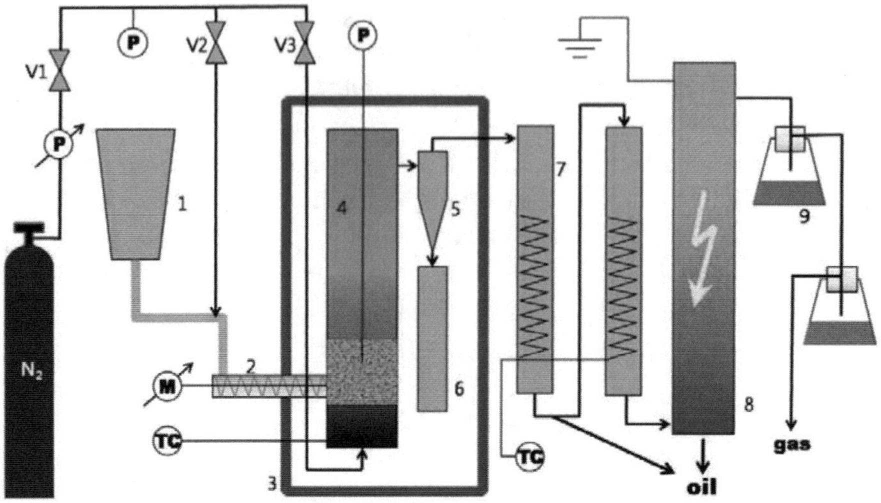

Figure 1. Schematic diagram of fluidized bed reactor (1: feeder, 2: screw feeder, 3: heater, 4: reactor, 5: cyclone, 6: char collector, 7: cooler, 8: electrostatic precipitator, 9: filter)

Analytical methods

Several physical and chemical analyses were performed to characterize biooil and biochar. The list of analytical methods for biooil is shown in Table 1. Elemental analysis and calorific value of biochars were measured with same equipment applied to biooil analyses. The surface morphology of the biochars was investigated by a field-emission scanning electron microscope (FE-SEM, SUPRA 55VP of Carl Zeiss).

Table 1. Analytical methods

Property	Method	Unit
Water content	Karl-Fischer titration	%(w/w, wet basis)
pH	pH meter	pH unit
Viscosity (@ 40°C)	Capillary type viscometer	cSt
Calorific value	Bomb type calorimeter (Parr 6400, Parr instrument company)	MJ/kg
Elemental analysis	Elemental analyzer (CHNS-932 Leco Corp.)	%C, %H, %N (w/w, wet basis)
Gas chromatography	Column type : DB 5 Dimensions : 30 m × 0.25 mm × 0.25 μm Injector : 250℃, split ratio 1:20, FID detector : 280℃ Oven : 45°C(4min. hold), 3 °C/min. to 280 °C (20min. hold)	

Results and discussion

Influence of pyrolysis temperature

Fast pyrolysis of yellow poplar wood meal was carried out in the temperature range of 400°C to 550°C at intervals of 50°C, and each yield of pyrolytic products (gas, biochar and biooil) is shown in Figure 2. This experiment revealed that the quantitative composition of the pyrolysis products was clearly affected by temperature. The yield of biooil, which was the target product of the fast pyrolysis of woody biomass, gradually increased up to 500°C, then slightly decreased at 550°C. The yield of biooil was maximized at approximately 68.5% at 500°C, based on the weight of yellow poplar. On the other hand, when the temperature of fast pyrolysis increased, the yield of non-condensable gases also continuously increased from 20.9% at 400°C to 28.2% at 550°C, while the formation of char decreased from 28.9% at 400°C to 7.9% at 550°C.

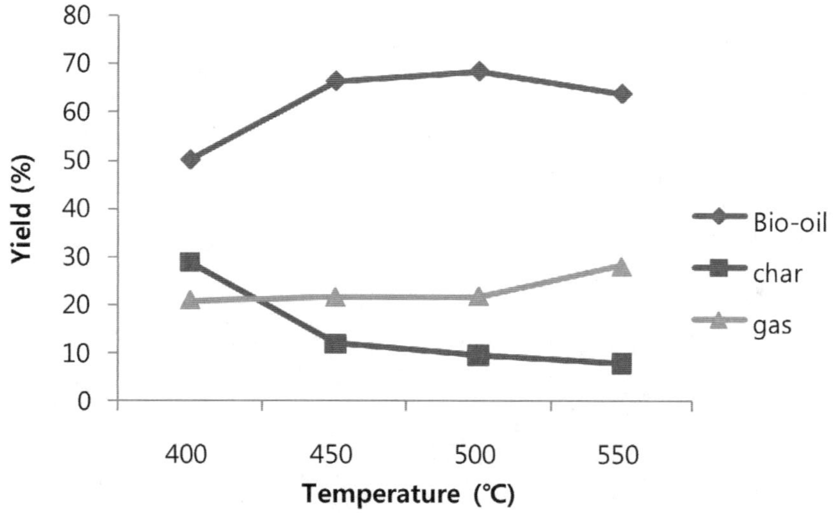

Figure 2. The distribution of product yields as a function of pyrolysis temperature (wet basis, residence time: 1.2sec)

According to previous studies [14-16], the highest yields of biooil from woody biomass feedstock were obtained at temperatures between 500°C and 520°C, which means that the energy required for biomass degradation to low molecular weight compounds was not sufficient below 500°C, but char formation was possible. In other words, a lower pyrolysis temperature prevents full decomposition of the biomass [7]. However, the components of biomass could decompose too much at temperature ranges over 500°C to be condensed into biooil. Higher pyrolysis temperatures cause secondary cracking of primary pyrolytic products, which leads to a decrease in biooil yield and an increase in gas yields [14, 15].

Influence of residence time

Residence time of pyrolytic products in the reactor is another critical parameter in the fast pyrolysis process. Pyrolysis was performed at different residence times, between 1.2s and 7.7s, which was adjusted by the nitrogen gas flow rate. The residence time is defined as the void volume of the reactor divided by the inlet gas flow rate at reactor conditions [8]. Four different nitrogen flow rates,

2.5ℓ/min, 5.0ℓ/min, 10.0ℓ/min, and 15.0ℓ/min, were chosen to examine the effect on pyrolytic product distribution. Each rate indicates 7.7s, 3.8s, 1.9s and 1.2s of residence time, respectively. In this case, the temperature was fixed at 500°C, at which the yield of biooil was the highest. The distribution of product yields, as a function of residence time at 500°C, is shown in Figure 3.

The yield of biooil was influenced by the residence time of the pyrolytic products. There was little difference in the yield of biooil at the residence times of 1.2s and 1.9s. However, when the residence time was prolonged to 3.8s and 7.7s, biooil production was clearly reduced to 61.8% and 51.6%, respectively. In the case of char and gas formation, however, their yields increased from 8.9% to 17.9%, and from 21.8% to 30.5%, respectively. This result could be explained by the residence time prolonging the primary degradation products to further exposure of secondary cracking reactions, thus converting the biomass to non-condensable gases and not biooil [7].

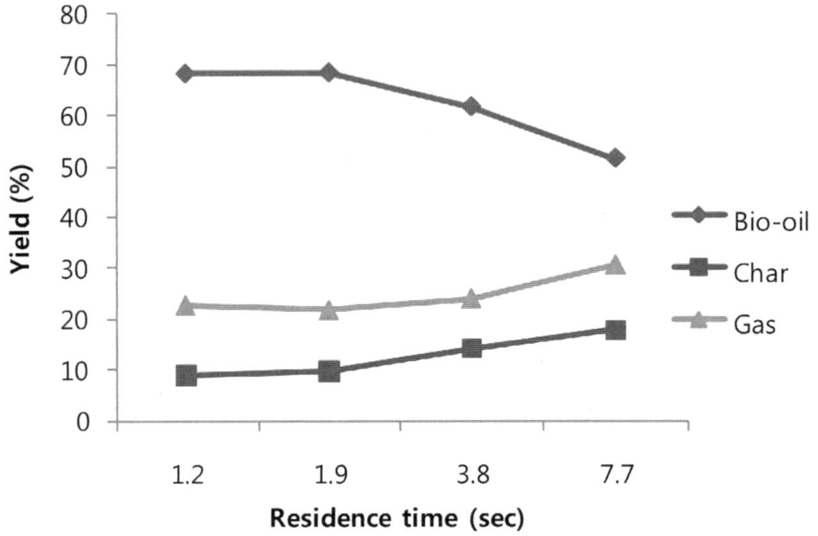

Figure 3. The distribution of product yields as a function of residence time (wet basis, pyrolysis temperature: 500°C)

Physicochemical properties of biooil

The water content in the biooil varied from 21.6 wt% to 42.6 wt%, depending on pyrolysis conditions. Considering the water content (approximately 8%) of the yellow poplar wood meal, the amount could be explained by the considerable amount of water formed during the fast pyrolysis of biomass, especially by a dehydration reaction [17]. Several kinds of acids, such as acetic acid and benzoic acid were formed from the wood components during fast pyrolysis, and pH values of the biooil were between 1.9 and 2.9 [17]. The heating values of the biooils, produced under different temperatures and residence times, were found to be between 10.4 and 17.2 MJ/kg. The relatively low heating values of the biooils were attributed to the high water content and oxygen-containing degradation products.

Characterization of biochar

Elemental composition and calorific value of the pyrolysis biochar was analyzed, and is presented in Table 2. As can be seen in table, the biochar was composed of more than 80% carbon in all experimental runs having a calorific value of 27.1~29.9MJ/kg, which is comparable to coal (approximately 30MJ/kg). The amount of carbon in biochar increased with decreasing pyrolysis temperature and increasing residence time. Like routine carbonization processes, the formation of biochar with a high carbon content and calorific value was favored at lower temperatures and longer residence times.

Table 2. Elemental analysis of biochar coproduced from fast pyrolysis of yellow poplar

Elemental analysis (wt%)	Residence time fixed at 1.9 sec				Temperature fixed at 500°C			
	400°C	450°C	500°C	550°C	7.7s	3.8s	1.9s	1.2s
Carbon	83.4	83.1	82.7	82.0	84.3	83.3	82.7	80.0
Hydrogen	2.0	1.4	1.1	0.8	1.6	1.3	1.1	1.0
Nitrogen	trace	trace	trace	trace	trace	trace	trace	trace
Higher heating value (MJ/kg)	29.9	29.0	28.4	27.8	29.8	28.9	28.4	27.1

A field-emission scanning electron microscope (FE-SEM) was applied to investigate the effect of thermal treatment on the biomass morphology. Figure 4 shows the FE-SEM images of feedstock and biochar obtained from pyrolysis. Significant structural differences can be seen when comparing both images. The biochar looked distorted, and its surface was rough, due to the thermal treatment.

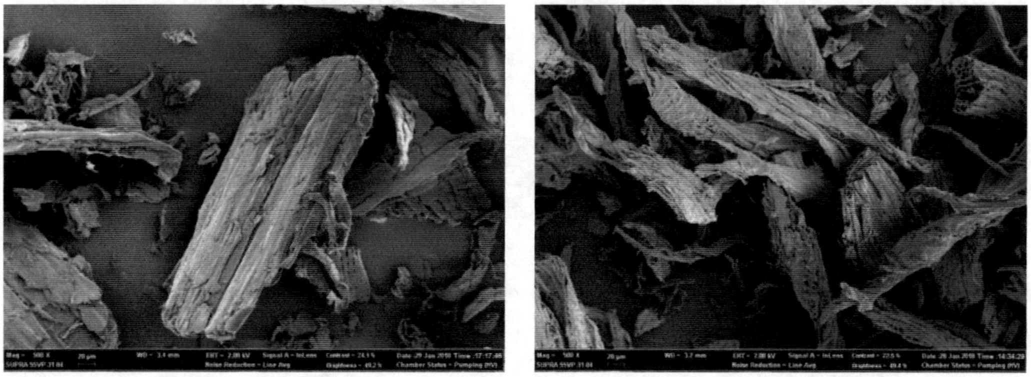

Figure 4. SEM pictures (×500) of yellow poplar wood meal (left) and biochar (right)

Conclusion

The yellow poplar wood meal was pyrolyzed to convert it into biooil and biochar under different experimental conditions using a fluidized bed reactor, and their physicochemical properties were characterized. The mass balance of pyrolytic products, biooil, biochar and gases, was significantly influenced by pyrolysis temperature and residence time. The maximum yield of biooil was 68.5%, based on a biomass weight at 500°C and with a residence time of 1.9s. Pyrolysis of yellow poplar wood for a prolonged residence time, as well as below 500°C, resulted in lower yields of biooil but activated the formation of co-products.

The yield of biochar was 8.0~28.9% on a mass basis. Biochar distribution was highly dependent on the pyrolysis conditions. The analytical results showed that the elemental composition of biochar was affected by experimental conditions. The carbon content of biochar increased with decreasing pyrolysis temperature and increasing residence time, which is similar to the carbonization mechanism.

References

1. Westerhof R.J.M., Kuipers N.J.M., Kersten S.R.A., Swaaij W.P.M. (2007) "Controlling the water content of biomass fast pyrolysis oil." *Industrial & Engineering Chemistry Research*, 46(26): p. 9238-9247.
2. Scott D.S., Majerski P., Piskorz J., Radlein D. (1999) "A second look at fast pyrolysis of biomass - the RTI process." *Journal of Analytical and Applied Pyrolysis*, 51(1-2): p. 23-37.
3. Bridgwater, A.V. (2003) "Renewable fuels and chemicals by thermal processing of biomass." *Chemical Engineering Journal*, 91(2-3): p. 87-102.
4. Putun, A.E. (2002) "Biomass to bio-oil via fast pyrolysis of cotton straw and stalk." *Energy Sources*, 24(3): p. 275-285.
5. Mullen C.A., Boateng A.A., Goldberg N.M., Lima I.M., Laird D.A., Hicks K.B. (2009) "Bio-oil and bio-char production from corn cobs and stover by fast pyrolysis." *Biomass and Bioenergy*, 34(1): p. 67-74.
6. Tsai W.T., Lee M.K., Chang Y.M. (2007) "Fast pyrolysis of rice husk: Product yields and compositions." *Bioresource and Technology*, 98(1): p. 22-8.
7. Xu R., Ferrante L., Briens C., Berruti F. (2009) "Flash pyrolysis of grape residues into biofuel in a bubbling fluid bed." *Journal of Analytical and Applied Pyrolysis*, 86(1): p. 58-65.
8. Piskorz J., Majerski P., Radlein D., Scott D.S., Bridgwater A.V. (1998) "Fast pyrolysis of sweet sorghum and sweet sorghum bagasse." *Journal of Analytical and Applied Pyrolysis*, 46(1): p. 15-29.
9. Park H.J., Park Y.K., Dong J.I., Kim J.S., Jeon J.K., Kim S.S. (2009) "Pyrolysis characteristics of Oriental white oak: Kinetic study and fast pyrolysis in a fluidized bed with an improved reaction system." *Fuel Processing Technology*, 90(2): p. 186-195.
10. Garcia-Perez M., Wang X.S., Shen J., Rhodes M.J., Tian F.J., Lee W.J. (2008) "Fast pyrolysis of oil mallee woody biomass: Effect of temperature on the yield and quality of pyrolysis products." *Industrial & Engineering Chemistry Research*, 47(6): p. 1846-1854.
11. Schnitzer M.I., Monreal C.M., Facey G.A., Fransham P.B. (2007) "The conversion of chicken manure to biooil by fast pyrolysis I. Analyses of chicken manure, biooils and char by 13C and 1H NMR and FTIR spectrophotometry." *Journal of Environmental Science and Health B*, 42(1): p. 71-7.
12. Uzun B.B., Putun A.E., Putun E. (2006) "Fast pyrolysis of soybean cake: product yields and compositions." *Bioresource and Technology*, 97(4): p. 569-76.
13. Zheng, J.L. (2007) "Bio-oil from fast pyrolysis of rice husk: Yields and related properties and improvement of the pyrolysis system." *Journal of Analytical and Applied Pyrolysis*, 80(1): p. 30-35.
14. Bridgwater, A.V. (1999) "An overview of fast pyrolysis of biomass." *Organic Geochemistry*, 30(12): p. 1479-1493.
15. Venderbosch, R.H, Prins, W. (2010) "Fast pyrolysis technology development." *Biofuels Bioproducts & Biorefining-Biofpr*, 4(2): p. 178-208.
16. Mohan D., Pittman C.U., Steele P.H. (2006) "Pyrolysis of wood/biomass for bio-oil: A critical review." *Energy & Fuels*, 20(3): p. 848-889.
17. Lu Q., Li W.Z., Zhu X.F. (2009) "Overview of fuel properties of biomass fast pyrolysis oils." *Energy Conversion and Management*, 50(5): p. 1376-1383.

(452c) Types of Carbon in Biochars From Slow Pyrolysis, Fast Pyrolysis and Gasification

Catherine E. Brewer, Center for Sustainable Environmental Technologies, Iowa State University, Ames, IA, Robert C. Brown, Center for Sustainable and Environmental Technologies, Iowa State University, Ames, IA and Klaus Schmidt-Rohr, Iowa State University, Ames, IA

Biochars, sustainably produced charcoals used for soil amendment and carbon sequestration applications, contain both organic (carbon) and inorganic (ash) constituents. The composition of the inorganic fraction is mostly dependent on the mineral matter present in the feedstock. The types and quantities of carbon in the organic fraction of the biochars, however, are highly dependent on processing conditions such as temperature, heating rate, reaction atmosphere (inert or oxic), and residence time. This study compares the types of carbons in biochars from the slow pyrolysis, fast pyrolysis, and gasification of switchgrass, corn stover and various hardwoods. Among the analytical techniques used to characterize the chars are CHNOS elemental analysis, proximate analysis, Fourier-transform infrared spectroscopy (FTIR), and quantitative ^{13}C solid-state nuclear magnetic resonance spectroscopy (NMR) using direct polarization magic angle spinning (DP/MAS) techniques. Three biochars from the fast pyrolysis of corn stover were also used to conduct a soil incubation study to measure carbon bioavailability based on CO_2 evolution rates. Overall findings suggest that biochars can be tailored to specific applications and that not all chars created from the thermochemical conversion of biomass will be suitable for use as biochars.

See more of this Session: Biorefinery - Thermochemical Conversion of Biomass I
See more of this Group/Topical: Forest and Plant Bioproducts Division

Product and Economic Analysis of Direct Liquefaction of Swine Manure

Mitchell James Minarick, University of Illinois, Urbana, IL, USA. Additional authors: Yuanhui Zhang, Zhichao Wang, Guo Yu, and Ted Funk.

Abstract

Hydrothermal processing of biomass is a technology that has shown promising results in treating waste and producing oil. A batch hydrothermal processing system that was previously developed at the University of Illinois at Urbana-Champaign was used to treat swine manure and it successfully converted up to 70% of swine manure volatile solids into oil and reduced manure chemical oxygen demand by up to 75%. A continuous reactor was developed and it resulted in similar conversion rates to the batch process and it indicated the potential for the process to be scaled-up. These results were also documented previously.

Following these results, a slurry pump was designed and built to enable longer test durations and lower repair costs, and it was used to conduct 24-hour continuous tests and produce quantities of oil suitable for more detailed analysis.

The hydrothermal process utilizes manure containing around 25% solids content which is not available in typical swine confinement facilities so a preliminary investigation of dewatering methods was conducted to determine the best dewatering method in terms of the resultant oil yields and the effects on fertilizer values. The best dewatering method appears to be a coarse filtering process or the use of a low-cost additive. A coarse filtering process followed by the hydrothermal process results in an oil product as well as a fertilizer product that retains the majority of its nitrogen value with a reduced level of phosphorus as compared to the unprocessed swine manure.

Energy and economic analyses of the dewatering and hydrothermal processes were conducted for swine manure as well as general analyses for other domestic biomass feedstocks. The feasibility of the conversion of swine manure was judged to be high, dependent mainly on the validity of the assumptions for the value of the oil product.

Introduction

Using waste materials to produce biofuel sidesteps the controversial food versus fuel debate and gives the production process a competitive advantage over methods using traditional feedstocks since they can be acquired much more cheaply, if not for free. It makes sense to research methods of biofuel production from waste streams alongside research of biofuels produced from more valuable feedstocks.

Waste Conversion technologies

The conversion of biomass to fuel generally follows one of two routes: biochemical or thermochemical. A summary of the primary biomass conversion technologies is provided by He (2000) in Figure 1 below.

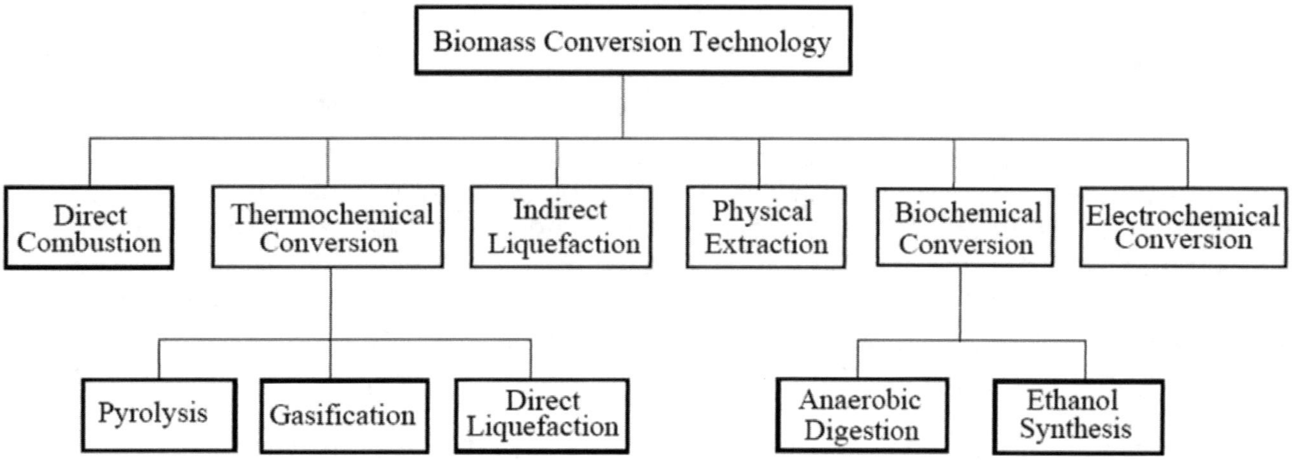

Figure 1. Biomass conversion technologies.

Direct liquefaction is a type of thermochemical conversion that has been studied extensively during the past 10 years at the University of Illinois. Similar to pyrolysis, direct liquefaction is well-suited for the conversion of livestock waste because it does not require a dry feedstock. Previous research teams headed by Dr. Yuanhui Zhang have concluded that direct liquefaction could be a viable solution for the disposal of swine manure, and they have attempted to determine the parameters that optimize the oil yield.

Ocfemia (2005) applied these parameters to a continuous reaction process to determine the feasibility of scaling the system up beyond the batch scale. His system had a capacity to process up to 48 kg of manure slurry per day, and successfully produced oil at equal or greater efficiencies than the batch process used by Brian He.

Water-separation considerations

The desired solids content of the feedstock for the hydrothermal process is around 25% according to He (2000). This is approximately the solids content contained in fresh (that is, taken directly from the confinement floor) swine manure. However, according to Fulhage (2001) most swine manure is handled as a liquid, utilizing shallow storage pits beneath slotted floors that allow the producer to apply manure to fields when it is most beneficial. This slurry has a higher water content than what is optimal for the hydrothermal conversion process (He, 2000) and since additional water translates into additional heating and processing costs, it is necessary to remove as much water as possible without affecting the oil yield.

For this technology to be useful on a farm, it is necessary to develop Ocfemia's continuous process (2005) into a working prototype that utilizes swine manure taken directly from the production facility. In a typical production facility, the manure as well as urine and spilt food and water falls through the slotted floor to the shallow-pit, where it is stored until it is pumped into an external tank for disposal every few weeks. Initial tests (Wang, 2007) showed that conversion into oil of the manure slurry as taken from the pit was impaired when compared to the conversion of manure taken from behind the pig, so it is necessary to determine the effect of different water separation methods on oil yields in order to dovetail thermochemical conversion technology with manure collection systems that are currently in place.

Methods

Water Separation Methods

As previously discussed, swine manure is commonly stored in a shallow pit, where it becomes diluted to less than 10% solids. The thermochemical process requires a feedstock with a solids content of around 20%, so a practical means for attaining higher solids content needed to be determined.

Evaporation is the simplest method of water separation, but also the most time consuming. The swine manure was simply stored in open containers with a fan circulating air over the tops of the containers. The solids content was measured periodically until it reached the desired level.

Coarse filtration was accomplished using a standard aluminum window screen as the filter medium. The screen had square openings measuring 1.13 x 1.30 mm, and it was used as a liner inside a round laundry basket that served as the filter holder. The manure from the pit was placed in the filter assembly and the laundry basket was rolled by hand until the flow of aqueous product out of the filter ceased.

The difficulties associated with this water separation method are discussed in section 3.3.4, but in summary fine filtration was defined as any filtration with a smaller pore size than the coarse filter (1.13x1.3 mm). The portion of the swine manure that was subjected to the fine filtration was the aqueous portion of the swine manure that was not already removed by the coarse filter.

For the low-speed centrifugation, a vertical-shaft solid-basket centrifuge manufactured by Lavin Centrifuge (model 12-413V) was used to subject the aqueous product from the coarse filtration to 2,250 Relative Centrifugal Force (2,250 RCF=2,250 times the force of gravity) for 5 minutes. The solids were collected and the aqueous portion was saved for further separation.

The aqueous solution remaining after the low-speed centrifugation was placed in capped 500mL Nalgene polycarbonate centrifuge bottles (Part #3140-0500, 69.5 mm diameter, 158.8 mm length, 13,700 maximum allowed RCF) and carefully weighed to make sure they would be balanced in the centrifuge. The centrifuge was a Dupont Sorvall centrifuge equipped with a GS-3 rotor and was used to subject the samples to 10,816 RCF for 15 minutes. The solids formed a dense layer on the bottom of the centrifuge bottles, so the aqueous solution was easily separated and both portions were saved.

Thermochemical Conversion Process

The thermochemical conversion was conducted using similar conditions as those used in the experiments by He (2000) and Ocfemia (2005). The temperature was the main control variable and was held at 305°C for 30 minutes of residence time in a Parr 2-L reactor. The oxygen was evacuated by pressurizing the reactor with 100 psi of nitrogen gas and then releasing the pressure. This was repeated three times before filling the reactor with 100 psi of nitrogen gas which was used as a process gas throughout the experiment.

These operating conditions resulted in a peak reactor pressure of approximately 1650 psi when the reactor was at its peak operating temperature. No catalyst was used during these tests.

Oil Analysis

The thermochemical process results in a raw oil product that is very crude and unrefined. The water content of the bio-oil was determined using ASTM standard D95-99 and subtracted from the oil yields. The toluene solubility was determined using ASTM standard D4072-98 and used to define a 'refined oil yield' which makes up roughly 35-45% of the raw oil. The refined oil yield gives an idea of both the quantity and quality of the oil and is a more robust measurement of the oil production.

Assumptions

Energy

The following assumptions were used when calculating energy balances:

The bio-oil has a heating value of 30,000 kJ/kg (Ocfemia, 2005).

The energy contained in the feedstocks before undergoing TCC was negligible or unavailable. In other words, it is assumed that the only energy input is heat via electricity, and the chemical energy in the feedstock is not considered. In most cases the feedstock had too high of a water content to be burned so this was an accurate assumption for the feedstocks considered.

The reactor is perfectly insulated. Although perfect insulation is not truly possible, the amount of heat loss through the reactor walls could be negligible with good insulation.

The internal energy values of the feedstocks are equal to those of water (i.e. the energy needed to heat/pressurize the feedstock is equal to that needed for water). This is a reasonable and conservative assumption because the feedstocks contained 75-95% water, and the specific heat of water is higher than most materials. For instance, the specific heat of water is over 4 times that of sawdust. Thus, more energy is needed to heat pure water to a given temperature than would be needed to heat the same amount of water mixed with sawdust. These figures for the change in internal energy are therefore higher than the actual values, and the actual energy needed to heat the different solutions should be less than (but fairly close to) what would be calculated using the following values for water:

Reaction conditions are approximately 100 bar (1450 psi) and 303°C,
Internal energy: U_f (303°C) = 1350.5 kJ/kg
U_f (22°C) = 92.32 kJ/kg
ΔU for reaction = 1258 kJ/kg
Energy used to pre-process the feedstock is negligible.

No heat is recaptured with a heat exchanger unless otherwise stated. When a heat exchanger is used in calculations, the effectiveness is assumed to be 75%.

Economic

For the following calculations (unless otherwise specified), the cost of electricity was estimated to be $0.05/kW-hr. If the reactor was only operated during certain times of the day, this is a reasonable estimate for industrial power usage, although in practice this value will vary.

Additionally, the value of crude oil varies greatly. It is reasonable to assume that the value of the bio-oil will be related to the value of crude oil since they are competitive products. Consequently, the value of the bio-oil can be expected to vary significantly based on the fluctuations in the price of crude oil, which can be volatile. For example, in just a 12-month span from winter of 2008 to winter of 2009 the price of oil varied from $40-$140. This variability certainly has an effect on the value of the bio-oil and on any alternative fuel for that matter.

For most of the calculations that follow, the price of oil is assumed to $50/barrel unless otherwise specified. If this price were assumed to be higher, it would result in more favorable results for the thermochemical process.

Estimating the actual value of the bio-oil is difficult. It is hard to find the true market value of an oil product which does not have an established market yet. For the purpose of calculation, the value of the refined bio-oil (i.e. the toluene soluble portion) was estimated to be 60% of the value of petroleum light crude oil. This value was chosen in an attempt to be conservative so as not to overestimate the value of the bio-oil. This value might well be higher since the crude oil is unrefined and the toluene-soluble portion of the bio-oil has already been subjected to one refinement step (hence the choice of the term 'refined oil yield' to describe it). This assumption is admittedly difficult to defend. Without an existing market, the value of the bio-oil will not be precisely known. However, taking 40% of the value off of the unrefined petroleum crude hopefully allows adequate room for this uncertainty.

Most petroleum chemists divide crude oil into four fractions: saturates, aromatics, resins and asphaltenes. Asphaltenes are, broadly speaking, the fraction of oil that are insoluble in heptane or pentane and soluble in benzene/toluene. Appleford (2004) analyzed the bio-oil from the thermochemical conversion processing of swine manure and found that the oil was primarily resins and asphaltenes as seen in Table 1 below where the first two columns of results represent crude oil from swine manure:

Table 1. Properties of bio-crude oil from swine manure vs. natural crude oil.

Fraction (%)	C-305-1500-60	N-305-1500-60	Pooled Sample	Natural Crude Oil*	Natural Bitumen*
Saturates	2.93	2.48	1.32	>30	15
Aromatics	1.84	1.67	1.60	>30	34
Resins	44.65	45.87	35.87	<20	34
Asphaltenes	44.73	43.39	54.81	<10	17
Water + Loss	5.85	6.59	6.40	varies	varies

As seen in Table 1, the bio-crude from the swine manure contains around 50% asphaltenes and around 35% resins. Therefore an alternative valuation for the bio-oil can be obtained by assuming the products will be used as asphaltanes rather than as a refined oil product. Asphalt prices, like oil prices, fluctuate dramatically. Table 2 shows the price of asphalt used for paving for 2007 and 2008 (California Department of Transportation, 2009):

Table 2. Monthly price of asphalt for 2007-2008.

Month-Year	$/ton	Month-Year	$/ton
Jan-07	$287.10	Jan-08	$497.80
Feb-07	$267.70	Feb-08	$434.40
Mar-07	$295.30	Mar-08	$510.00
Apr-07	$317.40	Apr-08	$504.10
May-07	$314.90	May-08	$567.70
Jun-07	$321.70	Jun-08	$645.30
Jul-07	$355.70	Jul-08	$716.90
Aug-07	$384.30	Aug-08	$630.40
Sep-07	$370.70	Sep-08	$547.40
Oct-07	$393.40	Oct-08	$486.10
Nov-07	$464.50	Nov-08	$288.10
Dec-07	$439.30	Dec-08	$199.40

Importantly, the use of the bio-oil as an asphalt would eliminate many of the expenses, technological hurdles, and uncertainties associated with refinement of the oil and finding a market for the finished product.

The net present value (NPV) of an investment is a standard method for using the time value of money to appraise long-term projects. Essentially it involves calculating the sum of the present values of future cash flows based on an assumed discount rate (or the expected interest rate of investment alternatives), and if the NPV is a positive number the investment is considered favorable to the investment alternatives. The NPV for a given set of cash flows can be calculated with a few assumptions. Labor costs could be limited to one employee at $35,000 per year. Maintenance costs can be estimated at $20,000 per year based on an assumed capital cost of $200,000. The annual discount rate (i.e. the rate of return of investment alternatives) is assumed to be 15% and the life expectancy of the equipment is 10 years.

Results

Figure 2 below shows the summarized results of the TCC tests using swine manure and a variety of different dewatering methods in terms of their refined oil yields. However, refined oil yield is not the only variable determining a successful test. Certainly energy input costs have a significant effect. For a feedstock with low solids content a significant amount of energy is spent heating and pressurizing water in addition to the solids. Manure slurry taken straight from the pit and sent to the reactor without any dewatering suffers from its high water content. Figure 3 below summarizes the relative processing costs associated with oil production based on the different dewatering methods.

Figure 2. Refined oil yields.

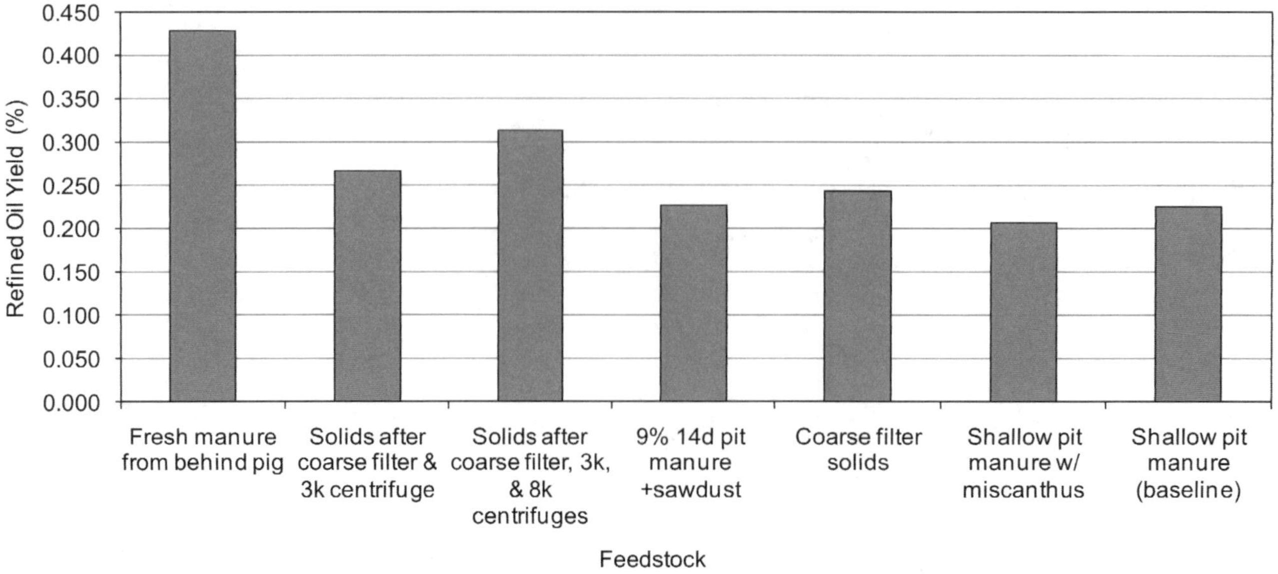

Figure 3. Relative processing costs per oil production.

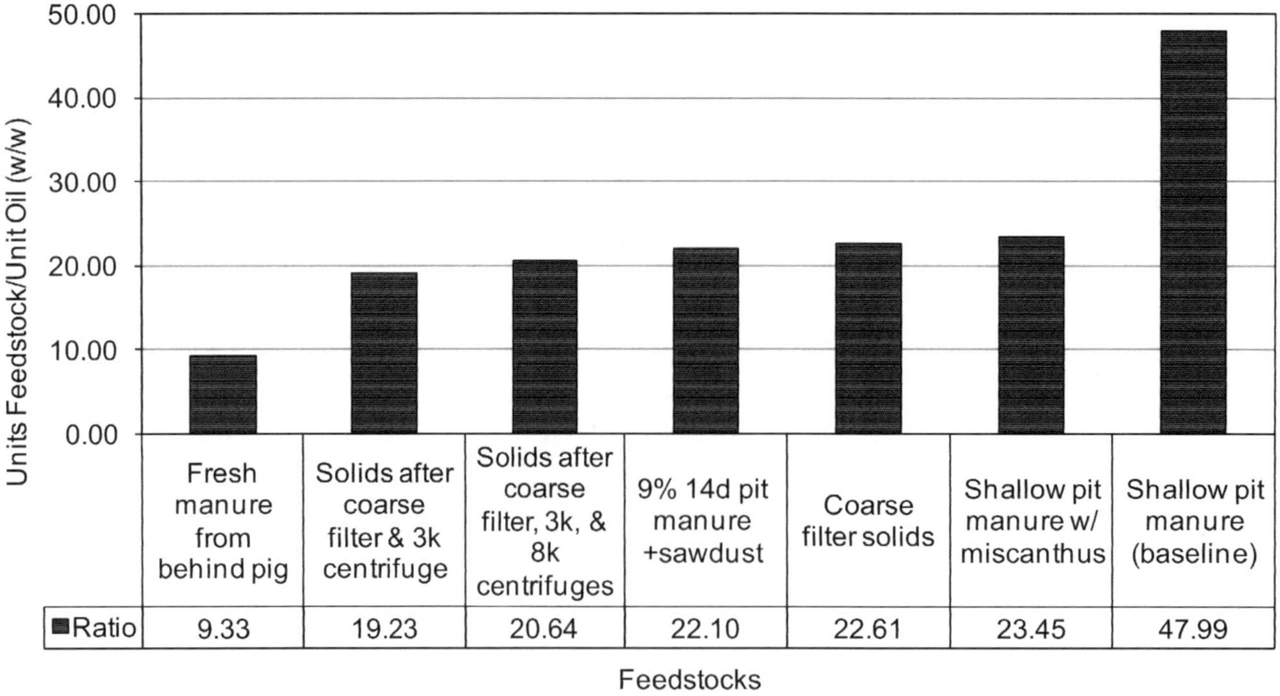

In summary, the best dewatering method appears to be mixing the slurry with additives such as sawdust or miscanthus if the additive is cheap or free. If the additive is not cheap, rough filtration is probably better.

Figure 4 below demonstrates the energy balances of the oil production process in terms of processing energy in versus oil-based energy out, both with and without a heat exchanger. The balances are favorable for all cases except the raw manure straight from the pit. It is important to remember that although most of the ratios seem to show a negligible energy gain, the main value added by the process is not due to an addition of net energy but rather a conversion of the form of the energy, from heat via electricity or combustion to a liquid fuel product.

Figure 4. Energy out versus energy in.

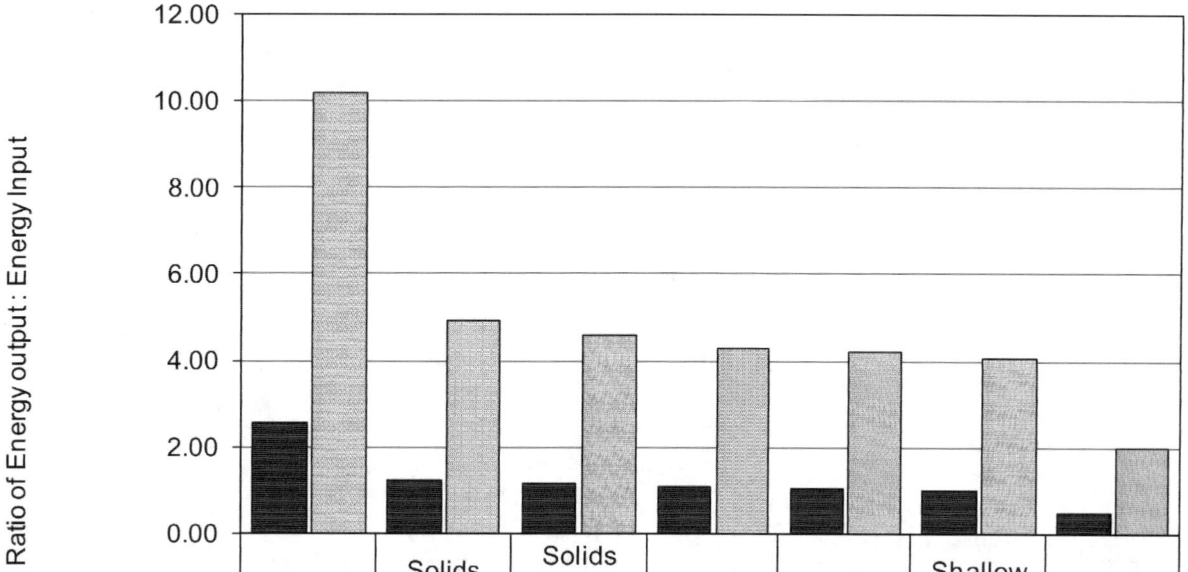

	Fresh manure from behind pig	Solids after coarse filter & 3k centrifuge	Solids after coarse filter, 3k, & 8k centrifuges	9% 14d pit manure +sawdust	Coarse filter solids	Shallow pit manure w/ miscanthus	Shallow pit manure (baseline)
■Energy output : energy input	2.55	1.24	1.16	1.08	1.05	1.02	0.50
□Energy out:in w/ heat exch.	10.22	4.96	4.62	4.32	4.22	4.07	1.99

Feedstocks

Figure 5 below shows that the fertilizer values of the aqueous products after the TCC test are not generally destroyed by the conversion process, with the exception of phosphorus content. The bars on the left side of each set of data represent the fertilizer values contained in the manure before processing. Nitrogen is fairly well preserved in most forms, except in Case 4 with the miscanthus as an additive. Potassium was retained well in all cases. This is an important piece of information because it shows that a farmer can process manure thermochemically to obtain bio-oil, and still retain over half of the original fertilizer value. Additionally, less phosphorus is applied to the field after the TCC process, lowering the risk of phosphorus runoff.

Figure 5. Fertilizer values remaining in aqueous solution.

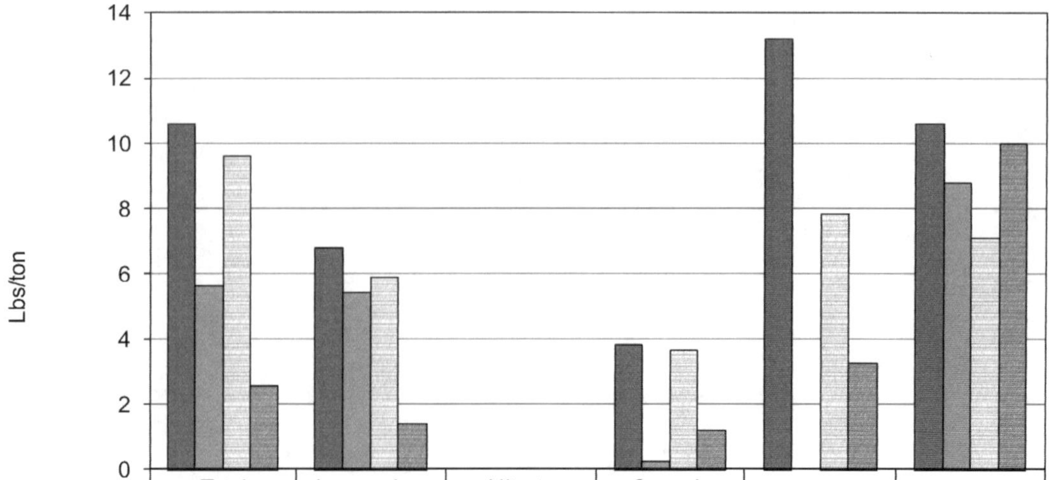

	Total Nitrogen	Ammonium Nitrogen	Nitrate Nitrogen	Organic Nitrogen	Phosphorus	Potassium
Raw Manure from pit, Case 1	10.60	6.80	0.00	3.80	13.20	10.60
Raw Manure after TCC, Case 2	5.64	5.41	0.00	0.23	0.00	8.80
Raw Manure after coarse filter & TCC, Case 3	9.61	5.90	0.00	3.66	7.84	7.09
Raw Manure w/ Miscanthus after TCC, Case 4	2.56	1.40	0.00	1.17	3.26	10.02

Components

Economics of Swine Manure Conversion

For an operation with 10,000 hogs, a feedstock supply that does not cost anything to acquire, and an empirically-determined (Wang, 2007) refined oil yield of 42.9% for manure from the floor behind the pig, 6,552 barrels of refined oil per year could be produced.

Assuming the value of refined oil is equal to 60% of the value of crude oil and a barrel of crude oil is worth $50, the annual gross income for the operation is $196,560. The cost to process this amount of biomass assuming a fresh manure solids content of 25%, an electrical energy cost of $0.05/kW-hr, and no heat recapture is $178,101 annually. Thus an annual income of $196,560-$178,101=$18,459 would be expected in this case study. This is probably not even enough income to pay for the associated labor costs, and a system like this is not economically feasible.

Factors that would decrease this amount include: oil prices lower than $50/bbl, refined oil worth less than 60% of crude oil, oil conversion rate of less than 42.9%, energy input costs more than $0.05/kW-hr, labor and capital costs, heat losses to the atmosphere, etc. Factors that would increase this amount include: oil prices higher than $50/bbl, refined oil worth more than 60% of crude oil, energy input costs less than $0.05/kW-hr, and heat recapture via a heat exchanger. Of these factors, the most significant changes would probably come in the estimates for energy input costs. If the operator was careful to run the reactor at hours of low electrical demand, it is conceivable that the cost of electricity would average less than $0.05/kW-hr. There are other energy input options as well. Natural gas was estimated to cost Midwest customers $1.11/therm or $0.0378/kW-hr during 2006 (Energy Information

Administration, 2007), although the efficiency of the boiler would not be 100% so that value would be increased slightly in a practical application.

The simplest way to reduce energy input costs, however, would be to install a heat exchanger to take some of the energy from the hot exit stream and use it to preheat the cool incoming feedstock.

When heat recovery assumptions are applied to the economic calculations, the numbers improve dramatically. The oil yield and gross income of $196,560 would remain the same, but the energy input costs are quartered due to the assumed energy recovery of 75%. The energy input costs become $44,525 annually. This drives the annual profit up from $18,459 to $152,035 before labor, capital, and maintenance costs and the NPV is then $486,996 which means the project is profitable (i.e. returns more than 15%) if the capital costs are less than the NPV. These results are very promising and would be even more compelling for larger hog farms because with larger volumes they would generate greater annual profits yet the labor and capital costs would remain virtually the same. This result is dependent on multiple assumptions but they were kept conservative and overall the feasibility is judged to be very high.

If the refined oil product was able to be used as an asphalt substitute, it could be conservatively priced at $200/ton. The gross income of the operation then becomes $241,098. This drives the annual profit up to $196,573 and the NPV becomes $710,522. To recap then, the NPVs for the manure from the floor are as follows:

- Manure from floor used to make oil: NPV=negative
- Manure from floor used to make oil, heat exchanger used: NPV=$486,996
- Manure from floor used to make asphalt, heat exch. used: NPV=$710,522

The benefits of using a heat exchanger and exploring the opportunities for using the bio-oil as an asphalt are apparent, and these modifications will be used in the remaining economic analyses.

NPVs using manure from shallow pit

The gross income using manure from a shallow pit and adding sawdust would be $323,479, however this is assuming that the additive does not cost anything. In reality, sawdust can cost up to $30 per ton in some areas, reducing the gross income by $137,394 to $186,085 per year. The operating costs would then be $141,815 and the gross profit is only $44,270, which is not even enough to pay for labor and maintenance expenses.

If the excess water was removed with rough filtration instead of using sawdust as an additive, then the gross income would be $42,201 with processing costs of $17,179 annually. The ratio of the income to the expenses is 2.46. This is a very acceptable ratio, but the gross profit of the TCC process ($42,201-$17,179=$25,022) is not even large enough to support a full-time employee. The high profitability ratio, however, indicates good potential for scaling up the system. If the thermochemical conversion system was mounted on a portable trailer and used to process the manure from 5 farms, the net profit would jump to $125,110 and the NPV would be $351,866.

To recap then, the NPVs for the manure from the shallow pit are as follows:

- Manure from pit, heat exchanger, asphalt product and $30/ton sawdust: NPV=negative

- Manure from pit, heat exchanger, asphalt product, rough filtration: NPV=negative after labor/maintenance costs are added, but the profitability ratio is high, meaning more volume is needed for profitability.
- Manure from pit, heat exchanger, asphalt product, rough filtration throughout 5 farms: NPV=$351,866

These results are very promising because they represent conservative calculations for the net present values of the types of system that would be used with existing production facilities without modification. The numbers would be even more compelling for larger hog farms, again because larger volumes would generate greater annual profits yet the labor and capital costs would remain virtually the same.

The primary downside risk to these NPV figures is the difficulty in determining an accurate value of the bio-oil. As stated previously, the refined bio-oil was sometimes valued at 60% of petroleum crude oil in these calculations, but that number is simply an estimate and not backed by firm data (since there really isn't any firm data extant). Other times the refined bio-oil was considered as a substitute for petroleum asphalt selling for $200/ton. This is probably more accurate than the other approximation since the oil product more closely resembles asphalt, but firmer data regarding the suitability of the bio-oil as a substitute for asphalt would solidify these estimates considerably.

References

Appleford, J. M. 2004. Analysis and optimization of the thermochemical conversion process to produce oil from biomass. Masters thesis. Urbana, IL: University of Illinois at Urbana-Champaign, Department of Agricultural Engineering.

ASAE Standards. 2005. D384.2: Manure production and characteristics. St. Joseph, Mich.: ASAE.

ASTM. 2004a. ASTM D95-99. Standard test method for water in petroleum products and bituminous materials by distillation. In: Annual Book of ASTM Standards. West Conshohocken, PA: Am. Soc. for Testing Materials.

ASTM. 2004b. ASTM D473-02. Standard test method for sediment in crude oils and fuel oils by the extraction. In: Annual Book of ASTM Standards. West Conshohocken, PA: Am. Soc. for Testing Materials.

ASTM. 2004c. ASTM D4072-98. Standard test method for toluene-insoluble (TI) content of tar and pitch. In: Annual Book of ASTM Standards. West Conshohocken, PA: Am. Soc. for Testing Materials.

California Department of Transportation. 2009. California statewide paving asphalt price index. Sacramento,CA. Available at: http://www.dot.ca.gov/hq/esc/oe/asphalt_index/PREVIOUS_YEARS/07-08_AC_INDEX.pdf.

Dong, R. 2007. Personal communication. Department of Agricultural and biological engineering. Urbana, IL.: University of Illinois at Urbana-Champaign.

Elliott, D. C., et al. 1994. Chemical processing in high-pressure aqueous environments: continuous-flow reactor process development experiments for organics destruction. *Industrial and Engineering Chemistry Research*. 33: 566-574.

Energy Initiative Administration. 2007. Natural gas annual 2006. Washington, D.C.: United States Department of Energy, Energy Initiative Administration.

Fulhage, C., and D. Pfost. 2001. Swine manure management systems in Missouri. Columbia, MO.: University of Missouri-Columbia Extension.

He, B. 2000. Thermochemical conversion of swine manure to produce oil and reduce waste. PhD dissertation. Urbana, IL: University of Illinois at Urbana-Champaign, Department of Agricultural Engineering.

Hrubant, G. R., R. A. Rhodes, and G. H. Sloneker. 1978. Specific composition of representative feedlot wastes: a chemical and microbial profile. Washington, D.C.: Science and Education Administration, U.S. Department of Agriculture.

Kreith, F., and D. Y. Goswami. 2004. The CRC handbook of mechanical engineering. 2nd edition. CRC Press. Boca Raton, FL.

Moran, M. J., and H. N. Shapiro. 2008. *Fundamentals of engineering thermodynamics*. 6th edition. John Wiley and Sons. Hoboken, NJ.

Ocfemia, K. 2005. Hydrothermal process of swine manure to oil using a continuous reactor system. PhD dissertation. Urbana, IL: University of Illinois at Urbana-Champaign, Department of Agricultural Engineering.

Ott, L. S., B. L. Smith, and T. J. Bruno. 2008. Advanced distillation curve measurement: application to a bio-derived crude oil prepared from swine manure. *Fuel* 87: 3379-3387.

Reece, M. 2004. Special issue paper 9: Densities of oil products. Energy Statistics Working Group Meeting, International Energy Agency. Paris, France.

Scurlock, J. 2009. Bioenergy feedstock characteristics. Oak Ridge, TN. Oak Ridge National Laboratory, Bioenergy Feedstock Development Program.

Speight, J. G. 2001. *Handbook of Petroleum Analysis*. New York: Wiley Interscience.

Sweeten, J. M. 1992. Livestock and poultry waste management: A national overview. In J. Blake et al., eds., National Livestock, Poultry and Aquaculture Waste Management, pp. 4–15. ASAE Publication 03–92. American Society of Agricultural Engineers, St. Joseph, MI.

USDA-ERS. 2009. Livestock, dairy, & poultry outlook. Washington, D.C.: USDA Economic Research Service.

USDA-NASS. 2008b. Quarterly hog and pigs report. Washington, D.C.: USDA National Agricultural Statistics Service.

USDA-NASS. 2009. Crop production 2008 summary, p. 13. Washington, D.C.: USDA National Agricultural Statistics Service.

USDOE and USDA, 2005. Biomass as feedstock for a bioenergy and bioproducts industry: the technical feasibility of a billion-ton annual supply, p. 7. Washington, D.C.: USDOE and USDA.

U.S. Environmental Protection Agency Office of Solid Waste. 1999. Biosolids Generation, Use, and Disposal in the United States. EPA530-R-99-009, p. 2.

U.S. Environmental Protection Agency Office of Solid Waste. 2008. Municipal Solid Waste in the United States: 2007 Facts and Figures. EPA530-R-08-010, p. 1.

Vogel, K. P., and R. A. Masters. 1998. Developing switchgrass into a biomass fuel crop for the midwestern USA. Lincoln, NE. University of Nebraska. Agricultural Research Service, United States Department of Agriculture.

Wang, Z. 2007. Unpublished data. Department of Agricultural and Biological Engineering. Urbana, IL.: University of Illinois at Urbana-Champaign.

Wang, Z. 2007. Personal communication. Department of Agricultural and Biological Engineering. Urbana, IL.: University of Illinois at Urbana-Champaign.

Yu, G. 2008. Unpublished data. Department of Agricultural and Biological Engineering. Urbana, IL.: University of Illinois at Urbana-Champaign.

(452e) Effect of Torrefaction on the Fast Pyrolysis Using Forest Resources

Jiajia Meng, Sunkyu Park, David Tilotta and Junyeong Park, Forest Biomaterials, North Carolina State University, Raleigh, NC

Bio-oil has a great potential for multiple uses including heat/electricity generation and chemical production. This thermochemical technology appears to be a practical, effective, and environmentally sustainable means of producing renewable fuels while simultaneously reducing emissions of greenhouse gases. In addition, pyrolysis has a key advantage over biochemical methods in that it is less sensitive to variation in feedstocks, which makes supply-chain development less challenging. However, the quality of pyrolysis bio-oil is not sufficient to directly produce biofuels and chemicals due to high oxygen content, high acidity, and poor long-term stability. In this study, we investigated the potential of integrating fast pyrolysis with torrefaction to produce high quality bio-oil from forest resources. The benefits of this integration along with the preliminary experimental results will be discussed.

See more of this Session: Biorefinery - Thermochemical Conversion of Biomass I
See more of this Group/Topical: Forest and Plant Bioproducts Division

LIQUID PRODUCTS OF LIQUID PHASE-PYROLYSIS

Nikolaus Schwaiger[1], Verena Mertlitz[1]; Kerstin Zahel[2], Peter Pucher[2], Edgar Ahn[2], Matthäus Siebenhofer[1],
[1]Graz University of Technology; Institut für Thermische Verfahrenstechnik und Umwelttechnik; Inffeldgasse 25/C/II 8010 Graz/Austria;
[2]BDI-BioEnergy International AG; Grambach/Austria

Ligno-cellulosic feed is expected to contribute significantly to production of liquified combustibles in future, because of the quantity and the variety of feed material. Therefore the pyrolytic degradation properties of spruce wood and its building blocks glucose and cellulose were investigated. The interaction between heat carrier, biomass and biomass products in the condensed liquid and vapour phase was investigated.

The process was carried out in a semi-batch reaction vessel under isothermal conditions at a temperature of T = 623 K and at ambient pressure. For optimum heat transfer the pyrolysis process was carried out in liquid heat carrier phase. Gaseous products as well as condensibles were removed on top of the reaction vessel. Condensation temperature was T = 20°C. After recording flux and composition the gaseous phase was incinerated in a flare.

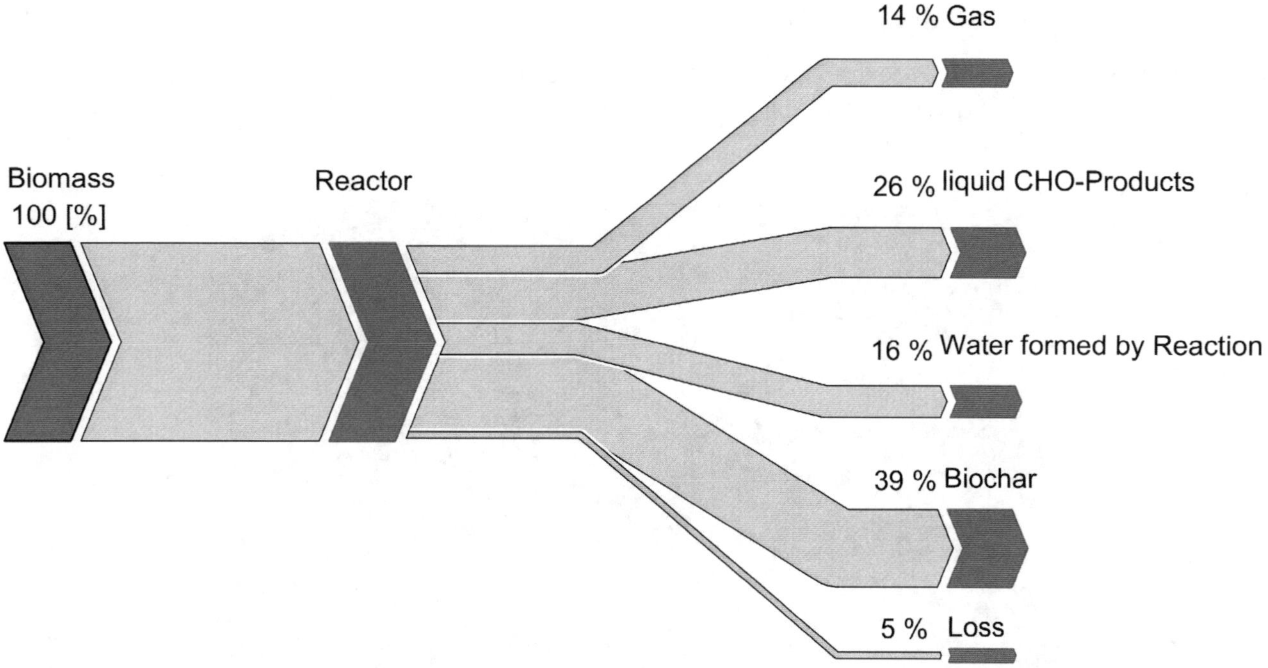

Figure 1: Biomass mass balance

The condensed products were analyzed by GC-MS and NMR spectroscopy as well as elemental analysis (CHNO).

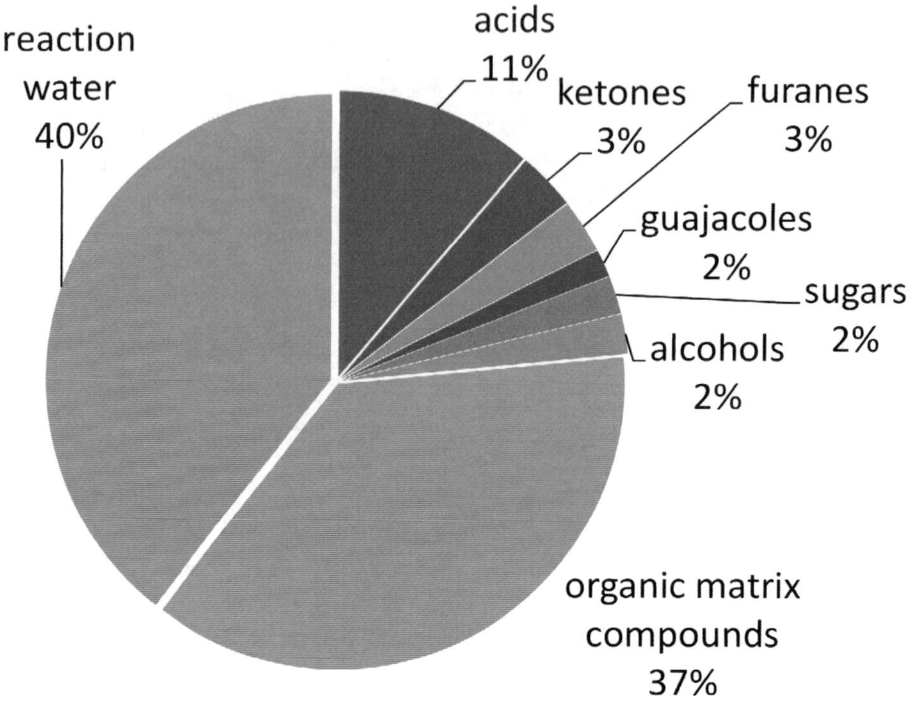

Figure 2: Composition of liquid products measured by GC-MS and Karl Fischer Titration

The solid product residues were pretreated by solid-liquid extraction and afterwards analyzed by elementary analysis, FTIR-ATR, Raman spectroscopy, scanning electron microscope and energy dispersive X-ray spectroscopy.

Figure 3: Scanning electron microscope photo of a Biochar cross section

Several operation parameters were investigated. Representatively the effect of particle size is briefly mentioned.

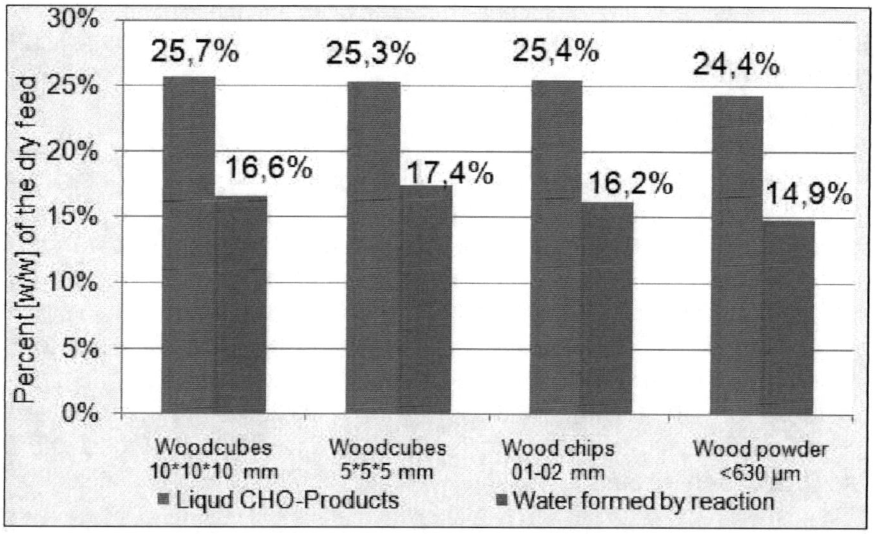

Diagram 1: Effect of feed particle size on the amount of CHO-products and the amount of water formed by thermal dehydration of softwood

Following the results shown in Diagram 1 variation of particle size does not have an effect on the amount of liquid CHO-products, when pyrolysing softwood. Water formation decreases for feed particles of smaller particles.

As shown in Diagram 2 the total amount of liquid CHO-products seems to be constant and independent of the feed particle size. The chemical composition of liquid phase pyrolysis products differs.

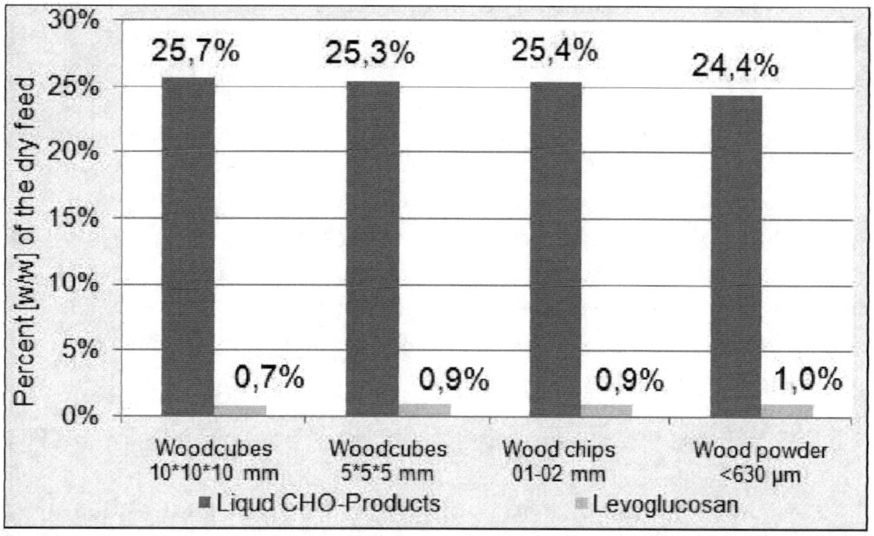

Diagram 2: Effect of feed particle size on non aqueous condesibles and products

Based on the detailed mass balance the effect of operation parameters on pyrolytic degradation of softwood, product formation and energy balance was determined.

Biobutanol Sustainable Development with Life Cycle Assessment

Steve Barr, Robin Jenkins, Bruce Vrana, Tom Xu, Michael Grady, Joseph Zaher
(DuPont, Wilmington DE)

Tyler Ames, Adam Schubert
(Butamax™ Advanced Biofuels LLC, Wilmington DE)

DuPont and BP have joined forces to form Butamax™ Advanced Biofuels LLC to manufacture the advanced biofuel biobutanol. Butamax™ combines the oil, gas and renewable energy capabilities of BP with the biotechnology discovery and process engineering capabilities of DuPont. In 2009, the Butamax™ joint venture was created to commercialize the biobutanol technology. Butamax™ is currently in the process of starting up a biobutanol demonstration facility in an existing BP operation at Saltend, Hull, north of London. DuPont and BP scientists and engineers are committed to making advanced biofuels and energy-efficient biofuel processes a reality.

Renewable feedstocks such as corn grain[1], wheat and sugar cane can be converted to fermentable sugars for the production of biobutanol. Starting with existing feedstocks allows for retrofitting existing ethanol plants to convert to biobutanol production. Additional feedstocks being pursued include macroalgae and cellulosic materials.

Biobutanol has a lower heating value of 27 MJ/liter, which results in fuel economy per volume more similar to gasoline than ethanol. An existing renewable supply of n-butanol comes from the acetone-butanol-ethanol (ABE) process. The ABE process is an expensive way to produce n-butanol due to the production of acetone and ethanol, which results in relatively low yields of n-butanol per bushel of corn. Compared to the other butanol isomers, n-butanol has a lower octane number and does not have ideal materials compatibility or toxicity. 2-butanol has similar properties to iso-butanol, but is believed to have less favorable production economics. Iso-butanol was selected because of its high octane number, distillation qualities, low vapor pressure, high compliance value in fuels, materials compatibility, low toxicity, and its ability to reach the targeted production economics.

Due to the lower oxygen content, biobutanol can be blended to a 16% volume versus the current 10% limit for ethanol in gasoline under an existing EPA waiver. This allows greater renewable fuel content with biobutanol while retaining compatibility with existing vehicles and infrastructure. Currently, ethanol has nearly reached the blend wall, where the maximum amount of ethanol possible is being blended within the limits of the waiver. The Renewable Fuel Standard (RFS2) requires increasing the volume of renewable fuel in gasoline blends through 2022[2]. Including biobutanol in gasoline fuel blends allows higher renewable fuel content within the existing and possible future waivers with the existing fuel infrastructure and vehicles. Ethanol requires increasing the number of flexible fuel vehicles on the road and the availability of E85 to meet the renewable fuel volume requirements of the RFS2.

Biobutanol creates value across the entire biofuels value chain. The first change or opportunity for value creation is direct access from the producer to the refinery where blending is now possible with biobutanol. The farmer will see growth and sustainable local demand of corn from the biobutanol producer. The biofuel producer will have access to greater demand

and higher returns versus ethanol today. The biobutanol volume produced will be 80% of the ethanol volume with a retrofit, using the same bushels of corn. Due to the higher energy content of biobutanol, a gallon is assigned 1.3 renewable identification numbers (RINS)[2]. This equates to the same fuel energy and RINS from biobutanol production as with ethanol production at the stated capacity of the plant. Use of biobutanol allows greater displacement of foreign crude with higher blends of butanol versus ethanol. This supports energy policy for greater energy diversity. The refinery can blend on-site and more precisely. The pennies saved per gallon equate to enormous value to a refinery. The lower Reid Vapor Pressure (RVP) of biobutanol creates value where pending ozone standards could disadvantage ethanol and management of state implementation plans (SIP) with butanol becomes more manageable. The lower RVP means the refinery is able to avoid special gasoline base blends and increase marginal blending opportunities, which leads to more value and more gasoline per barrel of oil. The terminal becomes more efficient with transport of fuel and requires no upgrading to distribute biobutanol. Biobutanol is pipeline capable and thus avoids the enormous cost and issues with creating a segregated ethanol pipeline. Biobutanol is also compatible with ethanol fuel blends, so it does not need to displace ethanol to enter the market. The retailer will have less risk and potential for higher margins. By incorporating biobutanol into the fuel mix, the retailer will avoid costs and liabilities associated with a separate canopy. This separate canopy would be necessary if the retailer were to provide the greater number of fuel blend options required when offering E10+ products (i.e. fuel blends higher than 10% ethanol). Alternatively, the retailer of E10+ product could reduce product choice like mid-range or premium blends which in many cases account for the only positive fuel margins for the retailer. The vehicle manufacturer has less risk from emissions versus use of E10+ blends, which can lead to issues with on-board diagnostics. Biobutanol is also more compatible for use across all engine types and sizes. The consumer will realize better fuel economy, lower emissions and supply convenience.

The process for producing biobutanol can be very similar to ethanol. Retrofitting an ethanol plant to produce biobutanol is believed to have minimal impact to production and capital investment.

Life Cycle Assessment (LCA) is the tool used to quantify the GHG emissions and fossil energy consumption throughout the supply chain for producing and using products. For biofuels, this includes fertilizer production, field soil emissions, tractor fuel, tractor emissions, transport of feedstock to the biorefinery, process energy use, transport of biofuel through distribution, combustion emissions from the vehicle, etc. LCA can be used to evaluate future technology options by utilizing process material and energy balances for input information. LCA is integrated into research and development, as shown in Figure 1, to enable innovative and competitive future technology options with quantifiable environmental benefits in the marketplace.

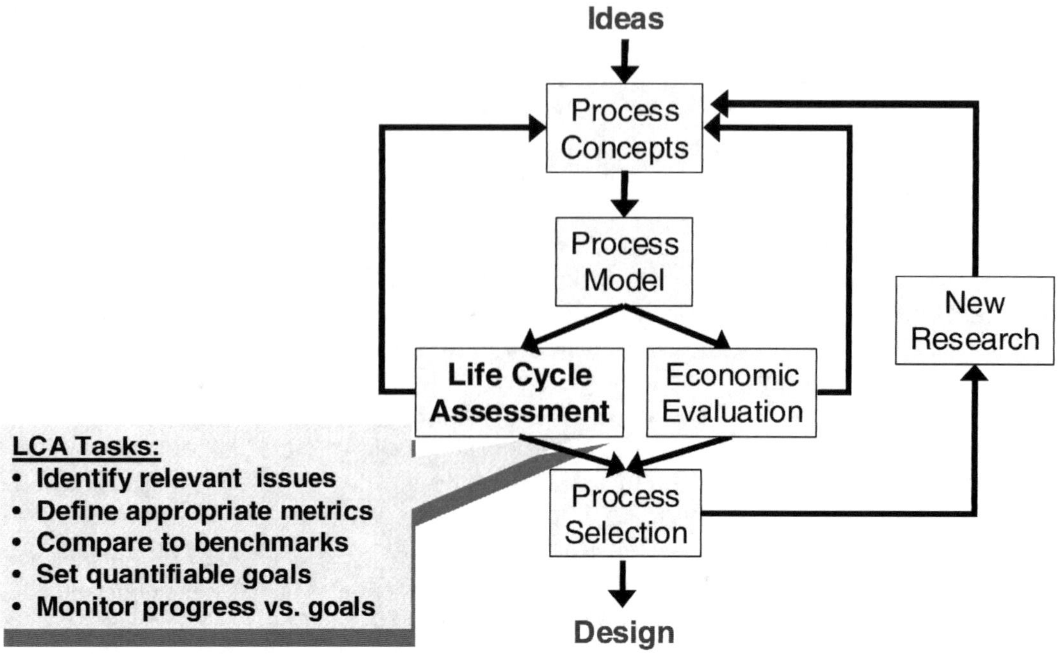

Figure 1: LCA in Process Development

In order to benchmark biobutanol with existing technologies, GHG results are compared to ethanol, conventional gasoline[3] and the traditional biological production of n-butanol using the ABE technology. In order to make a fair comparison, inputs for the different fuel options need to be aligned in the LCA models for comparison. Alignment means that life cycle inventory data and certain minor process assumptions are applied consistently among all of the cases being compared. For the ethanol, ABE, and biobutanol models, the same emission factors, or in other words life cycle inventory data, for corn, transport, natural gas and electricity input were used. In the analysis presented, the ethanol and biobutanol process models were also aligned to utilize similar process configurations and the same degree of heat integration to determine if there was a difference in the GHG results for ethanol and biobutanol.

As seen in Figure 2, preliminary results show that future technology options for producing biobutanol appear comparable to ethanol and favorable compared to conventional gasoline and butanol from ABE in terms of well-to-wheel (WTW) GHG emissions. The term WTW is used to describe impacts from the entire life cycle of a fuel, from extracting materials from the ground to combustion of the fuel in the vehicle. Since there is still no consensus in the scientific community regarding methodologies for accounting for GHG emissions from indirect effects, these results do not include any GHG emissions from indirect effects, such as land use change. The results indicate the relative GHG emissions to gasoline, based on GHG emissions per energy content of the fuel. The results for two ABE process options are shown in Figure 2. The "Current" result shows the GHG emissions for a typical ABE process currently operating. The "ANL" result shows the GHG emissions for a future ABE process based on an Argonne National Laboratory (ANL)[4] report detailing potential improvements to the ABE process, and may represent an optimistic view of the ABE process. Although not shown in this paper, the trend shown in Figure 2 for GHG emissions is the same as the trend would be for fossil energy consumption.

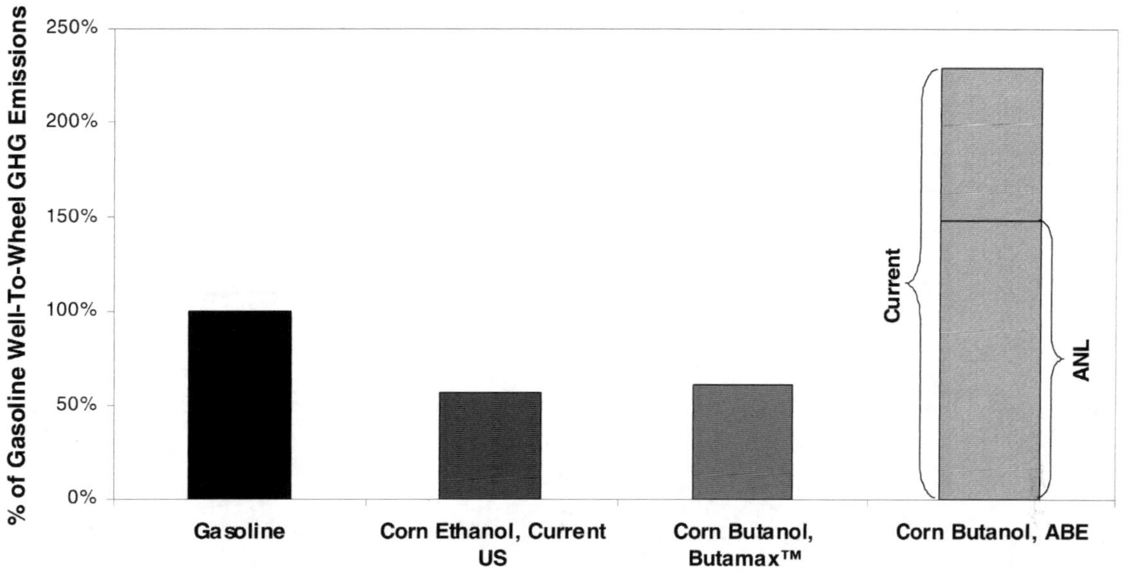

Figure 2: Corn Grain Biofuels: GHG Emissions

Figure 3 shows preliminary results regarding the relative contribution of different parts of the supply chain to the overall WTW GHG emissions. It can be seen that the corn grain production and process energy use account for nearly all of the GHG emissions for all three results shown. The credit is from sale of distillers' grains with solubles into the feed market, which displaces mainly corn grain and soybean meal in animal diets. The credit for the ABE process also includes the displacement of natural gas and gasoline. It is assumed that the acetone produced in the ABE process would be used as a boiler fuel displacing natural gas. It is not reasonable to assume that the acetone from the ABE process would displace fossil-based acetone since the world solvent acetone market is not large enough to support the co-product from multiple, large fuel ABE processes. The gasoline displacement in the ABE model is due to the small amount of co-product ethanol produced, according to the ANL process. Even with the additional credits, the much greater process energy consumption and lower yield from corn lead to greater life cycle GHG emissions for n-butanol from the ABE process compared to the biobutanol from the Butamax™ process.

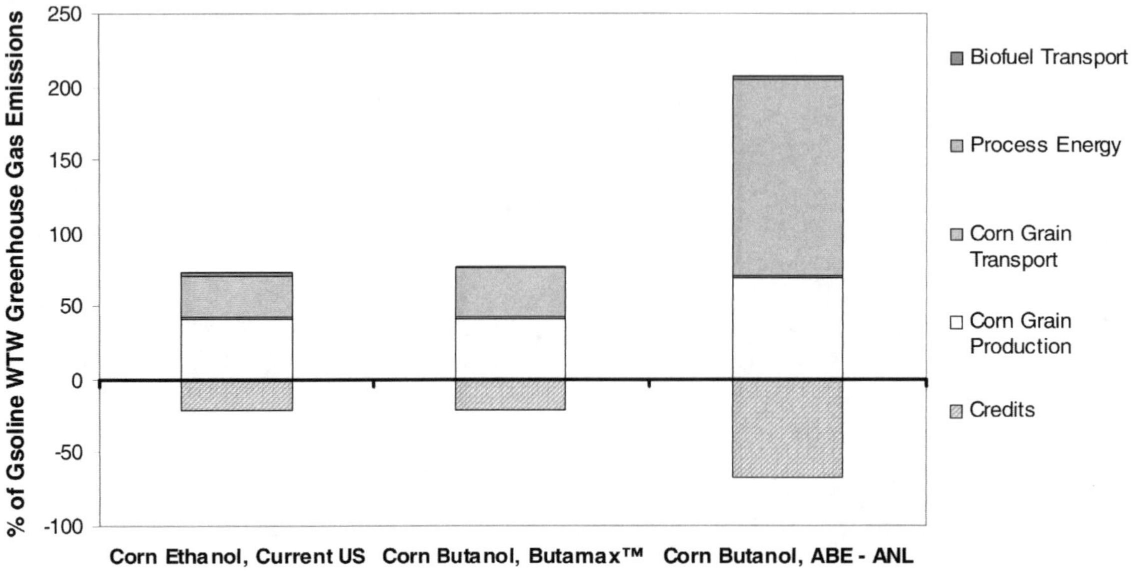

Figure 3: Corn Grain Biofuels: Relative Contribution to GHG Emissions

In the US market, ethanol is currently limited to a maximum volume concentration of 10% in gasoline. Based on the same limiting factor of oxygen content, biobutanol can be blended at a volume concentration of 16%. Because this allows a greater volume of similar LCA GHG emissions biofuel to be blended into gasoline, the blend has lower total GHG emissions. The effect on blend GHG emissions relative to conventional gasoline is shown in Figure 4.

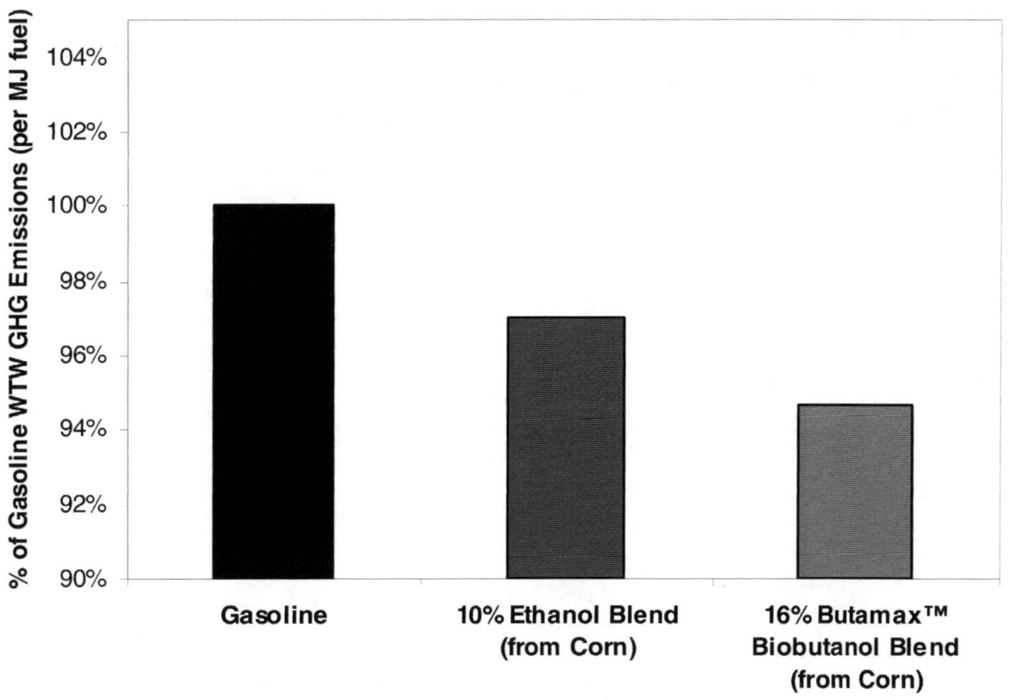

Figure 4: Corn Grain Biofuel-Gasoline Blend GHG Emissions

Sugar cane is an alternative feedstock for making biobutanol that has been modeled. The future technology options for producing biobutanol result in similar LCA GHG emissions to current high efficiency sugar cane ethanol production. If a sugar cane ethanol plant using azeotropic distillation for purification of ethanol were to be retrofitted to produce biobutanol, the biobutanol could be produced with lower GHG emissions per energy content of fuel due to the easier separation of biobutanol from water.

In conclusion, biobutanol is a fuel with advanced properties compared to ethanol. It allows greater volumes of biofuels to be blended into gasoline per gallon of fuel, without vehicle and infrastructure modifications. Biobutanol is similar to ethanol, and significantly lower than n-butanol from the ABE process, in regards to WTW GHG emissions. Future research and development efforts of Butamax™ will focus on second and third generation feedstocks, including cellulosic material and macroalgae.

References

1 Kim, S., B. Dale, and R. Jenkins. 2009. Life Cycle Assessment of Corn Grain and Corn Stover in the United States. The International Journal of Life Cycle Assessment 14: 160-174.

2 Environmental Protection Agency (2010), "40 CFR Part 80, Regulation of Fuels and Fuel Additives: Changes to Renewable Fuel Standard Program; Final Rule", Vol. 75, No. 58. (EPA-HQ-OAR-2005-0161-2642)

3 Farrell A.E., "Ethanol Can Contribute to Energy and Environmental Goals", Version 1.1.1, Updated July 13, 2006. EBAMM_1_1.xls.Transportation Models

4 Liu, J., M. Wu, M. Wang. (2009). "Simulation of the Process for Producing Butanol from Corn Fermentation". Ind. Eng. Chem. Res. 48: 5551-5557.

(508b) De-Oxygenation of Bio-Oil Via Electrocatalysis

Chungja Yang, Chemical Engineering, Michigan Technological University, Houghton, MI and Adrienne Minerick, Department of Chemical Engineering, Michigan Technological University, Houghton, MI

Bio-oil, a pyrolysis oil, has been recognized as a possible renewable energy source to supplement declining fossil fuel sources. Bio-oil offers sustainability, reduction of greenhouse gas emissions, and regional development. However, higher oxygen content (> 35%) in raw bio-oil is problematic because of reduced efficiency of energy recovery from the oil. Therefore, this research studies an upgrading approach involving the passage of electric fields through bio-oil to catalyze reduction reactions to reduce oxygenated hydrocarbons into higher energy hydrocarbons. This electro-catalytic process has the potential to overcome limitations of the solid catalytic method such as catalyst coking as well as high temperature and pressure operating conditions. Experiments were conducted in a custom designed electrolytic cell (E-cell) within which the complementary oxidation and reduction reactions occur at the electrode surface in the presence of the electric field. Chemical conversion via three types of electrocatalytic metals (Steel, Aluminum, and Nickel) and two different electric field strengths (5/3 and 200/2 [VDC/cm]) were studied. Two membrane separating the electrode chambers were compared: (1) an ion-selective semi-permeable nafion membrane and (2) dialysis membrane 10,000 MWCO. Experiment times of 30 min, 2 hr, and 5 hr were investigated. In addition, modifications of bio-oil solution were attempted to increase operating current within E-cell by addition of a metal catalyst, pH adjustment via NaOH, and electrolyte addition via NaCl. Temperature, pH, and current were monitored in real time via a LabVIEW/ National Instruments Data Acquisition system. Elemental, FTIR and GC/MS analysis were conducted to determine changes in Bio-oil composition. Electrocatalytic reactions with steel at 200 VDC for 5 hours showed the most significant oxygen reduction with 10% O (anode) and 27% O (cathode) of the original oxygen content. Furthermore, NaOH and NaCl bio-oil solutions achieved increased current (and thus reactions) between cathode and anode. FTIR and GC/MS results demonstrate that aliphatic ketones were converted to oxygen-reduced forms secondary aliphatic alcohols and that highly oxygenated hydrocarbons were converted to less oxygenated hydrocarbons. This process provides an alternative approach to in situ bio-oil upgrading.

See more of this Session: Alternative Fuels and Enabling Technologies II
See more of this Group/Topical: Fuels and Petrochemicals Division

Simulation of the Absorption of Gasoline Volatile Organic Compounds in Biodiesel

Rogelio Sotelo-Boyás, Nancy Yanine Gómez-Alvarado*
Escuela Superior de Ingeniería Química e Industrias Extractivas.
Instituto Politécnico Nacional. U.P.A.L.M., Edificio 7, México D.F. 07738. rsotelob@gmail.com

1. Introduction

The presence of volatile organic compounds (VOC's) in the atmosphere is a result of human activities, which is generated from car exhausts, use of solvents, industrial processes, agriculture, oil refining, as well as in the storage and distribution of petroleum distillates.

One of the main environmental problems is caused by VOC's when they are mixed with other air pollutants (e.g. NO_X) and react with sunlight, in such a way that they form ground-level ozone, which contributes to photochemical smog. In addition, VOC's may affect the degradation of the ozone layer and because of their toxicity they can have harmful effects on human health and natural ecosystems[1].

Among the VOC's with the highest concentration in the atmosphere of big cities, are the alkanes, mainly butane and pentane, which have a high ozone-forming potential (OFP)[2]. The presence of butane, pentane and hexane in the atmosphere may come from different sources, for example by evaporation of gasoline during storage and distribution, and in the ocean can come from oil spills, among other sources.

The recovery of VOC's can be accomplished by using solvents and equipment such as absorption columns, which would recover these VOC's to prevent further atmospheric pollution. Several researchers[2,3] have previously studied ways to reduce emissions of these compounds to the environment. In the recovery of volatile organic compounds, it has been usually proposed the use of solvents with specific properties such as low boiling point, high capacity for dissolved volatile organic compounds, insoluble in water, viscosity at room temperature no greater than 100 cSt. and a freezing point lower than -29°C.

Yamase et al.[3] have proposed the use of mineral oil in combination with phthalate esters, methyl ethyl phthalate esters, carboxylic acids, butyl benzyl-phthalate, butyl lauryl-phthalate, silicate, tetra-octyl-diethyl-phthalate. The solvent that these authors propose may contain aromatic hydrocarbons, which can be harmful to health and the environment[4]. Moreover, because of its organic nature, this type of solvent is not renewable and is difficult to degrade.

As an alternative, this paper proposes the use of biodiesel as a solvent in the recovery of normal alkane (paraffinic) VOC's from C_4 to C_6. Biodiesel has similar properties to those above mentioned, with the additional advantages of being non-toxic, biodegradable and renewable. Furthermore, due to its high calorific value, biodiesel can be recovered and reused as fuel for heating. A previous work[5] has studied the feasibility of biodiesel for the absorption of aromatic VOC's.

This work explores the potential for absorption of paraffinic VOC's in methyl oleate applied to the absorption of paraffinic hydrocarbons such as n-C_4, n-C_5 and n-C_6 through the simulation of the absorption process. The use of biodiesel is simulated on basis of a model compound, i.e., methyl oleate, which is usually the most abundant ester in biodiesel, and it is also known for having a high boiling point, low toxicity, low volatility and high solubility in organic compounds, which altogether make it an ideal solvent in absorption processes for the recovery of volatile organic compounds.

2. Methodology

The gaseous mixture containing the VOC's considered in this work is generated during the filling of 20000 Liters auto tanks in a gasoline distribution bulk terminal. The composition of this mixture is shown in Table 1. It contains about 70 mol % of air as nitrogen and oxygen.

Table 1. Composition of the gaseous stream fed in the absorption process

Component	*Mol composition, %*
n-Butane	2
n-Pentane	19
n-Hexane	10
Oxygen	15
Nitrogen	54

The properties of methyl oleate shown in Table 2 were determined by the group contribution method UNIFAC.

Table 2. Properties of methyl oleate

Property	Value
Molecular weight	296.49
Density, Kg/m^3	876.88
Boiling point, K	617.00
Critical temperature, K	764.00
Critical pressure, KPa	1280.00
Critical volume, $m^3/Kg\text{-}mol$	1.06
Compressibility Factor	0.21

The liquid-vapor equilibrium in systems of non-ideal behavior is predicted by the estimation of the activity coefficients in liquid mixtures with the UNIFAC model. This method evaluates the activity coefficient in terms of free energy in excess, the extra energy of the solution that makes it not ideal[6].

The affinity towards methyl oleate VOC's can be expressed by the activity coefficients at infinite dilution:

$$\gamma_i^{\infty} = \frac{H}{P_i^{sat}} \tag{1}$$

Both H and γ^{∞} allow the evaluation of the mutual affinity of the compounds. Henry's constant was determined from Henry's law in terms of partial pressure and mole fraction:

$$H = \lim_{x_i \to 0} \frac{P_i}{x_i} \tag{2}$$

Thus, the activity coefficient can also be determined by:

$$\gamma_i^{\infty} = \frac{P_i}{P_i^{sat} x_i} \tag{3}$$

The infinite dilution ratio of volatile organic compounds was estimated taking into account the vapor pressure of the different compounds at the same temperature. The estimation of activity coefficients of VOC's in methyl oleate, was determined with the simulation of an absorption column (Viz. Figure 1) fed on the top with methyl oleate as a solvent and on the bottom with the gaseous mixture containing the VOC's.

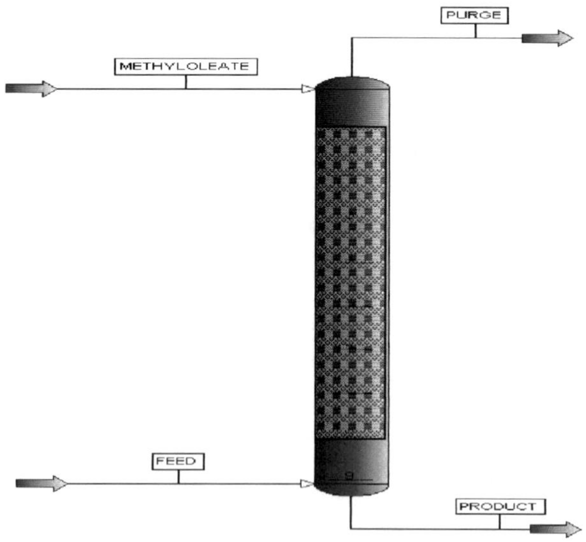

Figure 1. Absorption column

The biodiesel (as represented by methyl oleate) to hydrocarbons mass ratio was kept at 150 wt/wt. Thus, it was possible to assume infinite dilution of hydrocarbons in the methyl oleate.

3. Analysis of results

3.1 Activity Coefficients

Table 3 shows the estimated activity coefficients with PROII using UNIFAC method at various temperatures. The relative low values of activity coefficients of hydrocarbons indicate a good affinity with methyl oleate, and therefore it can be considered to be feasible as an absorbent for this kind of hydrocarbons.

In the case of butane, it can be observed an increasing trend in the values of the activity coefficients with increasing temperature, which indicates that n-butane can be desorbed more easily at high temperatures. The opposite is true with n-pentane and n-hexane, whose activity coefficients tend to decrease with increasing temperature. These data are therefore useful for the selection of appropriate temperatures for both absorption and desorption of VOC's in, or from methyl oleate, respectively.

Table 3. Activity Coefficients γ^{∞} of n-C_4, n-C_5 and n-C_6 in methyl oleate

Component	T_1 25°C	T_2 30°C	T_3 60°C	T_4 90°C
n-Butane	0.563	0.565	0.567	0.580
n-Pentane	0.665	0.662	0.66	0.650
n-Hexane	0.749	0.740	0.744	0.735

Due to the non-significant variation among the values of the activity coefficients from 25 to 90° C, it is possible to carry out the absorption process at low temperatures, i.e. close to ambient, by doing so, there would be a significant saving in energy and economics.

3.2. Simulation of an industrial absorption process

Once that methyl oleate has proven to be feasible as an absorbent of VOC´s, the simulation of an industrial absorption process was carried out. This consists of a 14 meter-column with a diameter of 1 m., packed with 25 mm.-metal Pall rings.

The composition of the feedstock is the same as shown in Table 1. The simulation was carried out with the objective of optimizing (minimizing) the flow of methyl oleate needed to obtain an optimal recovery of VOC´s, i.e. a minimum amount of n-butane, n-pentane and n-hexane in the purge.

The operative condition of the column are1 bar and 30°C. The flow of the gaseous feedstock is 1000 kg/h (ca. 23273 kmol/h).

After performing the simulaton, an optimum flow of methyl oleate was 17000 kg/h (57337 kmol/h). It was noticed that the larger the flow of methyl oleate, the larger was the recovery of hydrocarbons in the product stream. However, increasing the flow of methyl oleate it also means a higher cost, for example higher cost of pumping. A flow of 17000 kg/h was selected based on a large enough recovery and a low enough emission of hydrocarbons through the purge stream. In addition, a larger flow of methyl oleate does not significantly increase the recovery of hydrocarbons, but does increase the cost of pumping.

The compositions of the simulated streams are included in Table 4. It can be seen that methyl oleate absorbs over 99% of n-butane, n-pentane and n-hexane and therefore it can be argued that methyl oleate has an excellent absorption capacity for these VOC's. From these results, it can also be seen that the methyl oleate absorbs part of the air that is initially mixed with oil. This is probably due to the polar characteristic of the structure of ester.

Table 4. Simulation results of the industrial absorption column

Component/Stream	Feedstock	Biodiesel	Product	Purge
	Kmol/h	Kmol/h	Kmol/h	Kmol/h
n-Butane	0.3491	---	0.3491	$4.6x10^{-5}$
n-Pentane	4.4225	---	4.4225	$3.6x10^{-8}$
n-Hexane	2.2111	---	2.2111	$1.1x10^{-12}$
Oxygen	3.4216	---	0.0864	3.3352
Nitrogen	12.8717	---	0.2310	12.6407
Methyl oleate	---	57.3370	57.3369	$1.28x10^{-7}$
Total	**23.2730**	**57.3370**	**64.6370**	**15.9760**

Conclusions

The relative low activity coefficients obtained from the simulation with the UNIFAC method show the affinity of methyl oleate with n-paraffinic hydrocarbons such as butane, pentane and hexane in such a way that their absorption with biodiesel is possible at atmospheric pressure and temperature close to 30°C, and consequently is an energy efficient recovery process. It is therefore possible to use both economically and environmentally biodiesel as a solvent in the absorption of condensable light hydrocarbons.

References

1. Department of the Environment, P. T., Agriculture and Fisheries Volatile Organic Compounds. Accessed on April 2010: http://www.ingurumena.ejgv.euskadi.net/r49-434/es/contenidos/informacion/cov/es_7908/cov.html.

2. Belalcazar, L. C.; Clappier, A.; J.A., P.; Franco, J. F.; Behrentz, E. V., VOC´s Concentrations in the ambient air of Bogota City. Accessed on April 2010: http://www.kfpe.ch/projects/jeuneschercheurs/belalcazar.php.

3. Osamu Yamase, K.; Tetsuya lida, Y.; Ichizo Hara, H. Method or Purifying a Gas Containing Volatile Organic Materials. Patente US 4102983, 1978.

4. Ecologistas en Acción: Aceite De Motor. Accessed on April 2010: http://www.ecoloxistesasturies.org/Temas/Contaminacion/aceites_usados.htm.

5. Bay, K.; Wanko, H.; Ulrich, J. Absorption of Volatile Organic Compounds in Biodiesel: Determination of Infinite Dilution Activity Coefficients by Headspace Gas Chromatography. **2006**, 22-28.

6. Salazar, F. Estimación De Equilibrio Liquido Vapor Utilizando El Modelo UNIFAC. *Facultad de Ingeniería - Universidad Rafael Landívar: Electronic Bulletin No. 3.*, 2003.

(508d) Production of Green Diesel Via Decarboxylation of Free Fatty Acids

Elvan Sari, Craig DiMaggio, Manhoe Kim, Shuli Yan, Steven O. Salley and K. Y. Simon Ng, Chemical Engineering and Materials Science, Wayne State University, Detroit, MI

Green diesel is a next generation transportation fuel which emerged due to the need of renewable fuel replacement that is compatible with existing automotive technology,has high cetane number, low NOx emissions, and high heating value. Furthermore, the outstanding energy density of hydrocarbons as a fuel makes them a powerful transportation fuel option. It can be produced via a catalytic reaction involving hydroprocessing or/ and decarboxylation/ decarbonylation of the triglyceride . Advances in the production of green diesel have been scarce due to the lack of process knowledge. The literature indicates that decarboxylation of stearic acid with carbon supported catalysts is generally leading to higher raction rates especially with 5% Pd/C which showed the best performance in a semibatch reactor under helium atmosphere at 300 0C and 6 bar. However, it was concluded that catalyst deactivation has been inferred to occur via coking and/or CO poisoning. In this work decarboxylation of stearic and oleic in solvent was investigated under different atmosphere such as He, H2 and He-H2 mixture using carbon based catalyst. Batch experiments were conducted using a 500-mL stirred Parr high pressure reactor and 5% palladium on activated carbon support catalyst (Aldrich). Reaction conditions were 3000C and 220 psi. The liquid product was analyzed by gas chromatography using a Perkin Elmer Arnel Clarus 500 instrument equipped with a MXT-1 capillary column (60 m x 0.28 mm x 1.0 mm). Results show that saturated free fatty acids (FFAs) such as stearic acid can be decarboxylated to the linear alkanes having one carbon number less than the corresponding FFA under inert gas atmosphere with 99.9 % conversion. However, unsaturated FFAs such as oleic need to be hydrogenated before or during decarboxylation in the presence of hydrogen. The significance of this study is to show a better understanding of the processing conditions required, including catalyst support structure, optimization of the reaction conditions and especially the roles that hydrotreating of the catalyst plays, in order to further develop the current green diesel process.

See more of this Session: Alternative Fuels and Enabling Technologies II
See more of this Group/Topical: Fuels and Petrochemicals Division

(508e) Characterization and Catalytic Abatement of Emissions From Alternative Diesel Fuels Using a Benchtop Engine System

Gregory S. Bugosh, Rachel L. Muncrief and Michael P. Harold, Department of Chemical and Biomolecular Engineering, University of Houston, Houston, TX

Diesel fueled, compression ignition engines generally have higher efficiency, increased durability, and more torque than their gasoline powered spark ignition counterparts. This has led to their use in most heavy duty vehicles for transportation and construction, and an increasing presence in passenger vehicles. However, the emissions of diesel engines contribute to pollution problems, especially in urban areas. Of highest concern are the particulate matter (PM) and oxides of nitrogen (NOx) produced. Particulate soot has known carcinogens while ozone production from NOx and hydrocarbons is a respiratory irritant. In addition, the production of carbon dioxide (CO2) is also gaining importance because of its potential role in climate change. It is desirable to reduce the emissions without dramatically increasing overall cost. One approach to improvement is through modification of the fuel source. Diesel fuel must meet a large set of parameters; including viscosity, cetane number, and sulfur content. Engines have evolved to perform optimally on traditional petroleum-based fuels that meet those standards. Alternative forms of diesel fuel can meet or exceed those standards and differ from petrodiesel in other ways that result in improved performance. The development of new engines and emission systems, as well as the retrofit of these long-lived vehicles, must take into account the properties of the fuel.

In this study we evaluate the emissions of several types of diesel fuel and the effectiveness of catalysts to eliminate those species. We utilize a single cylinder, air-cooled, compression ignition, naturally aspirated, 5kW engine-generator with computer controlled load. The effect of several additives on emissions from the generator has previously been compared to chassis dynamometer testing of full size trucks to find a constant load condition which correlates to full scale transient heavy duty chassis dynamometer testing [An, H., Muncrief, R., Harold, M., Ismail, H., "Benchtop Engine System for Screening of Diesel Fuel and Additives for NOx Reduction," presented at SAE 2010 World Congress & Exhibition, April 2010, doi:10.4271/2010-01-1293.]. The benchtop system gives results faster and at less expense compared to full scale dynamometer testing. The experimental setup also includes the ability to switch between fuels without engine stopping, giving direct comparative results.

Several alternative diesel fuels, including Fischer-Tropsch fuel and biodiesel made from various feedstocks (soy, canola, and tallow), are compared to petrodiesel in terms of the emissions and fuel consumption. The petroleum-based fuel used as a baseline is Texas Low Emission Diesel (TxLED), which is the diesel fuel used in a 110 county region of Texas in order to reduce NOx emissions. Diesel fuel sold in this region must contain less than 10 percent by volume aromatic hydrocarbons and have a cetane number of 48 or greater. Alternative diesel fuel formulations can be considered TxLED if the Texas Commission on Environmental Quality (TCEQ) determines that they achieve similar or better reductions in NOx, PM, and hydrocarbons. The emission of NOx from the bench scale engine is determined by chemiluminescent NOx analyzer and also Fourier Transform Infrared Spectrometry (FTIR). Compared to the baseline fuel, NOx was found to decrease for Fisher-Tropsch diesel and biodiesel made from canola and tallow feedstocks, while an increase was found for biodiesel made from soy feedstocks. The chemical structure of the fuels, as well as parameters such as density, heat of combustion, and viscosity can help explain the differing performance.

We also will present the speciation of hydrocarbon emissions in the raw exhaust, including species like ethylene, propylene, formaldehyde, acetaldehyde, and 1,3- butadiene. Determining the spread of hydrocarbon species present in real exhaust is important because it can affect the catalytic activity of emissions control systems. Aftertreatment systems are often tested using simulated exhaust, which may contain only one hydrocarbon species. The benchtop engine setup gives the ability to test catalyst samples under real exhaust conditions and compare the efficiency to results obtained under simulated exhaust conditions.

See more of this Session: Alternative Fuels and Enabling Technologies II
See more of this Group/Topical: Fuels and Petrochemicals Division

(508f) Catalytic Conversion of Tall Oil to Green Diesel and Biodiesel

Kiran Pathapati, Stephen Dufreche, Rakesh Bajpai and Mark Zappi, Chemical Engineering, University of Louisiana at Lafayette, Lafayette, LA

The dependence of petroleum for nearly all of our transportation fuels fosters the importance and demand for renewable, biomass based transportation fuels. Biomass, the only current renewable source of liquid transportation fuel, has great potential to provide renewable energy. Biodiesel and Green Diesel can also be produced from the non-conventional, biologically derived feedstock such as tall oil. Crude tall oil(CTO) is a by-product of the Kraft Sulfate processing of the softwood for pulp and paper manufacture. Unsaturated C18 fatty acids, resin acids and unsaponifiables make up the principal constituents of the tall oil. CTO is upgraded by vacuum distillation to produce five major fractions. Catalytic cracking of these fractions along with CTO is performed to give green diesel, while acid catalyzed esterification leading to biodiesel. Results from conversion of tall oil to green diesel and biodiesel will be shown, as well as comparisons to traditional feedstocks.

See more of this Session: Alternative Fuels and Enabling Technologies II
See more of this Group/Topical: Fuels and Petrochemicals Division

(519a) Catalytic Conversion of Butanol to Olefins

Ivan C. Lee, Douglas A. Behrens and Jeff St Clair, US Army Research Laboratory, Adelphi, MD

Catalytic conversions of butanol, 2-butanol and isobutanol to olefins are investigated in a short contact time reactor. These butanol isomers were converted to 4-carbon olefins that are useful for synthetic fuel production.

See more of this Session: Catalytic Biomass Conversion to Chemicals
See more of this Group/Topical: Fuels and Petrochemicals Division

Extended Abstract # 186554

Microalgae, the third generation biofuel resource, have the potential to become a viable feedstock due to their high oil content and environmental-friendly nature (Gouveia and Oliveira, 2009). The major challenge, however, has been the high cost of recovering the oil from the microalgae prior to converting it into biodiesel. While there are many extraction methods such as solvent extraction, supercritical fluid extraction, ultrasonic extraction, and mechanical pressing; they require longer extraction times, large volumes of solvents, and are energy- and cost-intensive (Um and Kim, 2009; Vijayaraghavan and Hemanathan, 2009; Kanitkar, 2010). Unlike vegetable oils from various crops that can be extracted by crushing the oil seeds followed by a solvent extraction, releasing oil from algal cells is hindered by rigid cell wall structure. Thus, mechanical crushing seems to be an ineffective way to extract the oils from the algal biomass (Johnson and Wen, 2009). In addition, using solvents for extraction may involve expensive separation process. Microwave-assisted extraction or extractive transesterification could be an alternative to address the above concerns as the ability of the microwaves to penetrate through the cell wall structure may result in an efficient recovery of oils and lipids.

Microwave irradiation has been used in the past to extract the oils from the biomass, soils and vegetable feedstock (Barnabas et al. 1995; Kiss et al., 2000; Pan et al., 2002; Li et al., 2004). In microwave-assisted extraction, rapid generation of heat and pressure within the biological system forces out compounds from the biological matrix, producing good quality extracts with better target compound recovery (Hemwimon et al.,

2007). The rapid heating leads to localized high temperature and pressure gradients which assist in cellular wall degradation and enhanced mass transfer rates (Kanitkar, 2010).

A process which provides simultaneous oil extraction and transesterification is worthwhile to develop. Aresta et al. conducted thermochemical liquefaction using wet algal biomass and supercritical CO_2 extraction using dry algal biomass (Aresta et al., 2005). Both of the processes seem to be energy intensive by the reaction conditions they reported (thermochemical liquefaction conditions: 250-395°C for 1 hour and supercritical CO_2 extraction conditions: 50°C, 2.60 MPa for 7 hours). In a recent study, we have demonstrated the simultaneous extraction and transesterification (in-situ tranesterification) of the wet algal biomass in supercritical methanol conditions (Patil et al., 2010a). In a microwave-assisted extraction and transesterification process, as it has been demonstrated in many organic and biodiesel synthesis studies, it is anticipated that the reaction can be conducted at atmospheric pressures and temperatures merely close to the boiling point of methanol with much shorter reaction time (Leadbeater et al., 2006; Azcan and Danisman, 2007; Barnard et al., 2007; Refaat et al., 2008, Patil et al., 2010b).

Here, we propose a new, single-step extractive transesterification process that can extract the oils from the dry algal biomass and simultaneously convert them into algal biodiesel under the microwave irradiation. This method achieves a high degree of oil/lipid removal from the dry algal biomass and an efficient conversion of oils/lipids to biodiesel. It also reduces the reaction time and the solvent volume as compared with the separate lipid extraction and transesterification processes. The purpose of the present work is to investigate the effect of microwave irradiation on the simultaneous extraction and transesterification of algal oils to biodiesel from dry algal biomass. A response surface methodology technique was applied to optimize the effect of the catalyst concentration, dry algae to methanol ratio and reaction time. Results obtained from the RSM optimization study, GC-MS analysis, TLC analysis, and TEM analysis of raw material and final residue of the process are presented.

References:

Gouveia, L., Oliveira, A.C., 2009. Microalgae as a raw material for biofuels production. J. Ind Microbiol Biotechnol. 36, 269-274.

Um, B.H., Kim, Y.S., 2009. Review: A chance for Korea to advance algal-biodiesel technology. Journal of Industrial and Engineering Chemistry 15, 1-7.

Vijayaraghavan, K., Hemanathan, K., 2009. Biodiesel Production from Freshwater Algae. Energy & Fuels 23, 5448–5453.

Kanitkar, A.V., 2010. Parameterization of microwave assisted oil extraction and its transesterification to biodiesel. Master's thesis. Lousiana State University, Baton Rougue, LA.

Johnson, M.B., Wen, Z., 2009. Production of biodiesel from the microalga *Schizochytrium limacinum* by direct transesterification of algal biomass. Energy & Fuels 23, 5179-5183.

Barnabas, I. J., Dean, J. R., Fowlis, I.A., Owen, S.P., 1995. Extraction of polycyclic hydrocarbons from highly contaminated soils using microwave energy. Analyst 120, 1897-1904.

Kiss, G. A. C., Forgacs, E., Cserhati, T., Mota, T., Morais, H., Ramos, A., 2000. Optimization of the microwave-assisted extraction of pigments from paprika (*Capsicum annuum L.*) powders. Journal of Chromatography A. 889, 41-49.

Pan, X., Niu, G., Liu, H., 2002. Comparison of microwave-assisted extraction and conventional extraction techniques for the extraction of tanshinones from *Salvia miltiorrhiza bung*. Biochemical Engineering Journal 12, 71-77.

Hemwimon, S., Pavasant, P., and Shotipruk, A., 2007. Microwave-assisted extraction of antioxidative anthraquinones from roots of *Morinda citrifolia*. Separation and Purification Technology 54, 44-50.

Aresta, M., Dibenedetto, A., Carone, M., Colonna, T., Fragale, C., 2005. Production of biodiesel from macroalgae by supercritical CO2 extraction and thermochemical liquefaction. Environ Chem Lett. 3, 136-139.

Patil, P.D., Gude, V.G., Mannarswamy, A., Deng, S., Cooke, P., Munson-McGee, S., Rhodes, I., Lammers , P., Nirmalakhandan, N., 2010a. Optimization of direct conversion of wet algae to biodiesel under super critical methanol conditions. Bioresour. Technol. doi:10.1016/j.biortech.2010.06.31.

Leadbeater, N.E., Stencel, L.M., 2006. Fast, easy preparation of biodiesel using microwave heating. Energy & Fuels 20, 2281-2283.

Azcan, N., Danisman, A., 2007. Alkali catalyzed transesterification of cottonseed oil by microwave irradiation. Fuel 86, 2639–2644.

Barnard, T.M., Leadbeater, N.E., Boucher, M.B.; Stencel, L.M.; Wilhite, B.A., 2007. Continuous-flow preparation of biodiesel using microwave heating. Energy & Fuels. 21, 1777-1781.

Refaat, A. A., 2010. Different techniques for the production of biodiesel from waste vegetable oil. Int. J. Environ. Sci. Tech. 7 , 183-213.

Patil, P.D., Gude, V.G., Camacho, L.M., Deng, S., 2010b. Microwave-Assisted Catalytic Transesterification of Camelina Sativa Oil. Energy & Fuels 24, 1298–1304.

(519c) Compact Heat Exchange Reactor in Alcohol Production From Biomass

Zhijun Jia, Chart Energy & Chemicals, La Crosse, WI and Vijay Sethi, Western Research Institute, Laramie, WY

Chart's compact heat exchange reactor has been used in various applications especially for highly exothermic or endothermic reactions. Alcohol production from syngas for utilization of biomass is one of good applications to demonstrate the excellent features from the reactor since the reaction is very exothermic. A prototype of reactor consists of multiple reaction and cooling layers which are alternatively arranged. Each reaction layer contains multiple mini-channels in mm scale. Catalysts provided by WRI are packed into those mini-channels. The tested reactor is made of stainless steel to meet the requirements of the reaction conditions, e.g. 320C at 20 bars. The researchers have been continuously running the reactor for several weeks. The results indicated that the compact heat exchange reactor provides unique features to improve the process efficiency and productivity.

See more of this Session: Catalytic Biomass Conversion to Chemicals
See more of this Group/Topical: Fuels and Petrochemicals Division

(519d) Biojet Production From Vegetable Oil Hydroprocessing Using Supported Noble Metals, and Transition Metal Carbide and Nitride

Huali Wang, Manhoe Kim, Craig DiMaggio, Shuli Yan, Steven O. Salley and K. Y. Simon Ng,
Chemical Engineering and Material Science, Wayne State University, Detroit, MI

In recently years, many researchers are concentrating on developing biofuels from alternative and renewable sources to replace commercial petroleum products in the future. However, there are no known technically and economically feasible processes to produce a stand alone jet fuel from plant oils and animal fats within the scope of catalyst development, especially non-sulfided catalysts. Therefore, the research objective for this study is to synthesize, characterize and test various supported noble metal, metallic nitride and carbide catalysts for their activity in hydroprocessing of vegetable oils to produce jet fuel products. HY supported noble metal, transition metal nitride and carbide catalysts were prepared. Transition metal carbides and nitrides of molybdenum and tungsten were prepared by temperature-programmed reaction of the oxide precursor with a reactant gas (20% CH4/H2 for the carbides and 100% NH3 for the nitrides). The catalysts were tested for their activity in a high-pressure BTRS high pressure reactor system using soybean oil as the feedstock.

See more of this Session: Catalytic Biomass Conversion to Chemicals
See more of this Group/Topical: Fuels and Petrochemicals Division

Fractional catalytic pyrolysis of biomass for production of phenolic resins

F.A. Agblevor, A. C. Small*, A. B. Mullins*, O. Mante, J. Pereira, and S. Beis

Biological Systems Engineering, Virginia Tech, Blacksburg, VA 24061

*Luna Innovations Incorporated, 3157 State Street, Blacksburg, VA 24060

1. Abstract

Fractional catalytic pyrolysis is a process selectively targeting individual biopolymer components of the biomass to produce specific product slates. This method was used to screen several biomass feedstocks (switchgrass, corn stover, pinewood, hybrid poplar wood, oakwood) for adhesives production. Pyrolysis experiments were conducted at 400 to 450 °C and pyrolysis oil yields ranged from 30-50 wt%. The oils had very low molecular weight, low viscosity and pH of 3-4. The oils were used as substitutes for phenol in phenol /formaldehyde reactions. Both Novolac and Resol resins were prepared from the oils. For these reactions, there was no need for any pretreatments such as extractions, filtrations or neutralization (they were used as received). Excellent polymerization results were obtained and up to 95 wt% of the phenol could be replaced by the oils. The results were similar for all feedstocks except switchgrass oils which did not form good resins. All the oils were rich in both hydrocarbons and phenolics. However, the switchgrass oil had the highest hydrocarbon fraction.

2. Introduction

Petroleum-derived chemicals and materials play an integral role in the U.S. economy. However, increased reliance on imported sources of fossil fuels threatens our national security, economy and future competitiveness. Apart from fossil fuels, biomass is the only domestic, sustainable and renewable primary energy resource that can provide liquid transportation of fuels and organic chemicals. Biomass also supports a technology transition to a hydrogen economy through either gasification[1] or the production and refining of liquid intermediates such as ethanol, methanol, or bio-oil.[2] Over 98% of phenol is produced from the cumene process which is petroleum derived. However hardwoods, softwoods, and some annuals such as kenaf, wheat straw, and sugar cane bagasse contain phenolic moieties (lignin), useful in producing phenols and other aromatic compounds (Figure 1). The global phenol demand is projected to grow by 4% per year from 2008 to 2012 and the phenol demand in the USA is projected to reach over 2.13 million tonnes per year by 2011. The phenol market alone is greater than $10 billion, demonstrating high value.[3]

Rapid pyrolysis oils cannot be used directly as phenol substitutes in phenol formaldehyde (PF) resins and other applications because of the organic acids and the large fraction of carbohydrate decomposition products which are polar and water-soluble. For example, phenolic adhesives prepared from Tech Air™ pyrolysis oils were not effective as PF resins.[4] Chum et al. extracted the biomass pyrolysis oils first with ethyl acetate, then with water, and finally the ethyl acetate fraction was neutralized with sodium bicarbonate to remove the acidic components.[5] The ethyl acetate was evaporated and the product (phenol/neutral) was used as phenol replacements. The yield of the phenol/neutral fraction from the pyrolysis oil was 26% dry basis. The vortex pyrolysis reactor was used to produce the pyrolysis oils at 60% yield. The phenol/neutral fraction yield was about 15.6% on dry biomass basis. Although, this process was patented, it is yet to be commercialized. Extensive literature review on the production of renewable phenolic resins by thermochemical conversion of biomass showed that none of the phenolic production and fractionation techniques can substitute for 100% phenol content of resins compared to commercial formulations based on petroleum derived phenol without impacting the effectiveness[6].

Instead of the conventional rapid pyrolysis process for biomass conversion to pyrolysis oils followed by solvent extraction to produce phenol/neutral fraction and converting the phenol/neutral fraction to PF resins, we use the fractional catalytic pyrolysis method to produce biocrude oils. The fractional catalytic pyrolysis (FCP) is a novel method for in situ conversion of biomass components into a desirable product slate without post pyrolysis separation[7]. The FCP offers advantages to traditional biomass conversion in the form of significantly more stable products and much lower biocrude oil viscosities. Additionally, because the pH for the FCP process is higher than products from conventional routes, a neutralization step is not necessary thereby reducing man hours and materials required.[7] In fact, the FCP process is a single step process where biomass enters the reactor and leaves as three clean products: biocrude oil, biochar, and syngas. This single step process offers significant advantages over other conventional rapid pyrolysis methods. In this paper, we report direct substitution and characterization of up to 70 wt% of phenol with biocrude oils without any pretreatment. Additionally, we have produced materials with up to 95 wt% biooil substitution.

3. Materials and Methods

The biocrude oils used for the preparation of the resins were produced by the fractional catalytic pyrolysis process described elsewhere.[7] The feedstocks used were corn stover, oakwood, pine, switchgrass, and hybrid poplar wood. Phenol (99%), formaldehyde solution (37%), sodium hydroxide, and oxalic acid were used as received from Aldrich. Resol and Novolac resins were prepared from the biocrude oils and phenol. Additionally, a 100% petroleum derived phenol resin was prepared and used as a control.

Preparation of Resol Resins

Resol resins were synthesized from petroleum based phenol, oak, corn stover and switchgrass biocrude oils (Figure 2). All Resol prepolymers were prepared at a formaldehyde/phenol (F/P) ratio of 2.5. For example, a 50:50 oak biocrude oil/phenol Resol resin was synthesized using a formaldehyde/phenol (F/P) of 2.5 in a 5 wt% alkali solution. Oak biocrude oil (5 g) and phenol (5 g) were added to a 50 mL round bottom flask followed by 37% formaldehyde (20 mL) and 13.8% NaOH solution (6 mL). The flask was equipped with a water-cooled condenser and purged with nitrogen. The reactants were heated to 95°C and refluxed for 60 minutes followed by slow cooling to room temperature. The reaction mixture was left at room temperature for 24 hours after which 60 mL of methanol was added to dissolve the viscous reaction product. The solution was vacuum filtered through a 1 μm PFTE filter. The methanol was evaporated on a rotary evaporator to dryness. The solid residue was redissolved in methanol and evaporated under reduced pressure to give the final quantitative product. The product was characterized by DSC and TGA. The heat of cure was found to 47 J/G with two distinct cure temperatures at 128°C and 161°C. Resols from the corn stover and switchgrass biocrude oils were prepared and characterized using similar method, as was the 100% petroleum based phenol control material.

Preparation of Novolac Resins

Several preparations of Novolac resins were prepared from poplar wood biocrude oil. The following phenol to biocrude oil ratios were prepared: 70:30; 50:50; 30:70; and 5:95. Because the procedures were similar and the only variation was the ratios of phenol to biocrude oil, we describe the case of 50:50 phenol/biocrude oil ratio resin. The 50:50 Phenol/Poplar Biooil Novolac resin was prepared using a 1.22:1 mol to mol ratio of phenol to formaldehyde. Poplar based biocrude oil (3.4 grams), phenol (3.4 grams), 37% formaldehyde solution (4.7 grams), and oxalic acid (0.1005 grams) were combined in a 50 mL reactor. The flask was equipped with a water cooled condenser and nitrogen purge and heated to reflux (~95°C) for three hours followed by slow cooling to room temperature. The solution was left at room temperature for a total of 24 hours. After the cooling period two layers were observed in the flask. The bottom layer was dark in color and viscous while the top layer was clear and aqueous. Excess water was added to the flask and mixed and the layers allowed to phase separate over a period of two hours. The water layer was then decanted. Next, methanol (60 mL) was used to dissolve the viscous reaction product. The solution was vacuum filtered through a 1μm PTFE filter and the solution concentrated using a rotary evaporator. The product was quantitatively collected. DSC and TGA analysis of these materials was done on hexamethylenetetramine (HTA) cured resins. The material was mixed with 10 weight percent of HTA in a high speed orbital mixer and cured in an oven at 125 °C for 2 hours. DSC analysis of the material indicated a broad glass transition temperature of approximately 85 °C. TGA analysis indicates degradation onset at approximately 195 °C.

Epoxy Novolac Resin System

The synthesis of a poplar biooil based epoxy Novolac resin is a two-step synthetic process. The first step involves the synthesis of a Novolac resin as described above followed by epoxy functionalization of the resulting Novolac resin. For example, the 50/50 Novolac resin described above was epoxy functionalized by dissolving the poplar wood based Novolac resin (4 grams) in 100 mL of epichlorohydrin in a 500 mL reactor. The solution was heated to 80 °C and 30% sodium hydroxide solution was slowly added drop-wise. The rate of addition was slow enough to ensure that the pH of the reaction solution did not go above 8. Total addition time of sodium hydroxide solution was approximately eight hours. The addition was stopped and the reaction was left heating for another hour at 80°C followed by slow cooling to room temperature. The water/NaCl layer was then removed using a separatory funnel. The organic layer was filtered through a 1μm PTFE filter followed by vacuum distillation resulting in viscous oil. The epoxy equivalent weight was determined using ASTM D 1652-97 and found to be 155.8. The resin was cured using aminoethylpiperazine (AEP). The cure of the resin was examined using DSC analysis. The heat of cure was found to be 464.6 J/g with a cure temperature of 87 °C. The cured material was observed to have a glass transition temperature of 48°C.

4. Results and Discussion

Preparation of Resol Resins

Resol resins were synthesized from oak, corn stover, and switchgrass biocrude oils. All Resol prepolymers were prepared at a formaldehyde/phenol (F/P) ratio of 2.5. The prepolymers were synthesized by substituting the different biocrude oils for phenol in the formulations. A summary of the different Resols synthesized is given in Table 1. The amount of formaldehyde and 13.8% NaOH were held constant for each experiment. It was also interesting to note that when filtered through a 1-μm PTFE filter, only a negligible amount of residue was observed on the filter for all Resol preparations. Typical rapid pyrolysis oil resins leave considerable amount of residue after filtration. The difference could be the lower molecular of the biocrude oils as exemplified by M_n, M_w, M_z and M_p that were respectively 120, 257, 555, and 134. The polydispersity, M_w/M_n, was 2.136.

The resins were characterized by DSC analysis. The DSC curves for each formulation give the temperature range for the cure of the resins as well as the enthalpy of the cure. A second heating cycle reveals that all of these materials have broad glass transition values. The cure data is given in Table 1. Glass transition data is not given. A DSC curve comparing the cure of Resol Resin 2 to Resol Resin 4 is given in. As can be seen in Figure 3 the scan the enthalpy and temperatures of cure for the Resol blends are comparable for the two different types of materials. The control Resol shown also in Figure 3 gives a much larger cure enthalpy. Resol resins cure by condensation of methyloyl groups with phenol to form methylene bridges followed by the condensation of two methylol groups to form dibenzyl ether bridges. The second signal for the blended material can be attributed to the condensation of the dibenzyl ether bridges to methylene bridges resulting in the elimination of formaldehyde. The control Resol only shows one cure peak. The difference in the control and the blended samples as shown in Table 1 and Figure 3 could possibly be attributed to the difference in the types of linkages formed from introducing the biooil materials. The different sources of biooil thus may lead to different cure properties of the prepolymers.

Preparation of Novolac Resins

Novolac resins were synthesized from poplar, pine and switch grass biooils. All Novolac prepolymers were prepared by polymerization of phenol and formaldehyde (1.22:1 mol/mol ratio). The prepolymers were synthesized by substituting the different biooils for phenol in the formulations. A summary of the different Novolacs synthesized is given in Table 2. The amount of formaldehyde and oxalic acid were held constant for each experiment.

DSC analysis of these materials was done on hexamethylenetetramine (HTA) cured resins. Formaldehyde is a side product of the curing of these materials and as a result DSC analysis was only done on cured material. Attempts were made to analyze the cure of these materials using the DSC, but were unsuccessful because of the formaldehyde emission and the resulting noise in the baseline. The Novolacs were mixed with 10 weight percent of HTA in a high speed orbital mixer and cured in an oven at 125 °C for 2 hours. An example of the material resulting from the cure of Novolac Resin 3 is shown in Figure 4. Voids are evident in the cured material more so than the Resol resins. A DSC overlay curve for Novolac Resins 1, 2, 3 and control is shown in Figure 5. These materials have broad glass transitions. Thermal gravimetric analysis was also performed on cured and uncured materials. This data is shown in Figure 6. The degradation onset for uncured materials is between 90 °C and 100 °C, while degradation for cured materials is between 180 °C and 200 °C.

Preparation of Epoxidized Novolac Resins

Novolac Resins 2 and 4 as well as the Novolac control were converted to epoxy functionalized resins.[8,9] Novolac resin was taken up in epichlorohydrin and heated to approximately 80 ºC. Sodium hydroxide was then added while maintaining a narrow pH range. After reaction completion excess epichlorohydrin was removed under reduced pressure and epoxy equivalent weight was determined using ASTM D 1652-97. The materials were cured using aminoethylpiperazine (AEP) and analyzed using DSC. Cure data as well as glass transition temperatures for these materials are shown in Table 3. Clear castings were also made from these materials using small molds. Pictures of these parts are shown in Figure 7. The ability to produce clear castings from the biobased materials demonstrates their promise for future use in resins suitable for composites and adhesives.

5. Conclusions

The results to date demonstrate successful production of biocrude oil from a unique fractional catalytic fast pyrolysis process. The resulting biocrude oil produced is very clean and was readily converted into high biobased content (>70% with potential for 100% biobased content) resins such as Resol, Novolac, and Novolac epoxy based products. The versatility of the fractional catalytic fast pyrolysis (FCP) process on hard wood, soft wood, and grass based biomass feedstocks was evident with the production of high purity, storage stable biocrude oils from a single step process suitable for further processing. The biobased resins can be used in goods such as bonding pastes, wood adhesives, and high performance composites. From the results presented to date, up to 70% phenol can be substituted with biocrude oil in PF resins without modification of the reaction conditions. Up to 95% substitution may be possible without reaction condition modification, but it is unlikely 100% biocrude substitution is possible without significant reaction modification. Future work will focus on improving crosslinking and curing of the new biobased PF resins in order to improve T_g/HDT.

6. Acknowledgement and Disclaimer

This material is based upon work supported by the Department of Energy under Award Number DE-FG02-08ER86347. This paper was prepared as an account of work sponsored by an agency of the United States Government. Neither the United States Government or any agency thereof, nor any of their employees, makes any warranty, express or implied, or assumes any legal liability or responsibility for the accuracy, completeness, or usefulness of any information, apparatus, product, or process disclosed, or represents that its use would not infringe privately owned rights. Reference herein to any specific commercial product, process, or service by trade name, trademark, manufacturer, or otherwise does not necessarily constitute or imply its endorsement, recommendation or favoring by the United States Government or any agency thereof. The views and opinions of authors expressed herein do not necessarily state or reflect those of the United States Government or any agency thereof.

Figure and Tables

Figure 1. Wood, grass, and corn stover based biomass are converted to bio-oil, then to phenolic resin products such as Resol and Novolac epoxies.

Figure 2. From left to right – Control Resol, 50:50 Oak Biooil/Phenol Resol, 50:50 Corn Stover Biooil: Phenol Resol

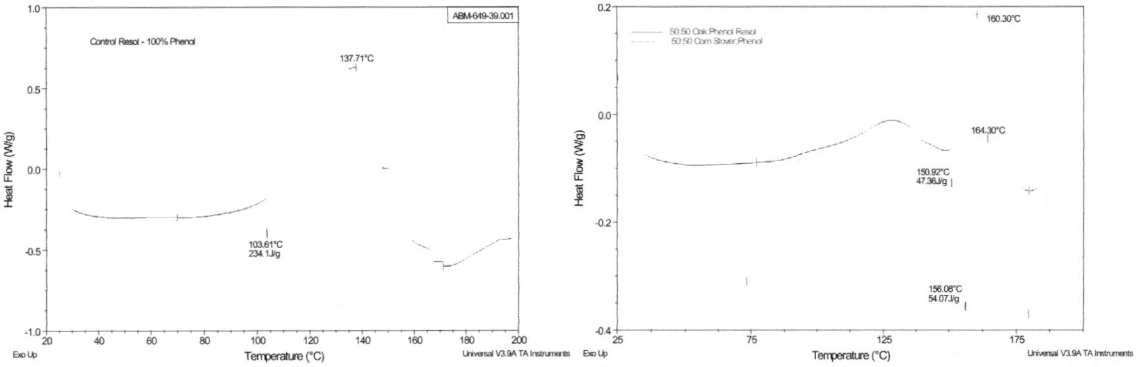

Figure 3. DSC Analysis of control Resol with 50:50 Oak Biooil/Phenol Resol and 50:50 Corn Stover Biooil: Phenol Resol

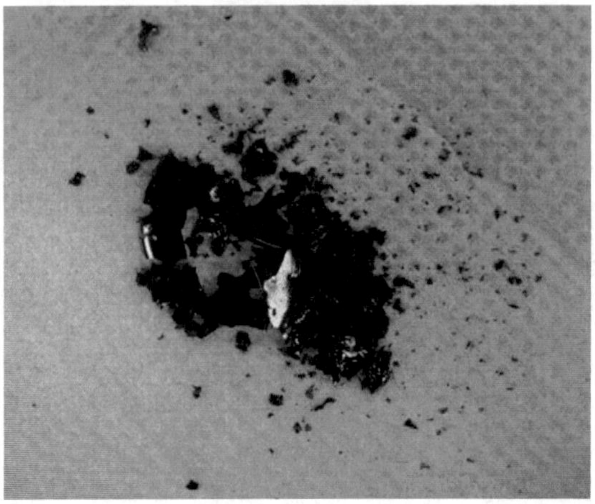

Figure 4. Cured 30:70 Phenol/Poplar Novolac. Material cured with 10% HTA

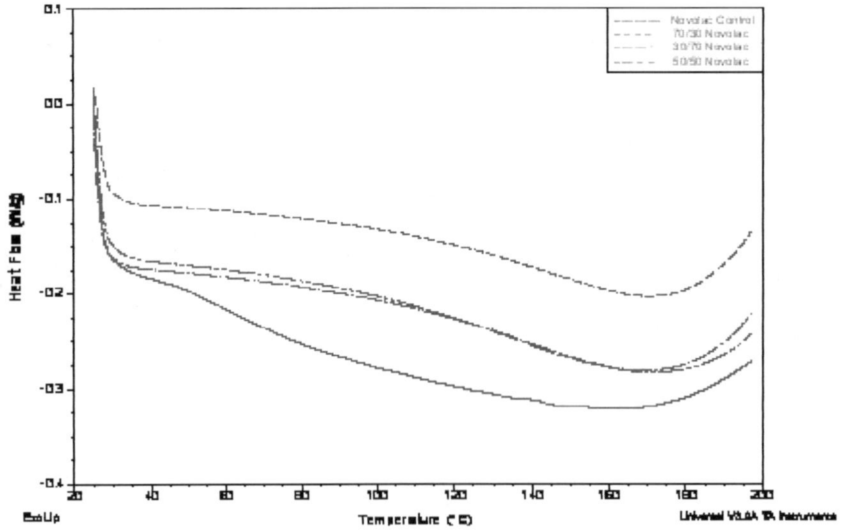

Figure 5. DSC analysis of cured Novolac resins. The glass transition for these materials is broad with the glass transition for the biobased Novolacs being higher than the control

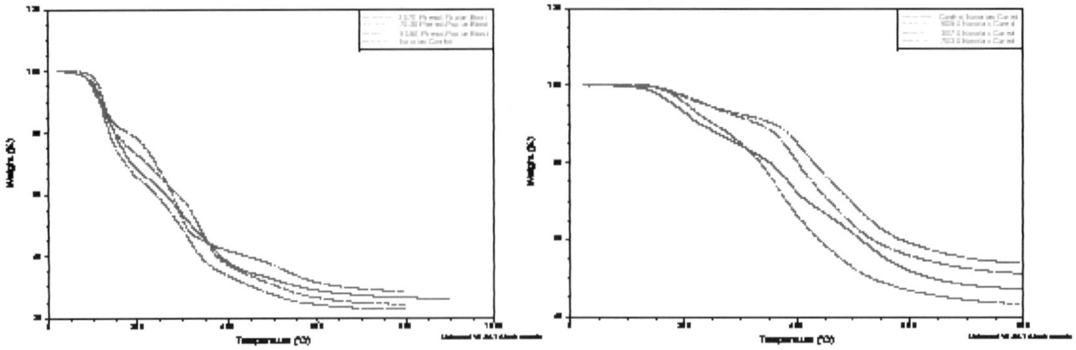

Figure 6. TGA curves showing degradation onset for uncured (left) and cured Novolac resins

Figure 7. Epoxy Novolac clear castings. The image on the left was made from the control epoxy Novolac, while the middle clear casting is made from poplar based epoxy Novolac. The image on the right was made from pine based epoxy Novolac. All samples were cured with AEP

Table 1. Cure data for Resol resins prepared from various biocrude oils

Resol Resin	Composition[a]	ΔH (J/g)	T_1 (°C)	T_2 (°C)
Control	Phenol (100%)	234	138[b]	No peak
1	Phenol:Oak biocrude oil (70:30)	20[c]	120[c]	149[c]
2	Phenol:Oak biocrude oil (50:50)	47	128	161
3	Phenol:Oak biocrude oil (30:70)	74	126	170
4	Phenol:corn stover biocrude oil (50:50)	54	138	164
5	Phenol:switchgrass biocrude oil (50:50)	129	141	No peak

a. Biooil was substituted for phenol these formulations. Formaldehyde and NaOH amount were held constant.
b. Only one cure peak observed for these materials.
c. Materials had partially cured at room temperature over a 48 h period. Resols must be stored t -20° C.

Table 2. Compositions of Novolac resins synthesized

Novolac Resin	Composition[a]
Control	100% Phenol
1	Phenol: Poplar biocrude oil (70:30)
2	Phenol: Poplar biocrude oil (50:50)
3	Phenol: Poplar biocrude oil (30:70)
4	Phenol: Pine biocrude oil (50:50)
5	Phenol: Switchgrass biocrude oil (50:50)[b]

[a] Biooil was substituted for phenol in these formulations. Formaldehyde and NaOH amounts were held constant

[b] Reaction was unsuccessful after two attempts

Table 3. Cure data and glass transition data for biobased epoxy Novolac samples

Epoxy Resin	Composition	ΔH (J/g)	T_1 (° C)	T_g (° C)
Control	100% Phenol Epoxidized Novolac	487	92	85
Poplar	50:50 Phenol/Poplar Biooil Epoxidized Novolac	465	87	48
Pine	50:50 Phenol/Pine Biooil Epoxidized Novolac	242	83	~ 30[a]

References

[1] Ni, M.; Leung, D.Y.C.; Leung, M.K.H.; Sumathy, K. "An overview of hydrogen production from biomass", *Fuel Processing Technology* **2006**, *87*, 461-472.

[2] For example: Multi-Year Program Plan, Biomass and Biorefinery Systems R&D (2007-2012), U.S. Department of Energy; Fernando, S.; Adhikari, S.; Chandrapal, C.; Murali, N. "Biorefineries: current status, challenges, and future direction", *Energy and Fuels*, **2006**, *20*, 1727-1737.

[3] Feng-Ru, Q., Global Market Analysis for Phenol, Fine and Specialty Chemicals, Issue 15, 2004.

[4] Elder, T. 1984. Effect of process conditions on the yield of pyrolytic products from southern pine. *Wood and Fiber Science*, 16(2), 169-179.

[5] Chum H., Diebold J., Scahill J., Johnson D., Black S., Schroeder H., Kreibich R. Biomass pyrolysis oil feedstocks for phenolic adhesives. In Hemingway, R.W., Conner, A.H. Brahham, S.J. (eds) Adhesives from Renewable Resources. ACS Symposium Series 385, American Chemical Society, Washington DC. pp.135-151.

[6] Effendi, A., Gerhauser, H., Bridgwater, A.V. 2008. Production of renewable phenolic resins by thermochemical conversion of biomass: a review. Renewable and Sustainable Energy Reviews, **12**, 2092-2116.

[7] Agblevor, F.A.; Beis, S.; Mante, O., Abdoulmoumine, N. "Fractional Catalytic Pyrolysis of Hybrid Poplar Wood", *Ind. Eng. Chem. Res.* **2010**, *49*, 3533-3538.

[8] Cherian, A.B, Thachil, E.T. *J. Appl. Polym. Sci.,* **100**, 457 (2006).

[9] Oyanguren, P.A., Williams, R.J.J. *Polymer,* **33**, 2376 (1992).

USE OF ANHYDRIDE BASED POLYMERS AS SCALE CONTROL ADDITIVES IN PRODUCTION OF HEAVY OILS
AIChE 2010 FALL PROCEEDINGS - Paper # 524a

Y. A. Roomi, K. F. Hussein and M. R. Riazi, Kuwait University, Kuwait*
**Corresponding author: RiaziM.com, email: mrriazi@gmail.com*

Introduction

Waterflooding is one of the procedures used in petroleum production. Recent studies show that even for high viscosity heavy oil fields waterflooding is the most inexpensive method of oil recovery in comparison with more expensive thermal EOR methods [1]. Many countries now contain significant amounts of heavy oils in their reserves [2,3]. During such processes inorganic chemicals in water may precipitate forming scales which may cause major problem in flow assurance and oil productivity. Precipitation of inorganics in water such as calcium sulfate ($CaSO_4$) or calcium carbonate ($CaCO_3$) is referred as scaling. In oil production by waterflooding as water approaches the vicinity of production well, initial pressures are diminished in which calcium sulfate solubility is rapidly reduced and scales are formed in the vicinity of wellbores causing restrictions in flow of oil from production well [4,5]. Inorganic scales are also a major problem in flow assurance in offshore oil and gas production which may lead to significant reductions in hydrocarbon production and costly workovers [6]. Researchers in recent years have studied mechanism of scale formation in hydrocarbon production wells [7-9].

The main method of preventing scale formation in oil and gas production or water desalination facilities is use of chemical inhibitors known as antiscalants or scale control additives (SCA). These additives are usually polymer type materials which are added in different dose rates to brine to combat scaling problems. The efficiencies of these antiscalant additives depend on the brine chemical constituents, chemical nature of antiscalant, design configuration, temperature and its reactivity with the brine. The exact chemical composition of a SCA is usually guarded secret of the manufacturer. The role or mechanism of antiscalants is still not clear in spite of their popular use [10]. Synthesis and properties of some of these polymer type SCAs have been investigated by both researchers and chemical manufacturers [11-17].

In this work we continue our previous research works on synthesis of some maleic anhydride based polymers as efficient antiscalants. Polymers based on maleic anhydride are particularly effective in sea water desalination and scale removal in oil production wells. The presence of both an electron-acceptor double bond and an anhydride moiety confers unique properties to maleic anhydride, providing an opportunity for application in chain and step-growth polymerization systems. Maleic anhydride is a widely available local material which is cost effective in comparison with other materials [18,19].

Synthesis and Experiments

Reaction schemes to produce monopolymer and copolymers from maleic anhydride (MA) in our laboratory are demonstrated in Figs. 1 and 2, respectively. Production of copolymer occurs at two step in which at first step n-octyl-acrylamide (AAO) monomer is produced as shown in Fig. 2.

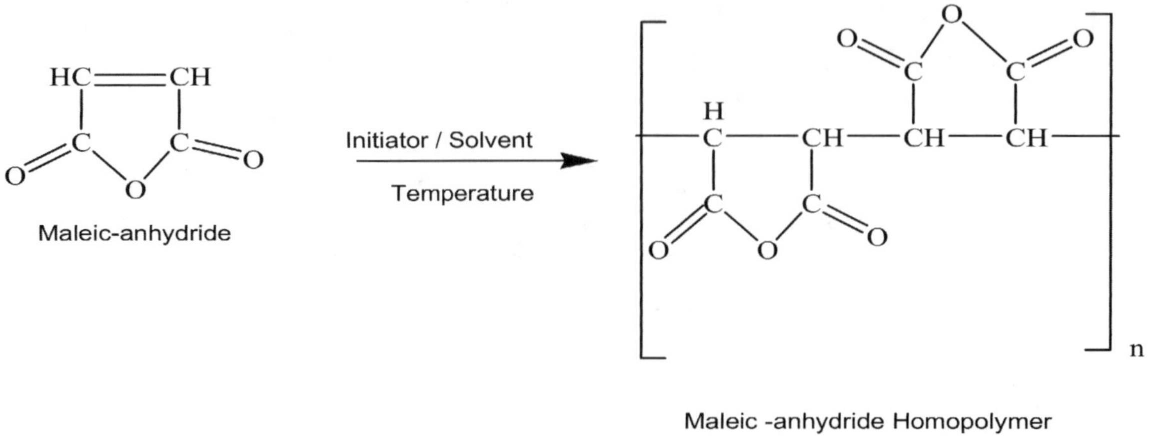

Figure 1. Homopolymer production from MA

Series of Low molecular weight maleic –anhydride homopolymers and copolymers synthesized in our laboratory are depicted in Tables 1 and 2. Maleic anhydride (Sigma, 99%) was recrystallized from benzene and dried under vacuum at room temperature. N- octyl amine (Aldrich, 98%) and acrolyl chloride (Aldrich, 96%) were used without further purification. BPO (Aldrich, 97 %) and AIBN (Aldrich, 97%) were crystallized from methanol prior to use. Molecular weight measurements of the monomers and copolymers were carried out by high temperature size exclusion chromatography (SEC). As shown in Table 1, when initiator composition was at 10 mol%, homopolymer with maximum molecular weight was produced. Table 2 shows the conditions at which maximum conversion can be achieved (50% MA, 50% AAO and 7.5% initiator).

Table 1: Conditions and Results for the Homo-polymerization of MA

Entry	Scale Inhibitor	MA (moles/liters)	Solvent	Initiator	Initiator composition, (mole %)	Conversion %	M_w[c]
1	YMR 41	0.247	O-Xylene	AIBN[a]	5	5.41	265
2	YMR 44	0.247	O-Xylene	AIBN	7.5	16.91	400
3	YMR 50	0.247	O-Xylene	AIBN	10	80.68	810
4	YMR 39	0.247	O-Xylene	AIBN	12	72.06	773
5	YMR 34	0.247	O-Xylene	AIBN	15	71.22	457
6	YMR 40	0.247	O-Xylene	BPO[b]	5	21.52	641
7	YMR 53	0.247	O-Xylene	BPO	7.5	60.64	706
8	YMR 47	0.247	O-Xylene	BPO	10	70.01	847
9	YMR 35	0.247	O-Xylene	BPO	12	30.31	405
10	YMR 31	0.247	Toluene	BPO	10	44.59	430

[a] AIBN = Azoisobutyronitrile; [b] BPO = Benzoyl peroxide; [c] Molecular weight obtained from SEC

Monomer Synthesis:

A 200 ml round bottom flask equipped with a glass stir bar was placed in an ice bath. N-Octyl amine and barium hydroxide (1%) were placed in the flask and stirred until the temperature equilibrated. Acrolyl chloride was added drop wise to the flask containing the n-octyl amine while continuously stirring. The mixture in the reaction vessel thickened after about 7 minutes indicating the formation of the monomer. The reaction mixture was then heated gradually up to 85°C, until HCl was eliminated. The resulting solution was whitish-yellow and turbid. Doubly deionized water was added to the vessel with vigorous stirring to separate the monomer. The procedure was repeated several times to eliminate excess reactants. The resulting material was dried by placing it in a 40°C oven. The waxy white colored final product was then stored at 0°C.

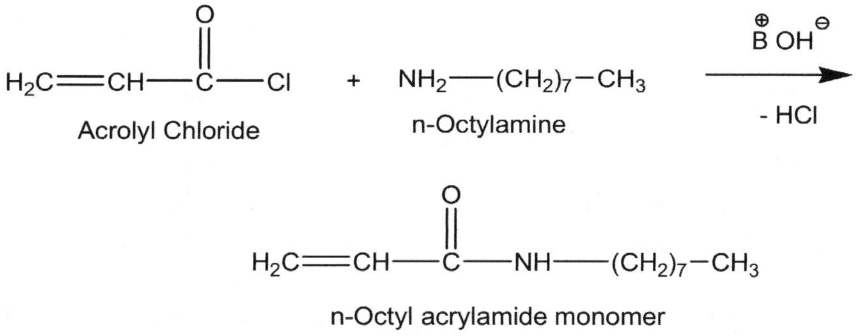

STEP- 1 : Reaction Scheme for preparation of n-Octyl acrylamide monomer (AAO)

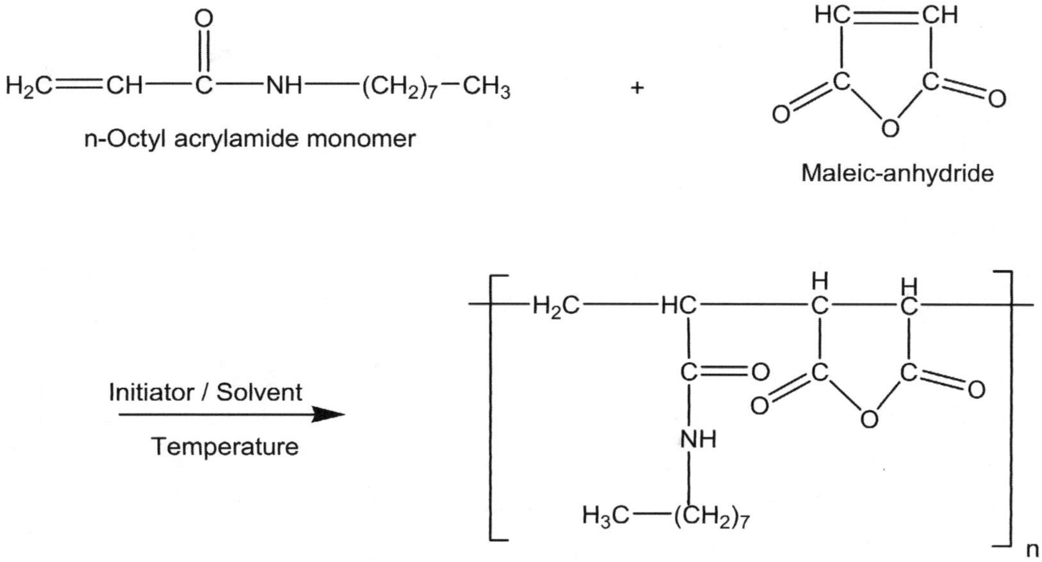

STEP- 2 : Reaction Scheme for preparation of Maleic-anhydride- n-Octyl acrylamide copolymer

Figure 2. Copolymer production in two steps

Copolymerization:

Copolymerization of MA with AAO was conducted with the objective of characterizing the reaction and to generate a series of low molecular weight co-polymers of varying composition. The reaction was conducted at four different concentrations. The samples were labeled as: MA / 30 AAO, MA / 50 AAO, MA / 60 AAO and MA / 40 AAO indicating 70:30, 50:50, 40:60 and 60:40 mol % of MA:AAO, respectively. A dry N_2 inert reaction atmosphere was maintained throughout all experiments. Two different initiators azo-isobutyronitrile (AIBN) and benzoyl peroxide (BPO), were used in the copolymerization study. The concentration of initiators (5, 7.5, 10 and 12 mol % of monomers) was varied to select and optimize their performance. The reaction temperature was maintained at the reflux temperature of o-xylene. The progress of the reaction was followed by measuring the disappearance of the monomer, which was conducted by taking out a small sample of reaction mixture every hour and determining its monomer content by LC and increase in molecular weight of the polymer by gel permeation chromatography.

A typical polymerization was carried out as follows. MA dissolved in 200ml of o-xylene was stirred and heated to reflux under a N_2 atmosphere until the temperature reached 70^0C. AAO and initiator dissolved in 50 ml of solvent were added over a period of five minutes. Refluxing was continued for a further 9 hours and the polymer was allowed to separate from the solution. Excess of solvent was distilled off and the oily layer so obtained was recrystallized in benzene to give the copolymer product. The series of n-octylacrylamide-maleic anhydride copolymers prepared by the free radical mechanism were analyzed by ^{1}H-NMR and ^{13}C-NMR spectroscopy to determine the copolymer composition [18].

Table 2: Conditions and Results for the Radical Copolymerization of MA and AAO in O-Xylene

S#	Scale Inhibitor	Composition of monomer mixture, mole%		Initiator	Initiator composition (mole %)	Conversion %	M_w/M_n	M_w[c]
		MA	AAO					
1	YMR-80	70	30	BPO[a]	7.5	49	3.89	1152
2	YMR-82	60	40	BPO	7.5	43.5	3.47	1345
3	YMR-68	50	50	BPO	7.5	52.5	1.89	1759
4	YMR-78	40	60	BPO	7.5	37.56	3.87	935
5	YMR-65	70	30	AIBN[b]	7.5	43.5	1.36	1658
6	YMR-72	60	40	AIBN	7.5	53.3	1.66	1646
7	YMR-74	50	50	AIBN	7.5	53	1.61	1890
8	YMR-75	40	60	AIBN	7.5	17	3.62	1469
9	YMR-67	50	50	BPO	5	40.2	1.63	1545
10	YMR-76	50	50	BPO	10	62	3.4	1364
11	YMR-81	50	50	BPO	12	50	3.1	1135
12	YMR-79	50	50	AIBN	5	33.25	3.78	1562
13	YMR-63	50	50	AIBN	10	53	1.67	1725
14	YMR-83	50	50	AIBN	12	43.5	3.86	1267

[a] BPO = Benzoyl peroxide; [b] AIBN = Azoisobutyronitrile; [c] Molecular weight obtained from SEC

Inhibition Efficiency of Synthesized Polymers

Synthesized polymers are all highly effective inhibitors of $CaCO_3$ and $CaSO_4$ precipitation. Precipitation tests for $CaCO_3$ were made with 0.1, 0.3 and 0.5 ppm while for $CaSO_4$ tests were conducted at 0.25, 0.5, 1, 2, 3, 5 and 7 ppm polymer concentrations. Virtually complete inhibition is achieved with 0.5 ppm polymer as it can be observed from Fig. 4 and Fig. 5 for $CaCO_3$. For homopolymers no significant difference in activity among the polymers is noted during this test as shown in Fig. 6, where as for copolymer series the inhibition percent increases with increase in molecular weight as can be observed from Fig. 7. Test methods used for inhibition efficiency of polymers were according to ASTM D 1126-02 and NACE TM 0374-95 [20,21]. To calculate the percentage inhibition of synthesized polymers following equation was used:

$$\%\ \text{inhibition} = \frac{[Ca^{2+}]_{antiscalant} - [Ca^{2+}]_{blank}}{[Ca^{2+}]_{original} - [Ca^{2+}]_{blank}} \times 100$$

where:

$[Ca^{2+}]_{antiscalant}$ = concentration of soluble Ca^{2+} in an antiscalant containing solution (mg/L)
$[Ca^{2+}]_{original}$ = concentration of soluble Ca^{2+} in original solution (mg/L)
$[Ca^{2+}]_{blank}$ = concentration of soluble Ca^{2+} in a blank solution (mg/L)

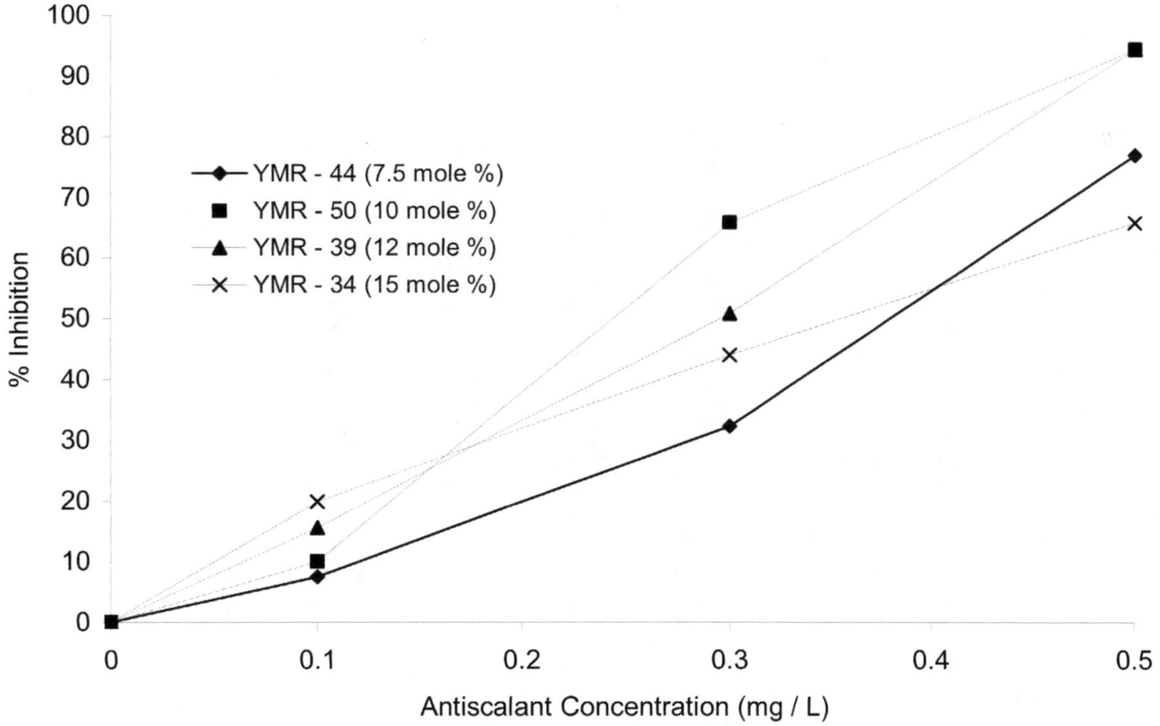

Figure 4: Inhibition% for $CaCO_3$ scales using synthesized homo-polymers obtained at varying concentration of AIBN initiator.

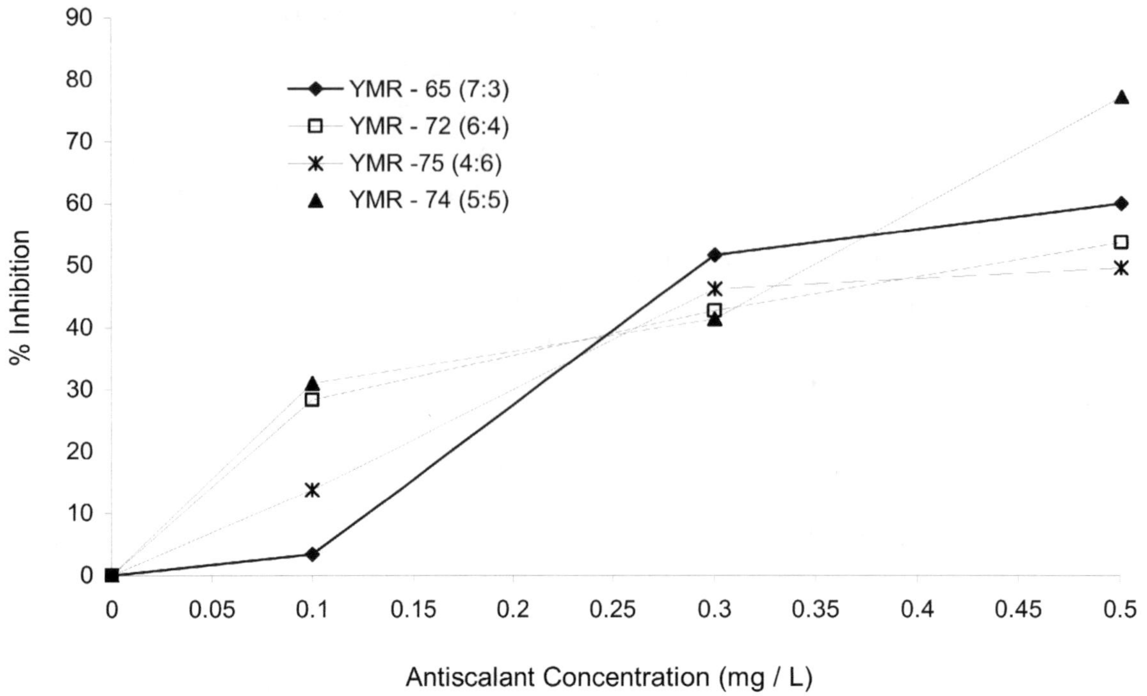

Figure 5: $CaCO_3$ Inhibition% for $CaCO_3$ scales using synthesized homo-polymers obtained at 7.5 mole % AIBN concentration and varying monomer (MA / AAO) ratio

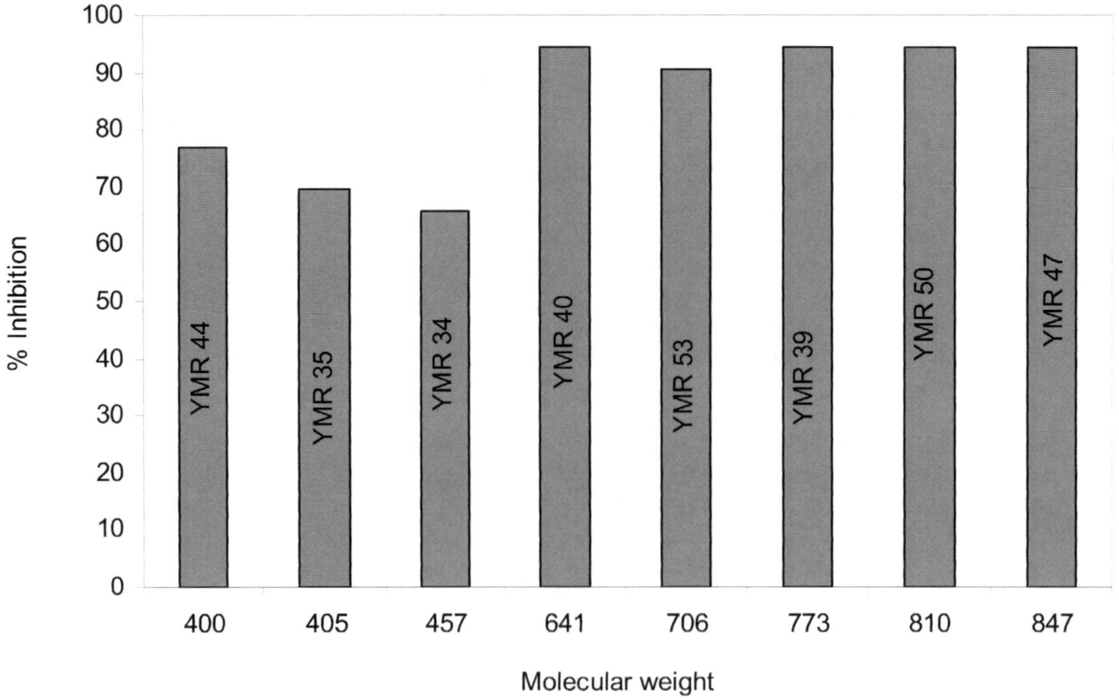

Figure 6: Inhibition % for $CaCO_3$ scales with molecular weight of MA homopolymers.

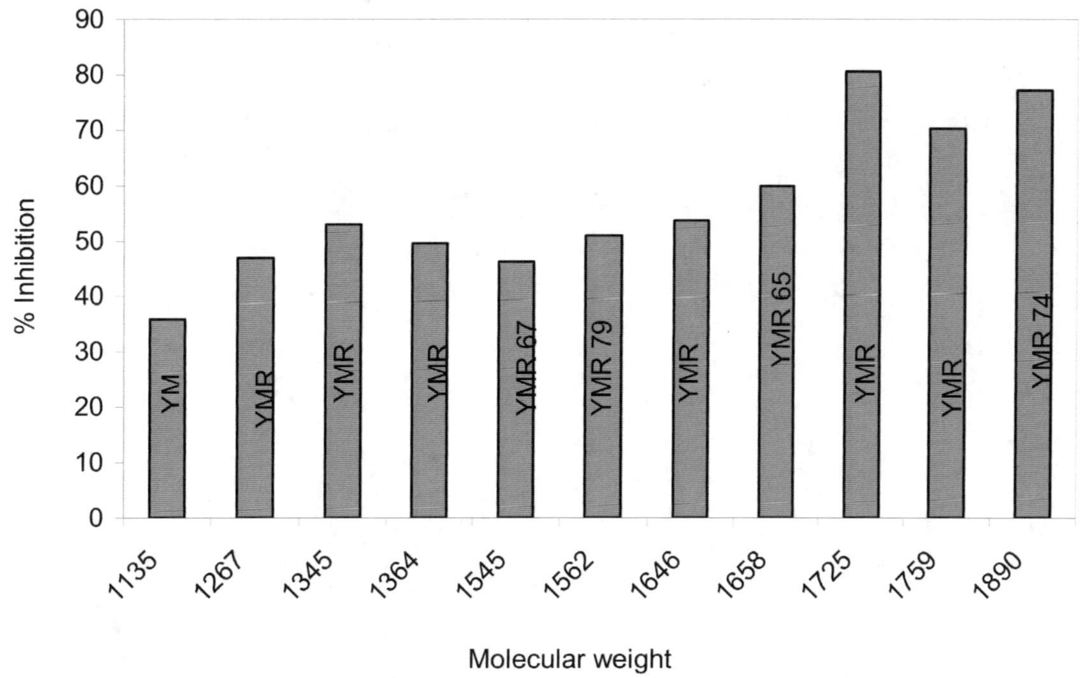

Figure 7: Plot showing increase in inhibition % for CaCO3 scales with molecular weight of the antiscalants (MA / AAO Copolymer)

To view the impact of produced antiscalants photographs of $CaCO_3$ scales before and after use of synthesized polymer are shown in Figs 8 and 9, respectively. Similar performances are observed for $CaSO_4$ and $Mg(OH)_2$ scaling.

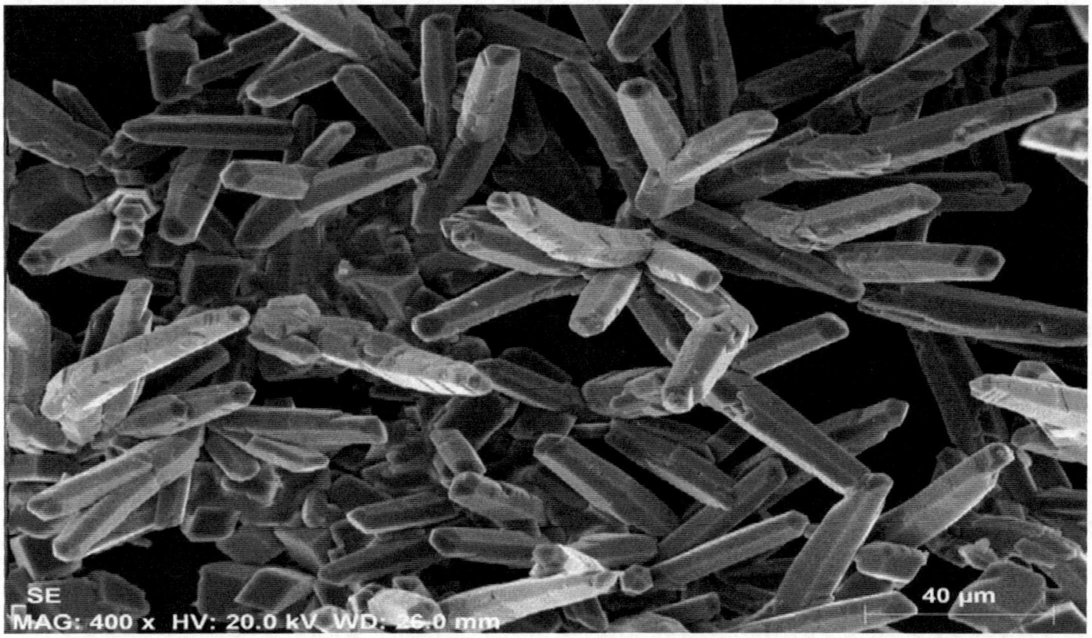

Figure 8: SEM photograph for the $CaCO_3$ in the absence of scale inhibitor (BLANK)

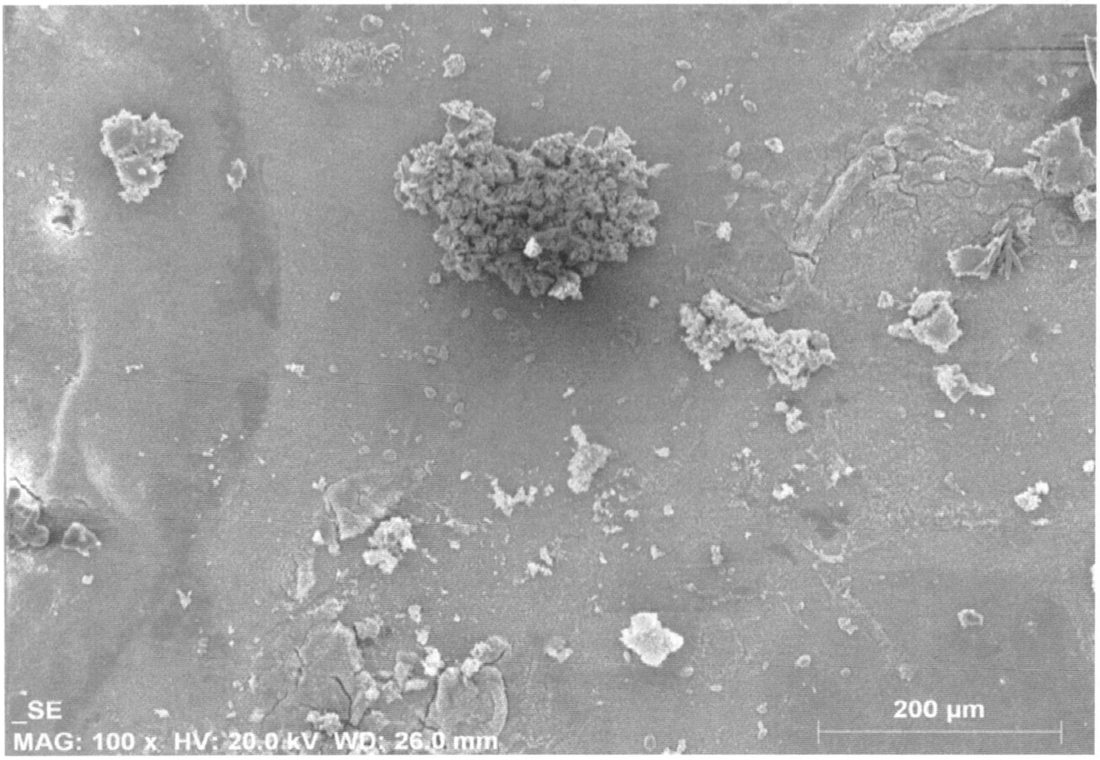

Figure 9: SEM photograph for the $CaCO_3$ in the presence of YMR-63

Conclusions

In this paper synthesis and production of series of antiscalants homo and copolymers based on maleic anhydride are presented. Produced polymers have been used as inhibitors for precipitation of calcium carbonate, calcium sulfate and magnesium hydroxide. Results of extensive experiments and testing show that synthesized polymers demonstrate very good efficiency and could be used as cost effective antiscalants in oil and gas production wells as well as in water desalination units.

Acknowledgment

The funding for this project was provided by the Research Administration of Kuwait University under Project No. RE 05/06 which is appreciated.

References

1. Mai, A., Kantzas, A. (2009), 'Heavy Oil Waterflooding: Effects of Flow Rate and Oil Viscosity,' *Journal of Canadian Petroleum Technology*, Vol 48, No. 3, pp. 42-51.
2. Riazi, M. R., (2010), 'Energy, Economy, Environment and Sustainable Development in the Middle East and North Africa', *International Journal of Oil, Gas and Coal Technology* (IJOGCT), Vol 3, No 3, pp. 201-240.

3. Riazi, M. R., (2005) "*Characterization and Properties of Petroleum Fractions,*" ASTM International, Conshohocken, PA, USA. (http://www.astm.org/mnl50.htm)
4. Oil recovery utilizing acidic organic phosphate scale inhibitors (1972), United States Patent 3703928 Website: http://www.freepatentsonline.com/3703928.html
5. Prevention of calcium sulfate scale in a waterflooding process (1971), United States Patent 3584686 Website: http://www.freepatentsonline.com/3584686.html
6. Gordon M. Graham, G.M., Eric James Mackay, E.J., and Sarah J. Dyer, S.J. (2002), 'The Challenges for Scale Control in Deepwater Production Systems: Chemical Inhibition and Placement', *NACE International*, Corrosion Conference, Paper No. 02316, April 7-12, Denver, CO, USA.
7. Salman, M., H. Qabazard, H., Moshfeghian, M.(2007),'Water scaling case studies in a Kuwaiti oil field', *Journal of Petroleum Science and Engineering*, Vol. 55,No. 1-2, January, pp. 48-55
8. Jamialahmadi, M. and Müller-Steinhagen, H. (2008) 'Mechanisms of scale deposition and scale removal in porous media', *Int. Journal of Oil, Gas and Coal Technology*, Vol. 1, Nos. 1/2, pp.81–108.
9. Fadairo, A.S.A. and Falode, O. (2009) 'Predicting tool for sulphate scale build-up around the wellbore in oilfield', *Int. Journal of Oil, Gas and Coal Technology*, Vol. 2, No. 4, pp.347 -364.
10. Dalvi, A.G.I., Kither Mohammad, N.M., Al-Sulami, S., Sahul, K. and Al-Rasheed, R., (2000) 'Role of chemical constituents in recycle brine on the performance of scale control additives in MSF plants', *Desalination*, Vol. 129, No. 2, pp.173-186.
11. Akbulut, U. and Hacioğlu, B. (1991) 'Electronitiated polymerization of maleic anhydride by direct electron transfer', *Journal of Polymer Science*, Part A: Polymer Chemistry, Vol. 29, No. 2, pp. 219-224.
12. '*Scale Inhibitors*' (1997) , Introduction of ACUMER series polymers, Rohm and Haas Company, http://www.rohmhaas.com , March 1997 Report, Taken from website: http://www.e-chemicalsource.com/bulletin/Acumer1000_1020.pdf
13. Shakkthivel, P., Sathiyamoorthi, R. and Vasudevan, T. (2004) 'Development of acrylonitrile copolymers for scale control in cooling water systems', *Desalination*, Vol. 164, pp.111-123.
14. Shakkthivel, P. and Vasudevan, T. (2006) 'Acrylic acid-diphenylamine sulphonic acid copolymer threshold inhibitor for sulphate and carbonate scales in cooling water systems', *Desalination*, Vol. 197, pp.179-189.
15. Li, H.-Y., Ma, W., Wang, L., Liu, R., Wei, L.-S., and Wang, Q. (2006) 'Inhibition of calcium and magnesium-containing scale by a new antiscalant polymer in laboratory tests and a field trial', *Desalination*, Vol. 196, No. 1-3, pp. 237-247.
16. Wang, C., Li, S.-P., and Li, T.-D. (2009) 'Calcium carbonate inhibition by a phosphonate-terminated poly(maleic-co-sulfonate) polymeric inhibitor', *Desalination*, Vol. 249, pp.1-4.
17. Tang, W., Yang, W., Yin, X., Ying Liu, Y., Yin, P. (2008) 'Investigation of CaCO3 scale inhibition by PAA, ATMP and PAPEMP', *Desalination*, Vol. 228, pp.55-60.
18. Al-Roomi, Y. M., and Hussain, K.F. (2005) 'Free Radical Solution Copolymerization of n-Octylacrylamide and Maleic Anhydride in a Non Polar Solvent; Synthesis and Characterization of a Difunctional Copolymer', *Polymer Bulletin*, Vol. 55, pp. 81-93.

19. Al-Roomi, Y. M., and Hussain, K.F. (2006) 'Homo-oligomerization of maleic anhydride in nonpolar solvents: A kinetic study of deviations from nonlinear behavior', Journal of *Applied Polymer Science*, Vol. 102, No. 4, pp. 3404 -3412.
20. ASTM International, West Conshohocken, PA, USA. http://www.astm.org/
21. NACE International (The *National Association of Corrosion Engineers*) Standards, Houston, TX, USA. http://www.nace.org/

(524b) A Novel Design for in-Flow Deposition Property Measurement

Mark Jemmett, Milind Deo, J. J. Magda, Husam El Gendy, Preston Montalvo and Dave Wagner, Chemical Engineering, University of Utah, Salt Lake City, UT

While wax deposition is a fairly well-known phenomena due to chemical potential gradients and solid-liquid equilibrium, effectively visualizing and measuring the thickness and strength of the deposition formed during flow without disturbing the crystalline structure is very difficult. Members of the department of Chemical Engineering at the University of Utah have constructed a flow loop that allows for non-invasive exploration of many facets of wax deposition. This loop includes an all-steel experimental heat exchanger where controlled deposition takes place under specified heat fluxes including "cold flow" – slurry conditions (below WPT), and zero heat-flux through the walls (isothermal) – and a clear acrylic plexiglass exchanger for controlled deposition and PIV visualization. Both sections are interchangeable, use pure ethylene glycol coolant, and also use high-resolution pressure transducers to measure pressure drop across the exchanger, allowing for accurate thickness measurements to be calculated using a modified Hagen-Poiseuille equation. In addition, the novel flow loop design incorporates a large, temperature-controlled reservoir to maintain fluid properties, as well as a particle size analyzer for use in determining particle size effect on deposition rate, strength, etc, and a conditioning loop to control these sizes. A method for determining in-line yield stress is also being developed so as to not break the crystals by transport out of the temperature controlled loop. The method currently considered is a pneumatic piston that fits flush inside the test section and will apply shearing stress until the deposition yields. Rheology of oils is extensively explored. Overall the flow loop design is unique, robust, and provides some of the most extensive information regarding the study of deposition available. Data generated shows measurement techniques within high accuracy of visual measurements. Novel design is also useful for any oil type, and is designed to handle water. Water effects to be explored after complete diagnostics and plain-oil exploration complete.

See more of this Session: Chemistry, Thermodynamics and Multiphase Flow of Complex Fluids Related to Flow Assurance II

See more of this Group/Topical: Topical D: Chemical Engineering in Oil and Gas Production and Other Complex Subsurface Processes

(524c) The Effect of Operating Temperatures and Solubility Curve of Paraffin On Wax Deposition

Zhenyu Huang[1], Shanpeng Han[1], Rainer Hoffmann[2] and H. Scott Fogler[1], (1)Chemical Engineering, University of Michigan, Ann Arbor, MI, (2)Herøya Research Centre Statoil ASA, Porsgrunn, Norway

Wax deposition in the subsea pipelines has been a major flow assurance problem for the petroleum industry. Flow loop experiments with different operating temperatures can be used to simulate the wax deposition at different axial positions of subsea pipelines. In this work, a series of flow loop deposition experiments are carried out based on different inlet temperatures and a model is applied based on the fundamentals of heat and mass transfer. The experimental results show that the wax deposit thickness increases with decreasing inlet temperature, which presents a contradictory trend from a previous study based on model oil [1]. To investigate this discrepancy, theoretical analysis using fundamentals of transport phenomena in the deposition model was carried out on the experiments of both the model oil and the crude oil. According to the deposition model, the temperature gradient at interface and the change in the solubility of wax at interface temperature are the two major elements to affect the growth of the wax deposit. By analyzing these two elements in the two aforementioned experiments, it is found that the discrepancy is due to the difference in the crude oil property and operating conditions. In conclusion, no generalization can be made to estimate on the effect of inlet temperature on wax deposition without considering the change in the temperature gradient of the particular experiment and the slope of the solubility curve of the oil.

[1] Hamid, O. B., Mehrotra, A. K. (2009) Energy & Fuels 23 3184-3194

See more of this Session: Chemistry, Thermodynamics and Multiphase Flow of Complex Fluids Related to Flow Assurance II
See more of this Group/Topical: Topical D: Chemical Engineering in Oil and Gas Production and Other Complex Subsurface Processes

(524d) The Propagation of Pressure and Wall Shear Stress in a Gelled Laboratory Pipeline with Application to the Flow Assurance of Paraffinic Crude Oils

Husam M. El Gendy[1], Mataz Alcoutlabi[1], Mark Jemmett[1], Milind Deo[1], Jules Magda[1], Rama Venkatesan[2] and Alberto Montesi[2], (1)Chemical Engineering, University of Utah, Salt Lake city, UT, (2) Chevron Technology Company, Houston, TX

There are many circumstances that lead to shutting down of oil pipelines for operational or emergency reasons in locations with ambient temperatures below the wax appearance temperature.Under these static conditions, paraffinic waxes in crude oils start to precipitate and form gels. Restarting the gelled pipeline, and eventually clearing the gel, requires the application of a pressure upstream of the gel that gives rise to a shear stress that exceeds the gel yield stress value.Pressure profile measurements and particle imaging velocimetry (PIV) experiments were performed both during gelation and after a sudden application of the restart pressure on a transparent pipeline containing a model waxy oil with measured rheological properties.

See more of this Session: Chemistry, Thermodynamics and Multiphase Flow of Complex Fluids Related to Flow Assurance II
See more of this Group/Topical: Topical D: Chemical Engineering in Oil and Gas Production and Other Complex Subsurface Processes

(577a) Biogas Conversion to Hydrocarbon Fuels Using Bromine Activation and Solid Reactants for Bromine Regeneration

Daniel J. Auerbach[1], Jennifer Blum[1], Jihong Cheng[1], Sagar B. Gadewar[2], Krishna Janmanchi[2], Vivek Julka[2] and Eric W. McFarland[3], (1)GRT Inc., Santa Barbara, CA, (2)GRT, Inc., Santa Barbara, CA, (3) Chemical Engineering, University of California, Santa Barbara, Santa Barbara, CA

GRT has developed a process for the selective partial oxidation of methane to produce liquid fuels and high value commodity chemicals. The process does not utilize a synthesis gas intermediate and is tolerant to common methane contaminants including the high concentrations of carbon dioxide present in biomethane.

Three steps are involved. The first step, "activation", uses bromine to activate a C-H bond in methane producing methylbromide and HBr.

$CH_4 + Br_2 \rightarrow CH_3Br + HBr$ (1)

In the second step, "coupling", the methyl bromide produced in activation is reacted over proprietary zeolite-type catalysts to form higher molecular weight hydrocarbons and HBr. Through careful design of the catalysts and choice of reaction conditions, GRT has shown that the products of the coupling reaction can be controlled to produce gasoline range molecules, aromatics for jet fuels, and a range of valuable chemicals. For "gasoline", an idealized equation for the overall reaction can be written as:

$8\ CH_4 + 8\ Br_2 \rightarrow 8\ CH_3Br + 8\ HBr \rightarrow "C_8H_{16}" + 16\ HBr$ (2)

Here "C_8H_{16}" is used to represent the mixture of actual hydrocarbon products making up gasoline. The third step, regenerates molecular bromine by oxidation of the HBr that is produced as a byproduct in both the activation and coupling reactions (reactions 1 and 2). This step is crucial since the process cannot tolerate loss of bromine. Perhaps the most cost effective small scale process option that GRT has developed for regeneration involves the use of supported metal oxide solid reactants that neutralize HBr and transform the metal oxide into a metal bromide. Bromine is regenerate by contacting the metal bromide with air, transforming the metal bromide back to an oxide.

$"C_8H_{16}" + 16HBr + 8MO \rightarrow "C_8H_{16}" + 8MBr_2 + 8H_2O$ (3)

$8MBr_2 + 4O_2 \rightarrow 8MO + 8Br_2$ (4)

GRT's initial solid reactant process configuration, referred to as the switched bed process, utilizes three or more fixed bed reactors containing the solid reactant operating in parallel so that while one bed is in HBr capture mode, another bed is in purge mode, and a third bed is in regeneration mode. By switching the role of the beds, quasi continuous operation is possible.

This talk will focus on the development of solid reactants that enable the switched bed solid reactant process to operate in an efficient and cost effective manner and on reactor design modeling to use data from small scale laboratory reactors to make design demonstration and commercial scale reactors.

The requirements on the solid reactant materials are quite daunting. 1. The solid reactant must have a large capacity for the capture and regeneration of Br2. The size of the solid reactant fixed beds required is of course inversely proportional to capacity. Practical considerations require a capacity of >2 mmol/g.

2. The capacity must be stable for thousands of capture/regeneration cycles. The large capacity requirement implies the material must be highly dispersed, i.e. present as nano scale particles. Sintering and deactivation with cycling is a serious problem.

3. The rates of capture of HBr and regeneration of Br2 must be high enough to give a sharp reaction front in the fixed bed reactor so that breakthrough of HBr and O2 can be avoided. High rates are also beneficial in that they enable faster cycle times and thus smaller amounts of solid reactant.

4. The material must have minimal or no degradation of the hydrocarbon products

5. The material must be packaged on a support with strength and mechanical integrity over cycles that are suitable for commercial operation.

We have developed synthesis strategies and methods for solid reactant materials that address these requirements. In addition we have developed a testing protocol for these materials to guide the synthesis and optimization of better materials. We initially use testing at atmospheric pressure to get data rapidly on a large number of materials. For the best materials, we then conduct laboratory scale tests at pressures up to 15 atm. to give data suitable for scale-up using kinetic data and reactor design models.

See more of this Session: Alternative Fuels and Enabling Technologies III
See more of this Group/Topical: Fuels and Petrochemicals Division

(577b) DuPont Danisco Cellulosic Ethanol (DDCE) LLC Evaluates Business Opportunities with Sustainability Assessments

Robin Jenkins[1], Carina Alles[1], Christopher Johnas[1], Bruce Vrana[1], Jamie Ginn[1], Susan M. Hennessey[2] and Stuart Thomas[3], (1)DuPont Engineering Research and Technology (DuET), Wilmington, DE, (2) DuPont Central Research and Development, Wilmington, DE, (3)DuPont Danisco Cellulosic Ethanol LLC, Itasca, IL

On January 29, 2010, one of the world's first cellulosic ethanol demonstration facilities commenced production. This DuPont Danisco Cellulosic Ethanol LLC (DDCE) joint venture facility is located in Vonore, TN and will convert corn cobs and dedicated bioenergy crops into ethanol. In 2008, DuPont and Genencor, a division of Danisco, formed the DDCE joint venture, with the mission to accelerate the development of commercial scale biorefineries, create value for the renewable fuels and agricultural industries, and lead the way toward a low-carbon economy (www.ddce.com). DDCE has made significant progress towards this goal and continues to use sustainability assessments as the business moves from the research and development to the commercialization phase.

Life cycle assessment (LCA) as a scientific decision support technique was integrated along side process development early in this biofuel research and development program. Over the years, the environmental benefits and consequences of process choices have been thoroughly evaluated. As DDCE moves towards commercialization, the tool of LCA is applied in new ways. LCA and general sustainability criteria are key considerations for DDCE business planning. By continuing to use LCA, environmental implications of various biorefinery integration and business options are better understood. Through the use of LCA, we see that cellulosic ethanol has the potential to provide a sustainable solution to the nation's growing concerns around energy security and climate change.

See more of this Session: Alternative Fuels and Enabling Technologies III
See more of this Group/Topical: Fuels and Petrochemicals Division

AGRICULTURE | YOUTH & FAMILIES | HEALTH | ECONOMY | ENVIRONMENT | ENERGY | COMMUNITIES

Investigation of ammonia inhibition on oleaginous yeast *Cryptococcus curvatus* growth

Yubin Zheng

Department of Biological System Engineering
Washington State University

WASHINGTON STATE UNIVERSITY EXTENSION

WASHINGTON STATE UNIVERSITY EXTENSION

Background

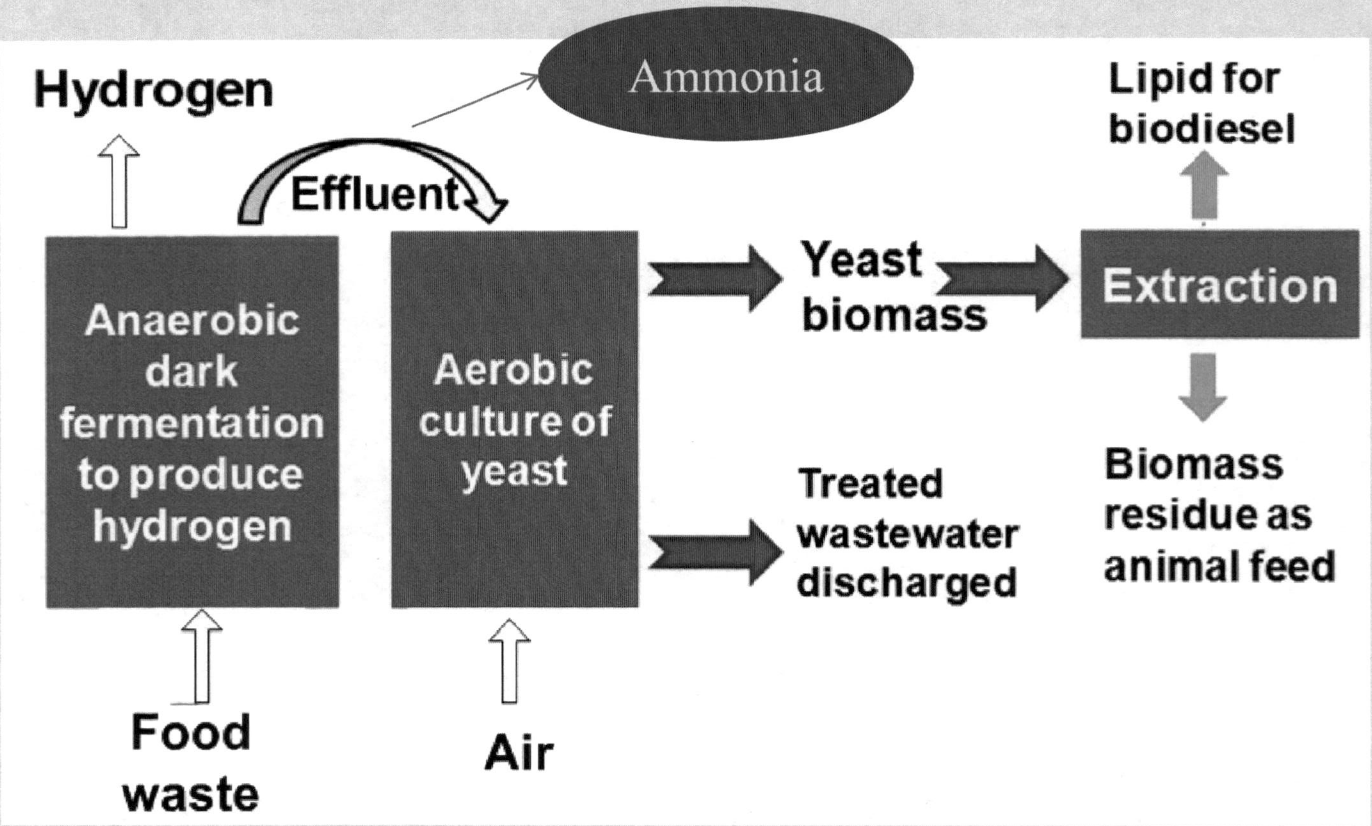

Effluent contains volatile fatty acids (**VFA**): acetic, propionic, butyric acids

Oleaginous yeast: >20% lipid/Dry weight

WASHINGTON STATE UNIVERSITY EXTENSION

Ammonia effects evaluation

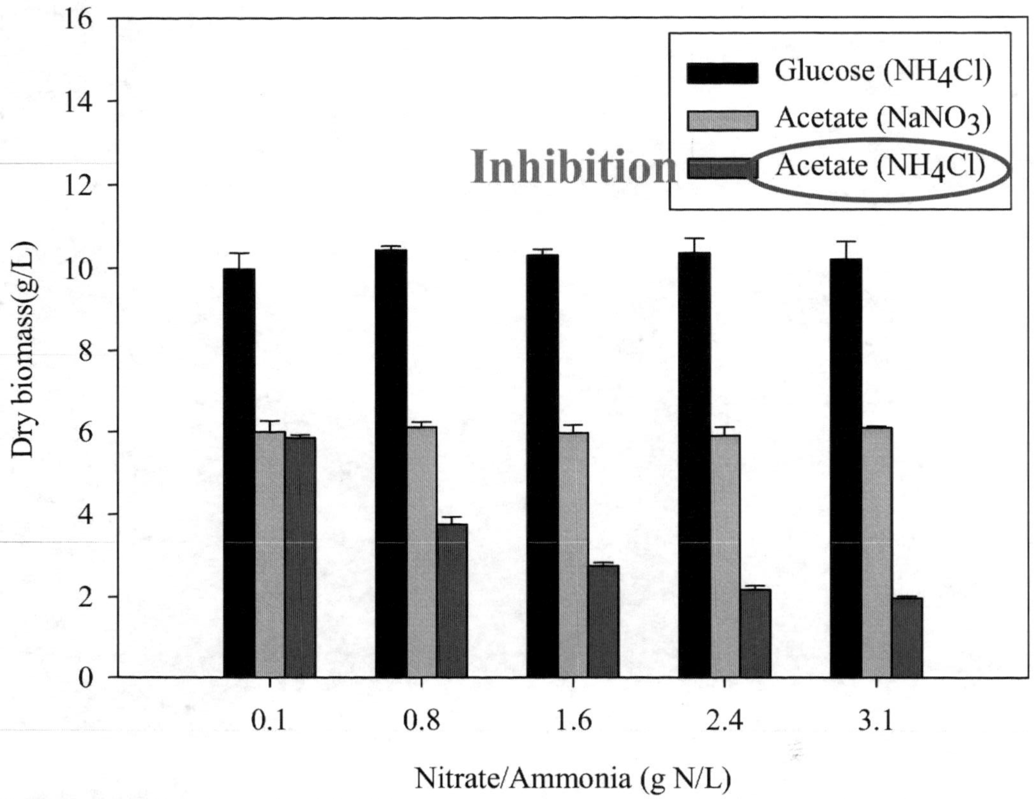

Dry biomass of *C. curvatus* versus different nitrate/ammonia concentrations. Ammonia concentrations ranged from 0.1 to 3.1 g N/L and the nitrate at the corresponding concentrations were used as the control. The culture conditions were temperature 30°C, rotational speed 200 rpm, initial pH 7.3. After 60 hours cultivation, the yeast biomass was harvested.

WASHINGTON STATE UNIVERSITY EXTENSION

Metabolic pathway (Nelson and Cox 2004)

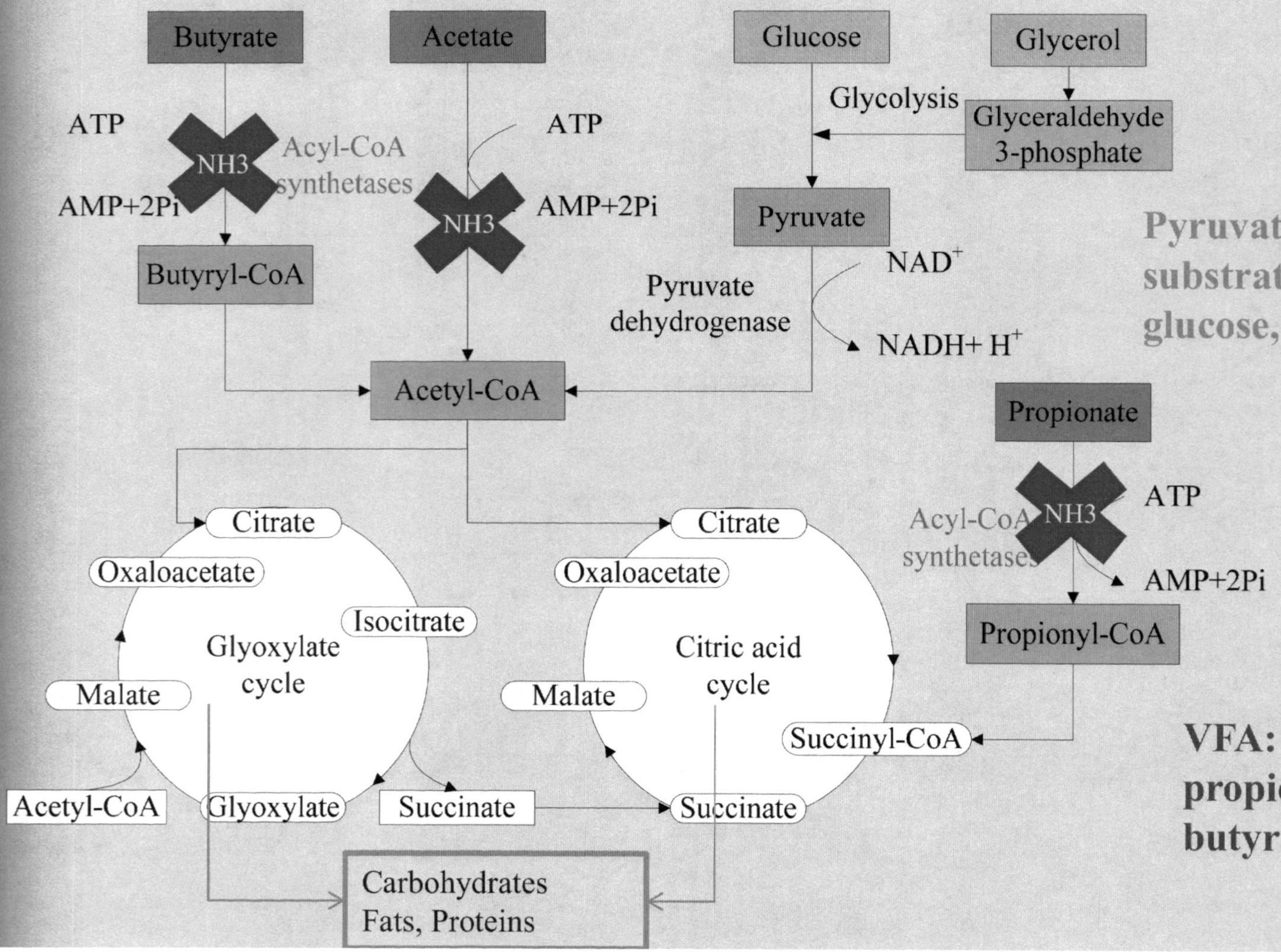

Pyruvate generating substrates (PGS): glucose, glycerol

VFA: acetate, propionate, butyrate

Washington State University Extension

Hypothesis

- ➢Ammonia may inhibit the reaction from VFA to acyl-CoA
- ➢Ammonia may not inhibit the reaction from pyruvate to acetyl-CoA
- ➢Ammonia inhibition is due to the suppression of the acyl-CoA synthetase activity

WASHINGTON STATE UNIVERSITY EXTENSION

Effects of ammonia on different substrates

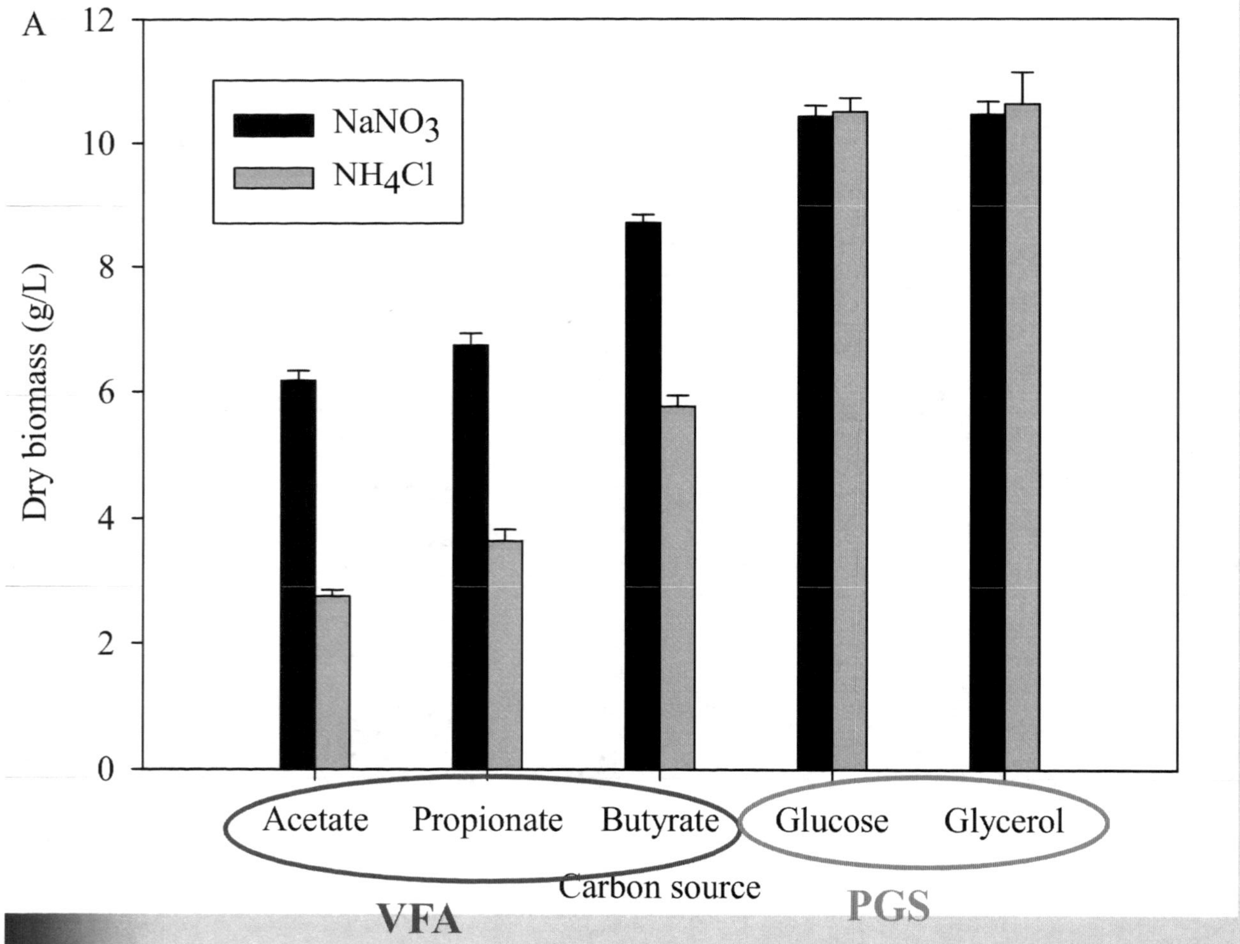

Ammonia effects on dry biomass of *C. curvatus* on different substrates. Ammonia concentration was fixed at 1.6 g N/L and nitrate at the corresponding concentration was used as the control. The culture conditions were temperature 30°C, rotational speed 200 rpm, initial pH 7.3. After 60 hours cultivation, the yeast biomass was harvested.

WASHINGTON STATE UNIVERSITY EXTENSION

Effects of ammonia on lipid production

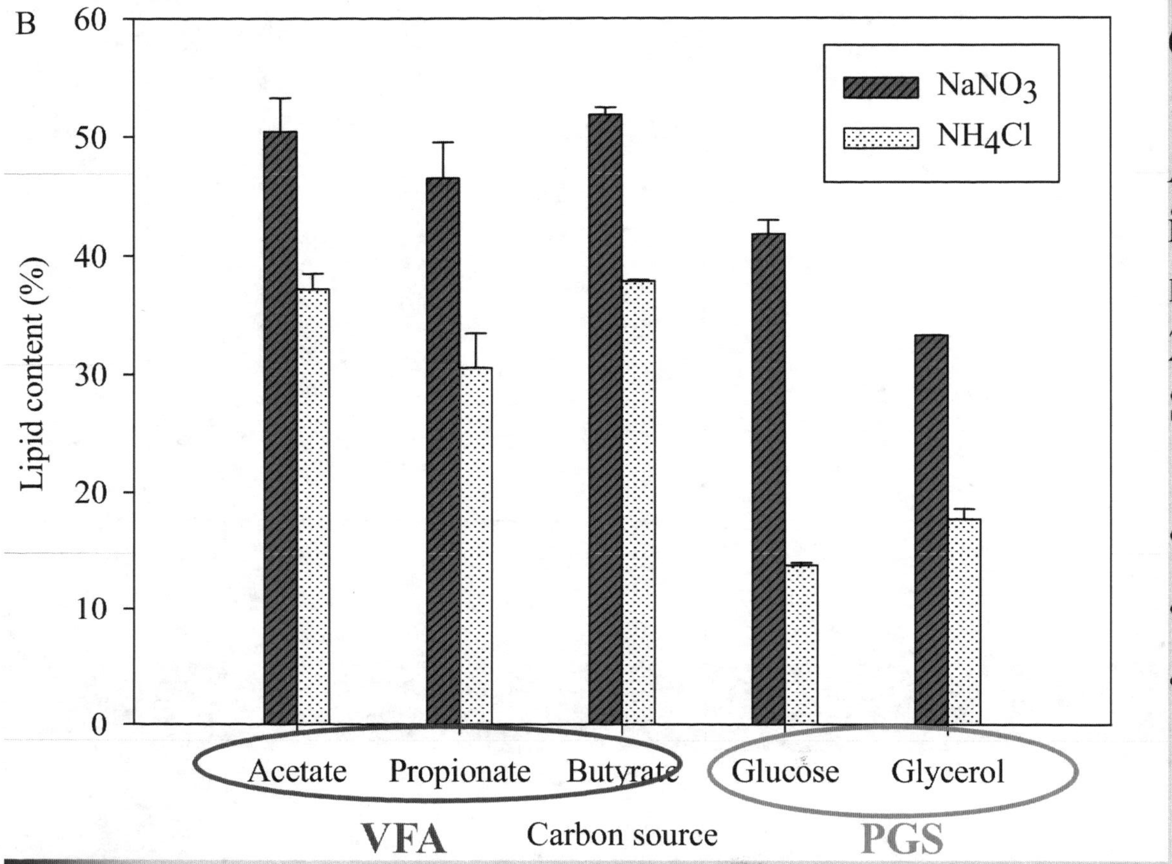

Ammonia may inhibit acyl-CoA synthetase activity

Acyl-CoAsynthetases is important for lipid metabolism (Ichihara et al. 2003; Parkes et al. 2006; Soupene et al. 2010; Waku 1992)

- Triglyceride synthesis
- Acyl chain elongation
- Phospholipid remodeling

WASHINGTON STATE UNIVERSITY EXTENSION

Effects of ammonia on acetyl-CoA synthetase (ACS) (*in vitro*)

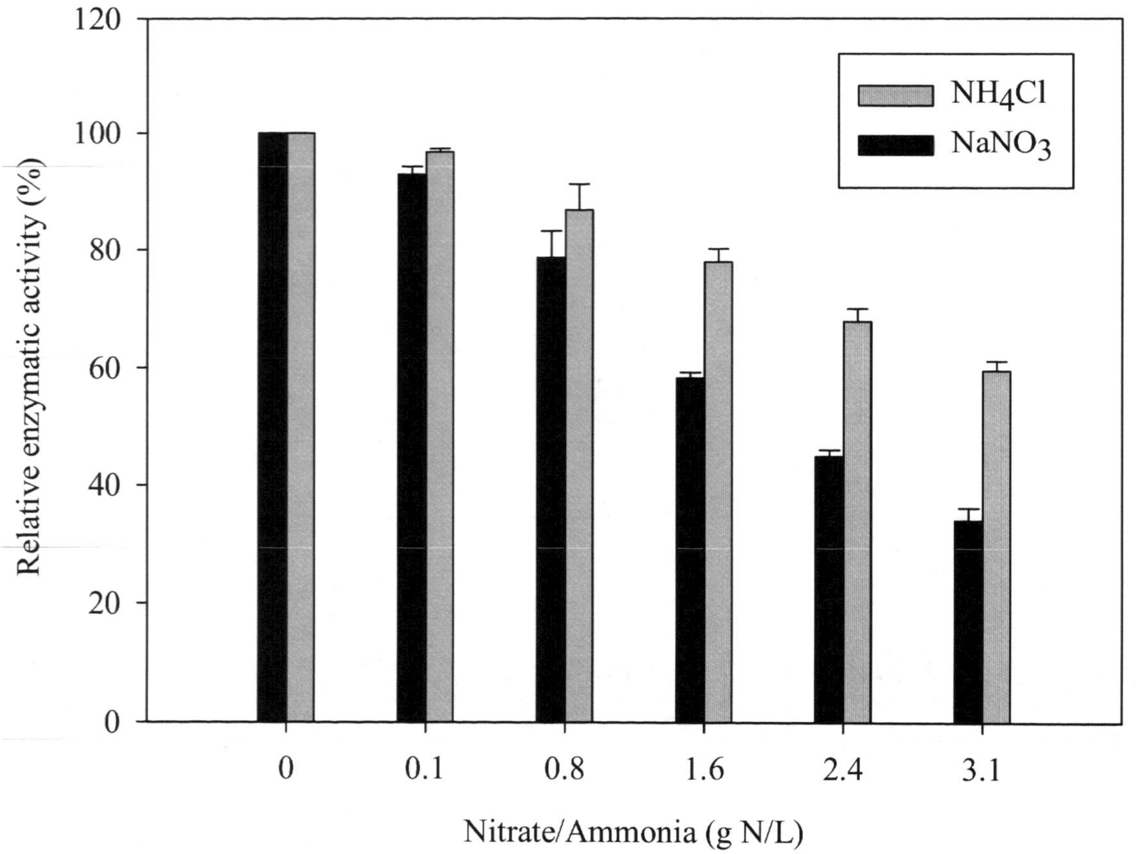

ACS activity versus different nitrate/ammonia concentrations. The enzyme activity was expressed in % of the control. The reaction mixture contained neither ammonia nor nitrate for the control. Ammonia/nitrate concentrations ranged from 0.1 to 3.1 g N/L for the experimental groups.

Effects of nitrate on ACS

Inhibition of nitrate on the ACS was more serious compared with ammonia but nitrate did not affect the yeast growth

- *In vitro* experiment
- Transportation of NH_4^+ and NO_3^- needs specific transporters (Siverio 2002; Kikeri et al. 1989; Knepper et al. 1989; Post and Jolly 1957)
- Intracellular NO_3^- concentration is low
- Free ammonia (NH_3) is a small, uncharged, lipophilic and membrane-permeable molecule (Boron and De 1976; Knepper et al. 1989; Schneider et al. 1996)
- NH_3 can diffuse passively into the cell to make the high intracellular ammonia concentration and cause inhibition on the ACS

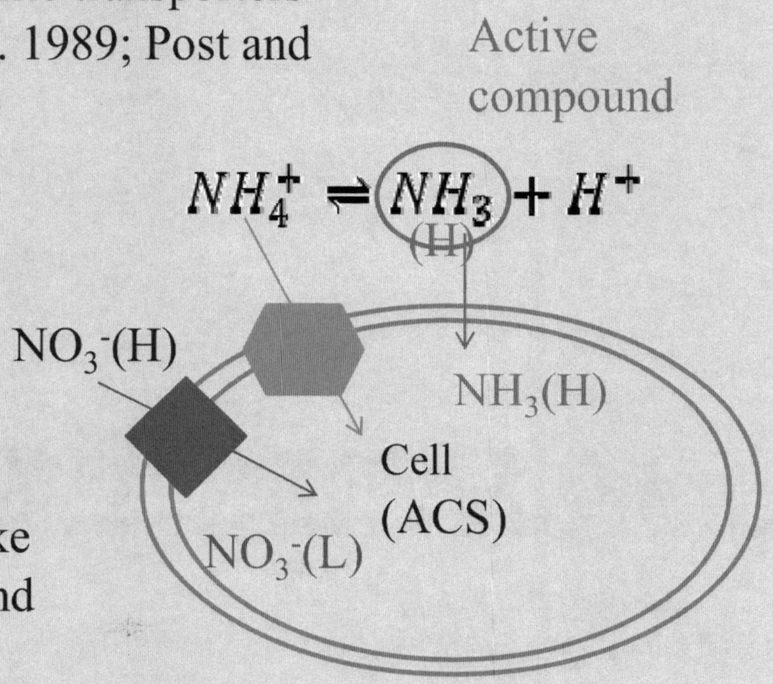

WASHINGTON STATE UNIVERSITY EXTENSION

Ammonia inhibition species investigation

- Two types of ammonia species in aqueous solution:

Ammonia ion (NH_4^+) and free ammonia (NH_3)

- The ratio is pH and temperature dependent (more NH_3 at higher pH at fixed temperature) (Emerson et al. 1975)

$$NH_3\% = 1 + \frac{10^{-pH}}{10^{-(0.09018 + 2729.92/T(K))}}$$

pH	6	7	7.3	8	8.3	9	10
NH3%	0.08	0.80	1.58	7.43	13.81	44.53	88.92

Washington State University Extension

Ammonia inhibition species (acetate)

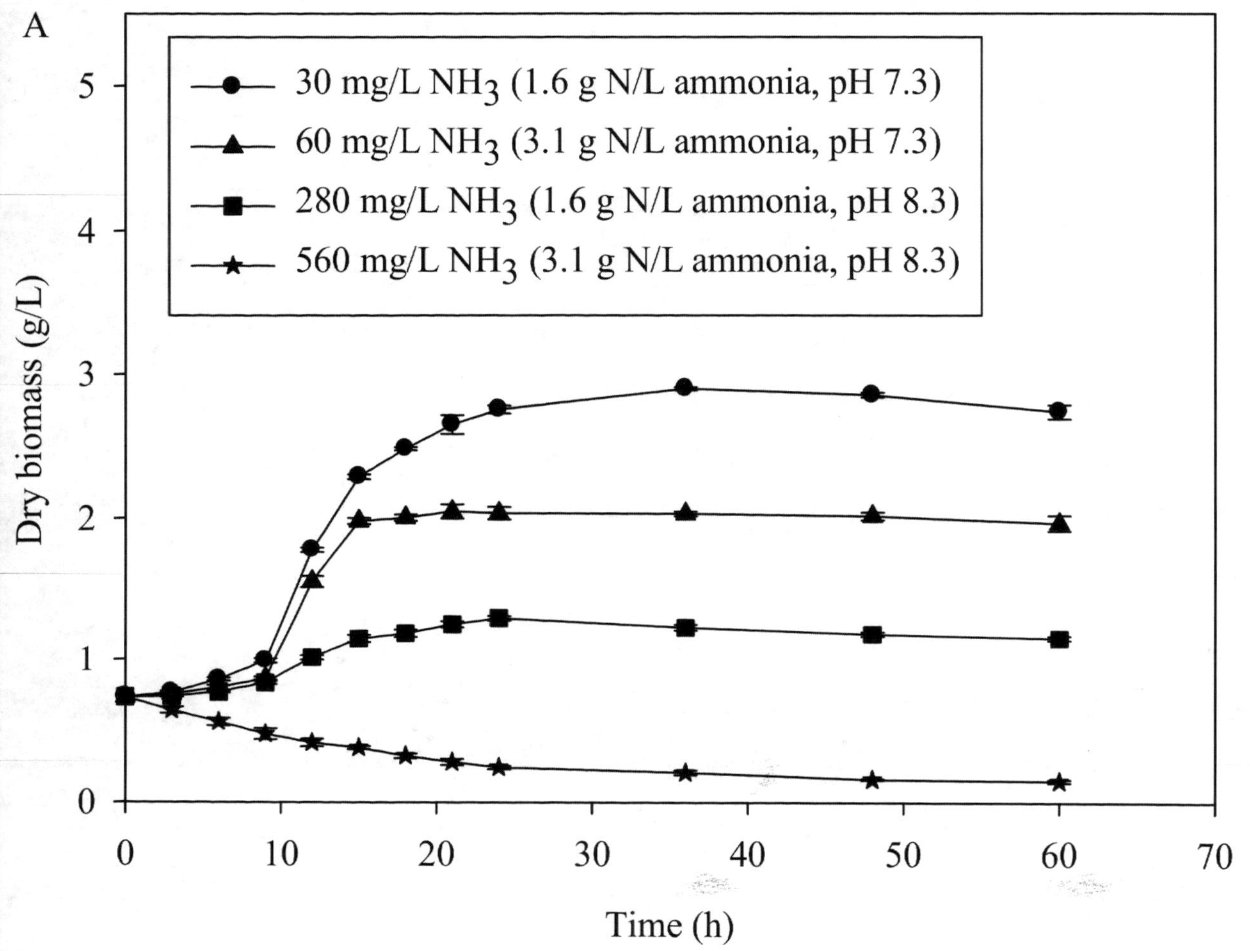

NH_3 is the active compound to cause ammonia inhibition on yeast growth

WASHINGTON STATE UNIVERSITY EXTENSION

Ammonia inhibition species (glucose)

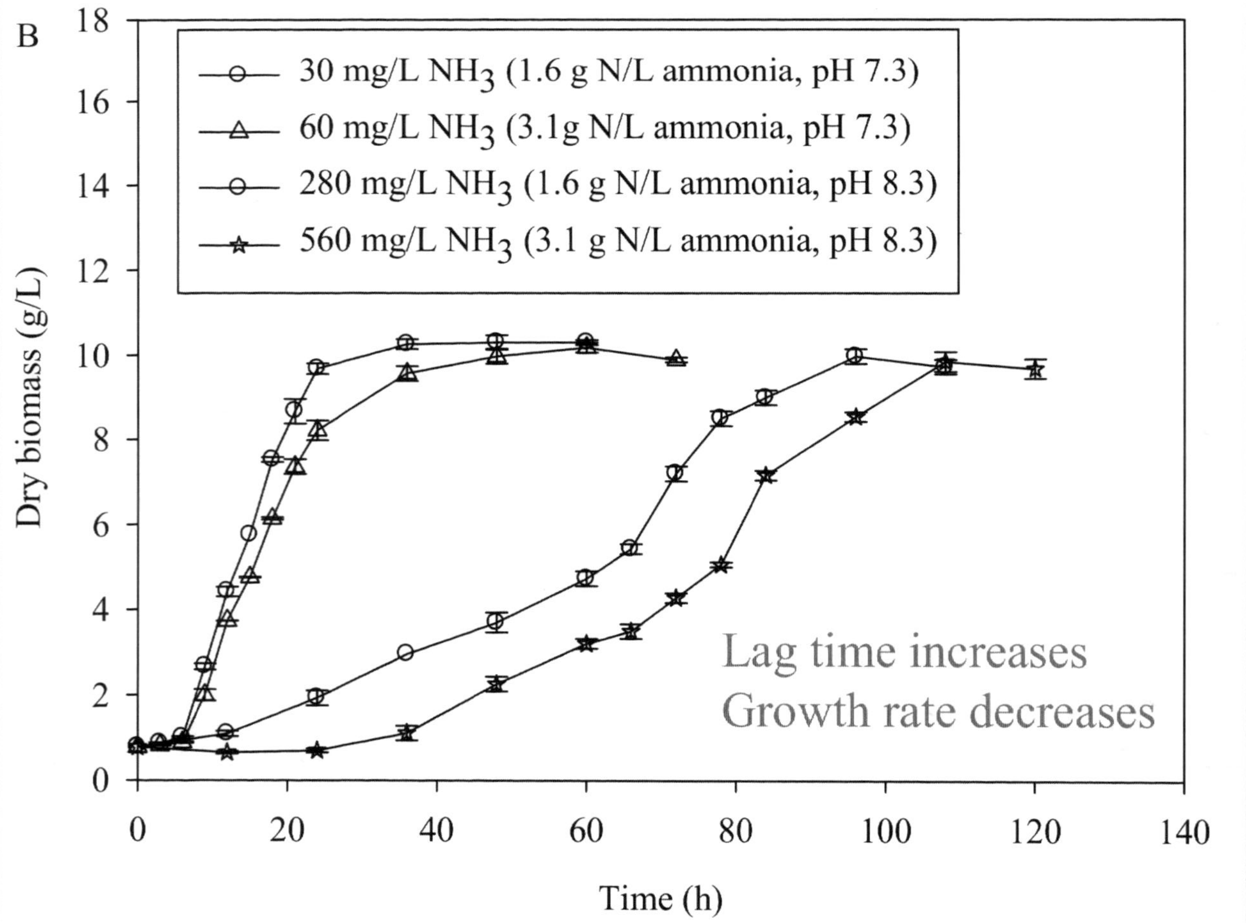

Washington State University Extension

Why?

- NH_3 can diffuse passively into the cell to cause inhibition on the acyl-CoA synthetase
- Acyl-CoA synthetase participates in gene expression (Takahashi et al. 2006)
- Acyl-CoA synthetase is the key enzyme for lipid metabolism and protein acylation acylation (Ichihara et al. 2003; Parkes et al. 2006; Soupene et al. 2010; Waku 1992)
- Gene expression, lipid metabolism and protein acylation are very active during the lag and exponential phase

WASHINGTON STATE UNIVERSITY EXTENSION

Conclusions

- Biomass production was inhibited by ammonia on VFA but not PGS
- Lipid production was inhibited by ammonia on both VFA and PGS
- NH_3 is the active compound to cause inhibition due to its membrane-permeable characteristics and suppression to acyl-CoA synthetases activity
- Inhibition of the acyl-CoA synthetases activity affects cell metabolism and results in the low biomass and lipid production

Washington State University Extension

Significance

- Provide a potential target for genetic modification to overcome this problem
- Help the resolution of ammonia inhibition in the pathway with VFA as a starting substrate (ref.):

•Anaerobic digestion for methane production (Demirel and Yenigun 2002)

•Phototrophic hydrogen production (Barbosa et al. 2001; Fang et al. 2005)

•Electricity production with microbial fuel cell (Liu et al. 2005; Oh and Logan 2005)

WASHINGTON STATE UNIVERSITY
EXTENSION

Acknowledgement

Dr. Shulin Chen

Zhanyou Chi

Jim O'Fallon

BBEL members

(577d) Comparison of Solvents for the Extraction of Algal Oil and Conversion to Biodiesel

Aaron Graham, Stephen Dufreche and Rakesh Bajpai, Chemical Engineering, University of Louisiana at Lafayette, Lafayette, LA

Fossil fuel demands are growing more and more each year and in order to supplement the ever increasing demand we need to develop new sources and types of fuel. One such fuel derived from animal fats or oils is known as Biodiesel. The primary constituents of biodiesel are fatty acid methyl esters. The goal of our research is to determine the viability of obtaining oil from algae cells, using the Accelerated Solvent Extractor (ASE), in order to produce Biofuels such as Biodiesel. The comparison of solvents will be measured against hexane, an industry standard. By comparing the efficiencies of these selected solvents we hope to achieve better extraction yields than pure hexane. Thus we hope to improve the industry standard and allow for a cheaper cost of production. Our research will compare the extraction efficiencies of different solvents and the saponifiable lipid yield obtained from the algae using these solvents. Further, the particle size of the dried algal biomass will be tested to try and determine an optimum particle size for yielding the largest amount of lipids. Another important parameter in the ASE extraction method is the static time. The static time is the amount of time the solvent is held in the cell at the beginning of each extraction. Differing values of time will be tested in order to determine the most efficient time that yields the most lipids.

See more of this Session: Alternative Fuels and Enabling Technologies III
See more of this Group/Topical: Fuels and Petrochemicals Division

(577e) Anaerobic Digestion of Glycerol

Adam Hoffman[1], Kenneth R. Muske[1], Metin Duran[2] and Cheol Park[2], (1)Department of Chemical Engineering, Villanova University, Villanova, PA, (2)Department of Civil and Environmental Engineering, Villanova University, PA

Waste glycerol is a byproduct from biodiesel production and its generation increases year after year. This study investigates the possibility of anaerobic digestion of glycerol waste, which can result in beneficial energy recovery, methane. Glycerol wastes from different biodiesel production processes including acidic and basic condition were tested and their results were compared with methane production from pure glycerol for any adverse impacts. The test has been done in bench scale serum bottle reactors with a total volume of 165mL. The results show that all tested concentrations recovered roughly 80% of theoretical methane production based on COD (Chemical Oxygen Demand) measurement and the sources of glycerol waste did not play an important role in overall results. After the completion of anaerobic digestion study, mathematical models were developed to predict a performance of anaerobic digestion of glycerol waste in a batch reactor. Even though the predictive models developed in the research may be valid only in the studied condition, the mathematical models may be able to yield predictive models for continuous reactors as well.

See more of this Session: Alternative Fuels and Enabling Technologies III
See more of this Group/Topical: Fuels and Petrochemicals Division

(577f) Culture of Microalgae Chlorella Minutissima for Biofuels Feedstock Production

Haiying Tang, Meng Chen, Nadia J. Abunasser, Danton Garcia, K.Y. Simon Ng and Steven O. Salley, Chemical Engineering and Materials Science, Wayne State University, Detroit, MI

Microalgae are among the most promising of non-food based biomass fuel feedstock alternatives. Algal biofuels production is challenged by limited oil content, growth rate, and economical cultivation. In this work, the effect of light source, light intensity, photoperiod, and N starvation on the growth and lipid content of Chlorella Minutissima were studied. Moreover, the fatty acid methyl ester (FAME) content and composition of Chlorella Minutissima was investigated with variable parameters. Red light-emitting diodes (LEDs), white LEDs, and fluorescent lights all were effective for algal growth. Increasing light intensity significantly resulted in more rapid algal growth, while increasing the period of light also significantly increased biomass productivity. N starvation significantly increased lipid content. Moreover, the different light source had effect of FAME composition of Chlorella Minutissima. The methyl palmitate (c16:0), methyl oleate (C18:1), methyl linoleate (C18:2), and methyl linolenate (C18:3) are the major components of Chlorella Minutissima FAME and makes Chlorella Minutissima biomass a suitable feedstock for biofuels production.

See more of this Session: Alternative Fuels and Enabling Technologies III
See more of this Group/Topical: Fuels and Petrochemicals Division

Enthalpy of Vaporization of Ethanol-Gasoline Fuel

Roman M. Balabin, Department of Chemistry and Applied Biosciences (D-CHAB), ETH Zurich, Zurich, Switzerland
Rustem Z. Syunyaev, Gubkin Russian State University of Oil and Gas, Moscow, Russia

Nations today often face divergent challenges in the form of climate change, air pollution, energy production, consumption security, and shrinking oil supplies. In response to these challenges, countries around the world have developed programs to support the use of clean fuels, including ethanol. The properties of gasoline have been altered in recent years to reduce motor vehicle emissions of carbon monoxide, photochemical smog precursors, and toxic organic air pollutants such as benzene. Changes have been made to sulphur, olefin, and aromatic contents, and to distillation properties of gasoline.

Presently, there is an increasing interest in adding oxygenated compounds to gasoline, because of their octane-enhancing and pollution-reducing capabilities. In the last several years, many interesting works on ternary, quaternary, or quinary systems that contain a synthetic reformate (hydrocarbon mixtures), an oxygenated compound (ethers or alcohols), and water, at approximately ambient temperatures, have been published. However, studies of systems that contain gasoline, an oxygenated compound, and water are not often found in the literature.

Among the oxygenated compounds, ethers and alcohols are the most important. Currently, among ethers and alcohols, ethanol has been receiving more attention.

Controversy has surrounded another major fuel change: the addition of methyl tert-butyl ether (MTBE). Presently, there is no requirement that a specific oxygenate be added to gasoline. However, the use of MTBE in gasoline was phased out in California at the end of 2002 due in part to concerns about surface water and groundwater contamination. Likewise, the United States Environment Protection Agency (USEPA) intends to significantly reduce the use of MTBE in gasoline nationwide. These decisions will lead to greater dependence on ethanol–gasoline blends and gasoline formulations that do not contain oxygenated compounds.

One of the major difficulties encountered in the use of alcohol–gasoline blends is their tendency to phase-separate on contact with small amounts of water.

Vapour pressure is one of the most important physical properties of gasoline mixtures because it defines the volatility of the mixtures. Vapour pressure can be determined by a variety of methods that are rather time-consuming. The refinery industry utilizes the Reid methods to determine vapour pressure, which is related to the gasoline performance characteristics and to its storage behaviour.

Evaporative emissions occur by a wide variety of mechanisms, including fuel spillage and vapour displacement during refuelling, venting of fuel tank vapours as ambient temperature changes, fuel evaporation from the engine compartment of parked vehicles due to residual engine heat, liquid leaks in vehicle fuel systems, and so on.

There is concern about increased evaporative emissions of volatile organic compounds (VOCs) when ethanol is blended with gasoline, because such blends tend to have higher Reid vapour pressure (RVP) than equivalent MTBE-blended fuel.

The thermodynamic properties of ethanol–gasoline mixtures have not been widely covered in the literature (with the exception of heat capacity), but these properties are necessary as they serve as basic thermodynamic data for the investigation of clean fuels. Enthalpy of vaporization is the most important thermodynamic constant that describes liquid–gas equilibrium.

The polarity of ethanol and nonpolarity of gasoline hydrocarbons are the reasons for the possible colloid state of ethanol–gasoline mixture. Particle size (the main parameter of any colloid system) was never reported for ethanol-containing fuel.

In this work, vapour pressure measurements of pure gasoline and its mixtures at three different temperatures (20, 40, and 60 °C) are presented. Enthalpy of vaporization is calculated using the Clausius–Clapeyron equation. Partial molar enthalpy of vaporization of ethanol and gasoline is also evaluated. The colloid state of ethanol–gasoline mixtures is also examined by scattered light correlation spectroscopy.

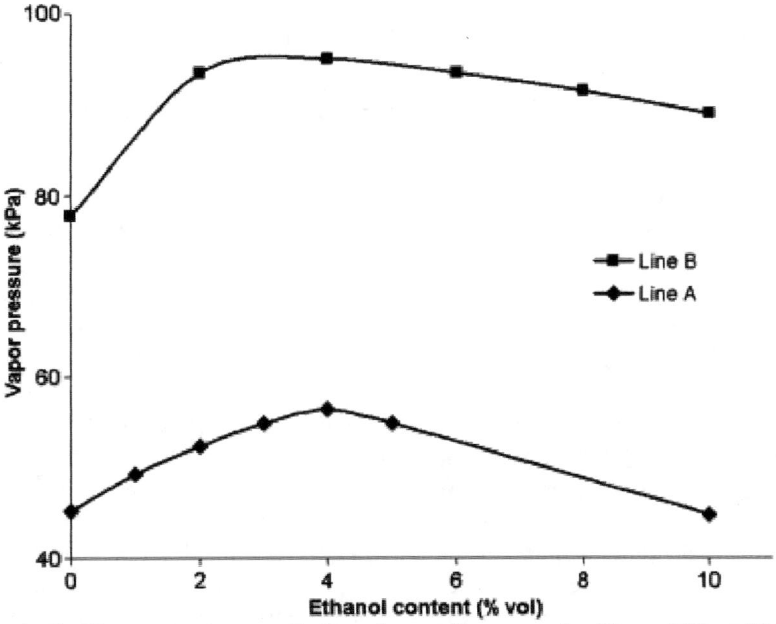

Figure 1. Vapour pressure of ethanol–gasoline samples (A and B) at 20 °C.

One of the most important thermodynamic properties of ethanol–gasoline system – enthalpy of vaporization – was studied. Values of partial molar enthalpy of vaporization were reported for the first time. They are found to be constant. These values can help normalize alcohol–gasoline blended fuel volatility and make this fuel more competitive than ordinary gasoline.

The dispersed structure of ethanol–gasoline mixture was shown. The most important parameter of any colloid structure, i.e., the radius of the dispersed particles, was reported. Its value is hundreds of nanometres and it dramatically depends on the concentration of ethanol. These data can help prevent phase-separation in ethanol–gasoline fuel caused mainly by the coalescence of alcohol–water drops and precipitation.

Changing the colloid structure of ethanol–gasoline systems can be an effective way to put in order the quality coefficients of the fuel.

Acknowledgements. Balabin Roman is grateful to ITERA International Group of companies for a nominal scholarship. The authors acknowledge the Yukos Oil Company for supplying gasoline and the corresponding data.

References

1. R. M. Balabin, R. Z. Syunyaev, S. A. Karpov (2007), "Molar enthalpy of vaporization of ethanol-gasoline mixtures and their colloid state" Fuel 86, 323.
2. R. M. Balabin, R. Z. Safieva (2007), "Capabilities of near infrared spectroscopy for the determination of petroleum macromolecule content in aromatic solutions" J. Near Infrared Spec. 15, 343.
3. R. M. Balabin, R. Z. Syunyaev, S. A. Karpov (2007), "Quantitative measurement of ethanol distribution over fractions of ethanol-gasoline fuel" Energy Fuels 21, 2460.
4. R. M. Balabin, R. Z. Safieva, E. I. Lomakina (2007), "Comparison of linear and nonlinear calibration models based on near infrared (NIR) spectroscopy data for gasoline properties prediction" Chemometr. Intell. Lab. Syst. 88, 183.
5. Z. Nan, Z.-C. Tan, and L. Sun (2004), "Investigation on Thermodynamic Properties of Ethanol + Gasoline Blended Fuel" Energy Fuels 18, 84.
6. N. Peschke, S. I. Sandler (1995) "Liquid-Liquid Equilibria of Fuel Oxygenate + Water + Hydrocarbon Mixtures. 1" J. Chem. Eng. Data 40, 315.
7. S. Hellinger, S. I. Sandler (1995) "Liquid-Liquid Equilibria of Fuel Oxygenate + Water + Hydrocarbon Mixtures. 2" J. Chem. Eng. Data 40, 321.
8. R. M. Balabin (2010), "Is quantum chemical treatment of biopolymers accurate? Intramolecular basis set superposition error" J. Chem. Phys. 132, 231101.
9. R. M. Balabin, R. Z. Safieva, E. I. Lomakina (2010), "Gasoline classification using near infrared (NIR) spectroscopy data: Comparison of multivariate techniques" Anal. Chim. Acta 671, 27.
10. R. M. Balabin (2010), "Intramolecular basis set superposition error as a measure of basis set incompleteness: Can one reach the basis set limit without extrapolation?" J. Chem. Phys. 132, 211103.
11. R. M. Balabin (2010), "The identification of the two missing conformers of gas-phase alanine: a jet-cooled Raman spectroscopy study" Phys. Chem. Chem. Phys. 12, 5980.
12. R. M. Balabin (2010), "Boryl substitution of acetaldehyde makes it an enol: Inconsistency between Gn/CBS and ab initio/DFT Data" J. Phys. Chem. A 114, 3698.
13. R. M. Balabin (2010), "Conformational equilibrium in glycine: Experimental jet-cooled Raman spectrum" J. Phys. Chem. Lett. 1, 20.
14. M. B. Gramajo de Doz, C. M. Bonatti and H. N. Sólimo (2003), "Liquid–liquid equilibria of ternary and quaternary systems with two hydrocarbons, an alcohol, and water at 303.15 K: Systems containing 2,2,4-trimethylpentane, toluene, methanol, and water, or 2,2,4-trimethylpentane, toluene, ethanol, and water" Fluid Phase Equilibria 205, 53.
15. M. B. Gramajo de Doz, C. M. Bonatti and H. N. Sólimo (2003) "(Liquid+liquid) equilibria of ternary and quaternary systems with two hydrocarbons, an alcohol, and water at T=303.15 K. Systems containing cyclohexane, benzene, methanol, and water" The Journal of Chemical Thermodynamics 35, 825.
16. Jasem A. Alkandary, Adel S. Aljimaz, Mohamed S. Fandary and Mohamed A. Fahim (2001) "Liquid–liquid equilibria of water + MTBE + reformate " Fluid Phase Equilibria 187-188, 131.
17. R. M. Balabin (2009), "Tautomeric equilibrium and hydrogen shifts in tetrazole and triazoles: Focal-point analysis and ab initio limit" J. Chem. Phys. 131, 154307.
18. R. M. Balabin (2009), "Conformational equilibrium in glycine: Focal-point analysis and ab initio limit" Chem. Phys. Lett. 479, 195.
19. R. Z. Syunyaev, R. M. Balabin, I. S. Akhatov, J. O. Safieva (2009), "Adsorption of petroleum asphaltenes onto reservoir rock sands studied by near-infrared (NIR) spectroscopy" Energy Fuels 23, 1230.
20. R. M. Balabin, E. I. Lomakina (2009), "Neural network approach to quantum-chemistry data: Accurate prediction of density functional theory energies" J. Chem. Phys. 131, 074104.
21. R. M. Balabin (2009), "Polar (acyclic) isomer of formic acid dimer: Gas-phase Raman spectroscopy study and thermodynamic parameters" J. Phys. Chem. A 113, 4910.
22. R. M. Balabin (2009), "Enthalpy difference between conformations of normal alkanes: Raman spectroscopy study of n-pentane and n-butane" J. Phys. Chem. A 113, 1012.
23. J. Chen, L.-P. Duan, J.-G. Mi, W.-Y. Fei and Z.-C. Li (2000) "Liquid–liquid equilibria of multi-component systems including n-hexane, n-octane, benzene, toluene, xylene and sulfolane at 298.15 K and atmospheric pressure" Fluid Phase Equilibria 173, 109.

24. F. Aiouache and S. Goto "Liquid–liquid equilibria of ternary 2M1B/2M2B–2M1BOH–H2O and quaternary 2M1B–2M2B–2M1BOH–H2O mixtures" Fluid Phase Equilibria187-188, 415.
25. R. M. Balabin (2008), "Enthalpy difference between conformations of normal alkanes: Intramolecular basis set superposition error (BSSE) in the case of n-butane and n-hexane" J. Chem. Phys. 129, 164101.
26. R. M. Balabin (2008), "Intermolecular dispersion interactions of normal alkanes with rare gas atoms: van der Waals complexes of n-pentane with helium, neon, and argon" Chem. Phys. 352, 267.
27. R. M. Balabin, R. Z. Safieva, E. I. Lomakina (2008), "Wavelet neural network (WNN) approach for calibration model building based on gasoline near infrared (NIR) spectra" Chemometr. Intell. Lab. Syst. 93, 58.
28. R. M. Balabin, R. Z. Safieva (2008), "Motor oil classification by base stock and viscosity based on near infrared (NIR) spectroscopy data" Fuel 87, 2745.
29. R. M. Balabin, R. Z. Syunyaev (2008), "Petroleum resins adsorption onto quartz sand: Near infrared (NIR) spectroscopy study" J. Colloid Interf. Sci. 318, 167.
30. R. M. Balabin, R. Z. Safieva (2008), "Gasoline classification by source and type based on near infrared (NIR) spectroscopy data" Fuel 87, 1096.
31. I. Hatzioannidis, E. C. Voutsas, E. Lois, and D. P. Tassios (1998) "Measurement and Prediction of Reid Vapor Pressure of Gasoline in the Presence of Additives" J. Chem. Eng. Data 43, 386.
32. R. A. Harley and S. C. Coulter-Burke (2000) "Relating Liquid Fuel and Headspace Vapor Composition for California Reformulated Gasoline Samples Containing Ethanol" Environ. Sci. Technol. 34, 4088.

(582a) Designing Reforming Catalysts to Condition Biomass Derived Syngas

Kim Magrini[1], Whitney Jablonski[2], Yves Parent[2] and Matthew M. Yung[2], (1)National Renewable Energy Laboratory, Golden, CO, (2)National Bioenergy Center, National Renewable Energy Laboratory, Golden, CO

One approach to the efficient thermochemical conversion of biomass to produce liquid fuels using indirect gasification utilizes a catalytic steam reforming process to remove and reform tars to produce clean syngas which additionally maximizes the use of the biomass carbon and hydrogen content. The catalysts used in the steam reforming reactor can comprise a significant portion of the overall process cost to produce biofuels. In order to reduce the catalyst cost and improve the overall fuel synthesis economics, highly active and stable reforming catalysts are required. To achieve this goal, we are using rational catalyst design to understand, design, and synthesize robust, tailor-made catalysts capable of handling biomass-derived impurities such as tars, polyaromatic hydrocarbons (PAHs), and H2S. Emerging reforming catalysts designed by industrial collaborators are also assessed for reforming performance. Results of catalyst design and reforming performance at laboratory and pilot scales are discussed.

See more of this Session: Catalytic Conversion of Renewable Resources to Synthesis Gases and Pyrolysis Oils
See more of this Group/Topical: Sustainable Engineering Forum

(582b) A DFT Study of Desulfurization and Tar Cracking of Gasifier Effluents Over Ceria-Based Rare-Earth Oxides

Adam D. Mayernick, Department of Chemical Engineering, Pennsylvania State University, University Park, PA, Michael Janik, Department of Chemical Engineering, Pennsylvania State University-University Park, University Park, PA and Kerry M. Dooley, Dept. of Chemical Engineering, Louisiana State University, Baton Rouge, LA

Cleanup of biomass gasifier effluents involves both the removal of sulfur-containing compounds such as H_2S and the cracking of tars. Mixed rare earth oxides of ceria containing lanthanides or transition metal additives are effective H_2S sorbents and tar cracking catalysts. We report density functional theory (DFT +*U*) results used to evaluate the H_2S adsorption capacity and tar cracking activity of pure ceria and ceria doped with a series of lanthanides and group VIIB-VIIIB transition metals. We use ab initio thermodynamics to construct phase diagrams for the surface layers of these rare earth oxides as well their corresponding sulfides and oxysulfides as a function of H_2S and O_2 partial pressures. Our results illustrate the relationship between oxide reducibility and H_2S sorbent capacity, and quantify changes in the surface activity for sulfur adsorption as a result of doping. We also evaluate the effects of transition metal additives on C-H and C-C bond activation, and discuss our results in the context of designing improved rare earth oxide catalysts for simultaneous desulfurization and tar cracking of gasifier effluents.

See more of this Session: Catalytic Conversion of Renewable Resources to Synthesis Gases and Pyrolysis Oils
See more of this Group/Topical: Sustainable Engineering Forum

(582c) Development of a Catalytic Reactor for Conditioning of Biomass Generated Producer Gas and Evaluation of Selected Commercial Catalysts

Ajay Kumar, Hasan K. Atiyeh, Danielle Bellmer and Raymond L. Huhnke, Biosystems and Agricultural Engineering, Oklahoma State University, Stillwater, OK

Gasification of biomass is one of the promising conversion technologies to produce fuels, chemicals and power from renewable resources. Removal of tar is one of the most challenging barriers for commercial-scale use of producer gas generated from gasification of biomass. Tar can be reduced to an acceptable level by cracking in presence of catalysts at high-temperature. This method can also increase carbon monoxide and hydrogen compositions which are low in biomass generated producer gas. In this project, we have designed and developed a catalytic reactor specifically for evaluating commercially-available catalysts for conditioning producer gas which is supplied from a lab-scale gasifier with a biomass feedrate of 3-7 kg/h. In our prior study, selected commercially-available catalysts were evaluated in decomposing toluene as a model tar compound. In this project, the catalysts and reaction conditions will be further evaluated using producer gas with tar generated by gasification of switchgrass as the feedstock. Results are expected to show the effects on selected catalysts, shape of catalysts, and reactor temperature on reducing tar and improving producer gas composition.

See more of this Session: Catalytic Conversion of Renewable Resources to Synthesis Gases and Pyrolysis Oils
See more of this Group/Topical: Sustainable Engineering Forum

(582d) Comparison and Characterization of Bimetallic, Ni-Based Catalysts for Conditioning of Biomass-Derived Syngas

Matthew M. Yung[1], Kimberly A. Magrini-Bair[2], Whitney S. Jablonski[2] and Jessica Olstad[2], (1) National Bioenergy Center, National Renewable Energy Laboratory, Golden, CO, (2)National Renewable Energy Laboratory, Golden, CO

Due to the decreasing availability of low-cost petroleum, efforts to develop technology for transportation fuel production from alternative and renewable sources are being actively researched. Biomass is one resource that has the potential for significant fuel production based on feedstock availability. Thermochemical conversion of biomass to syngas has the potential to be used as a cost-effective and renewable source of transportation fuels, displacing petroleum consumption for this purpose. In order to efficiently utilize the carbon in the biomass and protect downstream processing equipment, the light hydrocarbons and tars in the gas stream following gasification should be converted into additional synthesis gas through steam reforming. To this end, we have designed and evaluated several Ni-based tar reforming catalysts for the conditioning of biomass derived syngas. Catalysts of varying composition are evaluated through reaction experiments for hydrocarbon steam reforming in the presence of hydrogen sulfide, using a model syngas that is representative of the composition that would be created through biomass gasification. Catalyst characterization through several techniques (including XRD, TPR, SEM/EDS, and laser Raman spectroscopy) are also discussed.

See more of this Session: Catalytic Conversion of Renewable Resources to Synthesis Gases and Pyrolysis Oils
See more of this Group/Topical: Sustainable Engineering Forum

(582e) Autothermal Oxidative Pyrolysis of Alternative Hydrocarbon Feedstocks

Christine M. Balonek, Joshua L. Colby and Lanny D. Schmidt, Department of Chemical Engineering and Materials Science, University of Minnesota, Minneapolis, MN

Traditional pyrolysis units are heated externally, requiring large energy inputs, and produce gases, solid chars, and complex liquid mixtures. We have developed and demonstrated an oxidative pyrolysis reactor able to run autothermally with negligible solid char formation. The heat necessary for pyrolysis is generated internally, eliminating the heat transfer issues associated with external heating. This process can produce both synthesis gas and hydrocarbon products from alternative feedstock models of biomass and plastic wastes. Solid cellulose and polyethylene (PE) particles were fed individually with air to a Rh-Ce/γ-Al_2O_3 catalytic fixed-bed reactor. Catalyst temperature and product species were studied as a function of the inlet C/O ratio, the ratio of carbon moles in the solid feed to oxygen moles in the O_2 feed. At all C/O ratios reported, the solid feed was completely converted with negligible coke formation at millisecond time scales. Cellulose and PE pyrolysis produced large amounts of synthesis gas at low C/O ratios (0.5 to 1.0). Cellulose pyrolysis produced up to 20 % carbon selectivity to C_{3+} hydrocarbons at C/O of 1.3 (autothermal operation was unsustainable at higher C/O ratios). At similar conditions, PE pyrolysis had 50 % selectivity to C_{3+}, and selectivity continued to increase with increasing C/O ratio to 90 % at C/O = 3.4. This high initial yield of C_{3+} hydrocarbons from cellulose and PE pyrolysis demonstrates the potential for autothermal oxidative pyrolysis to utilize alternative hydrocarbon feedstocks in a fast, clean, and efficient way.

See more of this Session: Catalytic Conversion of Renewable Resources to Synthesis Gases and Pyrolysis Oils
See more of this Group/Topical: Sustainable Engineering Forum

Production of Bio-oil and Hydrogen from Sawdust via Pyrolysis and Catalytic Steam Reforming

Ebrahim Salehi, Jalal Abedi, Tom Harding, and Fakhry Seyedeyn Azad
Chemical and Petroleum Engineering, University of Calgary, Calgary, AB T2N 1N4, Canada
E-mail: jabedi@ucalgary.ca; Tel: 403-220-5594

ABSTRACT

Fixed-bed and fluidized-bed lab-scale reactors were developed to produce bio-oil using pyrolysis of biomass including sawdust and agricultural wastes such as wheat straw. The role of pyrolysis temperature, nitrogen gas flow rate, and condensation temperature on product yield was investigated. The chemical characteristics of the pyrolysis bio-oil and char obtained under the optimal operational conditions were identified by elemental and calorific analyses. The bio-oil was analyzed using chromatographic and spectroscopic techniques. Pyrolysis temperature had the most significant effect on the bio-oil yield. In addition, an experimental system was designed and constructed to make hydrogen from produced bio-oil in a catalytic steam reforming micro-reactor. Gas chromatographic technique was used to determine the concentration of hydrogen and the other components in produced flue gas from the catalytic micro-reactor. The effect on the hydrogen yield of Ni-based catalysts with different Ni loading was investigated.

Keywords: sawdust; fast pyrolysis; hydrogen; bio-oil, catalyst

1. Introduction

Concerns related to the depletion of fossil fuel reserves as well as environmental issues like global warming make hydrogen, in combination with fuel cells, an attractive alternative energy source. The conventional method of hydrogen production is steam reforming of fossil fuels which results in emission of green-house gasses. Renewable sources, such as biomass, may be used as alternative feedstocks for hydrogen production. Biomass for producing energy has a special place among all other renewable energy sources, and it is estimated to contribute 10-14% to the world's total energy supply [1-7].

To produce hydrogen from biomass, two main technologies are steam gasification and catalytic steam reforming of pyrolysis oil (bio-oil) [8]. The second method has an advantage over steam gasification. In this technique, the first step is to apply fast pyrolysis to bulky and low density biomass and thus convert it into an easy-to-transport intermediate called bio-oil which has a high volumetric energy density. The next step is to steam reform the produced bio-oil into a combustible gas mixture including hydrogen. In addition, bio-oil can be used as a feedstock in conventional petroleum refineries, as well as for the production of high value chemicals [5].

Pyrolysis of biomass is a process in which biomass is subjected to heat in an oxygen-free environment resulting in its conversion to char, liquid, and noncondensable gases. The steam reforming of the bio-oil can be simplified as the catalytic steam reforming of the oxygenated organic compounds (CnHmOk) [9-11], followed by water–gas shift reaction (WGS):

$$C_nH_mO_k + (n-k)H_2O \leftrightarrow nCO + (n+m/2-k)H_2$$

$$nCO + nH_2O \leftrightarrow nCO_2 + nH_2$$

Therefore, the overall process can be represented as follows:

$$C_nH_mO_k + (2n-k)\ H2O \leftrightarrow nCO_2 + (2n+m/2-k)H_2$$

This study aimed to develop first a fixed-bed, and then a fluidized-bed pyrolysis system to produce bio-oil from sawdust. Bio-oil was then steam reformed in another developed experimental apparatus using various Ni-based catalysts made in-house.

2. Experimental

2.1. Feedstock

The feedstock used for pyrolysis was sawdust provided by a waste wood recycling company. The sawdust was sieved into fractions of particle diameter < 590 μm. Before the experiments, the sample to be used was dried overnight at 100°C in an oven. A commercial bio-oil supplied by BTG Company, The

Netherlands, was used in catalytic steam reforming tests. Table 1 shows the elemental analyses and the calorific value of the biomass and bio-oil samples.

2.2. Fixed-bed pyrolysis

The fixed-bed pyrolysis system developed consisted of a stainless steel tubular reactor (1.9 cm I.D., 2.2 cm O.D., and 50 cm length) which was placed in a tubular electrical furnace. In a typical experiment, a sample of 10 g of sawdust was placed in the reactor and nitrogen was used as the carrier gas. After heating the reactor, the biomass particles decomposed to char, non-condensable gas and volatile materials. The gas and the volatile materials passed through two sequential condensers maintained at -10°C by a bath circulator. The schematic diagram of the fixed-bed pyrolysis system may be found in another publication [1].

Table 1. Characteristics of feedstocks

Properties	Bio-oil	Sawdust
Elemental analysis (wt%)[a]		
Carbon	45.83	47.37
Hydrogen	7.15	6.03
Nitrogen	0.87	2.40
Oxygen[b]	46.15	44.2
Heating value[c] (MJ/kg)	17.50	19.05
Water content wt (%)	23.10	-
Density (kg/m3)	1255.00	-

[a] dry basis
[b] by difference
[c] using bomb calorimeter

As discussed in our previous work [1], it was recommended that future tests be performed under fast pyrolysis conditions in a fluidized-bed system. Changing the pyrolyzer from the fixed-bed to a fluidized-bed system allows all feedstock particles to be heated faster and more homogenously which results in higher bio-oil yield and different quality.

2.3. Fluidized-bed pyrolysis

Figure 1 shows a picture of the developed fluidized-bed fast pyrolysis system. This system was designed to operate at temperature between 400-500°C, atmospheric pressure, and with a 100 g/hr biomass feed rate. The system comprised mass flow controllers to control and measure fluidizing gas (nitrogen), fluidized-bed reactor and plenum, biomass feed system, cyclone char collection system for char separation from gas stream, vapor condensation system for bio-oil recovery, instrumentation for data acquisition and control, and heating system to provide heat for the system.

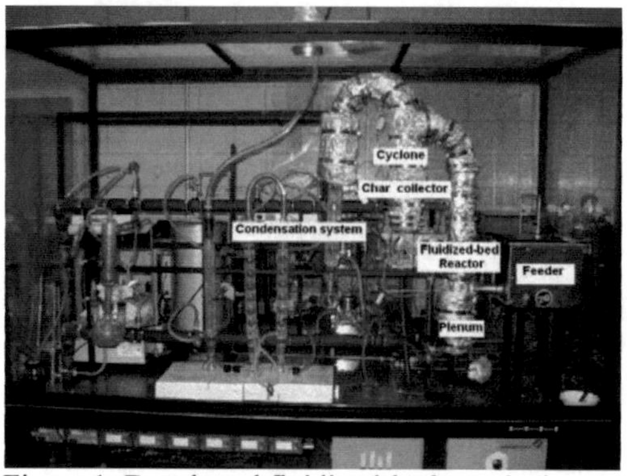

Figure 1. Developed fluidized-bed pyrolysis system

The reactor and the cyclone were sized to provide residence time of less than 1 second for gas phase before reaching the condenser. At the end, the product gas is cooled in a series of water-glycol glass condensers to recover bio-oil from gas steam.

2.4. Catalytic steam reforming

Because at the point of starting steam reforming experiments there was not enough produced bio-oil from the pyrolysis system, a commercial bio-oil was used. Figure 2 shows a picture of the developed catalytic steam reforming system. The heart of the system is the catalytic micro-reactor which is a stainless steel fixed-bed rector. Two syringe pumps, one for water injection and the other one for bio-oil injection, provide feeds to the reactor on top of the catalyst bed. Nitrogen is used as the carrier gas. Before running an experiment, catalyst is reduced in the medium of nitrogen and hydrogen (50:50 vol%) at the temperature of 850 °C.

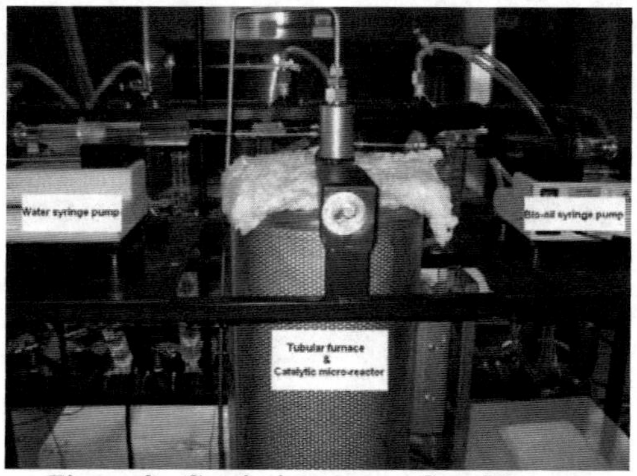

Figure 2. Catalytic steam reforming system

Ni-based catalysts which are very common for the steam reforming of natural gas were prepared in our lab in four different Ni loading (5.6%, 10.7%, 14.1%, and 18%) on alumina. Down stream of the catalytic steam reforming reactor, the product gases, after passing a glass condenser to recover excess steam, were analyzed by an on-line micro-GC capable of detecting H_2, N_2, CO, CO_2, CH_4, C_2H_6, C_3H_8, and C_4H_{10}.

3. Results and Discussion

In fixed-bed pyrolysis experiments, three operational parameters including reaction temperature, condensation temperature, and nitrogen gas flow rate were selected to investigate their effects on pyrolysis product yields. Maximum bio-oil yield of 45% could be achieved. GC-MS analysis was performed allowing detection of ten components in the produced bio-oil. Results indicated that bio-oil is a complex mixture of organic compounds containing a wide variety of aromatics and oxygenated compounds, such as carboxylic acids, phenols, and aldehydes [1].

Regarding the fluidized-bed pyrolysis system, more than 30 experiments have been done to date. Maximum bio-oil yield of ≈50% has been achieved. The char and gas yields were15% and 35%, respectively. During these initial tests challenges involving the char separation system and bio-oil recovery system have been gradually overcome. Now the main objective is to determine the optimal operational conditions to obtain the highest bio-oil yield and to investigate the effects of operational parameters on the bio-oil quality.

Catalytic steam reforming test results have shown yield of 73% for hydrogen on 14% Ni/Alumina. Compared to the rest of the Ni-based catalysts, the 14% Ni loading provided higher hydrogen yield while the temperature was set at 950 °C and steam to carbon ratio was 5. The main problem regarding catalytic steam reforming of bio-oil is coke deposition on the catalyst. In the next step, the plan is to investigate the effect of temperature and steam to carbon ratio on the hydrogen yield, and also to prepare noble metal catalysts for the bio-oil steam reforming to improve the hydrogen yield.

4. Conclusion

In this study, fast pyrolysis experiments of sawdust were carried out in fixed-bed and fluidized-bed pyrolysis systems. Bio-oil yields of 45% were achieved in the fixed-bed system and a 5% increase in yield to 50% was obtained in the fluidized-bed system. For the pyrolysis phase, the next step is to investigate the effect of operational parameters like biomass particle size, pyrolysis temperature, and vapor residence time on bio-oil yield and quality. Results of catalytic steam reforming show the hydrogen yield of 73%. Improving Ni-based catalysts by adding other metals like Mg, and preparing noble metal catalysts have potential to help prevent or at least decrease coke deposition.

Acknowledgement

Financial support from the Natural Sciences and Engineering Research Council of Canada (NSERC), the Agricultural Funding Consortium of Alberta, Alberta Innovates Bio Solutions, and the Institute for Sustainable Energy, Environment and Economy (ISEEE) at the University of Calgary are gratefully acknowledged.

References

[1] E. Salehi, J. Abedi, T. Harding, Energy & Fuels, 23 (2009) 3767–3772.
[2] N. Ozbay, A. E. Putun, E. Putun, Int. J. Energy Res. 30 (2006) 501.
[3] A. E. Putun, B.B. Uzun, E. Apaydin, E. Putun, Fuel Process. Techno. 87 (2005) 25.
[4] S.Sensoz, D.Angın, Bioresour. Technol. 99 (2008) 5492.
[5] H.J. Park, J.I. Dong, J.K. Jeon, Y.K. Park, K.S. Yoo, S.S. Kim, J. Kim, S. Kim, Chem. Eng. J., 143 (2008) 124–132.
[6] G. Chen, J. Andries, H. Spliethoff, Energ. Convers. Manage. 44 (2003) 2289.
[7] M. Krzesinska, M.B. Pilawa, S.Pusz, Biom. Bioenerg. 30 (2006) 166.
[8] A.C. Basagiannis, X.E. Verykios, Inter. J. Hydrogen Ener. 32 (2007) 3343 – 3355.
[9] A. C. Basagiannis, X. E. Verykios, Catalysis Today 127 (2007) 256–264.
[10] Z. Wang, Y. Pan, T. Dong, X. Zhu, T. Kan, App. Catal. A: General 320 (2007) 24–34.
[11] C. Rioche, S. Kulkarni, F. C. Meunier, J. P. Breen, R. Burch, App. Catal. B: Environmental 61 (2005) 130–139.

(583a) Performance Enhancement of Low Temperature Water Gas Shift Catalysts by Palladium-Iron and Platinum-Iron Interactions

Jorge Pazmino[1], Luis Bollmann[1], Vincent Kispersky[1], Damion Williams[1], Mayank Shekhar[1], Jeffrey T. Miller[2], Jeffrey W. Elam[3], Anil Mane[3], W. Nicholas Delgass[1] and Fabio Ribeiro[1], (1)Chemical Engineering, Purdue University, West Lafayette, IN, (2)Chemical Technology Division, Argonne National Laboratory, Argonne, IL, (3)Energy Systems Division, Argonne National Laboratory, Argonne, IL

We report on the effect of iron addition to Pd and Pt on three supports (Al_2O_3, ZrO_2 and CeO_2) as catalysts for the water gas shift reaction. Compared to a 2.3%Pd/Al_2O_3 catalyst, Pd/Fe samples with similar Pd loadings showed a promotion of the WGS rate per mole of Pd in the range of 35 to 235 times. On Pt/Al_2O_3 catalysts an increase in rate per mole of Pt (up to 25 times) was measured with the increase in Fe wt% loading from 0.6 to 11%. Interestingly, the addition of Fe to Pd supported on Al_2O_3, CeO_2 and ZrO_2 or to Pt supported on Al_2O_3 and CeO_2 erases the support and metal effects, i.e. although Pt has a turnover rate about 10 times higher than Pd on the three supports, the addition of Fe makes the rate similar on Pt and Pd catalysts on the three supports.

Extended X-ray absorption fine structure (EXAFS) and in situ diffuse reflectance infrared spectroscopy data showed that Pd and Pt particles were partially covered by a shell of Fe. In addition, for Al_2O_3 supported catalysts with Fe wt% loading between 1.8 and 13.5, the rate per gram of catalyst correlated with the amount of hydrogen consumed by titration of surface oxygen after the sample was exposed to O_2 at 100°C. These results suggest that reducible Fe species are active sites. X-ray photoelectron spectroscopy (XPS) data measured after reaction showed that some of the Fe is reduced to the zero valent form. We also found that the amount of Fe^0 correlates with the rate per gram. In line with these results, we propose that with the addition of iron the reaction takes place over new redox sites, mainly reduced Fe on top of the noble metal. As no Fe-M (M = Pd, Pt) alloy formation was observed by EXAFS, we believe that the role of Pd or Pt is the initial reduction of nearby Fe by H_2 spillover, and then the stabilization of a surface layer of reducible Fe. The promotion effect is therefore explained by the creation of more active redox sites as compared to those on Pd or Pt on Al_2O_3. This hypothesis is consistent with our kinetic observations on Pd and Pt on reducible supports like CeO_2 and ZrO_2, where the addition of iron showed minimal promotion mainly because these materials already provide enough redox sites for the reaction.

(583b) Determination of the Active Site for Nanoparticle Gold Water-Gas Shift Catalysts Using a Model Support

W. Damion Williams[1], Mayank Shekhar[1], Wen-Sheng Lee[1], Vincent Kispersky[1], Jun Wang[1], W. Nicholas Delgass[1], Fabio H. Ribeiro[1], Seung Min Kim[2], Eric A. Stach[2], Jeffrey T. Miller[3] and Lawrence F. Allard[4], (1)School of Chemical Engineering, Purdue University, West Lafayette, IN, (2) School of Materials Engineering, Purdue University, West Lafayette, IN, (3)Chemical Technology Division, Argonne National Laboratory, Argonne, IL, (4)Materials Science & Technology Division, Oak Ridge National Laboratory, Oak Ridge, TN

The rate per mole of gold for the water-gas shift (WGS) reaction was measured for a series of gold catalysts supported on model rutile TiO_2. The use of a model support provides an enhanced ability in resolving supported gold nanoclusters using Transmission Electron Microscopy (TEM). A physical model of the gold clusters as truncated cubo-octahedra was used to determine the fraction of atoms in a gold cluster that have a specific number of gold neighbors. The rate of each sample was then obtained by assuming a constant rate per active site on each cluster imaged by TEM and obtaining a value for the constant rate to best fit the experimental data. Results indicate that the number of active sites on each cluster is constant irrespective of the gold nanocluster size in the range of samples studied. DRIFTS and transmission FTIR experiments were used to show how different surface intermediates are related to the WGS rate under in situ conditions and during isotope switching experiments. The rate on a Au/TiO_2 catalyst with a 1.2 nm average particle size ($1.4x10^{-1}$ s^{-1}) is two orders of magnitude higher than the rate on an industrial $Cu/ZnO/Al_2O_3$ catalyst ($5x10^{-4}s^{-1}$) (moles CO reacted per surface mole of Au or Cu at 120°C, 6.8% CO, 21.9% H_2O, 8.5% CO_2, 37.4% H_2, and balance Ar) further emphasizing the potential practical and scientific interest of nanogold WGS catalysts.

See more of this Session: Catalytic Hydrogen Generation - General I
See more of this Group/Topical: Catalysis and Reaction Engineering Division

(583c) Deactivation of Gold-Ferrochrome Very Low Temperature Water-Gas Shift Catalysts

Carl R. F. Lund, Chemical and Biological Engineering, University at Buffalo, SUNY, Buffalo, NY and **Gaurav N. Vajani**, Dept. of Chem. & Bio. Engr., University at Buffalo, SUNY, Buffalo, NY

Gold-ferrochrome catalysts are active for water-gas shift at temperatures below 200 degrees Celsius, but they deactivate rapidly. The deactivation can be attributed primarily to two factors: sintering of the gold nanoparticles present in the catalyst, and deposition of carbon on the surface of the catalyst. High resolution transmission electron microscopy and X-ray diffraction characterization of fresh and used catalysts provide evidence for the former. The effect of reductive pretreatments using carbon-containing gases and carbon-free gases and in-situ gravimetric studies under reaction conditions provide evidence for carbon deposition. Additional characterization using BET, SEM and NEXAFS shows that the ferrochrome component of the catalyst is not altered significantly as the catalyst deactivates.

See more of this Session: Catalytic Hydrogen Generation - General I
See more of this Group/Topical: Catalysis and Reaction Engineering Division

(583d) Ultra-Low Temperature Water-Gas-Shift Reaction with Homogeneous Supported Ionic Liquid Phase (SILP) Catalysts

Sebastian Werner[1], Normen Szesni[2], Richard W. Fischer[3], Marco Haumann[1] and Peter Wasserscheid[1], (1)Chemical Reaction Engineering, Friedrich-Alexander University Erlangen-Nuremberg, Erlangen, Germany, (2)Catalytic Technologies R&D, Süd-Chemie AG, Bruckmühl, Germany, (3)Catalytic R&D, Süd-Chemie AG, Bruckmühl, Germany

Supported Ionic Liquid Phase (SILP) catalysts are new materials consisting of an ionic liquid, a metal catalyst and a porous support. The catalyst is dissolved in the ionic liquid which itself is dispersed as a thin film on the inorganic support. This application combines both the advantages of homogeneous and heterogeneous catalysis and thus bridges the gap between traditional homogeneous and heterogeneous catalysis. Especially continuous, gas-phase reactions are highly suited for this novel and innovative technology. An industrially important example is the water gas shift (WGS) reaction, by which hydrogen can be generated from carbon monoxide and water. Homogeneous WGS catalysts operate at milder temperatures than commercial heterogeneous systems. This can open new fields of applications such as distributed hydrogen production or fuel reforming. Since hydrogen production via WGS is an exothermic reaction, lower temperatures result in higher equilibrium conversions. In this contribution we present investigations of homogeneous metal complexes active in the water gas shift reaction which have been immobilized by the SILP technique. Because of that an optimization strategy for this multi-parameter catalysts system was developed as depicted in the figure focusing on improving stability, activity, robustness and removal of the induction period. We tested more than 50 catalyst precursors were tested as SILP WGS catalysts. These precursors include literature-known systems and newly developed catalysts. This detailed study showed, that catalysts bearing a ruthenium center exhibit the highest level of activity. So we selected RuCl3 as simple precursor for the further optimization steps. The ionic liquid severely influences the activity and stability of the SILP WGS catalyst by its changing physic-chemical properties. Especially the different solubilities of the substrates, the different coordination strength towards transition metal complexes and the basicity are important. In a set of experiments we tested more than 20 ionic liquids on their potential in the WGS. Indeed the final level of activity and the tendency of stripping out volatile ruthenium components can be tuned. In our experiments the ionic liquids [BMMIM]Cl and [BMMIM][OTf] turned out to be the most promising candidates for the optimized system. It is known in literature, that the activity of homogeneous water-gas shift catalysts is influenced by the basiscity of the reaction environment. In case of a SILP WGS catalysts basicity can be introduced via addition of an organic base or the support basicity. Our results also indicate a promising effect on activity by increased basicity. We found that either the support material or the addition of a base leads to this effect, which also can be combined. The highly porous support material based on Alumina turned out to be a significantly better choice than the Silica which was previously used, so it was selected for the optimized system. In order to remove the induction period, we carried out infrared spectroscopy measurements on the RuCl3 based SILP WGS catalysts. For this study, we washed out the catalyst-IL phase from the support material by ethanol and produced a sample for IR by removal of the solvent. In comparison of samples of catalysts that were just prepared and ones that were already prune to the reactive atmosphere against reference spectra of various ruthenium-chloro-carbonyl complexes. It turned out, that the induction period is possibly related to the formation of [Ru (CO)3 Cl2]2 from the RuCl3. So the dimeric ruthenium species was immobilized on Alumina and indeed no induction period was observed, so that the [Ru (CO)3 Cl2]2 was selected for the optimized system. Further experiments focused on the optimization of catalyst loading, ionic liquid loading and the addition of cocatalyst. With this building-block like approach the activity was increased by more than 2 orders of magnitude. The investigated SILP systems exhibit activities and stabilities exceeding those of homogenous systems reported in literature. The newly developed catalysts showed excellent restart behavior and long-term stability for more than 750h time-on-stream. Cross-tests in typical synthesis gas environments were also carried out at the Süd-Chemie AG labs successfully. The results indicate that SILP derived WGS catalysts may become a promising alternative to conventional heterogeneous systems, especially in distributed fuel cell setups or decentralized biogas applications.

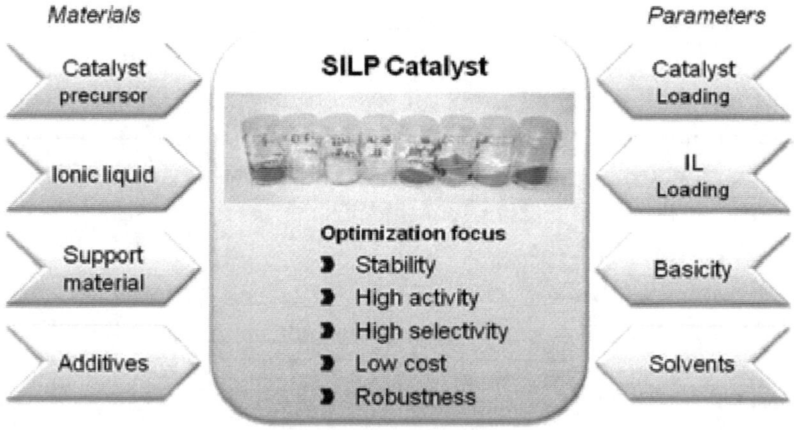

See more of this Session: Catalytic Hydrogen Generation - General I
See more of this Group/Topical: Catalysis and Reaction Engineering Division

(583e) Kinetic Studies On Au and Pt Catalysts Supported On Model Al2O3 and TiO2 for the WGS Reaction

Mayank Shekhar[1], W. Damion Williams[1], Wen-Sheng Lee[1], W. Nicholas Delgass[1], Fabio H. Ribeiro[1], Seung Min Kim[2], Eric A. Stach[2] and Jeffrey T. Miller[3], (1)School of Chemical Engineering, Purdue University, West Lafayette, IN, (2)School of Materials Engineering, Purdue University, West Lafayette, IN, (3)Chemical Technology Division, Argonne National Laboratory, Argonne, IL

The water-gas shift reaction (WGSR), CO + H2O → CO2 + H2 plays an important role in hydrogen generation and purification by CO removal. Our kinetic analysis of Pt/Al2O3, Au/Al2O3 and Au/TiO2 catalysts suggests that the WGSR rate scales with the total number of Pt surface atoms, whereas the rate scales primarily with perimeter Au atoms for Au/Al2O3 catalysts and, as reported in a separate paper, with coordinatively unsaturated corner Au atoms for Au/TiO2 catalysts. To determine the rate as a function of particle size, model supports with simple geometry and no porosity were used. Precise particle size distributions could then be obtained by TEM since all of the metal is exposed on the surface of the support.

The WGS rate for Pt/Al2O3 and Pt/TiO2 catalysts was found to have the form A exp (-85(kJ/mole)/RT) [CO]^0.1[H2O]^0.9[H2]^-0.6[CO2]^0. However, the rate expression for Au/Al2O3 and Au/TiO2 catalysts was found to be different from that for Pt catalysts with lower apparent activation energy at 10-50 kJ/mole, a higher order with respect to CO of 0.8 and a variable order with respect to H2O ranging from -0.3 to 0.8 and depending on the Au particle size and the support. Mechanistic interpretation of the rate data and descriptors for the catalytic activity will be discussed.

See more of this Session: Catalytic Hydrogen Generation - General I
See more of this Group/Topical: Catalysis and Reaction Engineering Division

(583f) Pt-Re Interaction Under Hydrothermal Environment In Aqueous Phase Reforming of Bioliquid for Hydrogen Production

Liang Zhang, Ayman Karim, David L. King and Yong Wang, Pacific Northwest National Lab, Richland, WA

Reforming of oxygenated hydrocarbons (sugars, sugar alcohols, polyols, etc.) in the aqueous phase (APR) provides a promising technology to produce hydrogen from renewable resources. Pt supported on carbon is selected as one of the potential catalysts due to Pt's high selectivity to C-C bond cleavage and carbon's hydrothermal stability, but suffers from low activity. Rhenium has proven to remarkably improve the catalyst activity upon addition to Pt. In an effort to understand the role of Re in the APR process, especially under hydrothermal environment, we have employed in-situ x-ray photoelectron spectroscopy (XPS), in-situ extended x-ray absorption fine structure (EXAFS) technique, and aberration-corrected STEM imaging and x-ray analysis to investigate the structure of a PtRe/C catalyst. Our results suggest a dual-functionality of the catalyst under working conditions: modification of Pt electronic structure for enhancement of conversion and formation of ReOx for additional competing reaction pathways through C-O scission.

See more of this Session: Catalytic Hydrogen Generation - General I
See more of this Group/Topical: Catalysis and Reaction Engineering Division

(611a) Opportunities for Oxyfiring to Reduce Upstream Life-Cycle Greenhouse Gas Emissions From Transportation Fuels

Kerry Kelly[1], Terry Ring[2], Jonathan Wilkey[3], Bernardo Castro[3], Adel F. Sarofim[3] and David W. Pershing[3], (1)Institute for Combustion and Energy Study, University of Utah, Salt Lake City, UT, (2) University of Utah, Salt Lake City, UT, (3)Chemical Engineering, University of Utah, Salt Lake City, UT

With the recent passage of the European Union's and California's low-carbon fuel standards and the potential for US climate-change legislation, the transportation fuel sector is under pressure reduce its greenhouse gas (GHG) emissions. Oxyfiring is a promising technology for reducing the CO2 footprint from this industrial sector, but it requires a significant amount of energy to generate oxygen. This study examines the potential for oxyfiring to reduce well-to-pump, life-cycle GHG emissions from the refining of conventional crude oil and the in-situ extraction and upgrading of oil sands. It will evaluate the importance of the potential reductions in the context of a vehicle's life-cycle GHG emissions. Specifically, the study considers the additional GHG emissions associated with the power required for air separation and CO2 handling; the fuel savings from oxyfiring compared to air firing; and the upstream GHG emissions associated with the additional fuel requirements.

See more of this Session: Novel Processes for Carbon Capture
See more of this Group/Topical: Sustainable Engineering Forum

(611b) A New Process for CO2 Capture with Ionic Liquids

Xiangping Zhang, Xiao Tian, Yansong Zhao, Yan Xu, Shaojuan Zeng and Suojiang Zhang, Institute of Process Engineering, Chinese Academy of Sciences, Beijing, China

Capturing CO2 from industrial gases is one of the important ways to mitigate the globe warming effective, and it is well known the technological obstacles do not exist since several technologies, such as absorption, adsorption, membrane separation, have been used commercially for many years. However, a big challenge we are meeting now is we need treat gigantic amount of the gases if we take the reduction of CO2 emission as the final objective, thus the huge energy consumption for CO2 capture has been regarded as the real obstacle to realize the CO2 capture with large scale.

In recent years, ionic liquid solvent has being regarded as a prospective solvent for CO2 capture, especially its potential advantage to reduce the energy consumption greatly. However, commercial running such new technologies requires us to consider: absorption with high efficiency, desorption with low energy consumption, low cost of the IL solvents, high stability of the solvents and excellent environmental performance after long term utilization.

In this work, a new alkanolamines-based ionic liquid, 2-[2-hydroxyethyl (methyl) amino] ethanol tetrafluoroborate (MDEABF4), were synthesized, which partly possesses the above required performance for capture CO2. The basic physico-chemical properties and the absorption performance of the new ILs solvent have been studied and a series absorption curves have been obtained. A new mechanism of CO2 capture was proposed through the in situ IR analysis. Furthermore, the ionic liquid-based composite solution consisting of 2-[2-hydroxyethyl (methyl) amino] ethanol (MDEA), MDEABF4, piperazine (PZ), and H2O, was investigated for CO2 capture. Moreover, we found that the new solvent has lower causticity than the amine solvent. A multi-scale simulation method was applied to simulate the CO2 capture process. Firstly, a priori prediction is employed to obtain thermodynamic data of the IL-containing system including activity coefficients, equilibrium vapor pressures and Henry constants with a QC continuum salvation model (CSM) COSMO-RS. Then, the interaction parameters in Electrolyte NRTL model are fitted to data from COSMO-RS to predict the physical properties and phase equilibrium data of the novel system over a wider range of concentration and temperature. Based on these parameters, process simulation was carried out with ASPEN plus. The simulation results show that the new process could save the regeneration energy consumption more than 30% compared with the amine method. Due to the above excellent properties of the new solvent and the new process, it would be an new generation solvent for CO2 capture.

See more of this Session: Novel Processes for Carbon Capture
See more of this Group/Topical: Sustainable Engineering Forum

(611c) CO_2 Capture with Ionic Liquid - Amine Solvents

Jason E. Bara, Chemical & Biological Engineering, University of Alabama, Tuscaloosa, AL

Solvents composed of ionic liquids and amines offer distinct advantages in post-combustion CO_2 capture processes when compared to conventional aqueous amine solvents. By blending ionic liquids with common (e.g. MEA, MDEA, etc.) and/or advanced amines, solvents can be readily formulated to capture at least 90% of CO_2 emissions from low pressure flue gas streams, such as those found at coal-fired power plants. Ionic liquids exhibit desirable properties for gas processing such as non-volatility and thermal and chemical stability. Associated benefits of ionic liquid - amine solvents include reduced energy usage, reduced solvent circulation rates, lower impact on COE, and smaller process footprints.

The discussion will focus on the physical and chemical properties of ionic liquid - amine solvents and process design considerations.

See more of this Session: Novel Processes for Carbon Capture
See more of this Group/Topical: Sustainable Engineering Forum

(611d) The Design of Reversible Ionic Liquids for Post-Combustion CO2 Recovery

Ryan Hart[1], Kyle Flack[2], Jackson Switzer[1], Amy Rohan[1], Manjusha Verma[2], Charles L. Liotta[2] and Charles A. Eckert[1], (1)School of Chemical & Biomolecular Engineering, Georgia Institute of Technology, Atlanta, GA, (2)School of Chemistry and Biochemistry, Georgia Institute of Technology, Atlanta, GA

The post-combustion recovery of CO2 from existing coal fired power plants presents an immense separation challenge in terms of scale and difficulty. Presented here are silylamine-based reversible ionic liquids that have been developed to overcome the problems associated with today's best available technology for the capture of CO2 from power plant flue gas – namely monoethanolamine (MEA). Over 600 coal fired power plants in the United States are currently producing the electricity to meet about 1/3 of the electricity demand for the nation. Regulations to limit the amount of CO2 released by these large point source emitters are imminent. In addition to the large number of existing facilities that release copious amounts of CO2, the flue gas streams themselves are technically challenging in terms of separating the CO2. Amine absorption represents today's best available technology for reducing the CO2 in power plant flue gas, but the process is costly due to the high energy requirement for regenerating the solvent. We are making modifications to the silyl amine-based reversible ionic liquid solvents in order to alter their physical and chemical properties favorably for application to CO2 recovery from power plant flue gas. We use FT-IR spectroscopy to determine physical and chemical absorption capacities. The information gathered is used to determine the thermodynamic properties (equilibrium constant, enthalpy and Gibb's free energy of reaction) for the reversible ionic liquid candidates for the purpose of developing a direct replacement for MEA-based absorption processes.

See more of this Session: Novel Processes for Carbon Capture
See more of this Group/Topical: Sustainable Engineering Forum

(611e) Surface Functionalized Liquid-Like Nanoparticle Ionic Materials for CO2 Capture

Youngjune Park[1], Kunyi Andrew Lin[1] and Ah-Hyung Alissa Park[2], (1)Earth and Environmental Engineering, Columbia University, New York, NY, (2)Earth and Environmental Engineering & Chemical Engineering, Columbia University, New York, NY

Because ionic liquids have a unique "self-designable" characteristic, the physical and chemical structure of ionic liquids can be modified for various research and industrial applications such as synthesis, biocatalysis, electrochemistry, and separation. Moreover, most of ionic liquids exhibit environmentally benign physicochemical properties such as negligible vapor pressure with excellent thermal stability and solvating property, which can significantly reduce the emission of solvent during the industrial processes. Recently, ionic liquids have been applied to CO2 capture and storage (CCS) technology. Particularly, task-specific ionic liquids which contain alkaline groups such as amines have been developed as an alternative option to the conventional amine-based solvents (e.g., MEA solvent). In this study, liquid-like ionic materials called nanoparticle ionic materials (NIMs) are designed and synthesized. NIMs are molecular analog of ionic liquid, which consist of inorganic nanoparticle and amine-functionalized corona. By tethering corona chains to nanoparticle cores, additional free volume is created with an engineered canopy structure. It has been suggested both theoretically and experimentally that greater the frustration between corona chains would lead to improved CO2 capture by NIMs. This entropy effect allows greater CO2 capture by NIMs compared to conventional ionic liquids. A series of analytical techniques including TEM, TGA, ATR FT-IR and NMR spectroscopy are employed to investigate the physical and chemical structures of NIMs during their synthesis and CO2 capture. Next, as the amine functional groups are added along the length of the corona chain, the CO2 capture capacity of NIMs is further improved. The effects of temperature, CO2 partial pressure and humidity on the CO2 capture capacity, selectivity and recyclability of task-specific NIMs are investigated. The findings from this study will provide valuable knowledgebase for novel organic-inorganic hybrid materials that can be applied to various environmental and energy systems including CCS with the optimized enthaphic and entropic controls.

See more of this Session: Novel Processes for Carbon Capture
See more of this Group/Topical: Sustainable Engineering Forum

(630a) Aqueous Phase Reforming Over Carbon Nanotube Supported Catalysts

Xiaoming Wang, Nan Li, Zhiteng Zhang, Lisa D. Pfefferle and Gary L. Haller, Department of Chemical Engineering, Yale University, New Haven, CT

Hydrogen production from aqueous phase reforming (APR) has been carried out on a large variety of different supports by many research groups. We have shown that carbon nanotubes (CNT), including both single-walled carbon nanotube (SWNT) and multi-walled carbon nanotube (MWNT), work as good catalyst supports for platinum (Pt) in APR, however, different metal-support interactions greatly affect the performance of the Pt/CNT catalysts. Previously we studied the effect of surface oxygen containing groups on the APR activity of Pt/CNT catalysts, and found that the competitive adsorption between water and polyol onto CNT is the key factor that governs the conversion of APR. In this study, different aspects of metal-support interactions are explored and correlated with APR activity.

Several factors affect the metal-support interaction in Pt/CNT catalysts. Larger particles behave more like bulk Pt and are less affected by the support, while the electronic structure of smaller particles is more affected by the CNT support. The particle size also affects the selectivity and TOF of APR. We have discovered that the binding energy of the CNT supported Pt particles are different than bulk Pt, and this binding energy change depends on the particle size, the radius of curvature of the CNT, as well as the location of particles inside/outside the CNT channel. The effects of particle size, CNT radius of curvature and particle location on APR activity have also been studied. The catalyst preparation method and the metal precursor also play important roles. As incipient wetness impregnation is the most commonly used method in research labs, which gives Pt particles with good dispersion, some methods such as solution reduction or mechanical mixing, give larger particles, but with less metal-support interaction, while other methods such as ion exchange give the catalyst more interaction with the support. Larger particles lead to larger turnover frequency, but not necessarily higher catalyst mass time yield. An optimal particle size is required for the best utilization of the precious metal.

See more of this Session: Catalytic Hydrogen Generation - General II
See more of this Group/Topical: Catalysis and Reaction Engineering Division

(630b) Nanoparticle Molybdenum Dioxide: a Highly Active Catalyst for the Partial Oxidation of Aviation Fuels

Oscar G. Marin Flores[1], Timothy Turba[2], Caleb Ellefson[2], Joe Breit[3], M. Grant Norton[2] and Su Ha[4], (1)Chemical Engineering, Washington State University, Pullman, WA, (2)School of Mechanical and Materials Engineering, Washington State University, Pullman, WA, (3)System Concept Center, Boeing Commercial Airplanes, Everett, WA, (4)School of Chemical Engineering and Bioengineering, Washington State University, Pullman, WA

Nanoparticle molybdenum dioxide (MoO_2) synthesized directly from a reducing ethylene glycol/water solution was found to be highly active for the partial oxidation of dodecane (a surrogate for Jet-A fuel) at 850°C, 1 atm and weight-hourly-space-velocities up to 10 h^{-1}. Even at these very high flow rates MoO_2 nanoparticle catalyst shows a remarkably high fuel conversion of > 90% with a hydrogen yield of > 70% and an exceptional coking resistance. Under similar environments, conventional Ni-based catalysts and commercial low surface-area MoO_2 quickly deactivate due to coking. In addition, nanoparticle MoO_2 also displays remarkable activity for the particle oxidation of real jet A fuel, with conversions >95% and H_2 yields around 50%, at O_2/C = 0.6. Our results demonstrate that, in its nanoparticle form, MoO_2 represents a very promising alternative to expensive noble metals for the reformation of various logistics fuels.

See more of this Session: Catalytic Hydrogen Generation - General II
See more of this Group/Topical: Catalysis and Reaction Engineering Division

(630c) Reaction Sequence for Syngas Production From Propane Using Ceria-Supported Pt and Ni Catalysts

Mayuri Mukka, Tapan K Das, Vijaya Bansode, Edwin L. Kugler and **Dady B. Dadyburjor**, Chemical Engineering, West Virginia University, Morgantown, WV

The partial oxidation of propane is a useful surrogate for the reforming of gasoline to produce hydrogen in-situ for transportation. Hydrogen production over CeO2-supported Pt was compared to the process using Ni/CeO2 catalysts. The aim of the study was to investigate the sequence of different reactions during propane oxidation over the two catalysts and to determine optimum methods of producing hydrogen in each case. The effect of space velocity on the reaction rates was evaluated by varying total inlet flow (100 to 400 SCCM) and catalyst weights. At 600 °C, six species (C3H8, O2, H2, CO, CO2 and C3H6) were detected at the reactor outlet. The Gaussian elimination process yielded four independent reactions. This resulted in ten sets (which would yield the six species). For each set, a material balance on the six outlet compositions measured obtained the extents of each of the four reactions in the set. The rate of each reaction in all the ten sets was calculated using a least-squares-regression technique. Sets were eliminated from consideration if they contained irreversible reactions with negative rates. To confirm the validity of sets containing dry reforming, steam reforming and water gas shift, these reactions were further carried out over the catalysts. Finally, the effect of weight hourly space velocity on the reaction rates for each of the catalyst was evaluated. This allows us to evaluate the relative importance of each reaction in each allowable set as a function of contact time.

The results indicate that, for Pt-based catalysts, hydrogen production is supported through the exothermic direct partial oxidation route at the lowest contact times. On the other hand, for Ni-based catalysts, hydrogen production can be maximized by using higher contact times, and it supports the endothermic steam reforming reaction.

See more of this Session: Catalytic Hydrogen Generation - General II
See more of this Group/Topical: Catalysis and Reaction Engineering Division

(630d) Catalytic Partial Oxidation of Butatnol for Hydrogen Production

Ivan C. Lee and Jeff St Clair, US Army Research Laboratory, Adelphi, MD

Catalytic partial oxidation of isobutanol was investigated for hydrogen generation. The rhodium coated foam catalyst was studied in a short contact time reactor. Role of rhodium in the selectivity of hydrogen production will be discussed.

See more of this Session: Catalytic Hydrogen Generation - General II
See more of this Group/Topical: Catalysis and Reaction Engineering Division

Low Activation Energy Dehydrogenation of Formic Acid at Ambient Temperature and Pressure

Kwong-Yu CHAN, Siu Wa TING, Jayashree BABY
Department of Chemistry
The University of Hong Kong

Nicole K. VAN der Laak (Herriot-Watt)
Shaoan Cheng (Zhejiang University)

Low Carbon Electricity and Transport

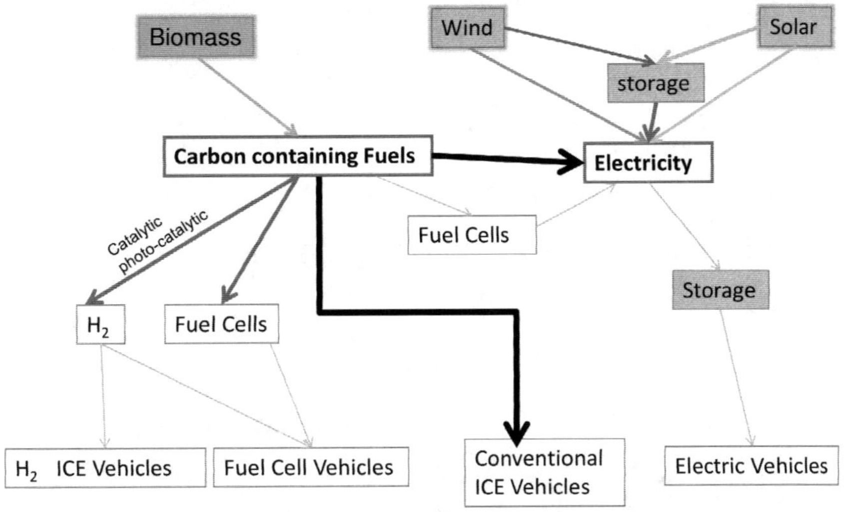

4/13/2010 HKU ICEE Workshop 2

Fueling a Fuel Cell

H_2 fuel cells give higher Power Density than Carbon containing Liquid Fuel Cells

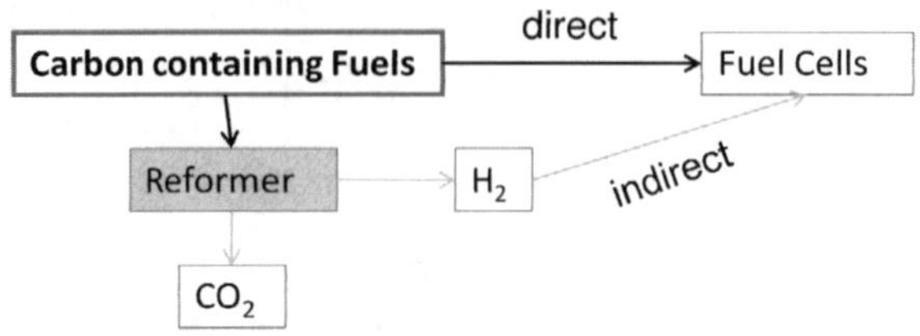

Reforming liquid fuels to produce H_2

$$HCOOH(l) \quad --slow--\rightarrow 2H^+ + CO_2 + 2e^-$$

ΔG_f° = -33.0 kJ

ΔH_f° = 31.2 kJ

$$CO_2(g) + \ H_2 \quad --fast--\rightarrow 2H^+ + 2e^-$$

10/12/2010 ECS Las Vegas 3

Problems in Reforming Fuel

- Heating for Thermal activation, > 200° C
- Presence of CO in product gas
- Additional weight, volume, thermal management to cell

H_2 Generation from formic acid

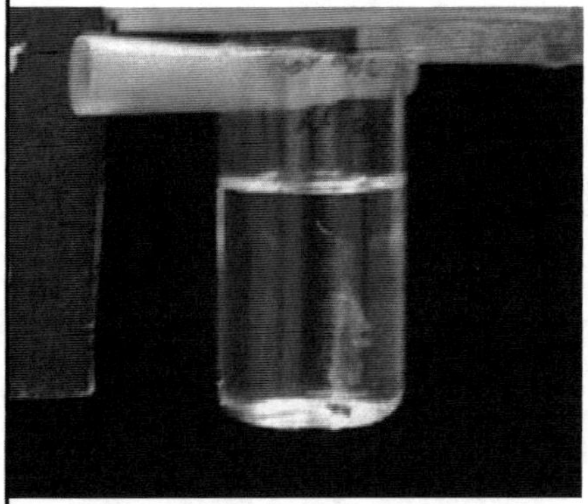

3 mg of Catalyst

10 mL 15% HCOOH in water

Room temperature

$HCOOH \longrightarrow H_2 + CO_2$

CO not detected by GC

US Patent application filed

Formic Acid as Fuel

- Hydrogen storage capacity of 4.3wt%
- A liquid at room temperature, 1.22g/cc.
  Corrosive, b.p. 100.4 °C
- Bio-derivable
- Carbon dioxide can be recycled by hydrogenation to formic acid
- Electrochemical reduction of CO_2

Catalyst: $PtRuBiO_x$

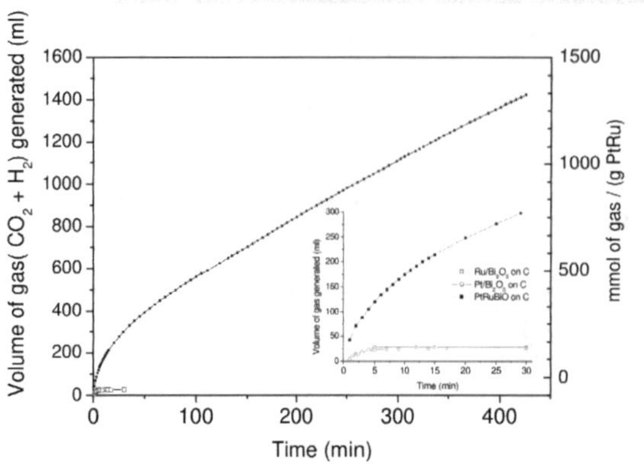

Decomposition of formic acid on Pt-Ru-BiOx/C (Pt/Ru/Bi ~ 2:1:4) at 80°C over 7 hours. inset: activity of initial 30 min for better comparison with Pt-Bi_2O_3 /C and Ru-Bi_2O_3 /C.

Chem. Comm. 2009 Ting et al.

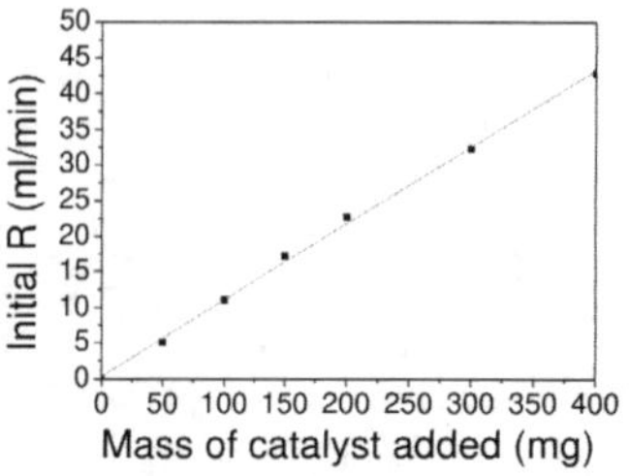

The initial rate of Pt-Ru-BiO_x on Vulcan XC 72 carbon (Pt/Ru/Bi ~ 2:1:4) towards formic acid (80 ml 15% v/v) decomposition at 80°C with different mass of catalysts.

Highly Selective Dehydrogenation

The gas chromatography spectrum of the standard gas containing 4.969% CO_2, 4.957% CO, 4.978% N_2, 4.990% O_2, 4.032% CH_4 and 4.141% H_2.

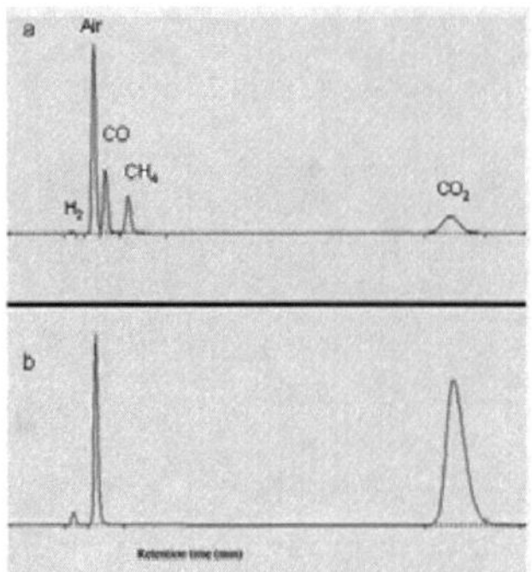

Gas chromatographs of (a) reference standard (b) the gas evolved from catalytic decomposition of formic acid.

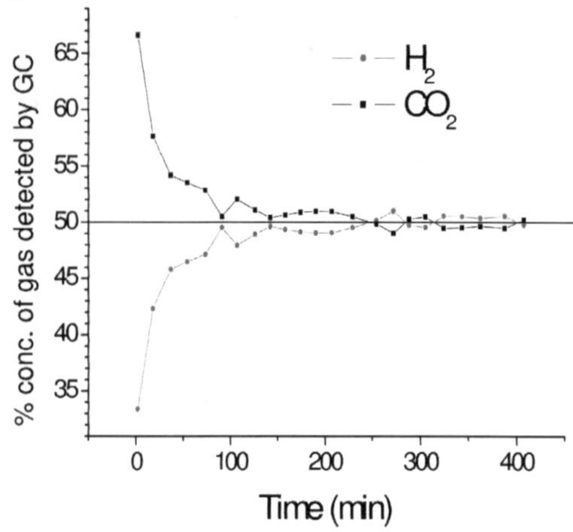

Synthesis

1. Reducing Pt and Ru precursors on Bi_2O_3 particles
2. Calcination of Colloidal PtRu plus $Bi(NO_3)_3$
3. Citric Gel Method
   - Pt^{4+} (aq) + Ru^{3+} (aq) + $Bi(NO_3)_3$ were dissolved in HNO_3
   - Citric acid (CA) was added (variable)
   - Dark brown solution was obtained
   - Vulcan XC 72 carbon (variable)was added to the solution
   - Heated at 100°C to remove water
   - Black powder was obtained and sintered at 850°C under Ar

Particle Size and Structures

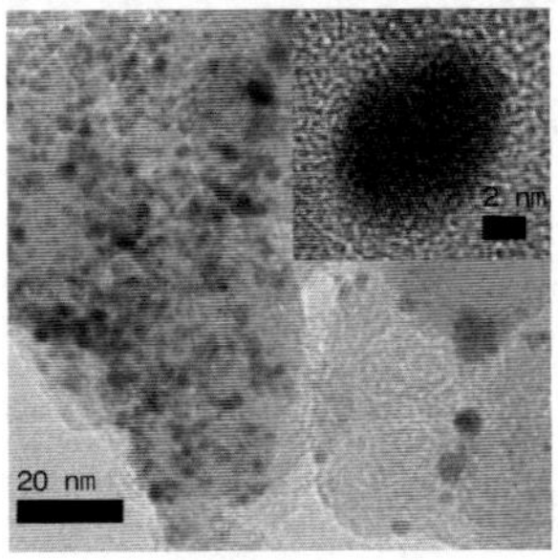

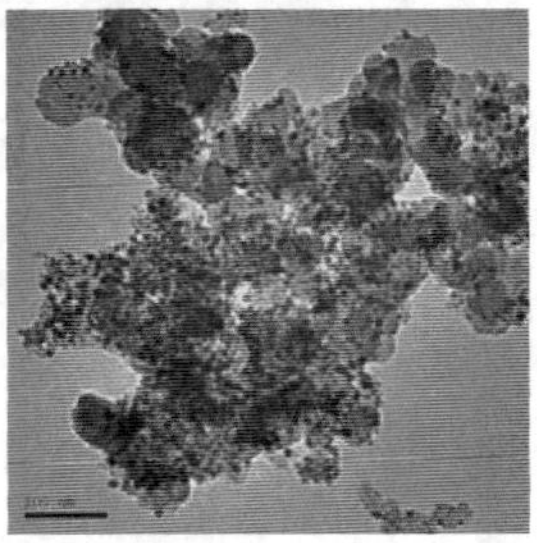

The Volume-area mean diameter $d_{VA} = \frac{\sum n_i d_i^3}{\sum n_i d_i^2}$

The Volume-area mean diameter of this sample was 26.5nm. $\frac{6 \frac{v_m}{a_m}}{d_{VA}}$

Dispersion D= $\frac{N_S}{N_T}$

TOF=312 hr^{-1} average over the first hour.

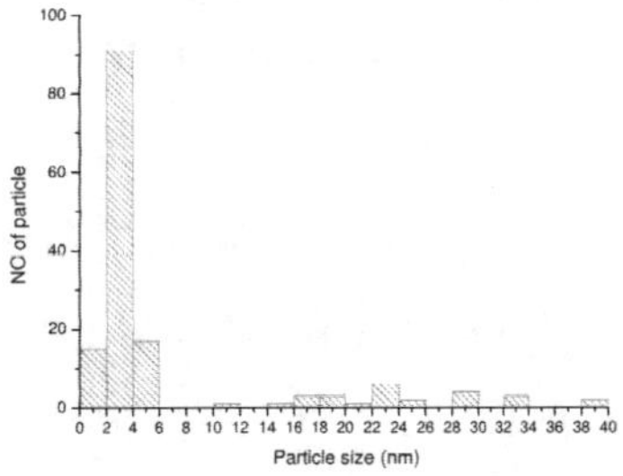

	Laurenczy and co-workers [5-7]	Beller and co-workers[8-10]	Fukuzumi & Suenobu [3,4]	Zhou et al. (2008) Chem Commun	HKU catalyst Pt/Ru/Bi/O on C
Reaction	Homogenous	Homogenous	Homogenous	Heterogeneous	Heterogeneous
Catalyst	Soluble Ru /TRRTS	Soluble Ruthenium phosphine complex $RuCl_2(PPh_3)_2$ soluble	Soluble Rhodium complex $Rh^{III}(Cp^*)(bpy)(H_2O)SO_4$ Cp*=pentamethyl-cyclopentadienyl	PdAu/C/CeO2	Pt/Ru/Bi on C
Reactant and solution content	HCCOH/HCOONa (9:1) 4M (aq)	HCCOH/ amine N,N-dimethyl hexaylamine	HCCOH/HCOONa optimum at pH 3.8	HCCOH/HCOONa 3:1	15% HCOOH 85% water
Additional constraints		Presence of amine Product gas required charcoal filter to purify	pH control to optimize rate		
Reaction temperature	25°C-120°C	25-40 °C	25 °C	92° C	20°C-80°C
Reaction pressure	1-220 bar	Ambiance	Ambiance	Ambiance	Ambiance
Atmosphere	Inert	Inert	Ambiance	Ambiance	Ambiance
% conversion	90-95%	93%	100%		>99%
Activation Energy kJ/mol	-	-	86.9		37.3

Homogeneous Catalysts of Soluble Ru Complexes

- High Turn Over Ratio, high selectivity to CO_2
- Stability? Air Sensitivity
- Cannot separate catalyst from Formic Acid once introduced
- Cannot use liquid level control to start/stop reaction
- Cannot remove excess water if formic acid feed is not 100%
- Need of Charcoal to Remove Trace Amines
- Low Solubility of catalyst needing large reactor

Literature Results

Formic Acid Decomposition in Gas Phase

Catalysts	Ref.	Temperature (°C)	Conversion (%)	Selectivity (%)	
				CO	CO_2
TiO_2	21	130	25.5	95.2	4.8
		180	73.3	98.1	1.9
Pt/ceria	17	210	53	-	48
SnO_2 (100)	1	427	95-99	34	66

Formate Decomposition

Metal oxide surface	Selectivity (%)	
	CO	CO_2
ZnO (0001)	20.6	79.4
TiO_2 (0001) – {114} faceted	67.8	32.2
TiO_2 (0001) – {011} faceted	73.7	26.3
SnO_2 (110)	71.4	28.6
CeO_2 (110)	81.8	18.2
CeO_2 (100)	85.5	14.5
ZnO_2 (100)	84.1	15.9

Electrochemical Reaction?

- Reaction proceeds without conducting support, PtRu on Bi_2O_3, titania
- Reaction does not depend on pH

Photocatalytic Reaction?

- Fast Reaction in the dark
- Some photocatalytic enhancement reported for homogeneous catalysts

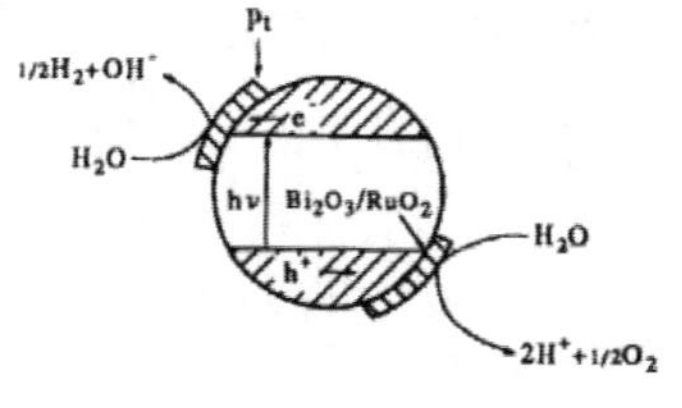

Scheme 1. Photodecomposition of water with Pt/Bi_2O_3/RuO_2.

Redox Reaction of "catalyst"?

$2Ru^{3+} + 3HCOOH \rightarrow 2Ru + 3CO_2 + 6H^+$

$Pt^{4+} + 2HCOOH \rightarrow Pt + 2CO_2 + 4H^+$

$2Bi^{3+} + 3HCOOH \rightarrow 3Bi + 6H^+ + 3CO_2$

$Bi_2O_3 + 6HCOOH \rightarrow 2\ Bi(OOCH)_3 + 3H_2O$

In experiments, HCOOH reacted >1000 X stoichiometric equivalents of "catalyst".

Dependence of Activity on Structure and Compositions

- $PtRuBiO_x$ can be a physical mixture
- Activity is various ratios of the ternary mixture
- No activity without Ru or Bi and Low activity without Pt
- Unidentified peaks in XRD
- Not uniform Particle Size distribution

XPS shows expected states of metals

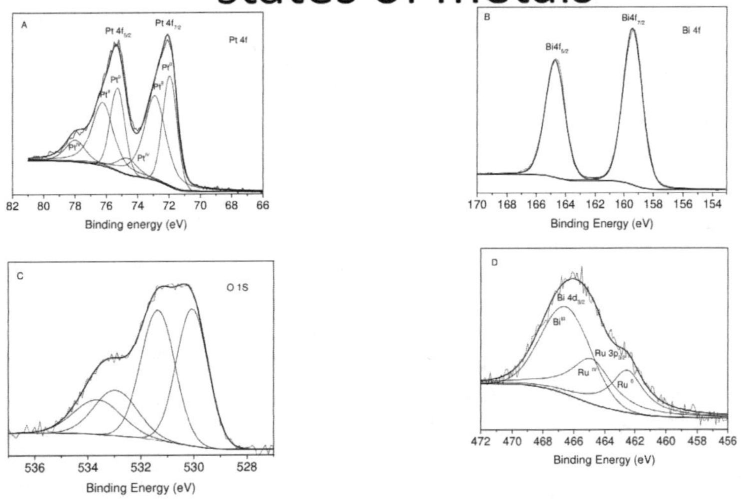

XPS spectra showing (A) Pt peaks, (B) Bi peaks (C) O peaks and (D) Ru and Bi Peak

Thermodynamics

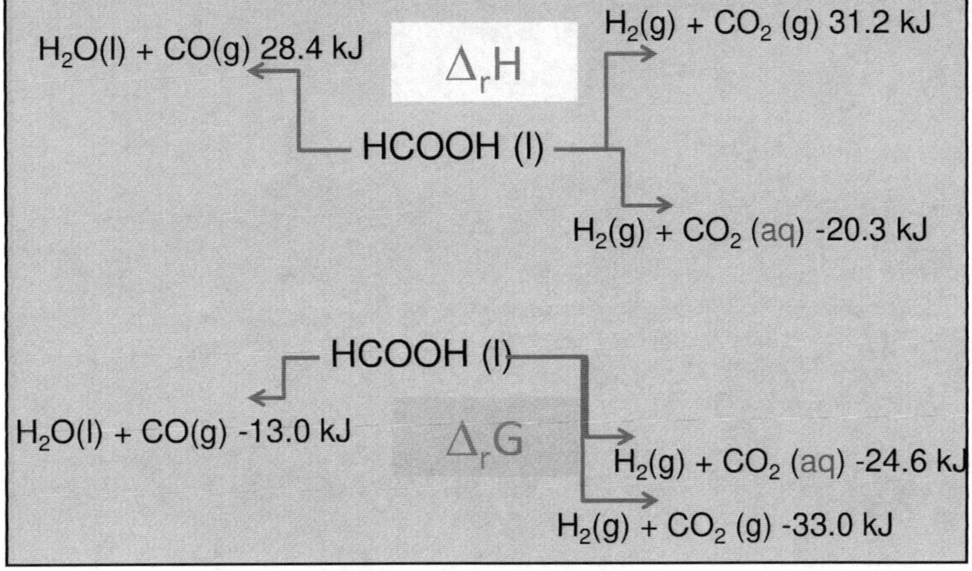

Mechanism?

HCOOH ⟶ intermediate I ⟶ H_2 +
intermediate II ⟶ CO_2

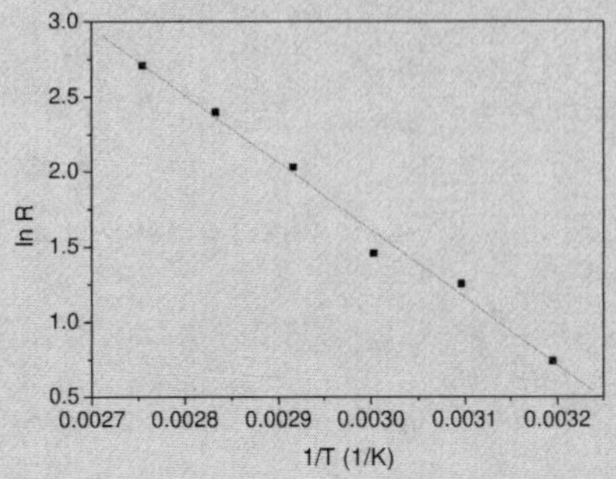

Catalyst	E_a (kJ/mol)
ZnO	125.5
Cr_2O_3	142.3
Cr_2O_3	157.3
Cr_2O_3	142.3
γ-Al_2O_3	142.3
Al_2O_3	188.3
γ-Al_2O_3	150.6
Al_2O_3	171.5
Al_2O_3	146.4
CaF_2	163.2
CaF_2	154.8
CaF_2	142.3
C	148.5
SiO_2	146.4
SiO_2	125.5
TiO_2	-
TiO_2	127.6
TiO_2	129.7
TiO_2 / 0.5% Cr_2O_3	125.1
TiO_2 / 1% Cr_2O_3	105.4
TiO_2 / 2% Cr_2O_3	104.6
TiO_2	194.6
V_2O_5	146.4
Cr_2O_3	154.8
MnO	-
Fe_3O_4	125.5
Cr_2O_3	105.9
Cr_2O_3	110.5
Cr_2O_3	132.6
Cr_2O_3 / 0.5 TiO_2	119.2
Cr_2O_3 / 1.0 TiO_2	123.8
Cr_2O_3 / 2.0 TiO_2	122.2
Cr_2O_3 / 5.0 TiO_2	124.7
Cr_2O_3 / 0.1 K_2O	85.8
Cr_2O_3 / 0.5 K_2O	60.7
Cr_2O_3 / 1.0 K_2O	46.0

Low Activation Energy 37.3 kJ/mol
Compared to literature values of 40 ~ 90 kJ/mol

Effect of [$HCOO^-$] and [HCOOH]

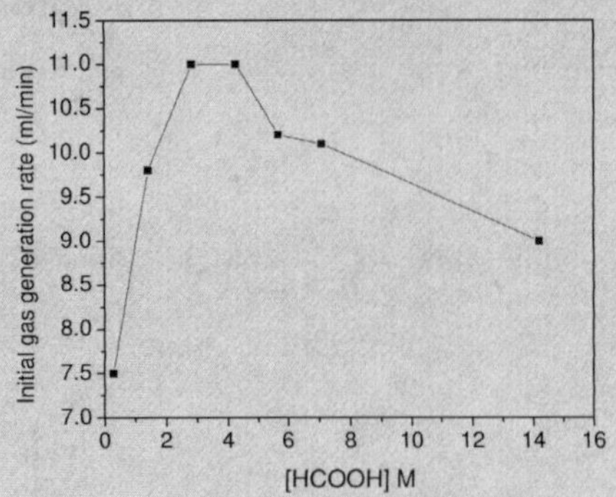

The initial rate of Pt/Ru/BiO on Vulcan-72 carbon (Pt/Ru/Bi ~ 2:1:4) towards formic acid decomposition at 80°C with different concentrations of formic acid.

Effect of [$HCOO^-$] and [HCOOH]

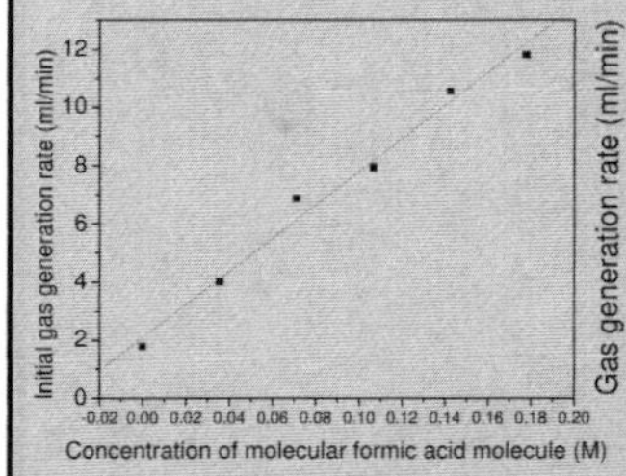

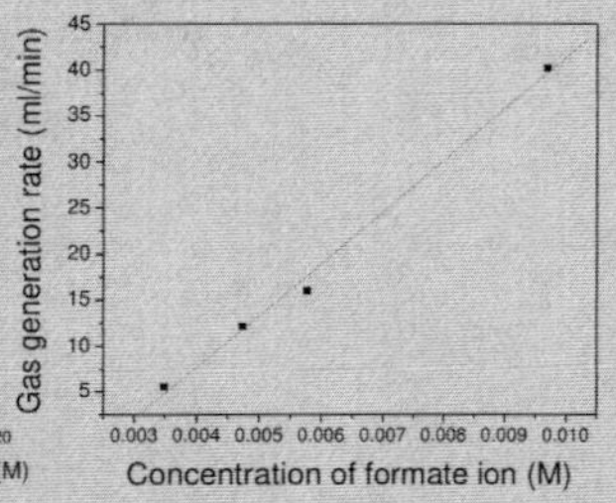

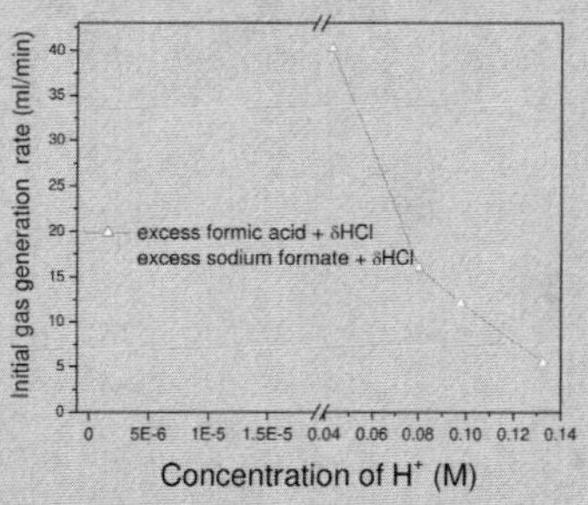

Rate of decomposition as a function of formic acid concentration. Formate ion is present in excess as sodium formate initially at 2.33M. Molecular formic acid was induced by addition of HCl and the x-axis shows the value calculated via ionic equilibrium with known K_a of formic acid.

Rate of decomposition as a function of proton concentration combining the data of experiments of Fig. S6 and Fig. S7 but with calculated proton concentration in the X axis. The results show no consistent dependence of formic acid decomposition on pH.

Roles of Catalyst components?

- Pt for cleaving C-H bond
- Ru for CO tolerance
- Bi_2O_3 for CO tolerance and CO_2 adsorption?

Low Temperature WGS

A novel effective hydration of carbon monoxide in liquid phase by a water-soluble ruthenium complex catalyst at moderate pressures in aqueous medium
Ram S. Shukla, Sharad D. Bhatt, Ravikumar B. Thorat and Raksh V. Jasra

$$CO + H_2O \xrightarrow{\text{Catalyst I}} HCOOH \xrightarrow{\text{Catalyst II}} CO_2 + H_2$$

Overall WGS ΔH – 42 kJ/mole

Catalyst I - [Ru(Hedta)Cl]K.2H2O

Catalyst II - $Pt/Ru/Bi_xO_y$

at ambient pressure and temperature.

Closed System under Pressure and Elevated Temperature

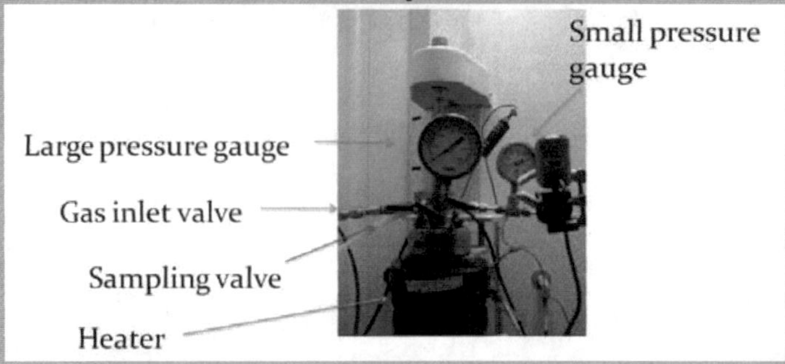

- Experimental details
  - 10 ml 15 % v/v formic acid mixed with 200mg C supported $Pt/Ru/BiO_x$ at RT
  - Reaction temperature: 353K - 413K
  - Time of the reaction and pressure change were recorded
  - The solution after reaction was titrated with 0.2 M NaOH
  - The product gas was analyzed by GC

Conclusions

- Hydrogen generation from aqueous Formic Acid observed on various catalysts at near ambient temperature with >99% selectivity to CO_2
- $PtRuBiO_x$ most active heterogeneous catalyst TOF much lower than some reported Homogeneous Catalysts
- Mechanism involves both molecular HCOOH and formate ion and max at 15% v/v formic acid in water
- Differential adsorption of CO vs CO_2
- Low activation energy
- Prospect of Low temperature WGS

Acknowledgments

- Research Grants Council of Hong Kong
- Sino Germany Collaborative Research Fund
- NSFC-RGC Collaborative Research Fund

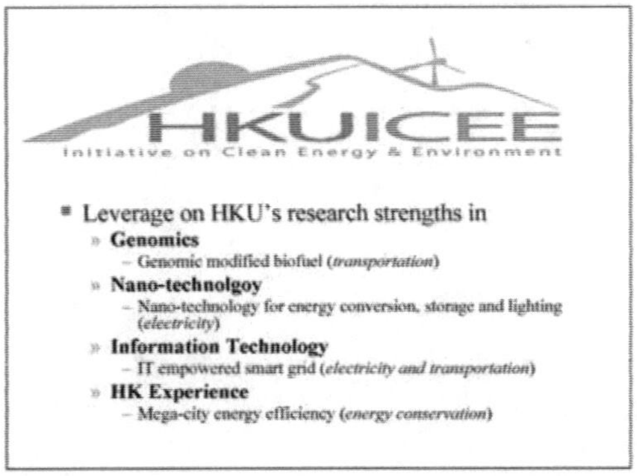

(630f) Partial Oxidation of Ethanol Over Platinum-Tin Catalysts for Hydrogen Generation

Kavi Geetharani Loganathan and Corey Leclerc, Department of Chemical Engineering, New Mexico Institute of Mining and Technology, Socorro, NM

The objective of this work was to investigate Pt-Sn bimetallic catalysts on various supports like PSZ (Partially stabilized zirconia), alumina and cordierite for partial oxidation of ethanol. The effects of each catalytic system on hydrogen selectivity, conversion of ethanol were studied. The selectivity to ethane and ethylene was found to be higher for pure Pt than Pt-Sn on PSZ support. Amongst the Pt-Sn catalysts, Pt-Sn-PSZ demonstrated hydrogen selectivity of 40% which is higher than on Pt-Sn-alumina and Pt-Sn-cordierite. The changes in selectivity was attributed to good active-metal support interactions which was characterized using temperature programmed technique and pulse-chemisorption. The reduction of platinum, tin and platinum-tin alloy were found to be favored on PSZ than the other two supports which was evident from Temperature programmed reduction profile. Pulse-Chemisorption shows that the active particle diameter was 25 nm on PSZ which is much smaller than alumina (270 nm) and cordierite (111 nm) . The metallic surface area achieved on PSZ was 18 m2/g metal which was higher than observed on alumina and cordierite. Overall, hydrogen selectivity, ethanol conversion was found to be better at all C/O ratios for Pt-Sn-PSZ than Pt-Sn- alumina and Pt-Sn-cordierite. As the second phase of the project plasma surface modification of the ceramic monoliths was achieved using nitrogen. TPR data shows that plasma surface modification introduced new phases of platinum and tin in the system which may contribute to definite changes in the product selectivities and conversion of ethanol.

See more of this Session: Catalytic Hydrogen Generation - General II
See more of this Group/Topical: Catalysis and Reaction Engineering Division

Breakdown of Polarization Losses in Button Sized SOFCs and a Prismatic Stack via Impedance Spectroscopy

*Ying Zhu, Wenhua H. Zhu, and Bruce J. Tatarchuk**
Center for Microfibrous Materials
Department of Chemical Engineering
212 Ross Hall, Auburn University, Auburn, AL 36849
** (334)844-2023, brucet@eng.auburn.edu*

Abstract

Impedance Spectroscopy has been widely used to study the contributions of the chemical and physical processes occurring in a SOFC to cell resistances. In this study, impedance data of both small button sized SOFCs and a full prismatic SOFC stack are obtained at various operating conditions. An equivalent circuit diagram is also derived and validated by varying cell temperature, load current and reactant concentration. Arrangement and values for the circuit elements in the equivalent circuit diagram are obtained by fitting the diagram to the experimental impedance data. The derived equivalent circuit diagram is mechanistically discriminated which and to what degree kinetics and mass transfer limit cell output. The performance of SOFCs in two different sizes is compared with each other to provide information for cell scale up. This work is also useful for performing stack maintenance and diagnostics and for designing process controls and power conditioning circuitry.

Introduction

Impedance Spectroscopy (IS) is a novel and powerful analytical method to assess the dynamic response and predict performance of fuel cells. It has been widely and successfully used to study cell performances and material characteristics of cell electrodes, electrolytes and interconnects. Also, it is useful to break down the cell losses into purely ohmic loss, diffusion loss, and other chemical or electrochemical losses. The most common and standard approach of IS[1] is to measure impedance directly in the frequency domain by superimposing a small AC signal on the DC output. Equivalent circuit (EC) diagram, a circuit of electrical elements which produces a similar load response to the cell under test, is utilized to explain, analyze and mechanistically discriminate the obtained impedance data. However, an EC diagram involving three or more circuit elements can be arranged in several ways, yet, to produce the same impedance spectrum[1]. Also, different combinations of circuit elements can yield the same impedance. It is easy to find a fitted EC diagram for the tested cell. The real challenge of IS analysis is to find the EC diagram which gives experimentally verifiable interpretations. The values of the plausible EC elements should perform consistency and certain trend under different test conditions.

Among all kinds of fuel cells, more and more attentions have been paid on Solid Oxide Fuel Cells (SOFCs), which possess higher energy efficiency than other power generation systems. Studies of SOFCs widely include[2] electrode and electrolyte materials; electrode catalysts; mechanisms and kinetics of reactions and processes undergone on electrodes, in electrolyte, and at three-phase interfaces; cell optimal designs, stability, degradations mechanisms, diagnostics and scale up; etc. Up till now, the most commonly used materials for SOFCs are yttria-stabilized zirconia (YSZ) electrolyte, nickel-zirconia cermet (Ni-YSZ) anode and lanthanum manganite ($LaMnO_3$) based cathode, due to their good stability, electrical conductivity, catalytic activity and compatibility with each others.

In this work, the IS measurements on the full button cells and the full stack cell are performed. The EC diagram is derived properly to analyze the performance of cells, and to determine the contributions of the cell processes to the cell resistances. Also, the impedance data of the button cell are in comparison with the stack to study cell scaling up.

Experimental

IS measurements were first applied to small scale button cells (Ceramatec, Inc.) using Solatron Analytical 1255A frequency response analyzer in conjunction with a Schlumberger SI-1286 potentiostat/galvanostat. As detailed in our previous works[3,4], the button cells consisted of an 8 cm^2 disc of zirconia based electrolyte (~170 μm thick), a 2 cm^2 (active area) nickel based anode, and a 2 cm^2 (active area) manganite based cathode. The button cells were fed with air at ~15 psig on the cathode side and mixtures of hydrogen and nitrogen bubbled through water on the anode side.

Most impedance measurements were performed near the Open Circuit Voltage (OCV) condition due to equipment limitations. In this work, DC current was applied to the cells under investigation. Impedance spectra were acquired at varying load current, cell temperature and anode feed composition. In each case, the system was perturbed with a sinusoidal wave with amplitude of 5 mV over a spectrum of 1.0 MHz to 10 mHz. Gas mixtures were altered using rotameters upstream from the water bubbler maintaining a total flow of 39 SCCM. Outlet gas compositions were measured with a molecular sieve column in an Agilent 3000A gas chromatograph and analyzed with Agilent's Cerity Network Data System software. Inlet compositions were approximately equal to those of the outlet at open circuit.

The second group of tests were performed on a five cell prismatic stack (Ceramatec, Inc.) using an FC350™ potentiostat (made by Gamry Instruments) in conjunction with a TDI-Dynaload® RBL488 programmable load. As described by Payne *et al.*[3], each cell in the stack consisted with a ~170 μm thick zirconia based electrolyte, a 61 cm^2 (active area) nickel based anode, and a 61cm^2 (active area) manganite based cathode. The stack was fed with 1.5 SLPM air and 2.2 SLPM anode mixtures of hydrogen, nitrogen, and water. Spectra were acquired for the cells on both ends (cell #s 1 and 5) and the middle cell (cell #3) in addition to the entire stack at varying hydrogen composition. Stack temperature was regulated to be 800°C. The sinusoidal waves superimposed on the stack were generated galvanostatically with currents producing maximum potential amplitude of 5 mV.

All spectra collected from either cell size were fitted to equivalent circuits in Gamry's Echem Analyst software using the Levenberg-Marquardt algorithm.

Results and Discussion

Button Cells

The effect of temperature on the cell performance is shown in Figure 1. Both the ohmic resistance and the polarization resistance decrease with increasing temperature. This trend is expected because higher temperature provides larger activation energy, thus lower the cell impedance. However, the half semicircle tail in the lowest frequency range merely changes with temperature either at OCV or at load. The change of polarization resistance with temperature is mainly dominated by the semicircle in the middle frequency range. Thus, the middle frequency impedance is ascribed to one or several electrode processes (thermally activated), while the lowest frequency impedance is contributed to one or several concentration processes (independent of temperature)[5], that is mass transfer processes (diffusion or gas conversion process), as what concluded in the former study[4].

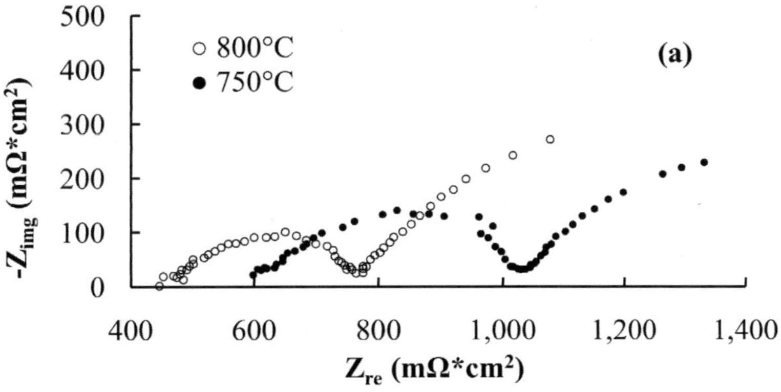

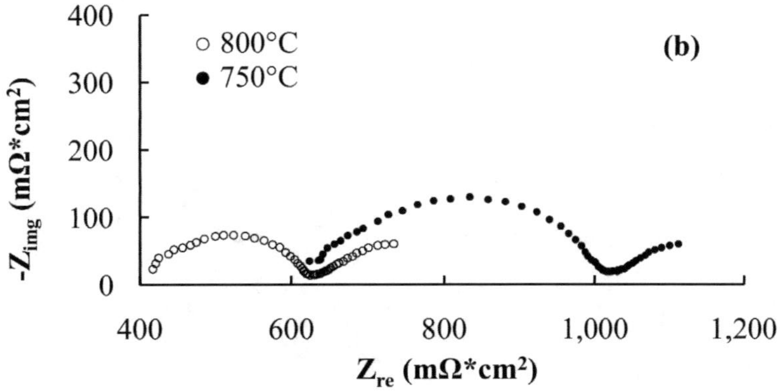

Figure 1. Full button cell impedance spectra obtained at anode gas composition of 97% H_2 + 3% H_2O. (a) at OCV; (b) at 175 mA/cm^2.

The effect of load current on the impedance performance is also shown in Figure 2. The half semicircle tail, ascribed to mass transfer processes, dominates the lowest frequency region, from 2 Hz to 0.02 Hz. Its impedance decreases while the load current density increases, especially dramatically when the load condition of the cell changes from at OCV to at 175 mA/cm^2 (Figure 2a, also shown in Figure 1). A bulk process most possibly contributes to the impedance in such low frequency region. Gas conversion over electrodes is the best explanation here. It has been recognized in several published studies, mainly included in anode measurements[5-8]. Gas conversion impedance does not reflect the anode performance, but is related to the setup geometry and the gas flow rate[6]. More measurements using varying anode gas composition and gas flow rate are necessary to verify the effect of gas conversion on the low frequency impedance.

The impedance spectra in Figure 3 were obtained at 800°C, with a DC current density of 175 mA/cm^2. The anode H_2 composition varied at 97%, 49% and 33%, all with 3% H_2O and balanced N_2. The highest frequency region of the spectra merely changes with H_2 concentration, however, the impedance arc in the frequency range between 200 Hz and 2 Hz increases significantly with decreasing H_2 concentration. The arc strongly dependent on anode H_2 composition is most likely ascribed to gas diffusion impedance.

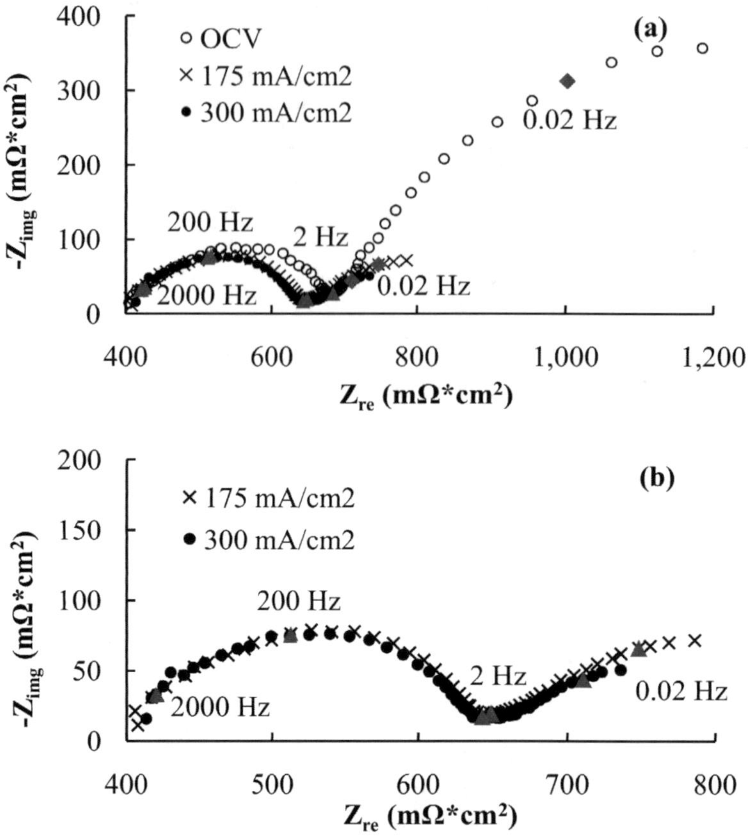

Figure 2. Full button cell impedance spectra measured at (a) 800°C, 49% H_2 + 3% H_2O + balanced N_2, at OCV, 175 mA/cm^2 and 300 mA/cm^2; (b) part of the impedance spectra in (a), providing only polarized conditions.

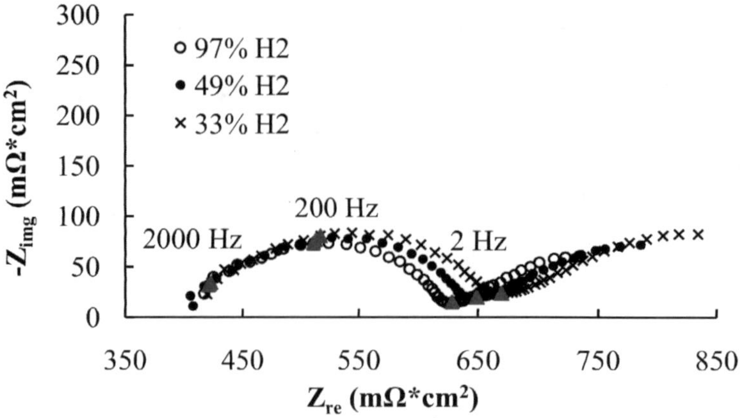

Figure 3. Full button cell impedance spectra measured at 800°C, 175 mA/cm^2, and varying H_2 composition on the anode side.

Up to 5 limiting processes have been reported for a full cell[8,9], but only three different arcs have be clearly separated in the analysis above. Since we did not test the effect of O_2 concentration on the cathode side on the cell performance, it is reasonable that there is one more arc has not been identified in this work, which appears to be related with O_2 adsorption process.

An EC (Figure 4) consisting with 4 limiting processes is used here to well model the full button cell (Figure 5). A finite diffusion element (FDE) is used to model the impedance in the frequency range around 2Hz. Constant Phase Elements (CPEs) have been used to replace the ideal capacitors in the sub-circuits ($R_{ad1}C_{ad1}$) and ($R_{ad2}C_{ad2}$).

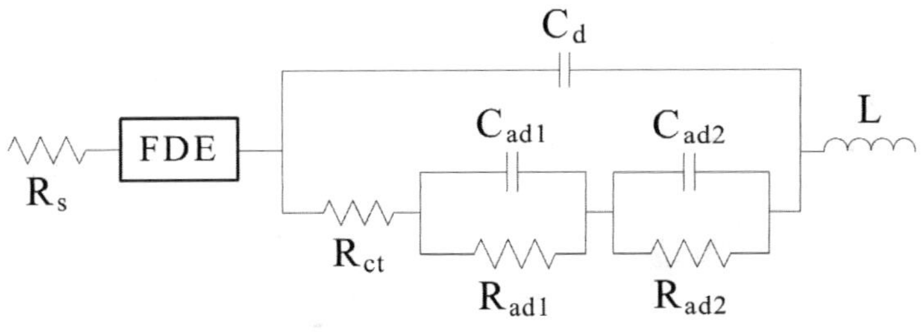

Figure 4. The EC diagram used to simulate full button cells.

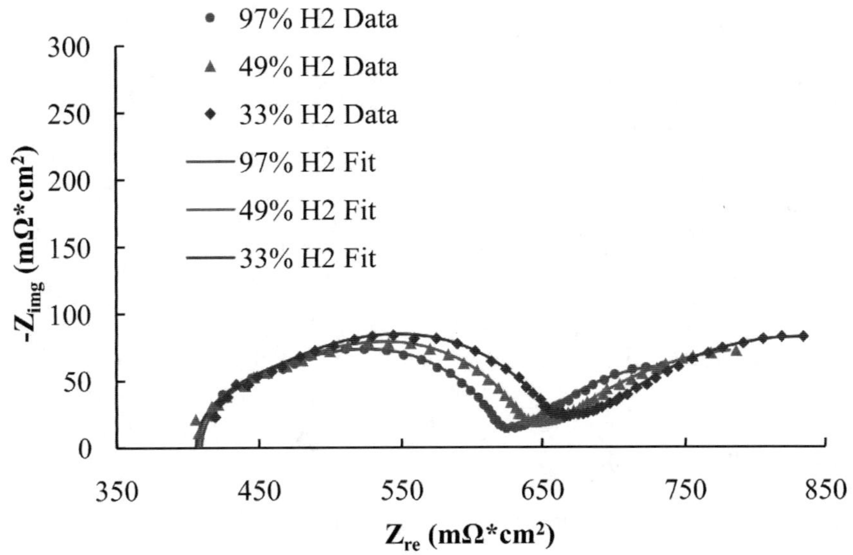

Figure 5. Impedance spectra of full button cell fitted with EC diagram shown in Figure 4, tested at 800°C, 175 mA/cm^2, and varying H_2 concentrations on the anode side. Solid points are experimental data, and lines are fitted data using circuit simulation.

Cell Stack

This EC is also used to model the prismatic cell stack analyzed below. The fitted EC values of cells in these two sizes are important to study cell scalability.

Figure 6 shows the impedance spectroscopy of the prismatic cell stack measured at 800°C, fed with 97% H_2 + 3% H_2O.

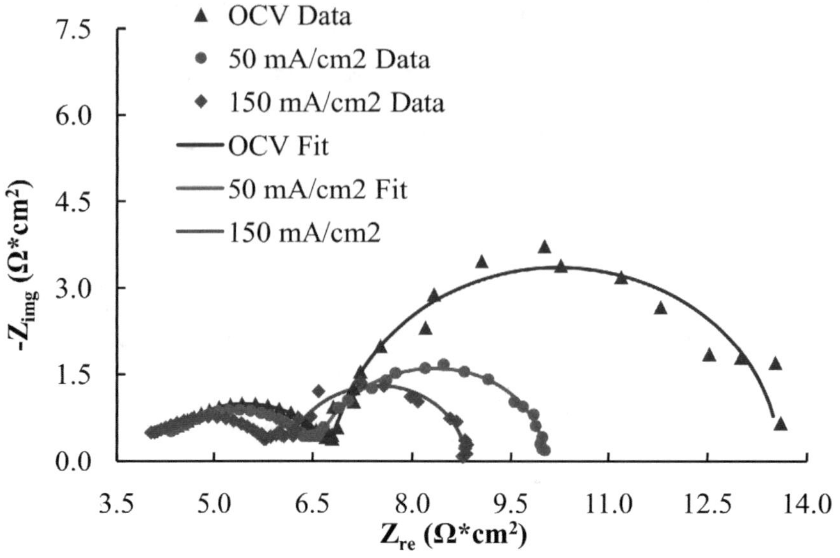

Figure 6. Prismatic cell stack measured at 800°C, fed with 97% H_2 + 3% H_2O on the anode, load current density varying at OCV, 50 mA/cm^2 and 150 mA/cm^2.

Cell Scaling

Each cell in the stack is similar to the button cells, but with wider, taller and thinner electrodes. Thus, ideally, the area specific impedance of the larger stack cells is approximately equal to that of the button cells; yet, non-uniformity among different processes in larger cells prevents this theory from being well observed. One of the most significant factors limiting ideal scalability is reactant distribution. The larger an electrode, the more locally variant are the reactant concentrations. Thus the reaction rate is more unevenly distributed in a larger cell.

The difference of the thickness of electrodes is also likely to impose large effect on the cell performance. Electrodes having different thickness may have different diffusion layer thickness, which changes the diffusivity of the electrode. Thus the impedance arc ascribed to diffusion process shifts its dominant frequency to another region.

Temperature distribution is another big issue in cell scaling. The five planar cells in the stack are not at the same temperature during the measurements. The one at lower temperature performs lower electrolyte conduction and diffusivity than others.

Conclusion

Impedance spectroscopy was applied to test and analyze the button sized solid oxide fuel cells and the prismatic cell stack at various operation conditions. An equivalent circuit diagram has been proposed to well fit the impedance data. The physical processes responsible for each impedance arc are validated by changing cell temperature, load current and anode H_2 concentration. The values of the EC elements for the full button cells and the prismatic stack are successfully obtained and compared with each other for further cell scaling considerations.

Acknowledgements

The authors would like to thank Dr. S. Elangovan, Mark Timper, Dennis Larsen, Lyman Frost, and Anthony Nickens at Ceramatec Inc. who helped make this project possible. We would also like to thank Dr. Robert U. Payne, and Dr. Donald Cahela else from CM3 group for their help and supports on this SOFC work.

References

1. Evgenij Barsoukov, and J. Ross Macdonald, Impedance Spectroscopy: Theory, Experiment, and Applications, 2nd Ed., John Wiley & Sons, Inc., 2005
2. Subhash C. Singhal, and Kevin Kendall, High Temperature Solid Oxide Fuel Cells: Fundamentals, Design, and Applications, Elsevier Ltd., 2003
3. Robert U. Payne, Wenhua H. Zhu, and Bruce J. Tatarchuk, "Mechanistic Discrimination of Circuit Elements in Solid Oxide Fuel Cell Equivalent Circuit Models", Proceedings of 43rd Power Sources Conference, Philadelphia, PA, July 7-10, 2008, pp. 327-330
4. Ying Zhu, Wenhua H. Zhu and Bruce J. Tatarchuk, "AC Impedance in Characterization of Solid Oxide Fuel Cell and an Interpretation of the Low Frequency Inductive Loop", 2009 AIChE Annual Meeting, Nashville, TN, Nov. 8-13, 2009, paper 89e
5. M. Brown, S. Primdahl, and M. Mogensen, "Structure/Performance Relations for Ni/Yttria-Stabilized Zirconia Anodes for Solid Oxide Fuel Cells", J. Electrochem. Soc., vol. 147, no. 4, pp. 475-485, 2000
6. S. Primdahl, and M. Mogensen, "Oxidation of Hydrogen on Ni/Yttria-Stabilized Zirconia Cermet Anodes", J. Electrochem Soc., vol. 144, no. 10, pp. 3409-3419, 1997
7. S. Primdahl, and M. Mogensen, "Gas Conversion Impedance: A Test Geometry Effect in Characterization of Solid Oxide Fuel Cell Anodes", J. Electrochem Soc., vol. 145, no. 7, pp. 2431-2438, 1998
8. Rasmus Barfod, Mogens Mogensen, Trine Klemensø, Anke Hagen, Yi-Lin Liu, and Peter Vang Hendriksen, "Detailed Characterization of Anode-Supported SOFCs by Impedance Spectroscopy", J. Electrochem. Soc., vol. 154, no. 4, pp. B371-B378, 2007
9. R. Barfod, A. Hagen, S. Ramousse, P. V. Hendriksen, and M. Mogensen, "Break Down of Losses in Thin Electrolyte SOFCs", Fuel Cells, vol. 6, no. 2, pp. 141-145, 2006

(641b) Activation Mechanism and Nanostructuring of Solid Oxide Fuel Cell Cathodes

M. Ali Haider and Steven McIntosh, Department of Chemical Engineering, University of Virginia, Charlottesville, VA

Perovskite structured oxides of series $La_{1-x}Sr_xMnO_{3-\delta}$ (LSM) are widely used as cathode for oxygen reduction reaction in a solid oxide fuel cell (SOFC). To study the electrode mechanism, a geometrically well-defined model system was developed using dense thin-film cathode. Polycrystalline nanostructured dense $La_{0.8}Sr_{0.2}MnO_{3-\delta}$ (LSM) thin-film cathodes with an average thickness of 600 nm were fabricated on yttria stabilized zirconia (YSZ) or gadolinia doped ceria (CGO) substrate electrolyte using spray pyrolysis. The polarization resistance of the cathode was observed to be reduced on initial application of voltage or current bias. In order to understand this initial activation, the reaction mechanism was probed. This include altering the electrode surface by doping with $La_{0.6}Sr_{0.4}FeO_{3-\delta}$ nanoparticles, characterizing changes in bulk microstructure before and after application of current and changing the nature of electrode-electrolyte interface by introducing $La_2Zr_2O_7$ impurities. Two different activation mechanisms were observed in the cathode. While applying current for a short duration (5 min) the activation was linked to changes in the surface activity, long term (16 hours) current application produced changes in bulk microstructure.

For surface doping of electrode, a reverse micelle (RM) based method was developed to synthesize highly active electrocatalytic nanoparticles of composition $La_xSr_{1-x}Co_yFe_{1-y}O_{3-\delta}$. The RM synthesis provides tight control over nanoparticle size and average particles size was changed from 14 nm to 50 nm by changing the water:surfactant ratio in the precursor RM solution. A noticeable difference in the sintering behavior of these nanoparticles was observed. While the bulk samples of the same perovskite materials sinter at high sintering temperatures, nanoparticles sinter into a pure perovskite phase at temperature as low as 823 K. The unit cell parameter (a=3.93 Å) for cubic perovskite nanoparticles was found to be greater than the bulk material (a=3.89 Å) of similar composition and structure.

See more of this Session: Fuel Cell Technology

(641c) Effect of Operating Temperature and Voltage On a SOFC Anode Fueled by Phosphorus Contaminated Coal Syngas

K. Channa R. De Silva[1], Brian J. Kaseman[2] and David J. Bayless[2], (1)Chemical and Biomolecular Engineering, Ohio University, Athens, OH, (2)Mechanical Engineering, Ohio University, Athens, OH

Among different types of fuel cells, solid oxide fuel cells (SOFCs) have the advantage of fuel flexibility where it can operate on fuels such as coal syngas and natural gas in addition to pure hydrogen. With the identification of potential to improve energy efficiency by employing an integrated gasification fuel cell (IGFC) system, research on trace contaminants in syngas that can deactivate SOFC Ni/YSZ anodes has gained significant attention. Thermodynamic analysis based on experimental data obtained from various gasifiers have revealed that phosphine (PH3) a phosphorus based gaseous contaminant can exist in concentrations in the region of 1.9 ppm in coal syngas. Such a fuel can be detrimental to electrochemical reactions occurring in a SOFC anode as nickel phosphides are formed. However the oxygen partial pressures at the anode that operates in a reducing atmosphere can be varied by current generated in the fuel cell. In addition phosphorus can react with available oxygen and form gaseous oxides. This research concentrates on varying oxygen partial pressures via changing operational voltage and monitoring electrochemical performance of the SOFC at temperatures above 800°C. Material tests such as SEM, EDXS, XRD and XPS will be performed on the SOFC anode following electrical tests, in order to investigate new phase formation and changes to surface morphology

See more of this Session: Fuel Cell Technology
See more of this Group/Topical: Fuels and Petrochemicals Division

(641d) Proton Conducting It-SOFC Utilizing Internally Steam Reformed Alcohol Fuels

Maria A. Azimova, Chemical Engineering, University of Virginia, Charlottesville, VA and Steven McIntosh, Department of Chemical Engineering, University of Virginia, Charlottesville, VA

The growing interest in proton conducting oxides is driven by the desire to reduce the solid oxide fuel cell (SOFC) operating temperature from >973 K to the intermediate temperature range (673 – 873 K). Fuel cells operating in this range would maintain the benefits of high temperature operation, such as increased cell and system efficiency, fuel flexibility, use of transition metal catalysts, CO and S tolerance and dry solid-state ion transport with no water management problems, while gaining the benefits of lower temperature operation - increased lifetime, reduced material costs and reduced startup/ shutdown time.

Perovskite structured oxides in the series $BaCe_{1-x-z}Zr_xY_zO_{3-\delta}$ (BCZY) have shown to provide technologically relevant proton conductivity in the target temperature range. The primary barriers to their application are stability in CO_2 containing atmospheres, low grain boundary conductivity, and the high sintering temperature required to produce dense electrolytes, typically >1973 K. In this study, we have utilized cobalt doping to lower this sintering temperature to <1698 K in $BaCe_{0.5}Zr_{0.4}(Y,Yb)_{0.1-x}Co_xO_{3-\delta}$.

The materials were synthesized using a modified Pechini procedure and were confirmed as phase pure cubic perovskites (space group Pm-3m) by X-ray diffraction. Density of >95% of theoretical was achieved by sintering at 1698 K or below. AC and DC conductivity measurements, performed in dry and humidified air, H_2 and Ar/N_2, demonstrate proton conductivities comparable with the best undoped proton conductors sintered at high temperatures. DC conductivity values are higher in humidified atmospheres; consistent with a proton-conducting mechanism. Proton conduction was confirmed by Nernst potential measurements, conducted in a dual chamber system as a function of pH_2 and pO_2 driving force and the materials were shown to be phase stable in the presence of hydrogen and steam. The upper limit for electrolyte applications was found to be between 5 and 10% cobalt doping on the B-site due to decreased stability and increased electronic contribution to the total conductivity above 5%.

Finally proton conducting SOFC were fabricated using a dual-layer tape-casting technique that allows flexible composition in the anode. The cells consisted of a 0.04 mm thick electrolyte supported on a 0.3 mm thick porous Ni/BCZY anode with an $La_{0.8}Sr_{0.2}CoO_{3-\delta}$/BCZY cathode. The cell performance was measured with humidified H_2 fuel and methanol/water fuel at 873 K. The maximum open circuit potential (OCP) of 1.04 V was achieved at 873 K in 3% humidified H_2 fuel with an accompanying peak power density of 89 mW/cm^2. The total cell polarization resistance was 0.83 Ω•cm^2. Impedance measurements showed decreasing polarization and ohmic resistance with increasing current density. Cell operation via internal steam reforming of methanol was demonstrated using a methanol/water feed with 3:1 and 2:1 steam-to-carbon (S:C) ratios; however, the OCP was reduced to 0.97 and 0.95 V and peak power density decreased to 65 and 58 mW/cm^2 respectively. This was primarily attributed to a lower H_2 concentration in the feed stream. Methanol reforming and carbon deposition rates were also assessed over Ni/BCZY catalysts for varying methanol/water S:C ratios.

See more of this Session: Fuel Cell Technology
See more of this Group/Topical: Fuels and Petrochemicals Division

A Hybrid Molten-Solid Oxide Direct Carbon Fuel Cell

Trung V. Nguyen* and Venkata R. Yarlagadda

Department of Chemical & Petroleum Engineering
University of Kansas, Lawrence, Kansas-66045, USA

*Corresponding E-mail: cptvn@ku.edu

Background:

Direct Carbon Fuel Cell (DCFC) could be the most efficient device for converting the chemical energy in a solid fuel to electricity. The concentrated carbon dioxide generated from this fuel cell can also be more effectively captured and sequestered. Direct usage of solid fuel in a DCFC requires a liquid electrolyte in order to facilitate ionic transport between the solid fuel and the electrolyte. Several studies have proposed bismuth oxide (Bi_2O_3) as a good electrolyte because of its high oxide ion conductivity. Also, electronic structure of Bi^{3+} includes lone pair of electrons which lead to the ionic mobility of oxide ions.

Until now bismuth oxide (Bi_2O_3) is doped with materials such as yttria, lanthanum, etc. to stabilize below its melting point facilitating its usage as an electrolyte in Solid Oxide Fuel Cells (SOFC). Present study involves the proposed use of Bi_2O_3 as a potential liquid electrolyte for a DCFC system. It is chosen because of its lower melting point and high ionic conductivity. In this study a four point conductivity measurement of molten Bi_2O_3 is done using an AC milli-ohmmeter. A DC multi-meter is also used to check whether molten Bi_2O_3 exhibits mixed conductivity.

Experimental Procedure:

A high form alumina crucible with platinum wires wrapped to a rectangular ceramic bridge present inside the crucible is used to make four point conductivity measurements. Platinum wires drawn out of the crucible are connected to AC milli-ohmmeter and DC multi-meter respectively in series. Figure 1 represents a schematic of the set up used for conductivity measurements.

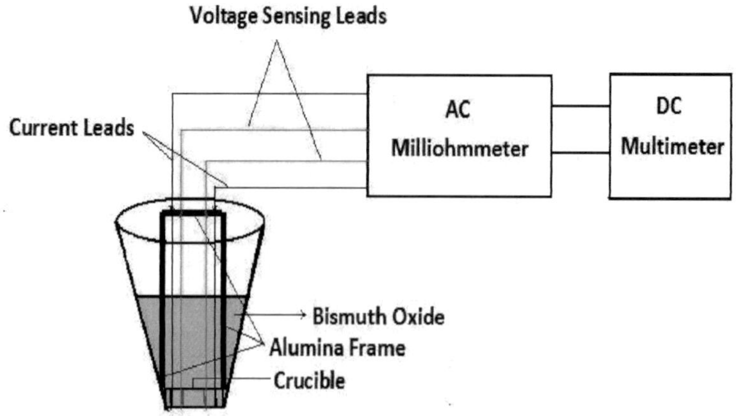

Figure 1: Setup for four point conductivity measurement

A high temperature oven is used to measure conductivity as a function of temperature. Data has been collected at 50°C intervals between 400-700°C and 25°C intervals between 700-900°C. At every interval, the oven is allowed to stabilize for 30 minutes before the readings are taken.

Results and Discussion:

Conductivity data from AC milli-ohmmeter and DC multimeter reveal that bismuth oxide is highly conductive in its molten state and it also acts as a mixed conductor above 600°C. From figure 2 one can see a sharp rise in both the ionic and electronic conductivity when bismuth oxide transforms from solid (α state) to molten state (δ state) around 725°C. Due to its high conductivity at temperatures around 900°C higher efficiency and power density can be achieved when compared to electrolyte systems such as molten carbonate, hydroxide and phosphoric acid which operate at lower temperatures.

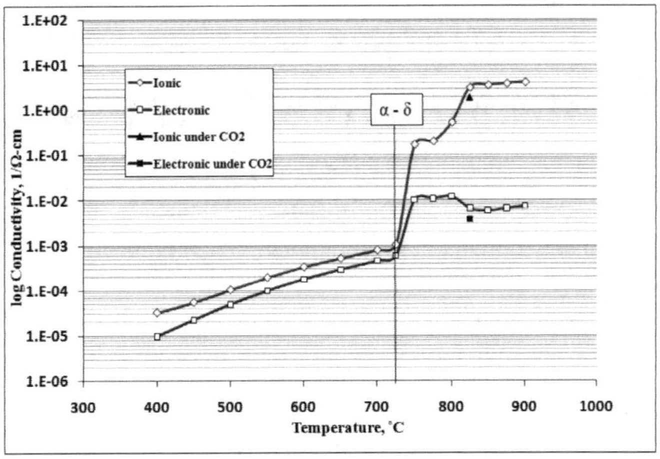

Figure 2: Ionic and electronic conductivity of Bi_2O_3

Since bismuth oxide is a mixed conductor using it in a DCFC will require placing a solid state ionic electrolyte between the cell cathode and the molten electrolyte to prevent shorting of the carbon anode and the molten electrolyte. YSZ/LSM is used as solid electrolyte/cathode and gold mesh is used as a current collector. Figure 3 shows a schematic of a hybrid molten/solid oxide DCFC used in this study.

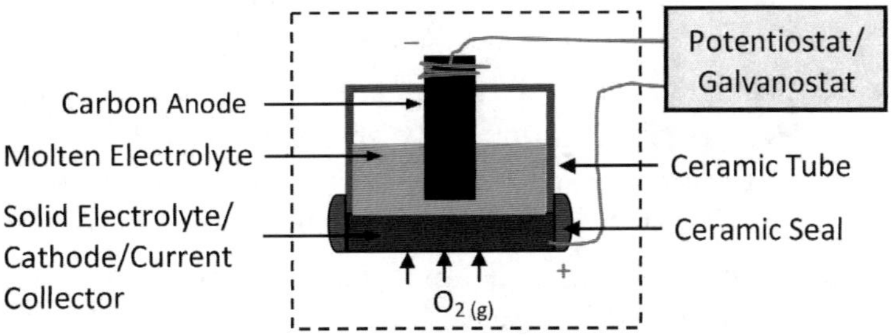

Figure 3: Hybrid molten/solid oxide DCFC

With carbon used as the fuel at the anode carbon dioxide is formed as a by-product. Resulting carbon dioxide may further react with Bi_2O_3 leading to possible formation of bismuth carbonate. In order to check whether bismuth carbonate hinders the performance of DCFC conductivity measurement using the set up in Figure 1 is done under CO_2 environment at 825°C. Figure 2 shows a small decrease in both ionic and electronic conductivity due to CO_2. Results regarding composition analysis of bismuth oxide crystals (both pure and CO_2 treated) and performance of hybrid molten/ solid DCFC will be discussed in this presentation.

References:

1. Sneh L. Jain, J. Barry Lakeman, Kevin D. Pointon, and John T. S. Irvine, "A Novel Direct Fuel Cell Concept," *J. Fuel Cell Science and Technology*, Vol. **4**, 280-282 (2007).
2. N. M. Sammes, G. A. Tompsett, H. Nafe, and F. Aldinger, "Bismuth Based Oxide Electrolytes-Structure and ionic Conductivity," *J. European Ceramic Society*, Vol. **19**, No. 10, 1801-1826 (1999).
3. P. Shuk, H. D. Wiemhofer, U. Guth, W. Gopel, and M. Greenblatt, "Oxide ion conducting solid electrolytes based on Bi_2O_3," *J. Solid State Ionics*, Vol. **89**, 179-196 (1996).
4. H. A. Harwig and A. G. Gerards, "Electrical properties of α, β, γ and δ phases of Bismuth Sesquioxide," *J. Solid State Chemistry*, Vol. 26, 265-274 (1978).

(641f) Platinum Reduction in the Cathode of a PEM Fuel Cell with Multiple Catalyst Layers

Debangsu Bhattacharyya, Department of Chemical Engg., West Virginia University, Morgantown, WV, M. Srinivasarao, Chemical Engineering, Indian Institute of Technology, Chennai, India, Raghunathan Rengaswamy, Department of Chemical Engineering, Texas Tech University, Lubbock, TX and Shankar Narasimhan, Department of Chemical Engineering, Indian Institute of Technology, Chennai, India

To make the polymer electrolyte membrane fuel cells (PEMFC) commercially viable, further reduction in its cost and improvement in its performance are required. The major cost component of a PEMFC is that of the Pt catalyst. More amount of Pt is typically used on the cathode side because of the sluggish oxygen reduction reaction. In this work, use of multiple catalyst layers (CL) is considered in place of traditionally used single catalyst layer for reduction in Pt loading without deteriorating the cell performance. Since the concentration of oxygen decreases in a CL from the diffusion medium-CL interface towards the polymer membrane, the CL adjacent to the diffusion medium should be of higher porosity than the other CLs. Similarly, the CL adjacent to the polymer membrane should contain more ionomer than the other CLs. Furthermore, liquid water should be removed without causing significant mass transport and/or ohmic losses. Therefore, the design parameters of a CL can be varied spatially for minimizing the Pt loading. However, such a continuously graded CL is difficult to manufacture due to lack of commercially available techniques and associated costs. As an alternative, a combination of layers can be synthesized where each layer is manufactured with different design parameters. It has been shown that such layered construction can result in a significant improvement in the cell performance1. In this work, the overall Pt loading is minimized by optimizing the design variables of each of the CLs. The design variables are chosen such that they can be achieved experimentally. For the optimization study, a detailed steady state model of a PEMFC cathode with multiple CLs is developed1,2. The model considers liquid water in all the layers. The catalyst layer microstructure is modeled as a network of spherical agglomerates. For improved water management, a thin micro-porous layer is considered between the gas diffusion layer (GDL) and the first CL. The fidelity of the model is also varied to study the effect of modeling various mechanisms on the values of the optimized decision variables. The tradeoff between the cell performance and the Pt loading will also be presented.

1. Srinivasarao,M; Bhattacharyya, D; Rengaswamy, R; Narasimhan, S; Performance analysis of a PEM fuel cell cathode with multiple catalyst layers, International Journal of Hydrogen Energy, doi:10.1016/j.ijhydene.2010.03.092, 2010

2. Srinivasarao,m; Bhattacharyya, D; Rengaswamy, R; Narasimhan, S; Parametric study of the cathode and the role of liquid saturation on the performance of a polymer electrolyte membrane fuel cell – A numerical approach, Journal of Power Sources, doi:10.1016/j.jpowsour.2010.03.003, 2010

EXPERIMENTAL INVESTIGATIONS OF THE EFFECT OF MICROPOROUS LAYER ON WATER MANAGEMENT IN POLYMER ELECTROLYTE MEMBRANE FUEL CELLS

Tak Cheung Yau [a], Xiaotao Bi [b], Jürgen Stumper [c]

a, b: Department of Chemical and Biological Engineering, University of British Columbia, Vancouver, British Columbia V6T 1Z3, Canada

c: Automotive Fuel Cell Cooperation Corporation, Burnaby, British Columbia V5J 5J8, Canada

Introduction

Water management has been an important issue in the operation of hydrogen PEM fuel cells. A certain level of water content in the membrane has to be maintained, otherwise membrane proton conductivity and hence cell performance can be adversely affected. On the other hand, flooding and blockage of oxygen access to the cathode catalyst sites would occur when the water produced from the cell reaction at the cathode is removed efficiently. To avoid flooding and dry-out conditions it is important to understand how the water is transferred in a fuel cell.

A useful parameter to study in this context is the water crossover rate, which is defined as the rate of water transported between anode and cathode through the membrane. In the literature there are different sign conventions, but here we define the flow of water from the anode to the cathode as positive. The water crossover flux is defined as the water crossover flow rate over a certain area of the membrane.

Atiyeh et al. (2007) experimentally measured the water transfer factor from membrane-electrode assemblies (MEAs) with and without microporous layers (MPLs). They found that the addition of an MPL at the cathode did not change the overall water transfer significantly. The same results were obtained by Janssen and Overvelde (2001) at a current density of 0.4 Acm^{-2}. At 0.6 Acm^{-2}, the difference was more pronounced, regardless of humidification conditions. The water crossover was found to be less if MPLs were applied. Dai et al. (2009) used water filled anode and a dry air cathode channel to measure water transfer to the cathode channel through the GDL. Contrarily, they found that the addition of an MPL could reduce water transfer to the cathode channel at dry air flow rates greater than 7.44×10^{-4} $mols^{-1}$. The authors claimed that at higher air flow rates more water was driven out of the membrane which caused a higher capillary pressure in micropores and hence decreased the transport efficiency through the GDL. Murahashi et al. (2006) used hydrophobic MPLs in both anode and cathode. Although they did not study directly the effects of the layer on water crossover, they found that when capillary pressure was included in their calculations, a lower overestimate in water crossover coefficient was observed. They claimed that capillary action increased the activity of water vapour at the membrane interface which in turn increased electro-osmotic drag and back diffusion coefficients and decreased water crossover from anode to cathode.

With no general agreement in experimental results, it would be of interest to study the effect of MPLs on water crossover in running fuel cells.

Experimental

MEA construction

4 pieces of catalyst coated Nafion NRE211 membrane with an area of 50 cm^2 were assembled with different gas diffusion layers. Two of them had MPLs on both anode and cathode, and the other two had MPLs on the cathode side only. Each set of identical MEAs served as replicates to check for reproducibility of the results. From this point on, the set of MEAs with micoroporous layer on the cathode would be denoted by "with MPL" and the other set would be denoted by "without MPL".

Water crossover measurement apparatus

The apparatus for water crossover measurement has been described in detail in Sauriol et al. (2009). The inlets of the fuel cell were humidified with syringe pumps water injection. The water content at the outlet was measured by an infrared analyzer. As a single phase stream was required for measurement, the outlet streams were diluted by preheated dry gas and the pressure was decreased to bring down the dew point before entering the analyzer. Accurate measurements in water content were made possible by the introduction of the 'calibration' steps. During these steps, water content 'standards' were made from humidified gas streams with known water content and known flow rate using the same syringe pump, dilution gas and flow meter as in 'normal' operations. The readings at the same infrared sensor were taken. So under the 'normal' mode, the outlet water content readings were referenced to the readings from 'calibration' steps and hence to the inlet syringe pump flow rate. The systematic errors of the equipment are cancelled out in the process of subtracting inlet from the outlet flow rates. During calibration tests, the operation of the cell was kept uninterrupted by humidification using backup syringe pumps. For each point on the curve, a constant current was withdrawn from the load bank and the cell voltage was measured, along with the water crossover rate determined from the measurement taken from the anode stream. Calibration steps were carried out for 20 minutes before and after the measurement of each point to minimize the effect of ambient temperature on water content readings. A minimum of 50 minutes of 'normal' experimental water crossover data were averaged to calculate the mean water crossover rate at steady state. The average water crossover flux was obtained by dividing the crossover rate by the cell active area, and in the following paragraphs 'water crossover flux' would simply mean this average value over the whole tested area in spite of the variation in water crossover flux at different areas of the cell.

Experimental conditions

Table 1 summarizes the conditions used in the polarization experiments. The temperature of the cell was kept constant by running a water coolant through the flow field. To minimize the effect of pressure on water crossover, both anode and cathode pressures were kept at 230kPa (absolute), while it was verified that the pressure gradient across the inlet and outlet of the cell was negligible. Each MEA was subjected to two different humidity conditions. A polarization curve was created by running the cell at 7 different current densities; and to avoid the possible effects of hysteresis, the order for different current densities were randomized for each run. The dry gas flow rates were constant under different current densities.

Table 1 Inlet conditions employed in polarization curve experiments.

Run	MEA #	With MPL?	Temperature (˚C)	Anode pressure (kPa)	Cathode pressure (kPa)	H_2 flow rate ($mols^{-1}$)	N_2 flow rate ($mols^{-1}$)	O_2 flow rate ($mols^{-1}$)	Anode RH (%)	Cathode RH (%)
1	1	No	70	230	230	4.46×10^{-3}	2.98×10^{-3}	7.81×10^{-4}	75	25
2	1	No	70	230	230	4.46×10^{-3}	2.98×10^{-3}	7.81×10^{-4}	75	75
3	2	No	70	230	230	4.46×10^{-3}	2.98×10^{-3}	7.81×10^{-4}	75	25
4	2	No	70	230	230	4.46×10^{-3}	2.98×10^{-3}	7.81×10^{-4}	75	75
5	3	Yes	70	230	230	4.46×10^{-3}	2.98×10^{-3}	7.81×10^{-4}	75	25
6	3	Yes	70	230	230	4.46×10^{-3}	2.98×10^{-3}	7.81×10^{-4}	75	75
7	4	Yes	70	230	230	4.46×10^{-3}	2.98×10^{-3}	7.81×10^{-4}	75	25
8	4	Yes	70	230	230	4.46×10^{-3}	2.98×10^{-3}	7.81×10^{-4}	75	75

Results and Discussion

Figure 1 shows the polarization curves for different conditions. The results presented are the average from two different MEAs with the same construction. Reproducibility was satisfactory with the difference in voltage being less than 31 mV between replicates. It could be seen that the addition of MPLs did give rise to a higher cell voltage at the same current density in the ohmic region, but the effect was not as significant compared to a higher cathode gas humidification.

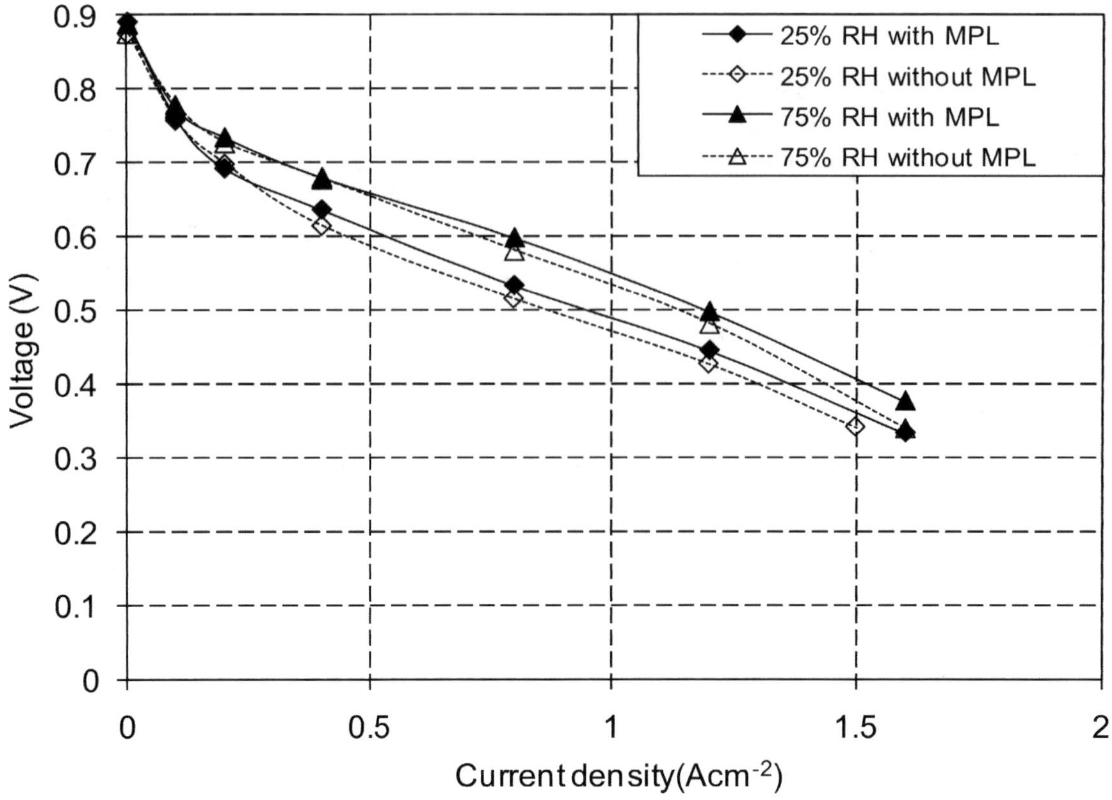

Figure 1 Experimental polarization curves at constant fuel and oxidant flow rates. Results are the average from two replicates. MEA construction and cathode RH are as indicated. T = 70 °C, P_A = 230 kPa, P_C = 230 kPa, $RH_{in,A}$ = 75%, H_2 flow rate = 4.46×10^{-3} mols^{-1}, N_2 flow rate = 2.98×10^{-3} mols^{-1}, O_2 flow rate = 7.81×10^{-4} mols^{-1}.

Figure 2 shows the corresponding water crossover values. Again the two replicates showed good reproducibility with a maximum difference in water crossover flux of 1.74×10^{-7} molcm^{-2}s^{-1}. Being the results of vapour phase diffusion only, the intercepts are clearly indicative of the RH gradients across the membrane for different runs: for cathode RH of 25% and anode RH of 75% at the inlet, water diffused from anode to cathode yielding a positive water crossover flux. When there was no difference in the feed RH between the two sides of the membrane, no water crossover could be observed. When the current density increased, the water generated by the electrochemical reaction at the cathode side of the membrane became an additional driving force for water going from the cathode to the anode which dominated over electro-osmotic drag. The trends in water crossover flux against current density were highly linear up to around 1.5 Acm^{-2} for 25% cathode RH and 1.2 Acm^{-2} for 75% cathode RH. This is consistent with what was found by Yau et al. (2010). The curve for 75% cathode RH slightly flattened out at high current densities, which indicated a drop in back diffusion in comparison to electro-osmotic drag. For a constant electro-osmotic drag coefficient with respect to water content (Zawodzinski et al. 1995), an increase in slope of the curve was probably caused by a decrease in concentration gradient across the membrane. It follows that the cathode interface of the membrane is saturated or very close to saturation so that back diffusion increased the water content at the anode interface and hence yielded a smaller concentration gradient. Even with negative values of water crossover it should be noted that the magnitude of back diffusion was still greater than that of electro-osmotic drag.

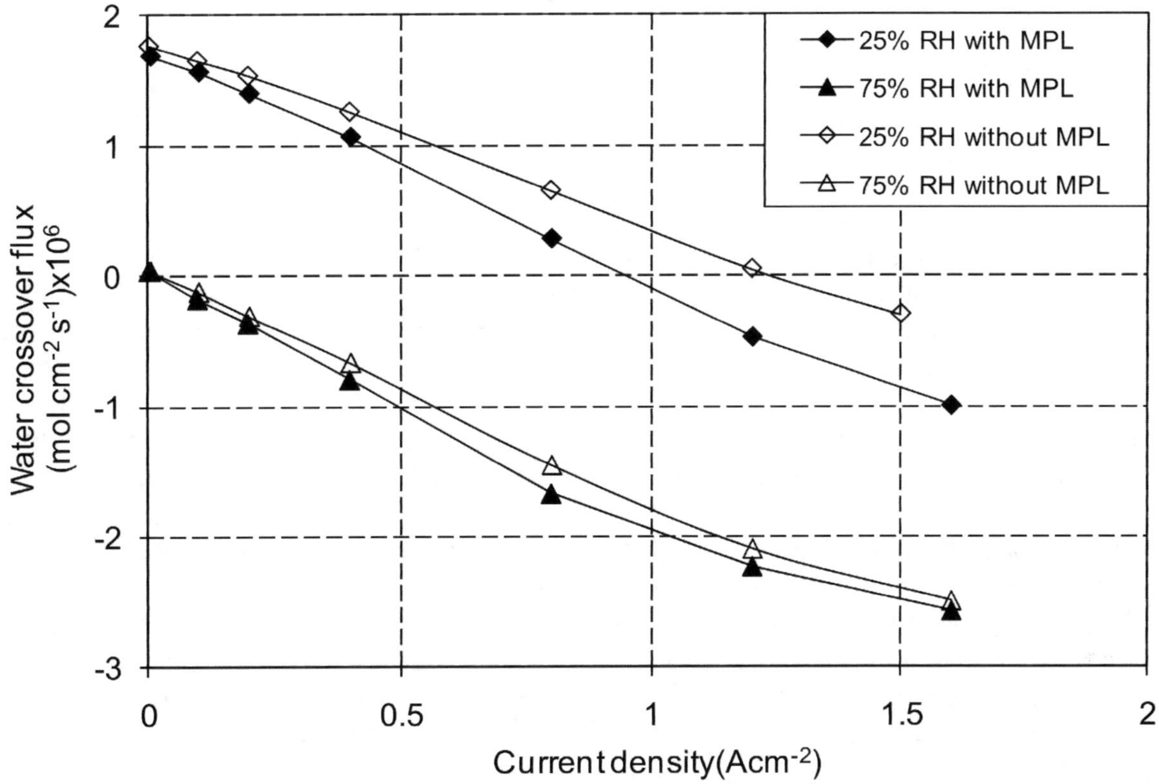

Figure 2 Water crossover flux (measured from anode) against current density. Negative values indicate water transfer from cathode back to anode. Results are the average from two replicates. MEA construction and cathode RH are as indicated. T = 70 °C, P_A = 230 kPa, P_C = 230 kPa, $RH_{in,A}$ = 75%, H_2 flow rate = 4.46×10^{-3} mols^{-1}, N_2 flow rate = 2.98×10^{-3} mols^{-1}, O_2 flow rate = 7.81×10^{-4} mols^{-1}.

By comparing the water crossover behaviour for MEAs with and without MPL, it is seen that adding an MPL makes the slope of the curve more negative for 25% cathode RH. By conducting a mass balance at the cathode interface, it could be shown that the slope is related to the distribution of generated water between anode and cathode channels:

$$J_{H_2O} + \frac{I}{2F} = J_{H_2O,cc} \quad [1]$$

$$-2F\frac{\partial J_{H_2O}}{\partial I} + 2F\frac{\partial J_{H_2O,cc}}{\partial I} = 1 \quad [2]$$

where J_{H_2O} is the anode-to-cathode water flux and $J_{H_2O,cc}$ is the water going from the cathode catalyst layer to the cathode channel. So the addition of MPL did enhance back diffusion of product water to the anode channel at 25% cathode RH. If the water crossover - current density relationship is regarded as linear, 27.4% of the water generated ended up in the anode channel without an MPL while that for the MEA with MPL would be 33.7%. However in the case of 75% cathode RH, the difference was a lot smaller. For the linear region up to 1.2 Acm^{-2}, 34.9% of the water generated was distributed to the anode without an MPL while the presence of MPL only increased this percentage to 37.4%. This is consistent with the findings by Atiyeh et al. (2007) that insignificant changes in water crossover occurred at anode and cathode humidities of 60% and 100%. In addition, it could be seen that the difference diminished at higher current densities.

A possible reason for the diminishing effect of back diffusion enhancement under high cathode RH is that the driving force for back diffusion is much lower, thereby minimizing the role of the MPL in changing the environment at the cathode catalyst layer, which is the source of water generation and back diffusion. This can be illustrated using the model proposed by Berg et al. (2004). At the cathode membrane interface (which is also the cathode catalyst layer in the model), the water desorption from the membrane to the cathode channel is proportional to the difference between the water content at the cathode membrane interface and the equilibrium water content with the cathode channel humidity. When the cathode RH is high, the equilibrium water content at the interface would also be high, according to Springer et al. (1991). Under the same current density, the water brought to the cathode membrane interface by electro-osmotic drag and generation from reaction would be the same, leading to similar water content at that interface. Therefore the driving force would decrease as a result of increasing cathode RH, as shown by simulation in Figure 3. With a smaller driving force for water transport, it would be less sensitive to the effect of MPL under high cathode RH.

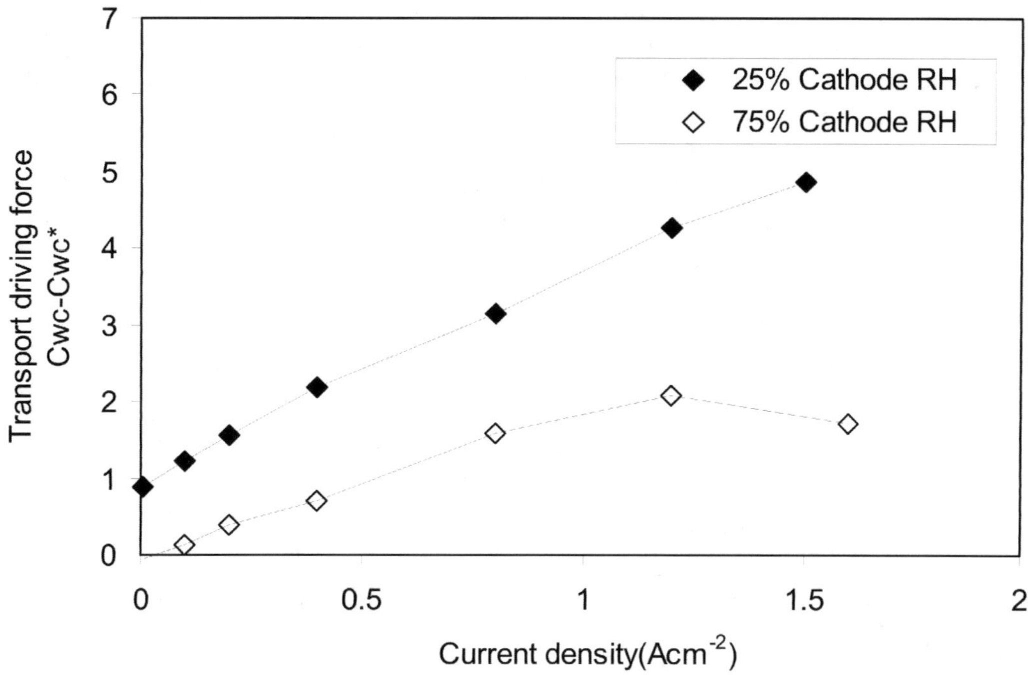

Figure 3 Driving force for water transport at the cathode membrane interface as predicted by the model by Berg et al. (2004). The conditions for simulation are the same as for the experiments without a MPL.

Conclusions

The effect of MPL under different humidity conditions has been studied using a well calibrated water crossover measurement apparatus with good reproducibility. The effect of MPL on polarization was positive, but was less significant than the increase in cathode RH. A general decreasing trend in water crossover flux with respect to current density was observed, and at a 75% cathode RH the decrease in flux became smaller at current densities greater than 1.2 Acm^{-2}. The addition of MPL at the cathode changes the slope of the curve to more negative, indicating increased distribution of produced water to the anode. The effect of MPL was less significant at 75% RH, which could be explained by a decrease in driving force of water transport at the cathode membrane interface under high RH conditions.

References

Atiyeh, H. Karan, K. Peppley, B. Phoenix, A. Halliop, E. and Pharoah, J. 2007. Experimental investigation of the role of a microporous layer on the water transport and performance of a PEM fuel cell. J. Power Sources 170: 111-121.

Berg, P. Promislow, K. St Pierre, J. Stumper, J. and Wetton, B. 2004. Water management in PEM fuel cells. J. Electrochem. Soc. 151: A341-A353.

Dai, W., Wang, H. Yuan, X. Martin, J. Shen, J. Luo, Z. et al. 2009. Measurement of water transport rates across the gas diffusion layer in a proton exchange membrane fuel cell, and the influence of polytetrafluoroethylene content and micro-porous layer. J. Power Sources 188: 122-126.

Janssen, G. and Overvelde, M. 2001. Water transport in the proton-exchange-membrane fuel cell: Measurements of the effective drag coefficient. J. Power Sources 101: 117-125.

Murahashi, T. Naiki, M. and Nishiyama, E. 2006. Water transport in the proton exchange-membrane fuel cell: Comparison of model computation and measurements of effective drag. J. Power Sources 162: 1130-1136.

Sauriol, P. Nobes, D. Bi, X. Stumper, J. Jones, D. and Kiel, D. 2009. Design and Validation of a Water Transfer Factor Measurement Apparatus for Proton Exchange Membrane Fuel Cells. J. Fuel Cell Sci. Technol. 6: 041014.

Springer, T. Zawodzinski, T. and Gottesfeld, S. 1991. Polymer Electrolyte Fuel-Cell Model. J. Electrochem. Soc. 138: 2334-2342.

Yau, T. Sauriol, P. Bi, X. and Stumper, J. 2010. Experimental determination of water transport in polymer electrolyte membrane fuel cells. J. Electrochem. Soc. 157: B1310-B1320.

Zawodzinski, T. Davey, J. Valerio, J. and Gottesfeld, S. 1995. The water-content dependence of electroosmotic drag in proton-conducting polymer electrolytes. Electrochim. Acta 40: 297-302.

In-Situ Measurement of Liquid Water Saturation Level in the Gas Diffusion Layers of a PEMFC

Trung V. Nguyen[a] and , Xuhai Wang[a1], Daniel S. Hussey[b], and David L. Jacobson[b]

[a]Chemical & Petroleum Engineering Department, University of Kansas, Lawrence, Kansas 66045, USA
[b]Physics Laboratory, National Institute of Standards and Technology, Gaithersburg, MD 20899, USA

I. Background

The gas diffusion layer (GDL), which is located between the catalyst layer and gas channel, can greatly affect the transport rates of reactant/product gases to and from the catalyst layers. During operation water generated in the cathode catalyst layer by the oxygen reduction reaction and electro-osmosis is transported out of the fuel cell through the cathode GDL to the gas flow channel. Excessive liquid water accumulation in the GDL can affect the transport rate of oxygen from the gas channel to the catalyst layer. Consequently, the ability to quantify the liquid saturation level in the GDLs of a PEM fuel cell is of great importance. It allows us to study the effect of the morphological and wetting properties of a GDL on the liquid water saturation level and the transport rate of liquid water in the GDL and, ultimately, the performance of the fuel cell. These measurements can also be used to validate the predictions of fuel cell models.

Currently only two in-situ methods can be used to measure the liquid water saturation level in a gas diffusion layer of a PEM fuel cell during operation: X-ray computed tomograph[1,2] and neutron imaging[3,4]. High-resolution x-ray is limited by the dimensions of the sample. High-resolution neutron imaging of both in-plane and through-plane water distributions in a GDL is possible; however, limited facility availability/accessibility is a major drawback

II. Methodology

To overcome these drawbacks, we have developed an approach that is based on a diagnostic tool developed by He *et al.*[5] This approach involves two experiments. **In the first experiment**, gravimetric and neutron imaging methods are used to obtain the correlations of liquid and gas permeability to liquid saturation levels. Figure 1 shows the relative permeability experimental setup with neutron imaging.

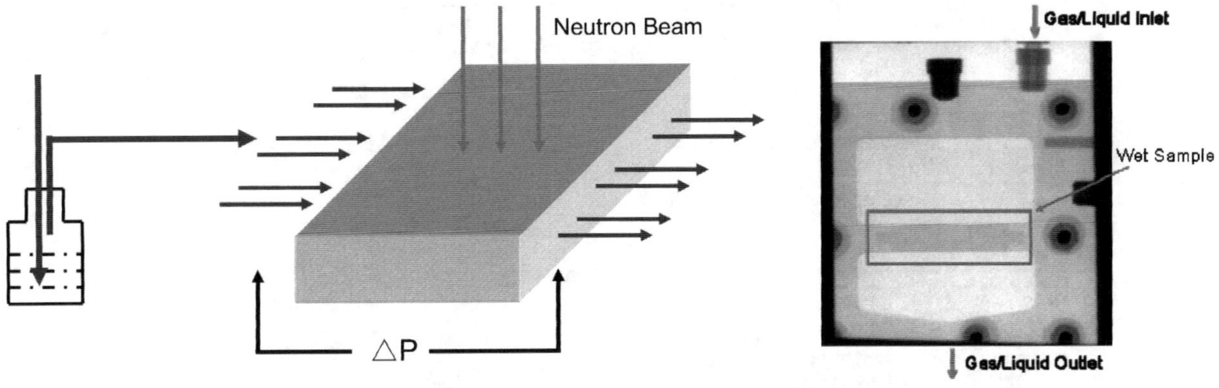

***Figure 1.** Schematic of relative permeability setup and neutron image of the fixture.*

To obtain the correlation of gas relative permeability and liquid saturation level, a GDL sample pre-saturated with liquid water is mounted in the fixture, water saturated gas at a controlled mass flow rate

[1] Current Address: Dept. of Chemical & Biomolecular Engineering, North Carolina State University, Raleigh, NC 27695

is flown through the sample, and the gas pressure drop across the sample is measured. In the neutron imaging approach, water content in the sample during the experiment is obtained from the changes in the intensity of the incident and transmitted neutron beam.[4] In the gravimetric method, the GDL sample is removed and weighed to determine the liquid saturation level. From the pressure drop, gas flow rate, saturation level, and viscosity of the gas, correlation of the gas relative permeability and saturation level is obtained ($k_{rg}(s) = -(v_g m_g \Delta x)/(\Delta P_g)$). To obtain the liquid relative permeability, a dry GDL sample is used, liquid water is flown through the sample, and the liquid pressure drop across the sample is measured. The liquid saturation level is obtained as in the gas relative permeability measurement. Note that these measurements are needed only once for each gas diffusion material.

The second experiment involves running a PEM fuel cell with a flow field that can be switched from serpentine flow mode to interdigitated flow mode and measuring the pressure drop across the GDL right after the fuel cell is switched from the serpentine mode to the interdigitated mode and at the end of the interdigitated mode. The pressure drop across the GDL right after the flow field is switched to the interdigitated mode correlates with the liquid water saturation level in the GDL during the serpentine mode. Differences between the pressure drop at the beginning and end of the interdigitated flow mode are used to determine how effective the interdigitated flow mode is in removing water from a GDL. The gas pressure drops measured in this experiment are converted first to gas relative permeability and then to liquid saturation levels using the correlations obtained from the gas relative permeability experiment.

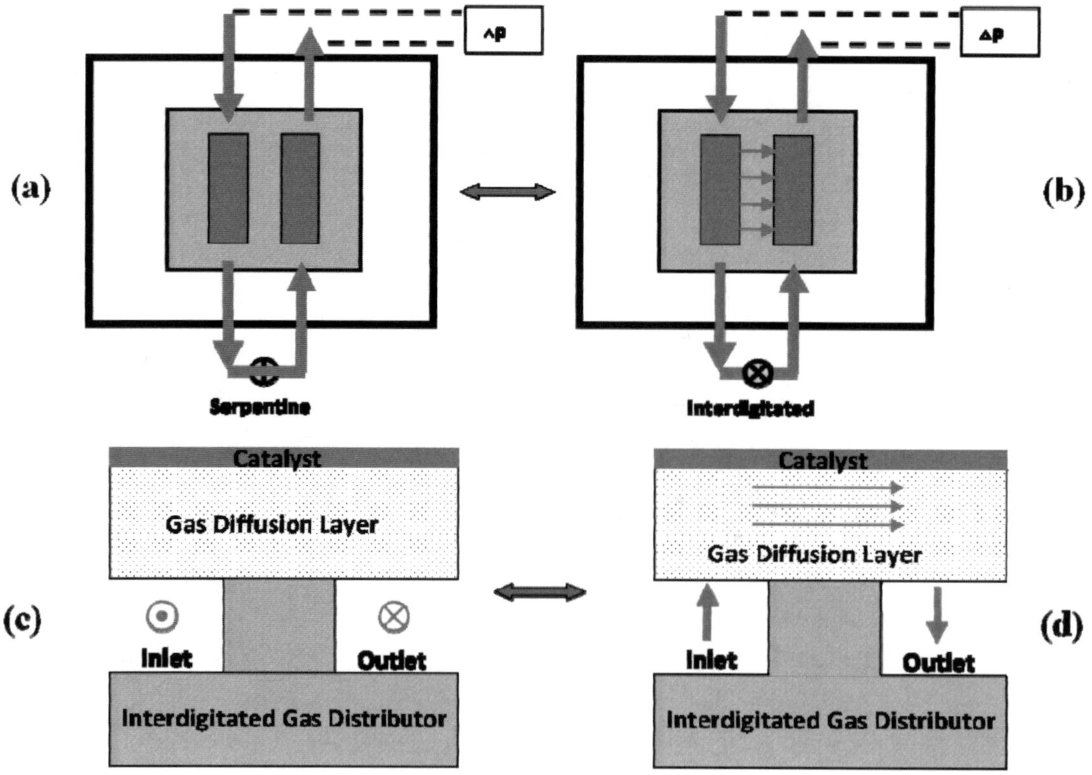

Figure 2. *Schematic of (a) serpentine and (b) interdigitated flow modes and cross-sectional views of (c) serpentine and (d) interdigitated flow fields. ⊙represents gas flowing into the surface, and⊗ represents gas flowing out of the surface.*

III. Results & Discussion

Gas relative permeability results of a proprietary gas diffusion material are shown in Figure 3. Note that for this material the 3rd-order power correlation often used in PEM fuel cell models does not fit the data very well. The discrepancy between the neutron imaging and gravimetric methods is attributed to the fact that some of the liquid droplets still attached to the edge of the sample at the outlet were imbibed back into the sample when the gas phase pressure was removed during each gravimetric measurement.

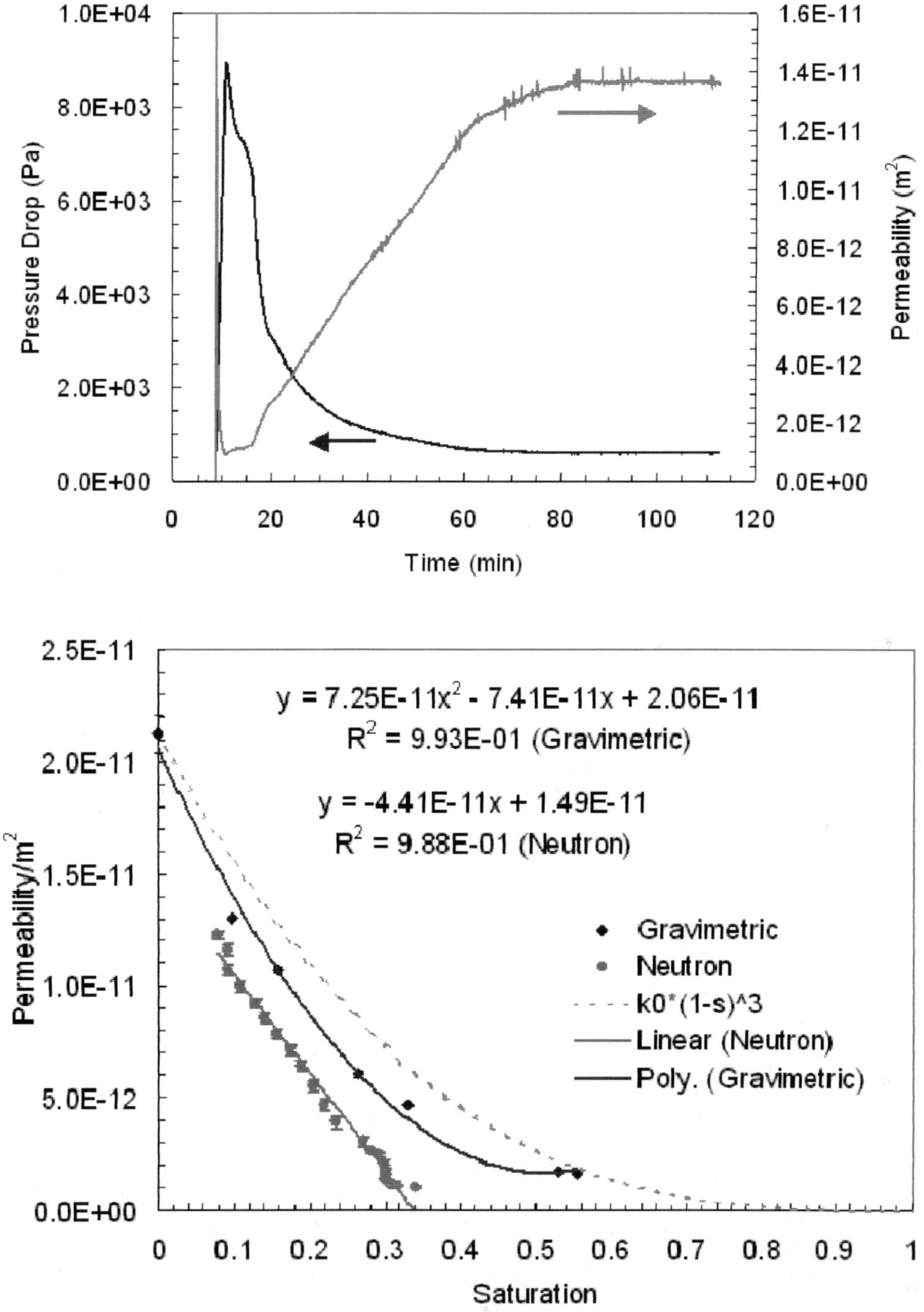

Figure 3. *Results from the gas relative permeability measurement*

Results from the saturation level study with a PEM fuel cell with a switchable cathode flow field for the same gas diffusion material are shown in Figure 4. The test conditions are 2/4 for anode/cathode stoichiometric flow rates, 1 bar for fuel cell outlet pressure, 70 °C for cell and anode/cathode humidifier temperatures.

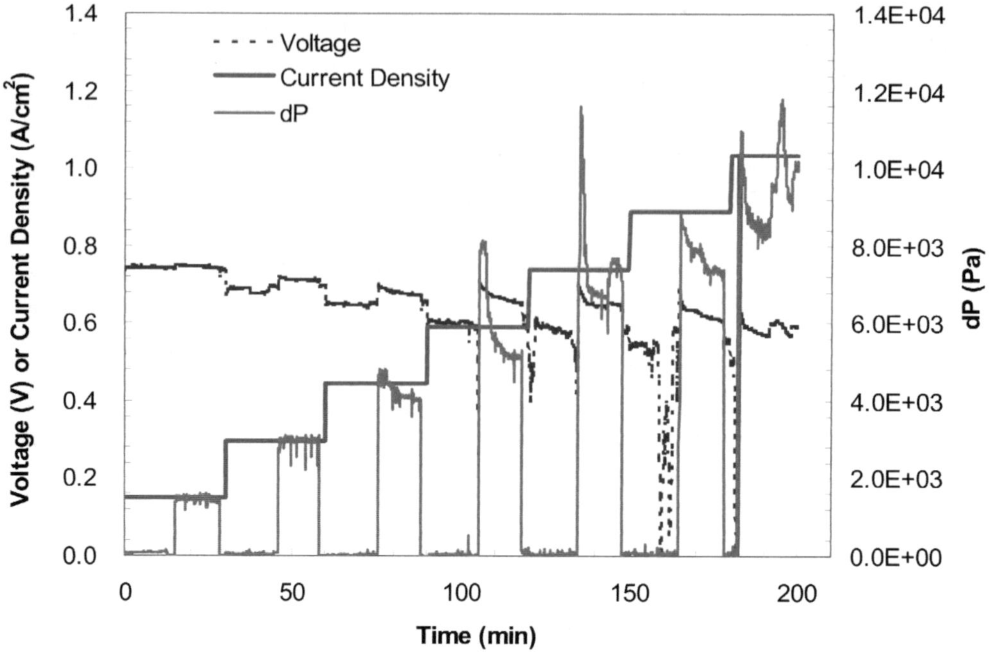

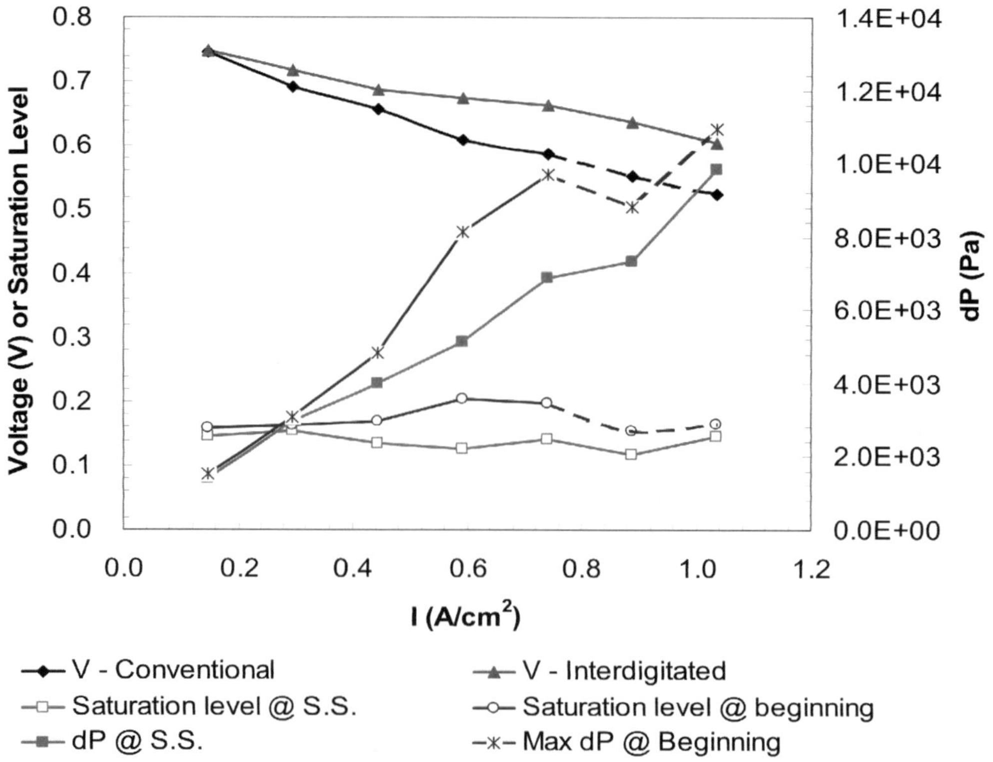

Figure 4. *Fuel cell test results*

From these results the followings observations can be made. First, the pressure drop in the serpentine mode is stable and negligible since the resistance in the flow channel is small. Second, in the interdigitated mode the pressure drop is stable at low current density but shows a decreasing trend at high current density. At low current density, little water accumulates in the porous GDL since the water generation rate by the oxygen reduction reaction and electro-osmosis is low. Hence, the water saturation levels under serpentine flow and interdigitated flow do not differ much. At high current density, liquid water accumulates in the porous GDL above the channels and rib during the serpentine flow mode resulting in reduced gas void volume. As the flow mode switches from serpentine to interdigitated, the pressure drop is initially high because of the high liquid water level in the GDL during the serpentine flow mode. As air flows through the GDL under the interdigitated mode, the shear force purges part of the liquid water accumulated in the GDL leading to a lower saturation level and lower pressure drop. Third, the fuel cell voltage under interdigitated mode is higher than that under serpentine mode. This higher performance can be attributed to higher reactant pressure or concentration and faster transport of the oxygen reactant to the catalyst layer in the interdigitated flow mode. Finally, the cell voltage under the serpentine mode became unstable at current density 0.9 A/cm^2 and higher because of excessive accumulated liquid water in the cathode GDL at these current densities. The critical saturation level at which the fuel cell begins to fail under serpentine flow mode appears to be around 0.2. Note that since air flow through a GDL under interdigitated mode can only sweep so far into the GDL in this arrangement, the actual liquid saturation level near the micro-porous layer could be much higher. Note also that in the current arrangement this technique cannot determine the liquid saturation levels in the MPL and the CL.

IV. Conclusions

This very effective method allows the liquid saturation level in a gas diffusion medium in a PEM fuel cell to be quantified during operation. This method has also been used to look at the effects of various fuel cell operating conditions such as stoichiometric flow rates, relative humidity, and temperature on the liquid water saturation level.

Acknowledgements

This work was supported by Toyota Motor Corporation and by the U.S. Department of Commerce, the NIST Ionizing Radiation Division, the Director's office of NIST, the NIST Center for Neutron Research, and the Department of Energy interagency agreement No. DEAI01-01EE50660.

References

1. P. K. Sinha, P. Halleck and C.-Y. Wang, *Electrochem. Solid-State Lett.*, **9**, A344 (2006).
2. W. Zhu, Z. W. Dunbar and R. I. Masel, *ECS Trans.*, **16**, 995 (2008).
3. A.B. Geiger, A. Tsukada, E. Lehmann, P. Vontobel, A. Wokaun and G.G. Scherer, *Fuel Cells*, **2**, 92 (2002).
4. D. S. Hussey, D. L. Jacobson, M. Arif, J. P. Owejan, J. J. Gagliardo and T. A. Trabold, *J. Power Sources*, **172**, 225 (2007).
5. W. He, G. Lin and T. V. Nguyen, *AIChE J.*, **49**, 3221 (2003).

(641i) AC Impedance Spectroscopy Analysis of Improved Proton Exchange Membrane Fuel Cell Operation Via Direct Inlet Gas Humidity Control

Li Tan[1], Doug McClelland[2] and **Anastasios Angelopoulos**[1], (1)Chemical and Materials Engineering, University of Cincinnati, Cincinnati, OH, (2)Mound Technical Solutions, Miamisburg, OH

Efficient management of water is a major factor toward improving operation of Proton Exchange Membrane (PEM) Fuel Cells. The amount of water in various PEM fuel cell components will impact a variety of microprocesses that involve the transport and reaction of protons, electrons, and molecular species critical to efficient fuel cell operation. Electrochemical impedance spectroscopy (EIS) may be used to help resolve and quantify the relative magnitude of these processes in the operating fuel cell. This study is aimed at using EIS to evaluate how a new method of precisely controlling fuel cell humidity may be used to optimize the performance of a PEM fuel cell. The method involves humidity sensing and control of the reactant gas streams immediately before entry into the fuel cell stack. The interaction between various relative anode/cathode inlet humidity levels (wet/wet, dry/wet, wet/dry, and dry/dry) and stoichiometry will be investigated at a given temperature. Data obtained from this study is compared and fitted to equivalent circuit models proposed in literature to test their validity and to quantify fundamental process parameters.

See more of this Session: Fuel Cell Technology
See more of this Group/Topical: Fuels and Petrochemicals Division

(674a) Application of Top-Down Ecological Control Principles to Maximize Algal Productivity in Open Pond Systems

Belinda S.M. Sturm, Environmental Engineering, The University of Kansas, Lawrence, KS, Val H. Smith, Ecology and Evolutionary Biology, University of Kansas, Lawrence and Frank Jerry deNoyelles, Kansas Biological Survey, Lawrence

Of the biofuel feedstocks investigated to reduce the nation's dependence on foreign oil, algae are the most promising due to their high lipid content, rapid growth rate, and ability to sequester CO2. Algae do not compete with existing food commodities, can grow on marginal lands not suitable for agriculture, and do not require large volumes of fresh water. Moreover, algae can remove nitrogen and phosphorus from nutrient-rich wastewaters, thus reducing eutrophication of receiving water bodies that cause gross disruptions of ecosystems, such as the Gulf of Mexico hypoxia. Algal production has been estimated to yield 3,200 to 14,600 gallons of oil/acre/year, but mass culture efforts have yielded 10-20 times less oil than expected. Outdoor bioreactors that are subject to natural variations and function as ecosystems have not mimicked pure laboratory reactors, and sustainable production of algae will require an abundant and nutrient-rich water source, which is unlikely to support the mass cultivation of pure cultures. This project incorporates ecological and reactor engineering concepts to develop a sustainable algal growth system in wastewater-fed, outdoor bioreactors. Four 2600-gallon continuous-flow reactors with a ten day hydraulic residence time were operated at the Lawrence, KS Wastewater Treatment Plant for six months in 2009. Top-down ecological control practices were applied to two of the reactors in order to increase and stabilize algal production.

The growth of algae growing in mixed-species, outdoor community cultures will be regulated by their supplies of macronutrients (nitrogen and phosphorus), carbon dioxide, and incident light. These algal species will vary significantly in size and shape; some of them will be small and edible by algae-eating zooplankton, while some of them will be large and inedible. The amount of lipid that can be produced from these algae will be constrained by zooplankton grazing in the bioreactor. During the past 25 years, a very large and globally-accepted body of literature has been developed which indicates that in natural lakes, algal biomass losses to zooplankton grazing can be minimized by including zooplanktivorous fish in the food web (top-down ecological control; see Fig. 2 from Smith et al. 2010). Likewise, algal biomass in outdoor bioreactors can be strongly affected by food web structure. Fishless bioreactors can be expected to develop large populations of large-bodied herbivorous zooplankton, thus exhibiting significantly lower algal biomass than would be predicted from their nutrient supply, relative to bioreactors that contain zooplanktivorous fish and smaller, less-efficient grazers.

One of our primary research goals was to determine whether the ecological concept of top-down control applies to microalgae grown in outdoor bioreactors. If this concept indeed applies, then careful control of food-web structure will allow for the design and operation of future large-scale outdoor bioreactors or ponds for algal biomass production – without the danger of algal population crashes due to predation.

Highlights of Results • Algae may be successfully cultivated in open pond reactors using wastewater effluent as an N and P source. • Top-down ecological control practices can decrease losses of algae to zooplankton predation. • The % lipid content of algae cultivated in the field is consistent with laboratory studies.

References Smith VH, Sturm BSM, Billings S, deNoyelles FJ. 2010. "Ecological aspects of algal biodiesel production". Trends in Ecology & Evolution, 25(5), 301-309.

See more of this Session: Advances in Algal Biorefineries I
See more of this Group/Topical: Sustainable Engineering Forum

(674b) Maximizing Productivity in Batch Reactors of Microalgae Nannochloropsis Gaditana

Ming Ren, Department of Chemical and Environmental Engineering, University of Arizona, Tucson, AZ and Kimberly Ogden, University of Arizona, Tucson, AZ

Biodiesel from algae is receiving more attention. Just recently, the United States Department of Energy (DOE) announced to fund around $85 million in the development of algae-based biofuels and advanced infrastructure-compatible biofuels. Compared to other terrestrial crops, such as soybean, sunflower, canola, palm, jatropha and others, algae has many advantages. It has high oil productivity, a short growth cycle and survives in a wide variety of water sources including high salinity and waste water.

Nutrients, salinity, pH, temperature, light intensity and aging of the culture can all affect on both lipid content and fatty acid profile. Usually nutrient stress is the easiest way to manipulate lipid composition and increase lipid content. In our research, microalgae *Nannochloropsis Gaditana* has been shown to have high lipid content (more than 40%) and quick growth rate (around $0.018 hr^{-1}$). Current work focuses on how the harvest time and nitrogen sources affect on the lipid content and composition.

The cell growth, sodium nitrate concentration, and lipid content and profile are monitored as a function of time. Decreasing nitrogen concentration does stimulate lipid content increase with sacrificing growth rate. In contrast, increasing nitrogen concentration does not increase lipid content. Maximum lipid content is obtained in mid-exponential phase and is relatively constant from this point through stationary phase. Major lipid compounds are palmitic acid (16:0) and palmitoleic acid (16:1), and lipid profile does change significantly with time.

See more of this Session: Advances in Algal Biorefineries I
See more of this Group/Topical: Sustainable Engineering Forum

(674c) Photobioreactor Productivity Enhancement Offered by Oscillatory Flow Mixing

Benjamin J. Taylor and Malcolm Mackley, Department of Chemical Engineering and Biotechnology, University of Cambridge, Cambridge, United Kingdom

Closed photobioreactors offer certain advantages over open algal growth systems in terms of consistent product output and purity but still suffer from other drawbacks. Limitations of many tubular photobioreactors were identified, and a new reactor was constructed to meet these challenges by utilizing the proven processing advantages of oscillatory flow mixing (OFM). The unique mass and energy transfer characteristics of OFM offer excellent bulk-to-wall turnover and suspension capabilities similar to that provided by turbulent flow but at significantly reduced input energy requirements. Growth experiments with the algal strain *Scenedesmus obliquus* demonstrated improved growth kinetics and sustained productivity within the laboratory-scale OFM reactor surpassing those from other bioreactor types. Computational fluid dynamics accurately predicted the suspension flow patterns obtained experimentally with OFM and offer insight into proper process control variables tailored to a wide range of algal species and reactor scales.

See more of this Session: Advances in Algal Biorefineries I
See more of this Group/Topical: Sustainable Engineering Forum

Effects of Total Inorganic Carbon on Growth of *Chlorella vulgaris*

Jinsoo Kim [a], Joo-Youp Lee [a*], Kaniz F. Siddiqui [b]
[a] *Chemical Engineering Program, School of Energy, Environmental, Biological, and Medical Engineering, University of Cincinnati, Cincinnati, OH, 45221-0012, U.S.A.*
[b] *The Metropolitan Sewer District of Greater Cincinnati, Cincinnati, Ohio 45204-2022, U.S.A.*

1. Introduction

Diminishing petroleum resources along with increased demand for petroleum by emerging economies are driving our society to search for new renewable and carbon neutral liquid transportation fuels. Biofuels such as bioethanol and biodiesel derived from agricultural and forestry residues and oil crops have been considered to be candidates for renewable liquid fuels. Recently, microalgae-derived biodiesel production has received great attention due to their ability to produce substantial amounts of triacylglycerols (e.g., 20–50% dry cell weight) [1], as a feedstock for biodiesel [2, 3], and to utilize a large fraction of solar energy (up to 10%) [4], and carbon dioxide (CO_2), and to grow at much faster rates than terrestrial biomass [5]. In addition, sugars generated as metabolites have potential to produce bioalcohols [6-10].

Microalgae utilize carbon dioxide (CO_2) as their carbon source and energy primarily from sunlight. They are photosynthetic, like terrestrial plants, but lack the various structures that characterize plants. It is estimated that there are 200,000–1,000,000 different microalgal strains in nature, but only ~30,000 strains have been described [11]. Among the strains, *Chlorella vulgaris* belongs to a genus of single-celled green algae, and is known as one of the fastest growing microalgae, and typically includes 14–22 % of lipid, 51–58 % of protein, 12–17 % of carbohydrates, and 4–5 % of nucleic acid [12]. Several studies on lipid production [13] and growth kinetics of *Chlorella vulgaris* have been conducted in terms of initial CO_2 concentration [14] and pH level [15, 16] in order to find optimum growth conditions for large-scale biofuels production. However, these previous growth kinetic studies did not focus on actual initial total inorganic carbon concentration used for algae growth in the aqueous phase. The solubility of CO_2 gas in water is very limited, and some of microalgae have the capability to convert bicarbonate ion into CO_2 for their carbon uptake with the help of an enzyme called carbonic anhydrase [17-20].

In this study, the initial total inorganic carbon concentration in the culture medium was controlled by using sodium bicarbonate ($NaHCO_3$) and CO_2 gas as a carbon source for photosynthesis of *Chlorella vulgaris* under the similar initial pH condition. This method is allowed to be controlled the initial total inorganic carbon concentration in culture medium due to the high solubility of the sodium bicarbonate in the water. In addition, the bubbled CO_2 was used for the decreasing pH and injected CO_2 gas was captured by the sodium bicarbonate.

2. Experimental

2.1. Preparation of Culture Medium

A culture medium was prepared by following the shuisheng-4 method [21, 22], and included ammonium sulfate ($(NH_4)_2SO_4$, EM Industries, Inc., Gibbstown, N.J.), boric acid (Allied chemical), cupric sulfate (J. T. Baker), calcium phosphate monobasic monohydrate, magnesium sulfate hetahydrate, sodium bicarbonate, potassium chloride, iron (III) chloride, potassium phosphate dibasic, manganese chloride, zinc sulfate, sodium molybdate (Fisher Scientific Company). The medium was mixed together with 200 mg/L of ammonia sulfate ($(NH_4)_2SO_4$), 30 mg/L of calcium phosphate monobasic monohydrate ($Ca(H_2PO_4)_2 \cdot H_2O$), 100 mg/L of magnesium sulfate hetahydrate ($MgSO_4 \cdot 7H_2O_3$) 80 mg/L, 25 mg/L of potassium chloride (KCl), 1.5 mg/L of iron (III) chloride($FeCl_3$), 10 mg/L of potassium phosphate dibasic (K_2HPO_4), and 1 mL of A5 liquid. Herein, A5 liquid was prepared by mixing together with 2.86 g/L of boric acid (H_3BO_3), 1.81 g/L of manganese chloride ($MnCl_2 . 4H_2O$), 0.222 g/L of zinc sulfate ($ZnSO_4 \cdot 7H_2O$), 0.391 g/L of sodium molybdate ($Na_2MO_4 \cdot 2H_2O$), 0.079 g/L of cupric sulfate ($CuSO_4 \cdot 5H_2O$). In addition, sodium bicarbonate ($NaHCO_3$) was not added in this step because the sodium bicarbonate was used as a inorganic carbon source in the next step.

2.2. Preparation of Different Initial Total Inorganic Carbon Concentration in Culture Medium

The initial total inorganic carbon in S1 culture medium was adjusted by adding sodium hydroxide (NaOH) until pH increased to 8.99 and bubbling CO_2 gas (≥99.5%, Wright Brothers, Inc., Cincinnati, OH) until pH decreased to 6.99. Then, the initial total inorganic carbon in S2-S5 culture mediums were controlled by adding different amount of sodium bicarbonate ($NaHCO_3$) (e.g. 100 mg of $NaHCO_3$/ 1L H_2O (S2, pH=7.05), 400 mg of $NaHCO_3$/ 1L H_2O (S3, pH=7.36), 1600 mg of $NaHCO_3$/ 1L H_2O (S4, pH=7.68) and 3200 mg of $NaHCO_3$/ 1L H_2O and bubbling CO_2 gas until pH decreased to pH=7 and the final pH was 7.05 (S2), 6.82 (S3), 7.04 (S4), and 7.08 (S5), respectively. Here, in the case of S2 culture medium, extra CO_2 gas was not bubbled after adding sodium bicarbonate ($NaHCO_3$) because the pH was close to pH=7.

2.3. Determination of Cell Density of Chlorella vulgaris

The cell density of *Chlorella vulgaris* was determined by measuring the optical density of a 15-mL sample at 682 nm [13] for every 24 hours by using UV-vis spectrophotometer (UV-1800, Shimadzu Scientific Instruments). The absorbance of UV spectrophotometer at 682 nm was calibrated by measuring the weight of dried *Chlorella vulgaris*. Then, the weight of dried biomass was obtained from the prepared calibration curve.

2.4. Determination of Total Inorganic Carbon Concentration

An acid-base titration method [23] was used to determine the inorganic carbon species in the aqueous phase. This titration method determines total inorganic carbon concentration (carbonate (CO_3^{2-}), bicarbonate (HCO_3^-) and aqueous carbon dioxide (CO_2* (aq))) using a 0.01 N hydrochloric acid solution as a titrant.

3. Results and Discussion

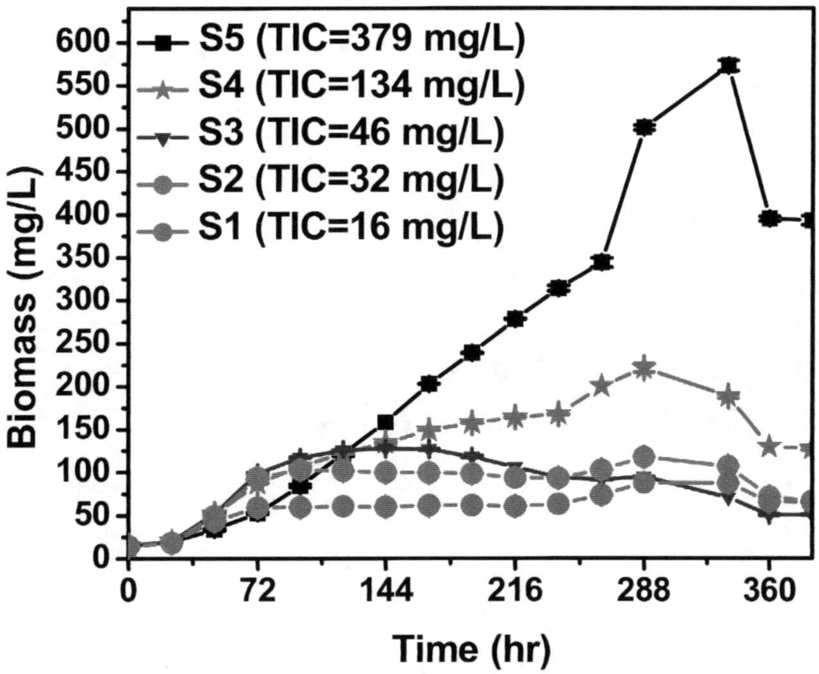

Figure 1. Growth kinetics of *Chlorella vulgaris* dependent on different initial total inorganic carbon (e.g. S1 (TIC=16 mg/L), S2 (TIC=32 mg/L), S3 (TIC=46 mg/L), S4 (TIC=134 mg/L), and S5 (TIC=379 mg/L).

The growth of *Chlorella vulgaris* and the consumption of total inorganic carbon are represented in Figures 1 and 2. At first, initial total inorganic carbon in S1-S4 culture mediums were 16 mg/L (S1), 32 mg/L (S2), 46 mg/L (S3), 134 mg/L (S4) and 379 mg/L (S5), respectively. The lag-phase of *Chlorella vulgaris* in S1 and S2 culture mediums was short (24 hrs) and the phase of *Chlorella vulgaris* in S3 and S4 culture mediums was modest (48 hrs) while a small amount of dissolved total inorganic carbon concentration in those mediums decreased during the period. Otherwise, the lag-phase in S5 culture medium showed the longest lag period (72 hrs) compared to other cases (S1-S4) because it might be derived from the high concentration of total inorganic carbon. During the exponential growth phase, the cell density of *Chlorella vulgaris* in S1-S5 culture mediums increased, and the dissolved total inorganic carbon dramatically decreased. Then, the growth of the algae in culture mediums stopped at 72 hrs (S1), 96 hrs (S2), 144 hrs (S3), 288 hrs (S4) and 336 hrs (S5), respectively. In addition, the maximum cell density decreased in the order of S5 (573.27±5.90 mg/L) > S4 (221.68±4.94 mg/L) > S3 (95.17±0.23) > S2 (118.06±0.23)> S1 (87.87±0.00). Therefore, it was noted that *Chlorella vulgaris* cultured in the medium with high total inorganic carbon concentrations can produce more algal biomass yield. However, the carbon consumption was delayed as its concentration increased, and too high a total inorganic carbon concentration resulted in longer lag-phase period. The pH of the culture medium continued to increase from the lag phase through the growth phase (Figure 2) as *Chlorella vulgaris* continued to produce hydroxyl ion (OH^-) during photosynthesis by the following reaction, $HCO_3^- \rightarrow CO_2 + OH^-$. For the culture mediums with moderate total inorganic carbon concentrations (i.e. S1–S3), as the uptake of total inorganic carbon levels off, the pH also reached a constant level.

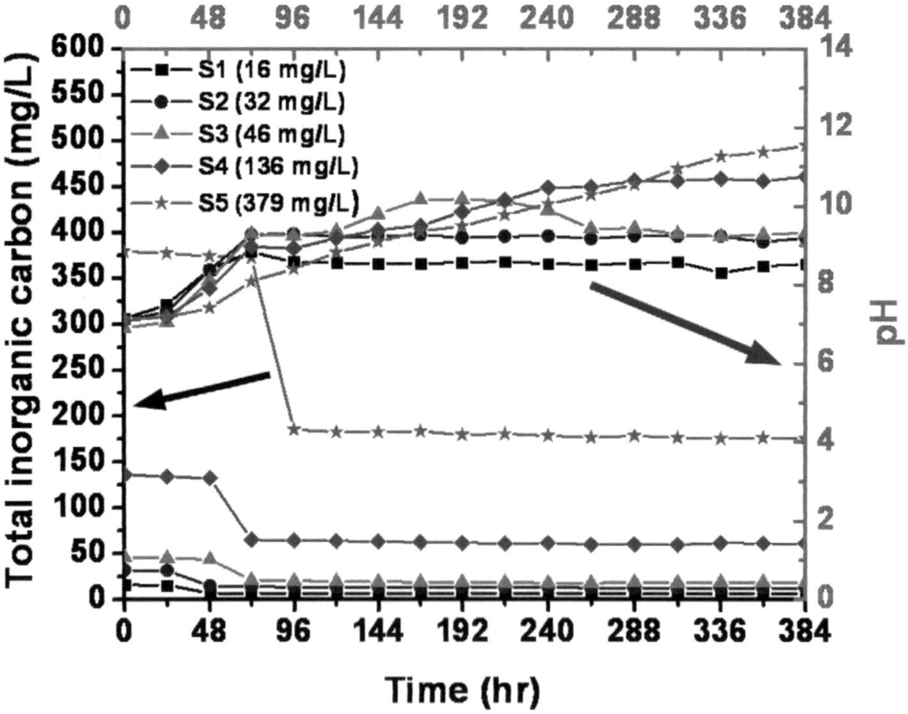

Figure 2. Change of total inorganic carbon concentration in culture mediums.

4. Conclusions

The effects of total inorganic carbon concentrations on the growth of *Chlorella vulgaris* have been studied using sodium bicarbonate and carbon dioxide gas as carbon sources for photosynthesis. The lag-phase period increased with an increase in inorganic carbon concentrations (i.e. 24 hrs for S1 and S2, 48 hrs for S3 and S4, and ~72 hrs for S5). The period of growth-phase increased with an increase in total inorganic carbon concentrations. However, a long lag-phase period was observed from the culture medium with an extremely high inorganic carbon concentration (i.e. S5). It was also observed that *Chlorella vulgaris* cultured in the medium with high total inorganic carbon concentrations can produce high algal biomass yield (i.e. S5 > S4> S3 > S2 >S1).

Acknowledgement

The This work was supported by the Metropolitan Sewer District of Greater Cincinnati (MSDGC) under Master Services Agreement No. 85X10431, Task Order No. 0210000209. The authors appreciate their financial support.

References

1. Hu, Q.; Sommerfeld, M.; Jarvis, E.; Ghirardi, M.; Posewitz, M.; Seibert, M.; Darzins, A., Microalgal triacylglycerols as feedstocks for biofuel production: perspectives and advances. *Plant Journal* **2008,** *54*, (4), 621-639.
2. Liu, Z.-Y.; Wang, G.-C.; Zhou, B.-C., Effect of iron on growth and lipid accumulation in Chlorella vulgaris. *Bioresource Technology* **2008,** *99*, (11), 4717-4722.
3. Hill, J.; Nelson, E.; Tilman, D.; Polasky, S.; Tiffany, D., Environmental, economic, and energetic costs and benefits of biodiesel and ethanol biofuels. *Proceedings of the National Academy of Sciences of the United States of America* **2006,** *103*, (30), 11206-11210.
4. Huber, G. W.; Iborra, S.; Corma, A., Synthesis of transportation fuels from biomass: chemistry, catalysts, and engineering. *Chemical Reviews* **2006,** *106*, (9), 4404-4498.
5. Chisti, Y., Biodiesel from microalgae. *Biotechnology Advances* **2007,** *25*.
6. Milne, T. A.; Evans, R. J.; Nagle, N., Catalytic conversion of microalgae and vegetable oils to premium gasoline, with shape-selective zeolites. *Biomass* **1990,** *21*, (3), 219-232.
7. Miao, X.; Wu, Q., Biodiesel production from heterotrophic microalgal oil. *Bioresource Technology* **2006,** *97*, (6), 841-846.
8. Dote, Y.; Sawayama, S.; Inoue, S.; Minowa, T.; Yokoyama, S.-y., Recovery of liquid fuel from hydrocarbon-rich microalgae by thermochemical liquefaction. *Fuel* **1994,** *73*, (12), 1855-1857.
9. Minowa, T.; Yokoyama, S.-y.; Kishimoto, M.; Okakura, T., Oil production from algal cells of Dunaliella tertiolecta by direct thermochemical liquefaction. *Fuel* **1995,** *74*, (12), 1735-1738.
10. Ginzburg, B.-Z., Liquid fuel (oil) from halophilic algae: A renewable source of non-polluting energy. *Renewable Energy* **1993,** *3*, (2-3), 249-252.
11. Sheehan, J.; Dunahay, T.; Benemann, J.; Roessler, P. *A look back at the U.S. Department of Energy's Aquatic Species Program—biodiesel from algae.*; National Renewable Energy Laboratory: 1998.

12. Becker, E. W., *Microalgae biotechnology and microbiology.* Cambridge University Press: New York, 1994.

13. Widjaja, A.; Chien, C.; Ju, Y., Study of increasing lipid production from fresh water microalgae Chlorella vulgaris. *Journal of the Taiwan Institute of Chemical Engineers* **2009,** *40*, (1), 13-20.

14. Azov, Y., Effect of pH on inorganic carbon uptake in algal cultures. *Applied and Environmental Microbiology* **1982,** *43*, (6), 1300-1306.

15. Goldman, J. C.; Azov, Y.; Riley, C. B.; Dennett, M. R., The effect of pH in intensive microalgal cultures. I. Biomass regulation. *Journal of Experimental Marine Biology and Ecology* **1982,** *57*, (1), 1-13.

16. Powell, E. E.; Mapiour, M. L.; Evitts, R. W.; Hill, G. A., Growth kinetics of Chlorella vulgaris and its use as a cathodic half cell. *Bioresource Technology* **2009,** *100*, (1), 269-274.

17. Badger, M. R.; Price, G. D., The role of carbonic anhydrase in photosynthesis. *Journal of Experimental Marine Biology and Ecology* **1994,** *45*, 369-392.

18. Kaplan, A.; Reinhold, L., CO2 concentrating mechanisms in photosynthetic microorganisms. *Annual Review of Plant Physiology and Plant Molecular Biology* **1999,** *50*, 539-570.

19. Moroney, J. V.; Somanchi, A., How do algae concentrate CO2 to increase the efficiency of photosynthetic carbon fixation? *Plant Physiology* **1999,** *119*, 9-16.

20. Giordano, M.; Beardall, J.; Reaven, J. A., CO2 concentrating mechanisms in algae: Mechanisms, environmental modulation, and evolution. *Annual Review of Plant Biology* **2005,** *56*, 99-131.

21. Zhou, Y. X.; Zhang, Z. S., *The toxicity methods in aquatic living.* Agriculture Press: Beijing, 1989.

22. Qian, H.; Chen, W.; Sheng, G. D.; Xu, X.; Liu, W.; Fu, Z., Effect of glufosinate on antioxidant enzymes, subcellular structure, and gene expression in the unicellular green alga Chlorella vulgaris. *Aquatic Toxicology* **2008,** *88*, (4), 301-307.

23. Fisher, R. B.; Peters, D. G., *Basic theory and practice of quantitative chemical analysis.* 3rd edition ed.; W. B. Saunders Company: Piladelphia, 1968.

(674e) Effects of CO_2 On South African Fresh Water Microalgae Growth

Edmore Kativu[1], Tonderayi S Matambo[2], Diane Hildebrandt[2] and David Glasser[2], (1)School of chemical and metallurgical engineering, University of the Witwatersrand, Johannesburg, Johannesburg, South Africa, (2)Chemical and Metallurgical Engineering, University of the Witwatersrand, Johannesburg, Johannesburg, South Africa

South Africa obtains over 80% of its energy needs from fossil fuels. Latest research indicate that South Africa emits over 222 000 000 tones of carbon dioxide (CO2) per year and it is ranked number eight as the world's CO2 emitter. Unabated or controlled emission of CO2 and other green house gases has largely contributed to detrimental global warming and adverse climatic change. Microalgae are capable to convert the hazardous CO2 into valuable biomass. A high CO2 tolerating microalgae was collected from Johannesburg Zoo Lake. Batch cultures were grown on 1000 ml Erlenmeyer flasks. CO2 was bubbled into microalgae culture media. The media contained optimal nutrients and under optimal photoperiod and light intensity and controlled pH. Concentrations of 100 %, 50 %, 25 %, 10 %, and 5 % CO2 and total gas flow rates of 20, 50 and 100 ml/min were used. Final concentrations were made up to 100 % by mixing with nitrogen gas. The aim of this study is to find out the effective CO2 concentrations that gives optimal microalgae growth. Microalgae have at 50% carbon in its dry mass and when all the other nutritional and environmental conditions are not limiting carbon is the main limiting factor. When atmospheric air was supplied at 50 ml/minute the growth rate was 0.1151 per day and it increased drastically by almost 5 folds when 100% CO2 was supplied at 50 ml per minute. At a gas flow rate of 20 ml/min 10 % CO2, growth rate optimal growth rate was 1.42 per day and a dry biomass yield of 809.96mg/l after 12 days. At a gas flow rate 50 ml/min 5% CO2 and the highest growth rate was 1.993 per day and an overall biomass yield of 1200mg/l and the average pH was 6.36, at 100% CO2 the growth rate was 1.265 per day and dry biomass yield of 469.81 mg/l and the average pH was 5.12. When gas flow rate was increased to 100 ml/minute 5 % CO2 the growth rate slightly reduced to 1.30 per day and dry biomass yield of 1000 mg/l, while at 100% the growth rate was 0.34 and dry biomass yield of 279.47 mg/l and the average pH was 5.09. Higher flow rates and higher concentrations resulted in slightly reduced growth due to low pH. These investigations indicate that the species under study is CO2 tolerant.

See more of this Session: Advances in Algal Biorefineries I
See more of this Group/Topical: Sustainable Engineering Forum

(674f) Investigation of Chlorella Microalgae Cultivation On Anaerobically Digested Manure

Amarjeet S. Bassi, Chemical & Biochemical Engineering, The University of Western Ontario, London, ON, Canada, Gureet Chandok, Chemical and Biochemical Engineering, The University of Western Ontario, London, ON, Canada, Shaikh Razzak, Chemical and Biochem Engg, The University of Western Ontario, London, ON, Canada and Peter Schnurr, Chemical and Biochemical Engg, The University of Western Ontario, London, ON, Canada

Laboratory and pilot scale studies were carried out to investigate the cultivation of the microalgae *Chlorella vulgaris* on anaerobically digested manure from a local farm. The effects of light, nutrient load and CO_2 to air ratios on cell growth and lipid accumulation were investigated. Both batch and continuous studies were carried out. Rates of CO_2 sequestration were also measured. The results showed that the photo-bioreactor configuration was appropriate to cultivate microalgae on a large on farm operation. Only a small fraction of CO_2 supplied was effectively taken up as microalgae biomass, indicating that using microalgae only for CO_2 uptake is not effective.

See more of this Session: Advances in Algal Biorefineries I
See more of this Group/Topical: Sustainable Engineering Forum

(683a) Carbon Formation and Catalyst Deactivation 3D Simulations of Hydrogen-Producing Reactions in a Fixed-Bed Reactor

Mohsen Behnam[1], Anthony G. Dixon[1], Michiel Nijemeisland[2] and Hugh Stitt[2], (1)Chemical Engineering, Worcester Polytechnic Institute, Worcester, MA, (2)Johnson Matthey, Billingham, United Kingdom

To produce hydrogen as a clean fuel and a new source of energy, catalytic reactions are widely used in industry [1]. The deactivation of catalysts is an important problem in industry, such as steam reforming and the catalytic dehydrogenation of alkanes, which are both strongly endothermic reactions. Different publications reported on the deactivation of steam reforming of methane (SRM) as a significant source of hydrogen production [2] and propane dehydrogenation (PDH) as a commercial and interesting reaction [3] to understand the effect of deactivation on the reactor situations and performance generally. The local carbon deposition on catalysts can cause particle breakage and strongly decrease reaction rates. Catalyst deactivation in heated tubes removes the heat sink and can result in local hot spots that weaken the reactor tube. This is particularly a problem for a low tube-to-particle diameter ratio (N) fixed bed reactor, where a large fraction of the catalyst particles are located next to the heated tube wall. Computational fluid dynamics (CFD) has been reviewed as a suitable tool to simulate packed bed tubes [4]. Previous works [5] have introduced the coupling of gas flow and resolved species and temperature gradients inside pellets by CFD for SRM and PDH without considering deactivation. In this work, the focus is on using CFD to obtain better comprehension of the role of catalyst deactivation on reactions, mass and heat transfer of the gas-phase fluid flow around and inside of cylindrical catalyst particles. SRM and PDH reactions were simulated by CFD to study the interaction between local flow and reaction/deactivation. CFD simulations of flow, heat transfer, diffusion and reaction were carried out using the commercial CFD code Fluent 6.3 in a 3D 120-degree periodic wall segment of an N=4 tube. The mesh used boundary layer prism cells at both the inside and outside particle surfaces and at the tube wall, with tetrahedral cells in the main fluid volume. These reactions were represented in the solid particles using user-defined scalars to mimic species transport and reaction, with user-defined functions supplying reaction rates. Diffusion in the particles was modeled by Fick's law using an effective diffusivity, given by Hite and Jackson's approximation of the Dusty Gas Model [6]. Catalyst activity is related to the coke accumulation with a suitable initial coking expression and causes decrease of reaction rates with deposition of carbon. Carbon deposition can be calculated by first running a CFD simulation at base case (undeactivated) conditions, then accumulating carbon production rates over a 60-second interval to obtain the local carbon build-up. This is then inserted into the activity factor of the catalyst, which is used to modify the fresh catalyst reaction rates to give the reaction rates after 60 s. The CFD simulation is then run for a further 60-second time period to obtain increased values of carbon accumulation and the process repeated. The CFD simulations show local details of carbon laydown both on the surface of, and inside near-wall catalyst particles. The transient development of particle internal gradients and carbon accumulation were studied for the early stages of deactivation. Carbon concentration is initially strongest close to the surface and in the high temperature regions of the catalysts and affected by the wall heat flux. Deactivation of the endothermic reactions causes a slow increase in the average catalyst temperature. The results present a decrease of the product mole percentages at the outlet with time.

References:

[1] Chen, Z.; Yan, Y.; S.E.H. Elnashaie, S. Catalyst deactivation and engineering control for steam reforming of higher hydrocarbons in a novel membrane reformer. Chemical Engineering Science 59 (2004) 1965 – 1978.

[2] Snoeck, J.; Froment, G.; Fowles, M. Kinetic Evaluation of Carbon Formation in Steam/CO2-Natural Gas Reformers. Influence of the Catalyst Activity and Alkalinity, International Journal of Chemical Reactor Engineering, Vol. 1 [2003], A7.

[3] Jackson, S. D.; Stitt, E. H. Propane dehydrogenation over chromia catalysts: micro-catalysis, deactivation, macro-kinetics, and reactor modelling. Current Topics in Catalysis 2002, 3, 245.

[4] Dixon, A. G.; Nijemeisland, M.; Stitt, E. H. Packed tubular reactor modeling and catalyst design using computational fluid dynamics. Advances in Chemical Engineering 2006, 31, 307.

[5] Taskin, M.E.; Troupel, A.; Dixon, A. G.; Nijemeisland, M; Stitt, E. H. Transport and Reaction Interactions for cylindrical particles with strongly endothermic reactions. Ind. Eng. Chem. Res., submitted.

[6] Hite, R.; Jackson, R. Pressure gradients in porous catalyst pellets in the intermediate diffusion regime. Chemical Engineering Science 1977, 32, 703.

See more of this Session: Catalytic Hydrogen Generation - General III
See more of this Group/Topical: Catalysis and Reaction Engineering Division

(683b) Molybdenum- Zeolite Catalysts for Conversion of Light Hydrocarbons to Fuels and Chemicals

Lucia M. Petkovic, Daniel M. Ginosar, Kyle C. Burch and Harry W. Rollins, Interfacial Chemistry, Idaho National Laboratory, Idaho Falls, ID

Recent reports on methane dehydro-aromatization (Tessonnier et al. Catal. Comm. 8 (2007) 1787) have shown that careful catalyst preparation allowed for increased activity while completely suppressing deactivation by coke deposition. These findings are positioning the dehydro-aromatization of methane as an environment friendly reaction which could be used to convert both raw methane feedstocks and light hydrocarbons byproducts, which otherwise would be flared, into valuable aromatic hydrocarbons such as benzene and toluene and fuel cell grade (CO-free) hydrogen. In addition, the dehydro-aromatization reaction does not require additional materials and can easily use nuclear, solar, or other heat sources. These factors make it a potentially advantageous alternative in support of the development of hybrid energy systems. The research reported here will present the comparison of catalyst preparation method, i. e., vaporization-reaction of molybdenum trioxide and incipient wetness with a molybdenum salt, on catalyst activity and stability. Temperature programmed oxidation and surface area analyses were used to identify potential routes to improved catalyst performance and regeneration alternatives.

See more of this Session: Catalytic Hydrogen Generation - General III
See more of this Group/Topical: Catalysis and Reaction Engineering Division

(683c) Improvement of Catalytic Hydrogen Production From Methane Using Dense Ba0.5Sr0.5Co0.8 Fe0.2Ox Ceramic Membranes

Sedigheh Faraji[1], Karen Nordheden[2] and Susan Stagg-Williams[2], (1)Chemical Engineering Department, California State University, Long Beach, Long Beach, CA, (2)The university of Kansas, Lawrence, KS

Dense oxygen-permeable ceramic membranes have been explored as an alternative oxygen source for carbon dioxide reforming of methane to syngas (CO+H2). In this study, the effects of various parameters on syngas production over a dense Ba0.5Sr0.5Co0.8Fe0.2Ox (BSCF) membrane have been investigated. These parameters include catalysis type, reaction temperature, membrane thickness, pH, and feed concentration. Although Pt-Ni/Al2O3 catalyst shows high methane conversion in the presence of the BSCF membrane at 800°C, the activity of this catalyst is low at 600°C. Pt-Ni/CeZrO2 bimetallic catalyst demonstrates superior performance compared to Pt-Ni/Al2O3 catalyst at 600°C. The BSCF membrane can reduce the apparent activation energy of CO2 reforming reaction by changing the reaction pathway. The BSCF membrane with low thickness demonstrates a higher methane conversion and H2:CO ratio than BSCF membrane with high thickness because membrane oxygen flux is inversely proportional to thickness. Varying the pH of the precursor solution during membrane preparation has no significant effect on the oxygen flux or the reaction. The CH4:CO2 feed ratio significantly affects the hydrogen production over the BSCF membrane. Altering the CH4:CO2 ratio has a direct impact on the oxygen flux, which in turn can influence the reaction pathway. These studies suggest that Pt-Ni/CeZrO2 catalyst might be suitable for low-temperature hydrocarbon conversion reactions over low-thickness BSCF ceramic membranes.

See more of this Session: Catalytic Hydrogen Generation - General III
See more of this Group/Topical: Catalysis and Reaction Engineering Division

(683d) Cluster-Expansion-Based Modeling of the Coverage Dependence of Adsorbate Binding at a Metal Surface

David J. Schmidt, Chemical and Biomolecular Engineering, University of Notre Dame, Notre Dame, IN, William F. Schneider, Department of Chemical and Biomolecular Engineering, University of Notre Dame, Notre Dame, IN, Wei Chen, Material Science and Engineering, Northwestern University, Evanston, IL and Christopher M. Wolverton, Materials Science and Engineering, Northwestern University, Evanston, IL

The binding of adsorbates to surfaces is often characterized using mean field models. This approach treats adsorbate binding energies as constant independent of coverage and neglects and correlations between the locations of adsorbates. In this work, we use lattice-based cluster expansions to model the configuration-dependent energetics of oxygen adsorption on the Pt(111) surface. The cluster expansions were parameterized using DFT calculations of 49 distinct configurations, the convergence of the expansion with cluster and configuration size evaluated, and the relative contributions of electronic and relaxation effects to adsorbate interactions evaluated. The cluster expansion was used in concert with Monte Carlo simulations to model adsorbate chemical potential as a function of surface coverage and temperature. The binding energy was found to be roughly linear with respect to the surface coverage, consistent with earlier studies. Ordered states were found at ¼, ½, and 2/3 ML coverage. Binding energy distributions were calculated using the cluster expansion and snapshots from the Monte Carlo simulations. Binding energy distributions of ordered structures can be identified by the presence of large spikes at specific binding energies, while disordered structures were identified by their low and broad distributions.

See more of this Session: Catalytic Hydrogen Generation - General III
See more of this Group/Topical: Catalysis and Reaction Engineering Division

(683e) H2 Generation On Low Content Pt Supported Over Nano Size Mesoporous TiO2

Wei Shao[1], Luzheng Zhang[2], Robert L. Lee[2] and Xiaohua Lu[3], (1)Materials Engineering, New Mexico Institute of Minging and Technology, Socorro, NM, (2)Petroleum Recovery Research Center, New Mexico Institute of Minging and Technology, Socorro, NM, (3)State Key Laboratory of Materials-oriented Chemical Engineering, Nanjing University of Technology, Nanjing, China

Production of hydrogen from renewable resources (e.g., biomass) is considered an effective approach to mitigating the environmental problems caused by pollutant emissions from fossil fuels. There is a great deal of interest in Pt/TiO2 because of its promising catalytic prospects in applications of hydrogen production from biomass. However, one important problem should be overcome is that CO as a by-product that will significantly impair the catalytic performance even the presence of carbon monoxide (CO) is as low as only 5-10 ppm. In the present paper, we aimed to fabricate CO poisoning resistant Pt/TiO2 catalyst by doping Pt onto nano-size mesoporous TiO2 and investigate its H2 generation performance in steam reforming reactions.

Nano-size mesoporous TiO2 and 0.5, 1.0, and 2.0% wt Pt-doped, nano-size mesoporous TiO2 were synthesized by a hydrolyzing followed by sintering process. Adsorption isotherms of carbon monoxide (CO) on the synthesized mesoporous TiO2 and on Pt/TiO2 were measured at temperatures 308K, 323K, 348K, 373K, 398K, and 423K and pressures up to 700 mmHg. The 0.5% wt Pt/TiO2 catalyst resulted in better CO adsorption than those which were tested on the nano-size mesoporous TiO2 support. The amounts of CO adsorbed with 1.0% wt and 2.0% wt Pt/TiO2 catalysts were found to be larger than those on nano-size mesoporous TiO2. All measured adsorption isotherms of CO were correlated to fit adsorption models Sips, Toth, and UNILAN in the experimental temperature/pressure range. Sips and Toth models were found to be more accurate for presenting CO adsorption on Pt/TiO2. It suggests that the 0.5% wt Pt/TiO2 catalyst have potentially good CO poisoning resistivity within a wide temperature range. Steam reforming of glucose and methanol under different experimental condition were utilized to generate H2 on the synthesized Pt/TiO2. And for comparison purposes, 5% Pt/Al2O3 was applied as well. It is found that the synthesized low content Pt/TiO2 is more cost-effective.

See more of this Session: Catalytic Hydrogen Generation - General III
See more of this Group/Topical: Catalysis and Reaction Engineering Division

(683f) Sulfur Tolerant Doped Fe/Ce Catalysts for the High Temperature Water Gas Shift Reaction – TPR and Mossbauer Spectroscopic Study

Krishna Reddy Gunugunuri[1], Manasa Sridhar[1], Punit Boolchand[2] and Panagiotis Smirniotis[1], (1) Chemical and Material Science Engneering, University of Cincinnati, Cincinnati, OH, (2)Electrical and Computer science Engneering, University of Cincinnati, Cincinnati, OH

To stabilize the formation of Magnetite phase during the activation and to promote the $Fe3+ \leftrightarrow Fe2+$ redox couple during the water gas shift reaction, in our previous study we introduced a variety of metal ions (M = Cr, Mn, Co, Ni, Cu, Zn, and Ce) into iron oxide (spinel lattice) and screened for effectiveness for a high-temperature water–gas shift reaction. Among the various catalysts tested Fe/Ce and Fe/Cr spinels with an atomic ratio of 10:1 exhibited the highest activity. Between the Fe/Ce and Fe/Cr catalysts, Fe/Ce catalyst exhibits stable and higher activity in the presence of sulfur compared to Fe/Cr catalysts. In the present study, we investigated the effect of foreign metals as a dopant on high temperature water gas shift activity of Fe/Ce catalyst. For this purpose, various metals like Co, Mo, Zr, Hf, and Cr were doped into Fe/Ce catalyst by using ammonia assisted coprecipitation method. The prepared catalysts were characterized by using TPR, Raman Mossbauer spectroscopic techniques and evaluated for WGS reaction in the temperature region of 400-550 oC and at a steam to CO ratio of 1, which the conditions typically used in our membrane reactor. A relatively high space velocity 60,000 h-1 was used in this study. To verify the sulfur tolerance the experiments were also carried out in the presence of H2S. These interesting results will be presented in the presentation.

See more of this Session: Catalytic Hydrogen Generation - General III
See more of this Group/Topical: Catalysis and Reaction Engineering Division

(690a) Influence of Sulfated Zirconia-Doped Nafion Composite Membrane On Direct Methanol Fuel Cell Performance

Shilpa Beravelli, Cynthia Rice York, Dalton York and Akshay S Bauskar, Chemical Engineering, Tennessee Technological University, Cookeville, TN

Fuel cells are an attractive alternative to many energy conversion/storage devices. In recent years, due to the system simplicity, easy handling of liquid methanol and high specific energy (3800kcal/L)compared to H2 (658kcal/L at 360atm) direct methanol fuel cells (DMFCs) are being recognized as one of the most attractive power sources for portable devices such as cellular phones, personal digital assistants, laptop computers, etc. Nafion membrane doped with superacid Sulfated Zirconia (S-ZrO2) was synthesized to enhance performance. The properties of composite membrane are compared with that of undoped Nafion membrane prepared by the same recasting technique and commercially available Nafion115 membrane. Conductivity results revealed enhancement in proton conductivity of composite membrane. This confirms that the additive, in virtue of both its acidity and hydrophilicity, promotes proton transport. Fuel cell with composite membrane delivered, power density (120mW/cm2) when compared to Nafion115 membrane power density (90mW/cm2).

See more of this Session: Fuel Cell Technology II
See more of this Group/Topical: Fuels and Petrochemicals Division

(690b) Understanding Mass Transfer Limitations in the Anode Porous Catalyst Layer of Direct Formic Acid Fuel Cell

Rajavardhan [Raja] Reddy Nagolu[1], Vinten Diwakar[1], Pedro Arce[2] and Cynthia A. Rice-York[2], (1) Department of Chemical Engineering, Tennessee Technological University, Cookeville, TN, (2) Chemical Engineering, Tennessee Technological University, Cookeville, TN

Direct Formic Acid Fuel Cells (DFAFC) are a promising alternative to hydrogen and alcohol based fuel cells and are attractive alternatives for small portable applications for example laptops, cell phones, etc. Among the available types of fuel cells, formic acid (HCOOH) has emerged as a promising choice due to better oxidation kinetics, high theoretical open circuit potential, limited crossover of fuel from anode to cathode through the Nafion membrane and reasonable power density at room temperature. The anode porous catalyst layer must provide efficient transport of HCOOH from the Gas Diffusion Layer (GDL) to the surface of the catalyst and at the same time the products of HCOOH oxidation reaction (water and carbon dioxide) should be efficiently removed from the catalyst layer since product accumulation in the system can lower the efficiency and performance of the cell by blocking surface sites. Also, the performance of the anode catalyst depends on the parameters of the intrinsic chemical kinetics of formic acid oxidation and physical parameters such as electrode thickness, surface area, effective diffusion, concentration of formic acid and temperature.

In this contribution, a diffusive-kinetic-based model is developed for the porous anode catalyst layer of DFAFCs to study mass transfer limitations that may affect the distribution of fuel and removal of products. An effectiveness factor approach is introduced to find the effects of mass transfer and kinetic parameters on Pt-Ru catalyst utilization. This model is developed in stages from simple physical situations that capture kinetics and pore geometry to more complex and realistic ones. The analysis is used to provide guidelines for experimental conditions that help the transport of reactants and products in the layer. The need and effect of pore formers to facilitate effective transfer of reactants and products through the Catalyst Layer will also be analyzed and discussed. Details about the formulation and preliminary results will be discussed. Pore size, concentrations, diffusion, and pore shape are among the key parameters studied. Since the effectiveness factor is a "macro-scale" approach, comparisons with more rigorous microscopic models will also be presented as validation.

See more of this Session: Fuel Cell Technology II
See more of this Group/Topical: Fuels and Petrochemicals Division

(690c) Quantitative Measurement of Single Electrode Performance within An Operating Fuel Cell

Matthew S. Naughton[1], Fikile R. Brushett[1] and Paul J.A. Kenis[2], (1)Department of Chemical & Biomolecular Engineering, University of Illinois at Urbana-Champaign, Urbana, IL, (2)Chemical and Biomolecular Engineering, University of Illinois at Urbana-Champaign, Urbana, IL

Fuel cells hold promise as alternative power sources due to their ability to bypass Carnot efficiency limitations by directly converting chemical energy into electrical energy. However, the high costs of catalysts and membranes, as well as component durability issues, have barred widespread implementation. Alkaline fuel cells (AFCs) have gained increasing attention due to superior electrode kinetics in alkaline media, compared to acidic media, which enables the use of non-noble metal catalysts [1]. However, analysis of individual electrodes within an operating fuel cell is typically hampered by the reliance on overall measures such as maximum power density and open circuit potential. Standard electrochemical analysis, such as RDE (ring disk electrode) experiments, commonly characterizes catalyst activity and stability under conditions which are very different from those found in an operating fuel cell (e.g., 0.1 M KOH vs. 1-7 M KOH).

Previously, we have developed a pH-flexible flowing electrolyte microfluidic fuel cell, which uses an external reference electrode to individually analyze cathode and anode performance. This microfluidic configuration combines the versatility of a traditional three electrode cell with the conditions found in an operating fuel cell. To date, typical analyses have mainly consisted of qualitative comparisons between electrodes as a function of operating conditions and/or catalyst type. These analyses compared inward or outward shifts for a given electrode and the onset of mass transport limits. Quantitative comparisons have been restricted to maximum power / current density and open circuit potential, which do not utilize the advantages of an external reference electrode.

Here, we present a method systematically analyzing the performance of individual electrodes within an operating fuel cell. This in-situ single electrode analysis enables detailed quantification of kinetic, ohmic, and mass transport losses as compared to a reference electrode. The effects of varying catalysts, reactant delivery, and electrolyte compositions are determined. These detailed measurements can be used to better understand the impact of local environments and electrode preparation techniques on the performance and durability of fuel cell systems.

[1] Brushett et al., Journal of the Electrochemical Society, 2009, 156, B565

See more of this Session: Fuel Cell Technology II

(690d) Scaling and Manufacturing of Laminar Flow-Based Fuel Cells

Adam S. Hollinger, Department of Chemical and Biomolecular Engineering, University of Illinois, Urbana, IL, Larry J. Markoski, INI Power Systems, Morrisville, NC and Paul J.A. Kenis, Department of Chemical & Biomolecular Engineering, University of Illinois at Urbana Champaign, Urbana, IL

Since the development of laminar flow-based fuel cells (LFFCs), many different configurations, fuels, and media have been successfully established [1,2]. More recently, our group has created a similar microfluidic fuel cell as an analytical platform for catalyst and electrode characterization and optimization [3]. For LFFCs to be successfully integrated into a fuel cell-based power source, LFFCs must be scaled out and manufactured effectively. The power output of a single-channel cell is inadequate for most power applications, but multichannel cells and stacked cells can address these power needs. To date, most LFFCs have been proof-of-concept unit cells. Here we will demonstrate a fourteen channel cell at various fuel concentrations and flow rates [4].

Presently, membrane-based fuel cells are assembled manually in layer-by-layer fashion followed by clamping the layers between metal plates. Scaling to larger fuel cell systems typically involves the stacking of more and more layers, requiring heavier and heavier clamping constructions to keep the stack together without leaks. Here we will demonstrate an alternative approach to fuel cell manufacturing which utilizes patterned polymer layers to eliminate the heavy metal clamping structures needed in traditional fuel cell stacks. This approach results in much smaller, lighter structures and enables thinner cell stacks, which is attractive for portable electronic equipment.

[1] E. Kjeang, N. Djilali and D. Sinton, Journal of Power Sources, 2 (2009) 353-369.

[2] F.R. Brushett, R.S. Jayashree, W.P. Zhou, P.J.A. Kenis, Electrochemica Acta, 2009, 54 (27), 7099-7105.

[3] F.R. Brushett, W.P. Zhou, R. S. Jayashree, P.J.A. Kenis, Journal of the Electrochemical Society, 2009, 156 (5), B565-571.

[4] A.S. Hollinger, R.J. Maloney, R.S. Jayashree, D. Natarajan, L.J. Markoski, P.J.A. Kenis, Journal of Power Sources, 2010, 195 (11), 3523-3528.

See more of this Session: Fuel Cell Technology II
See more of this Group/Topical: Fuels and Petrochemicals Division

(690e) Metabolic Engineering Strategies for Microbial Fuel Cells

Adam C. Baughman, Chemical & Biological Engineering, Rensselaer Polytechnic Institute, Troy, NY and Lealon L. Martin, Chemical and Biological Engineering, Rensselaer Polytechnic Institute, Troy, NY

We apply of a novel framework for the topological modeling of reaction networks (TMA) for the rational redesign of microbial fuel cell devices. We demonstrate the ability of our approach to quantitatively capture all metabolic activities of the chosen anode microbe agent (Pseudomonas Putida), as well as all electrochemical processes external to the microbe, using a compartmentalized modeling approach within a single framework. We show that a rigorous examination of microbial metabolism and its participation within the fuel cell can reveal targets for genetic redesign of the microbe improving both the overall bioremediation and elecrical poser generation of the fuel cell device. We close by discussing aspects of practically constructing the redesigned fuel cell devices identified using our modeling approach.

See more of this Session: Fuel Cell Technology II
See more of this Group/Topical: Fuels and Petrochemicals Division

(705a) The Synergy Between Energy Applications and Nanomaterials

Randy L. Vander Wal, Energy and Mineral Engineering, Pennsylvania State University, University Park, PA

Energy and Nanomaterials are intimately related. Most energy processes occur at the interface. Nanomaterials, when used as interfacial modifiers have the potential to significantly alter the energy landscape. Although carbon nanomaterials have been explored for a suite of energy applications, to-date the approach has been based on replacing bulk materials, not as interfacial modifiers. The work presented here will highlight this latter approach by showing results for a suite of energy applications from the author's own work.

Experimental details will be presented at the presentation with regards to the synthesis and utilization of the nanomaterials. Our applications include energy conversion, generation, storage, efficiency, conservation and control.

Nanoscale materials are redefining the relation between material composition, size and properties. Chemical properties (e.g. reactivity) and physical properties (e.g. surface area) become a strong function of size at the nanoscale. Applications illustrated include the following selected examples from the author's work: catalysis, composite materials, energy storage, sensors, thermal management, and tribology. The key concept is that the nanomaterials serve as interfacial modifiers. Since most energy related processes are dominated by interfacial reactions, nanomaterials have the potential to dramatically affect energy conversion rates and magnitudes. Examples follow.

Catalysis is central to particulate and NOx after-treatment systems. A prime example of reducing materials to nanoscale and realizing new properties is catalysis by nanoscale gold. Au nanoparticles supported on oxides such as CeO2, TiO2 and Fe2O3 offer ambient temperature oxidation of CO, volatile organic compounds (VOCs) and potentially exhaust hydrocarbons.

Lightweight materials will reduce weight significantly yielding substantial benefits in fuel efficiency with reduced emissions. Composites with substantial gains in Young's modulus, tensile strength, and EM shielding may be realized in polymeric composites using carbon nanotubes as an interfacial modifier rather than bulk filler [1]. The purpose of the foil is to serve as a gas impermeable barrier layer within the polymer composite.

Advances in energy storage include batteries, ultra-capacitors. These can support lighting, appliances, a starter, cooling fans, transmission and hydraulic systems, fuel and air handling systems and ultimately enable hybrid systems. Towards these goals, substantial gains in Li ion battery cathode and anode materials have been realized using carbon nanofibers, coating processes and including elements such as tin and silicon. Modifications of the CNT surfaces increases the Li ion capacity beyond the theoretical limit of normal graphite. As an illustration, CNTs were directly fabricated for this end-application use upon ultra-fine SS mesh. Significantly no harvesting, purification or processing (using binders) was required. The CNTs could be used directly as synthesized [2].

With regards to overall system integration and control, sensors will play a prominent role [3]. With ultrahigh surface exposure relative to bulk material, nanoscale materials are exceedingly sensitive to gas adsorption. Exploitation of nanoscale properties will lead to new NOx sensors and in-cylinder oxygen sensors. Examples include catalyst coated metal oxide semiconductors capable of ambient temperature operation.

Thermal management will benefit from nanofluids. Nanofluids can increase thermal conductivity and reduce radiator and heat exchanger size. Carbon-based nanofluids using nano-onions and carbon nanotubes have increased water conductivity by ~ 20% [4]..

Lubrication is critical to many engine components and powertrain systems. Nanolubricants can bridge the gap between fluid and solid materials [5]. As additives with liquids or greases, synergistic properties may be realized, particularly in boundary-phase lubrication. Improved coating formulations and properties can reduce or eliminate fretting and pitting. Results with nanocarbons show superior performance relative to graphite, diamond like carbon (DLC) and even Teflon.

Nanoscale materials are redefining the relation between material composition, size and properties. Their applications to energy processes include selected examples from the author's work. Selected technologies include catalysis, composite materials, energy storage, sensors, thermal management, and tribology. Interfacial modification serves to affect changes in energy conversion rates, and transfer magnitudes. This appears to be a rather universal concept and suggests R&D directions for nano-engineering of interfaces. Funding through The Penn State Institutes for Energy and the Environment (PSIEE) and the Pennsylvania Keystone Innovation Starter Kit (KISK) is gratefully acknowledged.

(1) Vander Wal, R. L., and Hall, L. J., Advanced Engineering Materials 2004, 6, 48-52. (2) Luo, Y., Vander Wal, R. L., and Scherson, D. A., Electrochemical and Solid State Letters 2003, 6, A56-A58. Vander Wal, R. L., Mozes, S. D., and Pushkarev, Nanotechnology 2009, 20, 105702-10. (3) Hunter, G. W., Vander Wal, R. L., Liu, C. C., Xu, J. C., and Berger, G. M., Sensor and Actuators B 2009, 138, 113-119. (4) Vander Wal, R. L., Mozes, S. D., and Pushkarev, Nanotechnology 2009, 20, 105702-10. (5) Street, K. W., Miyoshi, K., and Vander Wal, R. L., "Application of Carbon Based Nano-particles to Aeronautics and Space Lubrication", in Superlubricity, Chapter 19, pp 311-340. Edited by Drs. Jean-Michel Martin and Ali Erdemir, Elsevier, Amsterdam (2007). (also NASA/TM-2007-214473).

2007). (also NASA/TM-2007-214473).
See more of this Session: Nanowires II: Energy Conversion and Storage Applications
See more of this Group/Topical: Nanoscale Science and Engineering Forum

(705b) Molecular Scale Chalcogenide Nanowires for Thermoelectric Conversion

Genqiang Zhang, Gautam G. Yadav, Haoran Yang and **Yue Wu**, School of Chemical Engineering, Purdue University, West Lafayette, IN

PbTe and Bi2Te3 are the best candidates for thermoelectric conversion at temperature close to room temperature and 500 K, respectively. Herein, we developed a facile solution phase method for the synthesis of ultrathin PbTe and Bi2Te3 nanowires with diameters of about or less than 10 nm using the ultrathin Te nanowires as the in-situ templates. The phase transfer from Te to PbTe or Bi2Te3 is accomplished through the injection of Pb and Bi precursor solution to the Te nanowire solution.By fabricating themoelectric devices using these nanowires with diameters less than 10 nm, reduced thermal conductivity and enhanced Seebeck coefficient have been observed to improve the thermoelectric figure of merits.

See more of this Session: Nanowires II: Energy Conversion and Storage Applications
See more of this Group/Topical: Nanoscale Science and Engineering Forum

(705c) Intense SERS Effect From a Single Metallized Silicon Nanowire

Ajay Agarwal[1], Effendi Widjaja[2], Mohamed Khalid Nizamudin[2], Shaik Mohamed Salim[2] and **Marc Garland**[2], (1)Institute of Microelectronics, A*STAR (Agency for Science, Technology and Research),, Singapore, Singapore, (2)Process Science and Modelling, Institute of Chemical and Engineering Sciences, Singapore, Singapore

Metallized silicon (Si) nanowires are created by just by dipping the nanowires into an aqueous deposition solution for several minutes. Upon dipping into solution, metal ions were reduced and deposited on the top of the Si nanowire. Surface silicon oxide was simultaneously dissolved and removed by hydrogen fluoride (HF), so that the deposition reaction was sustainable and controllable. The deposited silver (Ag) nanoparticles (NPs) uniformly covered the silicon nanowires. Also Cu, Pd, Co, Au, and Pt were deposited to metallize the silicon nanowires in a similar manner. Applications of the new nano-composite materials were explored. When the Ag coated silicon nanowire was tested as a Surface-Enhanced Raman Scattering (SERS) substrate, an extremely strong signal was observed. The significantly enhanced SERS effect appears to be associated with the close packing of the neighboring NPs that self-assemble along the highly curved nanowire [1].

[1] C Fang, A Agarwal, E Widjaja, M Garland, SM Wong, L Linn, M K Nizamudin , S M Salim, and N Balasubramanian Chem. Mater. 2009, 21, 3542–3548

See more of this Session: Nanowires II: Energy Conversion and Storage Applications
See more of this Group/Topical: Nanoscale Science and Engineering Forum

(705d) Oriented Single-Crystalline TiO2 Nanowires On Titanium Foil for Lithium Ion Batteries

Bin Liu[1], Da Deng[2], Jim Yang Lee[2] and Eray S. Aydil[1], (1)Department of Chemical Engineering and Materials Science, University of Minnesota, Minneapolis, MN, (2)Department of Chemical and Biomolecular Engineering, National University of Singapore, Singapore, Singapore

Rechargeable lithium ion batteries (LIBs) are important for powering mobile or portable electronic devices and for storing energy in intermittent renewable power production systems that rely on the wind and the sun. The three key components of an LIB are the anode, the cathode and the electrolyte. During the charging and the discharging of an LIB, lithium ions insert the anode and the cathode materials, respectively. The ideal host electrode materials must have large surface-to-volume ratios and must be able to repeatedly accept and reject a large number of lithium ions without significant degradation after many charging-discharging cycles. Nanostructured electrodes provide high electrode-electrolyte interfacial area, fast lithium ion diffusion, and can accommodate large strains. Specifically, one-dimensional (1D) nanostructures such as nanorods, nanowires and nanotubes are emerging as a new class of electrode materials for LIBs because they can provide short ion diffusion lengths, improved electron transport and high surface-to-volume ratios. The use of aligned 1D nanostructured materials on a current-collecting substrate (e.g., copper foil or titanium foil) as the electrode could have additional benefits. Firstly, easy electrolyte diffusion into thicker electrodes through the open space in 1D nanostructured arrays could reduce the internal resistance for high capacity electrodes and thus improve power performance. Secondly, aligned 1D nanostructures could provide direct electrical pathways for enhanced charge transport. Finally, the direct connection of the 1D nanostructured electrode material to the current-collecting substrate eliminates the use of binders and conducting additives, which not only simplifies the battery fabrication process but also reduces the battery weight. In this presentation, we describe a simple and environmentally benign three-step hydrothermal method to grow oriented single-crystalline TiO2-B and/or anatase TiO2 nanowire arrays on titanium foil over large areas. We show that these nanowire arrays are suitable for use as the anode in lithium-ion-batteries; they exhibit specific capacities ranging from 200-250 mAh/g at 0.3 C rate (here 1 C is defined based on the theoretical capacity of anatase 168 mAh/g) and good capacity retention at high charge-discharge rates and over as many as 200 charging-discharging cycles. These promising properties are attributed to both the nanometer size of the nanowires and their oriented alignment. The comparable electrochemical performance to existing technology, improved safety, and the ability to roll titanium foils into compact three-dimensional structures without additional substrates, binders or additives suggest that these TiO2 nanowires on titanium foil are promising anode materials for large scale energy storage.

See more of this Session: Nanowires II: Energy Conversion and Storage Applications
See more of this Group/Topical: Nanoscale Science and Engineering Forum

(705e) Thermoelectric Characterization of Suspended Single Silicon-Germanium Alloy Nanowires

Julio A. Martinez[1], Tom S. Picraux[2], John P. Sullivan[1] and Brian S. Swartzentruber[1], (1)Center for Integrated Nanotechnologies, Sandia National Laboratories, Albuquerque, NM, (2)Center for Integrated Nanotechnologies, Los Alamos National Laboratory, Los Alamos, NM

The use of nanowires for thermoelectric energy generation has gained momentum in recent years as an approach to improve the figure of merit (ZT) due in part to larger phonon scattering at the boundary resulting in reduced thermal conductivity while electrical conductivity is not significantly affected. Silicon-germanium (SiGe) alloy nanowires are promising candidates to further reduce thermal conductivity by phonon scattering because bulk SiGe alloys already have thermal conductivity comparable to reported Si nanowires1. In this work, we show that thermal and electrical conductivity can be measured for the same single nanowire eliminating the uncertainties in ZT estimation due to measuring the thermal conduction on one set of wires and the electrical conduction on another set. In order to do so, we use nanomanipulation to place vapor-liquid-solid boron-doped SiGe alloy nanowires on predefined surface structures. Furthermore, we developed a contact-annealing technique to achieve negligible electrical contact resistance for the placed nanowires that allows us, for the first time, to measure electrical and thermal properties on the same device. We observe that thermal conductivity for SiGe nanowires is dominated by alloy scattering for nanowires down to 100 nm in diameter between the temperature range 40-300 °K. The estimated electronic contribution of the thermal conductivity as given by the Wiedemann-Franz relationship is about 1 order of magnitude smaller than the measured thermal conductivity which indicates that phonons carry a large portion of the heat even at such small dimensions.

1.Li D. et al. Thermal conductivity of Si/SiGe superlattice nanowires. APL 2003.

This work was performed at the Center for Integrated Nanotechnologies, a U.S. Department of Energy, Office of Basic Energy Sciences user facility at Los Alamos National Laboratory (Contract DE-AC52-06NA25396) and Sandia National Laboratories (Contract DE-AC04-94AL85000).

See more of this Session: Nanowires II: Energy Conversion and Storage Applications
See more of this Group/Topical: Nanoscale Science and Engineering Forum

(705f) Copper Templated Platinum Nanotubes as Oxygen Reducing Electrocatalysts

Shaun Alia, Yanqi Zhang, Qian Xu, Kurt Jensen, Christian Contreras and Yushan Yan, Department of Chemical & Environmental Engineering, University of California, Riverside, Riverside, CA

Platinum supported on carbon (Pt/C) is commonly utilized in proton exchange membrane fuel cells (PEMFCs) as an oxygen reducing electrocatalyst. The small particles are used due to the high surface area and thus high mass activity in the oxygen reduction reaction (ORR). The carbon support provides spacing for the platinum particles and prevents agglomeration. The use of Pt/C, however, creates concerns with respect to catalyst durability and specific activity loss due to alterations in the coordination number and crystal planes of the surface platinum atoms. Platinum nanotubes (PtNTs) with a thickness of 5 nm have previously been synthesized to improve the durability and ORR activity of PEMFC catalysts.[1]

Previous PtNT catalysts were synthesized by galvanic replacement with silver nanowires, which were formed by the ethylene glycol reduction of silver nitrate. Current studies have investigated the effect of the template material on PtNT activity. PtNTs have been synthesized by galvanic replacement with copper nanowires, which were formed by the sodium hydroxide and hydrazine reduction of copper nitrate. While the PtNTs have been synthesized under similar conditions, differences are observed in surface area and ORR activity. An alteration in ORR specific activity, or the ORR capability per unit area of available platinum, is of particular concern in this study. Improvement to ORR specific activity has been observed in catalysts templated from copper nanowires, principally due to changes in the coordination number and crystal planes of surface platinum and a decrease in atomic spacing created during the templating process. From these findings, optimization of the template material can be utilized to further increase the ORR activity of electrocatalysts synthesized by the galvanic replacement method.

Optimal ORR activity has been explored with the synthesis of platinum thin-coated copper nanowires and platinum thin-coated palladium nanotubes, synthesized in subsequent galvanic replacement reactions to create a surface layer of platinum instead of alloys. Catalysts have been characterized for ORR activity and durability to demonstrate the benefit of PtNTs to conventional PEMFC catalysts.

[1] Z. W. Chen, M. Waje, W. Z. Li, Y. S. Yan, Angewandte Chemie-International Edition 2007, 46, 4060.

See more of this Session: Nanowires II: Energy Conversion and Storage Applications
See more of this Group/Topical: Nanoscale Science and Engineering Forum

The Production Of Alkanes From Algae

Liane S. Carlson, Michael Y. Lee, Chukuemeka A.E. Oje, Arthur Xu, Stuart W. Churchill
Department of Chemical and Biomolecular Engineering, University of Pennsylvania

The potential of farmed algae as a biofuel has been recognized for some time because 1) its cultivation does not directly affect the food supply, 2) its productivity of lipids per acre far exceeds that of any other agricultural commodity, 3) it grows by absorbing CO_2, and if converted to a fuel and burned, is CO_2-neutral, and 4) the lipids can be converted to biodiesel or conventional fuel-range hydrocarbons. The purpose of our investigation has been to evaluate the technical and economic feasibility of an integrated algae-to-fuel process that incorporates several recently proposed technologies. This evaluation is based on the production of 20,000 BPD (226,000 lbs/hr) of n-alkanes since this is comparable to the daily output of a small refinery hydrotreater unit.

This analysis differs from prior ones in two major respects. First, it incorporates a new method of lipid extraction under development by OriginOil, Inc. rather than the conventional process of extraction with hexane. Second, the processing of the lipid has the objective of producing n-alkanes from triglycerides, as opposed to producing fatty acid methyl ester (FAME) biodiesel. Since the n-alkane product from this process is comparable to petroleum-based diesel fuels, it can be readily incorporated into existing energy infrastructures as a diesel blending stock or as a feedstock for other processing units in the refinery.

The conversion of algae to useful biofuels by means of this new process and in accord with this new concept can be divided into three modules: 1) algae cultivation, 2) lipid extraction, and 3) lipid conversion to alkanes. For each of the three modules, there are a number of possible approaches, each with a high degree of uncertainty. We attempted to evaluate the several methodologies for each module separately, and to formulate an integrated process of the most promising options.

Location. The selection of a site for an algae-to-alkanes plant is critical because of the need for 1) isolation from a population center, 2) a large relatively flat area with sufficient and relatively constant insolation, 3) a moderate year-around temperature, 4) a nearby coal-fired power plant whose stack gas contains a sufficient quantity of CO_2, 5) a sufficient source of saline water, 6) a nearby petroleum refinery as a market for the alkanes and as a source of hydrogen, and 7) a means of transportation of the lipids or alkanes thereto.

Sites that might meet these requirements were identified in Arizona and California, but East Texas appeared to be more promising overall. As a representative location that meets all of the listed requirements we propose an undeveloped area near the small town of Thompsons in East Texas. Thompsons, which is about 25 miles southwest from Houston, 30, 60, and 60 miles northward from three petroleum refineries on the Gulf Coast, and on a branch of the BNSF railway, receives insolation varying from the equivalent of 112 W/m^2 in January to 250W/m^2 in June, so the plant can be operated year-round [1]. A near-by coal-fired power-plant generates up to 2500 MW and produces sufficient CO_2 in its stack gas to meet the needs for algae cultivation. In addition, saline water for use in the cultivation process can be drawn from an underground aquifer and reused.

Cultivation. Three promising approaches for cultivation were identified. The first, which utilizes open-air "raceway" ponds, does not require much energy or attention but is hampered by contamination and low-growth rates. The second, which utilizes a compact photo-bioreactor with cultivation of the algae inside transparent tubes enclosing complex internals designed to optimize the rate of growth, requires an accounting for the concentration of carbon dioxide, free exposure to light, an expeditious rate of mass transfer from the gas-phase to the liquid phase, and the possible use of nitrogen starvation to maximize lipid production at the expense of proteins. The third is a hybrid that utilizes a simple infrastructure modeled after agricultural practices combined with the ability to control certain parameters of the environment by means of cultivating the algae inside polyethylene tubes. The Simgae™ Algal Biomass Production System developed by the Diversified Energy Corporation was chosen as a prototype of this methodology. *Nannochloropsis sp.*, a marine species with a high lipid content of 31- 68% (dry wt.) and other favorable qualities, was chosen as the strain of algae [2].

With a 50% recycle, a residence time of nearly 48 hours on average is required to double the amount of algae under the chosen conditions [3]. The corresponding length of the 6-in tubes is 1250 ft with a flow rate of 0.023ft/s. Each field consists of 100 reactor beds, each with 16 tubes, and, together with manifolding and paths, occupies 33 acres [4]. 1120 such fields, totaling 36,960 acres, were required for the chosen rate of production of algae.

Nutrients in suspension and solution in the saline water are combined with the recycled algae at the inlet. Stack gas pumped from the power plant by hydraulic induction and mixed with air to reduce its CO_2 content from 32 to 6 wt% is introduced every 300 ft along the tubing. The traces of SO_2 and NO_x in the stack gas enhance the growth of the algae. The O_2 produced by the growth of the algae is released by relief valves every 300 ft.

An overview of the proposed process is shown in Figure 1, the layout of the cultivation field in Figure 2, and of a single element thereof in Figure 3.

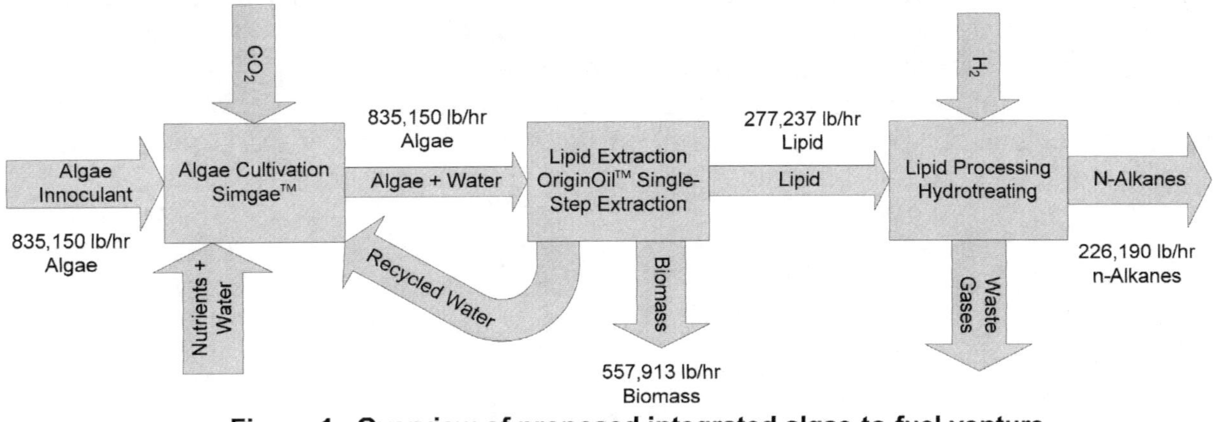

Figure 1. Overview of proposed integrated algae-to-fuel venture.

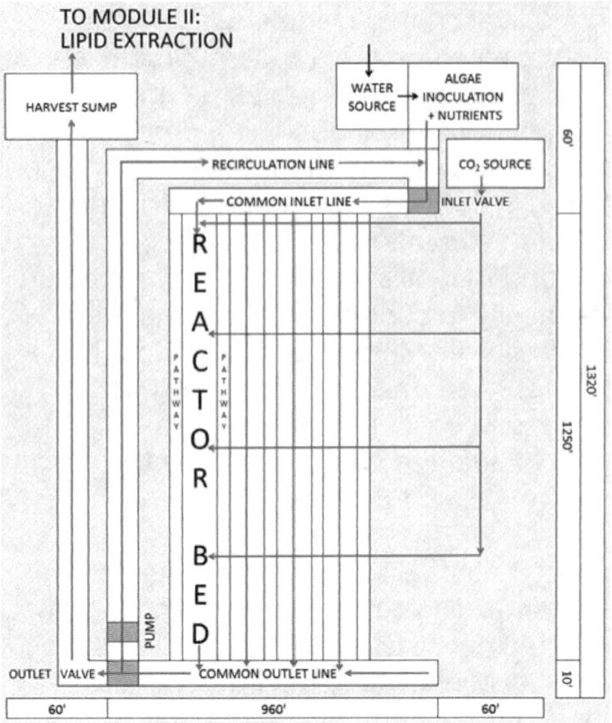

Figure 1. Schematic of cultivation field.

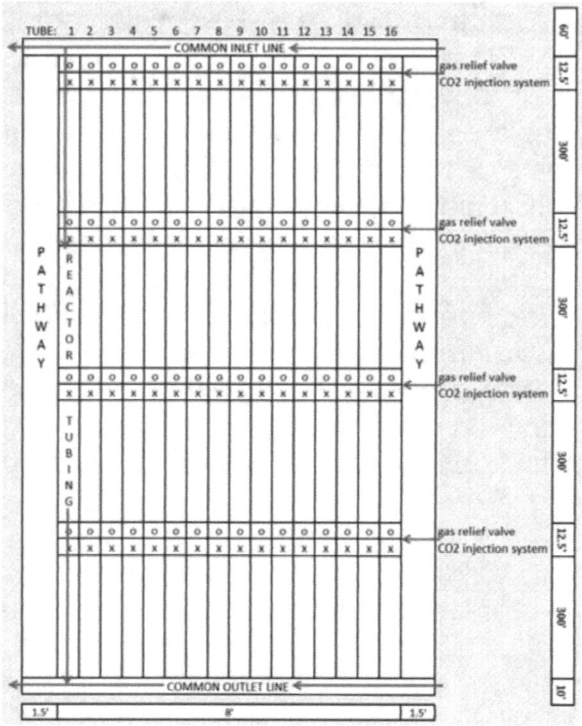

Figure 2. Schematic of a single reactor bed.

Lipid extraction. Historically, the desirable triglyceride fraction has been separated from the rest of the raw algae (proteins, carbohydrates, phospholipids, nucleic acids, and water) through extraction with hexane. A new process called Quantum Fracturing^{TM} by OriginOil, Inc., for which a patent has been applied but not yet issued, claims a combination of pH modification, microwaves, and ultrasonic pulses can rupture the wall of the algae cells while in slurry form, and allow the lipids to escape, resulting in a 90% reduction in energy consumption [5]. The resulting slurry can then be separated through gravitational settling into a lipid layer, an aqueous layer, and a wet layer of biomass. The lipids are transported to a petroleum refinery for conversion into alkanes, the aqueous fluid is recycled, and the biomass is centrifuged and drum-dried to reduce the residual water below 10 wt% [5]. It may then be sent to a nearby market for livestock feed or other uses. This process is shown schematically in Figure 4, the conjectured characteristics are compared with those for the conventional process in Table 1, and the estimated electrical consumption in Figure 5.

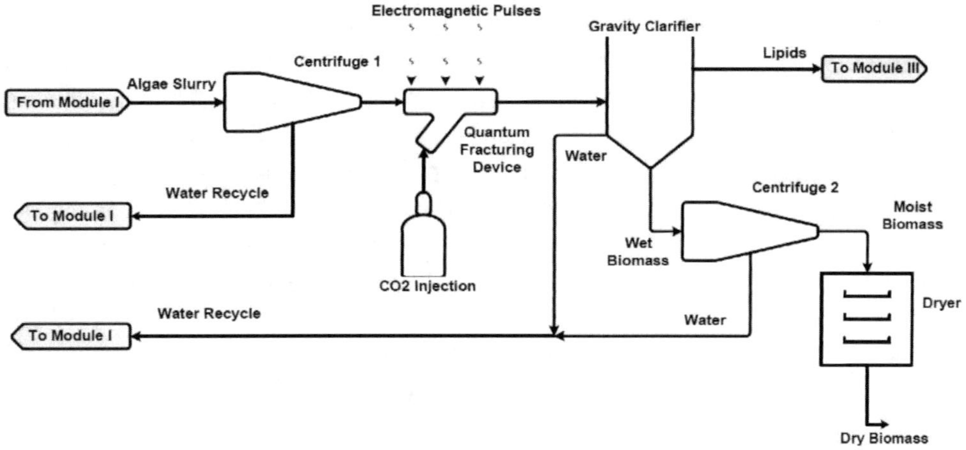

Figure 3. Schematic of proposed lipid extraction process based on Quantum Fracturing™ by OriginOil

Table 1. Comparison between conventional hexane extraction and Quantum Fracturing™ by OriginOil

	Conventional hexane extraction	OriginOil process
Separate by	Hexane	Gravity
Dewatering	Before separation	After separation
Lipid Yield	70 – 95%	85 – 97%
Energy Use (per lb lipid)	8.05 kWh	1.30 kWh

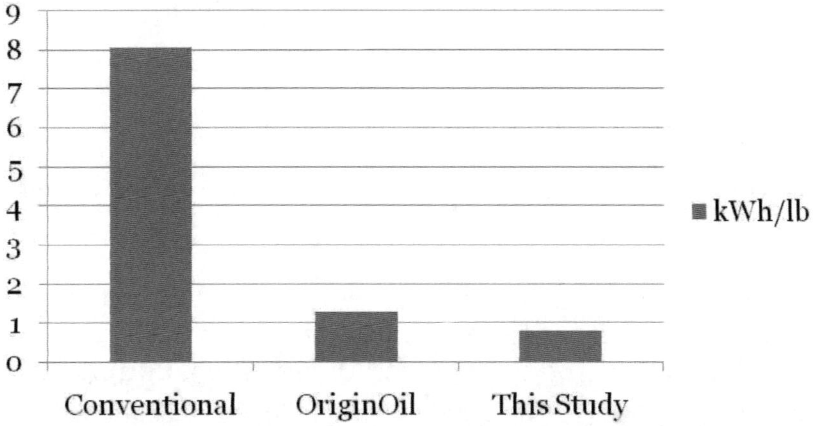

Figure 4. Comparison of estimated electrical consumption per pound of algae produced.

Lipid processing. Because the catalytic hydrotreating process required to convert the lipids to alkanes is similar to those employed in hydrocarbon processing and requires a source of hydrogen, it is proposed to carry out this step at a petroleum refinery rather than at the site of cultivation and extraction. The most feasible form of transportation to the refinery is by rail in tank cars. Since railway shipments are intermittent while lipid production and processing are continuous, it is proposed to provide for seven days of storage in tanks at the site of production and for two days of storage in tanks at the refinery.

The lipids react with hydrogen at 350°C and 50 bars in downward flow through packed beds of granules impregnated with a $NiMo/Al_2O_3$ catalyst [6]. In the first step, the triglycerides are hydrogenated and broken down into C_3 hydrocarbons and components of free fatty-acid.

These intermediates are then converted into straight-chain alkanes by three competing reaction pathways: decarboxylation, decarbonylation, and hydrodeoxygenation [7]. Figure 6 is a simplified schematic of the catalytic hydrotreating process. The product is separated by means of a high-temperature flash drum, a low-temperature, three-phase flash drum, a steam-stripper, and an amine scrubber with intermediate heat exchange. The alkane product from this process is comparable to petroleum-based diesel fuels. Thus it can be readily incorporated into the existing energy infrastructure as a diesel blending-stock or as a feedstock for other processing units in the refinery.

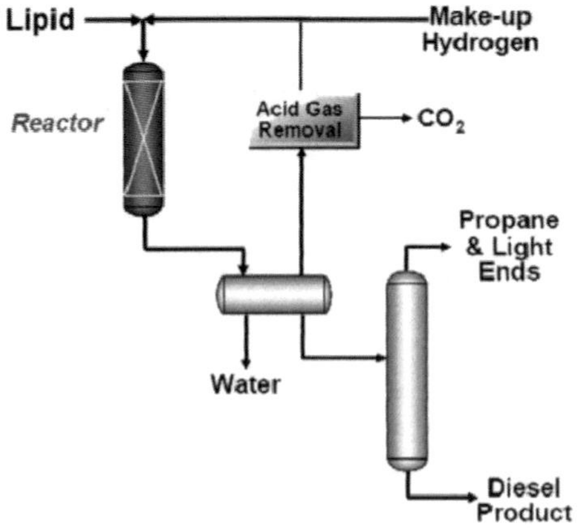

Figure 5. Simplified schematic of catalytic hydrotreating process.

Economic analysis. The profitability of the algae-to-fuel process described herein depends on a more accurate determination of costs and sources of revenues than is possible at the moment. Figure 7 shows a crude estimation of feedstock and utility costs per gallon of n-alkane, and Figure 8 a crude estimate of potential revenue. It is particularly difficult to estimate the utility costs because of great uncertainty of the electricity requirements in the Simgae™ and OriginOil™ processes.

Algae-to-fuel economics can be improved by developing higher-value uses of the algae biomass. Biomass which constitutes 55% of the algae, has possible use in pharmaceuticals and chemicals, and as a source of energy. These products are more valuable than livestock feed. As indicated by Figure 8, the sale of byproducts might exceed the revenue from the fuel oil. Carbon credits from potential cap-and-trade programs could be an additional source of revenue.

The bad news has been saved for last. The capital cost, which is primarily for the purchase of the 36,960 acres required for cultivation, is estimated to be roughly $2.6 billion. Despite the encouraging aspects of the potential income relative to the operating costs, the huge outlay for land suggests that consideration should be given to all possible means of increasing the rate of growth of algae and thereby reducing the required acreage before proceeding.

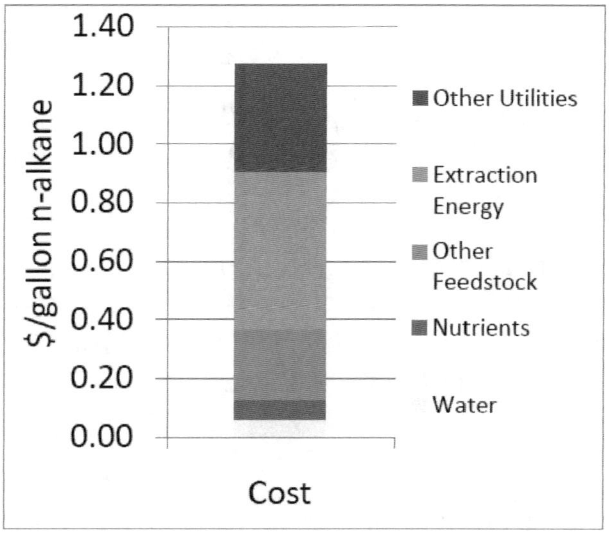

Figure 6. Estimated feedstock and utility costs of the proposed integrated algae-to-fuel venture.

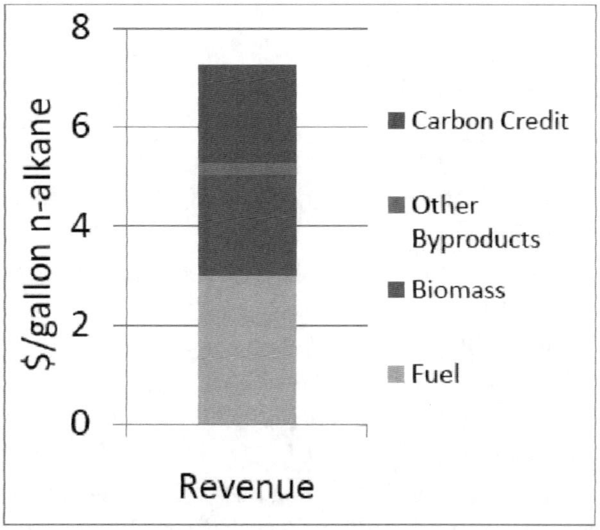

Figure 7. Projected revenue from the proposed integrated algae-to-fuel venture.

Conclusions. This manuscript describes a possibly improved process for the production of fuel from algae. Although encouraging both technically and economically, the analysis identifies several challenges that must be overcome before it merits commercialization. Future work should focus on these challenges.

Some of the challenges posed by the use of the hybrid SIMGAE™ design are maintenance of the proper temperature in the cultivation system, makeup water, the resistance of algae strain to invasion from other species, the environment impact of the algae farm, and most importantly, containing capital and operating costs. Nutrient consumption and costs, electrical requirements, CO_2-injection requirements, and most of all, land requirements are unresolved.

Some of the challenges in terms of oil (lipid) recovery from the algae include dewatering methods, lipid purification, energy costs, and optimizing the return from the residual biomass. The OriginOil™ lipid extraction process claims to reduce significantly the energy costs compared to those for conventional extraction processes. The most innovative aspect of this process is that dewatering is not necessary before lipid separation. However, the Quantum Fracturing™ technology from OriginOil™ is still in the development stage and is not a proven technology and is apparently not yet available for licensing.

The lipid processing step is probably the least risky of the three steps of the integrated algae-to-fuel process. Although its use to break up triglycerides is innovative, the fundamental technology behind catalytic hydrotreating is practiced in petroleum refineries worldwide. When adapting this technology for a lipid extraction process, it is important to focus on the catalyst selection and hydrotreater design for a lipid feedstock as opposed to a crude-oil feedstock.

It is important to optimize the value of biomass byproducts in order to maximize overall revenue, but the key challenge for this venture is to increase the rate of growth of the algae per acre. The possibilities include improvements by means of genetic engineering of the algae and by means of the physical and optical design of the reactors.

References.

1. Solar Radiation Data Manual for Flat-Plate and Concentrating Collectors. NREL. Web. 4 Apr. 2010. http://rredc.nrel.gov/solar/pubs/redbook/PDFs/TX.PDF.
2. Zittelli, G. Chini, et al. “Production of eicosapentaenoic acid by Nannochloropsis sp. cultures in outdoor tubular photobioreactors.” Journal of Biotechnology (1999) 299-312.
3. Hassania, J. (Diversified Energy). “Simgae – Low Cost, Broad Application – Algal Biomass Production System.” NETL 2008 Conference Proceedings. 18 Sept 2008.
4. Cloud, G. et al., US2008/0311649 A1. “Pressurized Flexible Tubing System for Producing Algae.” 18 Dec 2008. XL Renewables.
5. OriginOil presentation at the World Biofuel Market Conference, 15-17 March 2010
6. UOP/Eni Ecofining™ Process for Green Diesel Fuel. Web. 20 April 2010. http://www.uop.com/renewables/10010.html
7. Huber, George W., et al. “Processing biomass in conventional oil refineries: Production of high quality diesel by hydrotreating vegetable oils in heavy vacuum oil mixtures.” Applied Catalysis A: General, Volume 329, (2007), 120-129.

(718b) An Economic Evaluation of Lipid Production by Cultivation of Algae – A Critical Path Analysis

Ramalingam Subramaniam[1], Cunwen Wang[1], **Stephen Dufreche**[1], Mark Zappi[1], Barbara C. Benson[2] and Rakesh Bajpai[1], (1)Chemical Engineering, University of Louisiana at Lafayette, Lafayette, LA, (2) Bioprocessing Research Laboratory, University of Louisiana, Lafayette, LA

This presentation will deal with evaluation of heterotrophic growth of oleaginous algae on carbohydrates available from different sources and compare its economics of lipid production from this route with that from autotrophically growing algae. Energy efficiency during the different processes, light delivery issues, cell harvesting and recovery of lipids from the cells will be considered in identifying the effect of different processing steps on the unit cost of production of lipids.

See more of this Session: Advances in Algal Biorefineries II
See more of this Group/Topical: Sustainable Engineering Forum

(718c) An Integrated Algal Culture System for Biofuel Production Based On Mixotrophic and Heterotrophic Growth Modes

Shulin Chen, Zhanyou Chi, Yubin Zheng and Ben Lucker, Biological Systems Engineering, Washington State University, Pullman, WA

The commercial reality of algae fuel production depends upon innovations ingeniously incorporating engineering and algal biology and efficient utilization of resources. The Bioprocessing and Bioproduct Engineering Laboratory (BBEL) at Washington State University has developed a novel algal culture system that (1) incorporates the dual-trophism capability of some algae species into a two-stage culture process, and (2) integrates heterotrophic and phototrophic processes for year around fuel production in cold climate regions. In this system, the first stage utilizes organic carbon as substrate for algae fermentation, growing seed cells at high efficiency for inoculating to open ponds for phototrophic culture. This system can significantly improve productivity and offers a strategy for contaminant control. In parallel to the mixotrophic process, additional algae production can be accomplished using organic carbon derived from waste materials and agricultural residues. Systematic research efforts have been made in BBEL to further the science discovery and technology development for refining the system. This paper will present a summary of the related progresses. The information will provide new insights for biomass production for algae-based biofuel systems.

See more of this Session: Advances in Algal Biorefineries II
See more of this Group/Topical: Sustainable Engineering Forum

(718d) Biofuel Feedstock Production From Microalgae Grown in Municipal Wastewaters

Sage R. Hiibel, Mark S. Lemos and John C. Cushman, Biochemistry & Molecular Biology, University of Nevada Reno, Reno, NV

Microalgae offer great potential as a second-generation biofuel feedstock. Microalgae are attractive because they require minimal nutrient inputs, can capture atmospheric CO2, and can be grown in fresh, brackish, or salt water. Additionally, they do not compete with current food, feed or fiber feedstock needs, and can be grown on non-arable lands. However, limited water supplies and nutrient sources can significantly add to microalgae production costs.

Currently, municipal wastewater treatment facilities spend millions of dollars each year to remove abundant nitrogen and phosphorus from wastewater. In lieu of traditional treatments, microalgae can be used to effectively and efficiently remove these compounds, while simultaneously producing biofuel feedstocks. Preliminary research has demonstrated that some microalgal species can be grown successfully on centrate, the liquid fraction remaining after dewatering anaerobically digested wastewater sludge.

This project evaluates the potential for growth on centrate of over forty microalgae strains from fresh, brackish, or salt water genera. Generally, freshwater strains were able to grow in higher centrate concentrations, with two species, *Neochloris pseudostigmata* and *N. conjuncta*, demonstrating growth with up to 40 *vol%* centrate. Nine of eighteen salt water *Dunaliella* spp. also demonstrated growth in centrate levels exceeding 25 *vol%*. The use of UV sterilization for centrate sterilization was also evaluated. UV dosage requirements were significantly higher than those used in traditional drinking water treatment plants due to coloration and high TSS/TDS levels.

The lipid profiles of strains with high centrate tolerance were determined and compared against growth on defined maintenance media using GC-MS. The major lipid compounds were found to be palmitic (16:0), oleic (18:1), and linoleic (18:2) acids for all strains grown on either centrate or their respective maintenance medium.

See more of this Session: Advances in Algal Biorefineries II
See more of this Group/Topical: Sustainable Engineering Forum

(718e) Two-Step, Catalyst-Free Biodiesel Production From Wet Algal Biomass

Robert Levine, Tanawan Pinnarat and Phillip E. Savage, Chemical Engineering, University of Michigan, Ann Arbor, MI

Worldwide energy demand, particularly for high-density liquid transportation fuels, continues to rise as populations grow and become more affluent. Concurrently, concern over climate change, declining petroleum reserves, and national security has encouraged the use of biodiesel, a mixture of fatty acid alkyl esters commonly derived from the transesterification of vegetable oils. Recent interest in using oleaginous microalgae as a non-edible biodiesel feedstock has grown considerably, largely on the promise of high oil yields (1,000 to 100,000 L ha-1 y-1), the opportunity to capture CO2 from flue gases, and the ability to cultivate it on abandoned or unproductive land using brackish, salt or wastewaters instead of freshwater. Despite recent advancements and substantial public and private investment, large-scale algal biofuel production has not reached cost parity with the petroleum products they seek to replace. While energy inputs and costs must be reduced at each stage of algal biofuel production, harvesting, dewatering, and lipid processing remain significant challenges.

Instead of attempting to remove lipids from cells in an aqueous suspension or convert them directly to biodiesel in the presence of water, we propose a novel two-step, catalyst-free biodiesel production process involving intracellular lipid hydrolysis coupled with supercritical in-situ (trans)esterification (SC-IST/E). In the first step, wet algal biomass (ca. 80% moisture) is reacted at subcritical water conditions to hydrolyze intracellular lipids, conglomerate cells into an easily filterable solid which retains the lipids, and produce a sterile, nutrient-rich aqueous phase. In the second step, the wet fatty acid (FA)-rich solids are subjected to SC-IST/E with ethanol to produce biodiesel in the form of fatty acid ethyl esters (FAEEs). In this paper, we demonstrate that it is possible to carry out in-situ hydrolysis of cellular lipids in wet algal biomass, retain those lipids within a filterable solid, and then produce biodiesel using supercritical ethanol. In an effort to maximize lipid productivity and efficient carbon substrate utilization, Chlorella vulgaris was grown phototrophically in a bubble column reactor and then heterotrophically. Wet algal biomass contained about 50% total lipids as FAEEs and was used in hydrolysis reaction as a fresh paste or dried and rehydrated. The effects of hydrolysis processing time and temperature were investigated and lipid conversion to FA and smaller glycerides was quantified. Esterification reaction conditions and alcohol loading were investigated as predictors of biodiesel yield and biodiesel was analyzed for impurities in accordance with standard methods (ASTM 6584/EN14105 and EN14103).

See more of this Session: Advances in Algal Biorefineries II
See more of this Group/Topical: Sustainable Engineering Forum

(718f) The Sustainable Conversion of Algal Sugars Into Butanol

Jamie Hestekin[1], **Thomas M. Potts**[2], Jianjun Du[1], Robert R. Beitle[1] and Edgar C. Clausen[1], (1)Ralph E. Martin Department of Chemical Engineering, University of Arkansas, Fayetteville, AR, (2)Chemical Engineering, University of Arkansas, Fayettville, AR

Algae is an attractive feedstock since it can grow on marginal land, has a very high growth rate, and takes up nutrients as it grows so it has the potential to clean wastewater while making fuel feedstock. However, although much research is taking place on converting the algal lipids to biodiesel, there are few studies on converting algal carbohydrates to alcohols. The algal carbohydrates are mainly starches and simple sugars so they are much easier to pretreat and ferment than cellulosic sugars. Our group is working on taking a two step butanol process and continually converting algal sugars into butanol. The first step does an acidgenesis, with Clostridium Tyrobutyricum, for the conversion of sugars to organic acids, specifically butyric acid. The second step takes the butyric acid and converts it into butanol via Clostridium Beijerinckii. Although this is an anaerobic fermentation, we have used membrane processes to make both steps in the two-step process continuous and have modeled the reaction steps in the first and the second step. The yield of this process can go from 6:3:1 butanol, acetone, and ethanol, respecitively to greater than 90% butanol by optimizing both steps.

See more of this Session: Advances in Algal Biorefineries II
See more of this Group/Topical: Sustainable Engineering Forum

AUTHOR INDEX

AUTHOR INDEX

AUTHOR INDEX

AUTHOR INDEX

AUTHOR INDEX

AUTHOR INDEX